Advances in
CATALYSIS

VOLUME **52**

Advances in
CATALYSIS

VOLUME **52**

Editors

BRUCE C. GATES
University of California
Davis, California, USA

HELMUT KNÖZINGER
University of Munich
Munich, Germany

Associate Editor

FRIEDERIKE C. JENTOFT
University of Oklahoma
Norman, Oklahoma, USA

AMSTERDAM • BOSTON • HEIDELBERG • LONDON
NEW YORK • OXFORD • PARIS • SAN DIEGO
SAN FRANCISCO • SINGAPORE • SYDNEY • TOKYO
Academic Press is an imprint of Elsevier

Academic Press is an imprint of Elsevier
Radarweg 29, PO Box 211, 1000 AE Amsterdam, The Netherlands
32, Jamestown Road, London NW1 7BY, UK
30 Corporate Drive, Suite 400, Burlington, MA 01803, USA
525 B Street, Suite 1900, San Diego, CA 92101-4495, USA

First edition 2009
Copyright © 2009 Elsevier Inc. All rights reserved.

Notice
No responsibility is assumed by the publisher for any injury and/or
damage to persons or property as a matter of products liability,
negligence or otherwise, or from any use or operation of any methods,
products, instructions or ideas contained in the material herein. Because
of rapid advances in the medical sciences, in particular, independent
verification of diagnoses and drug dosages should be made

Library of Congress Cataloging-in-Publication Data
A catalog record for this book is available from the Library of Congress

British Library Cataloguing in Publication Data
A catalogue record for this book is available from the British Library

ISBN: 978-0-12-374336-7
ISSN: 0360-0564

For information on all Academic Press publications
visit our web site at books.elsevier.com

Printed and Bound in Hungary
09 10   10 9 8 7 6 5 4 3 2 1

Working together to grow
libraries in developing countries

www.elsevier.com | www.bookaid.org | www.sabre.org

ELSEVIER   BOOK AID International   Sabre Foundation

CONTENTS

Contributors   ix
Preface   xi

**1. Applications of Photoluminescence Spectroscopy to the
     Investigation of Oxide-Containing Catalysts in the Working State   1**

     Masakazu Anpo, Stanislaw Dzwigaj, and Michel Che

     1. Introduction   3
     2. Characterization of Catalytically Active Sites by PL Spectroscopy   4
     3. Characterization of Acidic and Basic Sites by Means
         of Probe Molecules and Photoluminescence Spectroscopy   21
     4. Photoluminescence Investigations of Photocatalytic
         Processes Involving Bulk Semiconductor TiO_2-Containing Photocatalysts 23
     5. Effect of Pressure on Photoluminescence Spectra   28
     6. Effect of Temperature on Photoluminescence Spectra   32
     7. Use of Photoluminescence Spectroscopy in
         High-Throughput Experiments   34
     8. Conclusions and Outlook   35
     Acknowledgments   38
     References   39

2. Structural Characterization of Operating Catalysts by Raman Spectroscopy   43

     Miguel A. Bañares and Gerhard Mestl

     1. Introduction   45
     2. Advanced Raman Spectroscopy for Characterization
         of Samples in Reactive Environments   47
     3. Raman Scattering   48
     4. Raman Investigations of Catalysts Under Controlled Environments   65
     5. Perspective and Conclusions   113
     Acknowledgments   115
     References   115

**3. Ultraviolet–Visible–Near Infrared Spectroscopy in Catalysis:
     Theory, Experiment, Analysis, and Application Under
     Reaction Conditions   129**

     Friederike C. Jentoft

     1. Introduction   132
     2. Theoretical Background   134
     3. Experimental: Optical Accessories for UV–vis–NIR
         Spectroscopy in Transmission Mode   148

4. Experimental: Optical Accessories for UV–vis–NIR
   Spectroscopy in Reflection Mode   149
5. Cells for Measurements in Controlled Environments   160
6. Data Acquisition and Analysis   167
7. Examples   176
8. Conclusions   203
Acknowledgments   205
References   205

4. X-Ray Photoelectron Spectroscopy for Investigation of Heterogeneous Catalytic Processes   213

Axel Knop-Gericke, Evgueni Kleimenov, Michael Hävecker, Raoul Blume,
Detre Teschner, Spiros Zafeiratos, Robert Schlögl, Valerii I. Bukhtiyarov, Vasily
V. Kaichev, Igor P. Prosvirin, Alexander I. Nizovskii, Hendrik Bluhm, Alexei
Barinov, Pavel Dudin, and Maya Kiskinova

1. Introduction   216
2. History of XPS Applied at Substantial Pressures   218
3. Description of XPS Apparatus   221
4. Interaction of CO with Pd(111)   229
5. Dehydrogenation and Oxidation of Methanol on Pd(111)   234
6. Ethene Epoxidation Catalyzed by Silver   240
7. Methanol Oxidation Catalyzed by Copper   247
8. CO Oxidation Catalyzed by Ru(0001)   256
9. Outlook   266
Acknowledgments   267
References   267

5. X-ray Diffraction: A Basic Tool for Characterization of Solid Catalysts in the Working State   273

Robert Schlögl

1. Introduction   274
2. Objectives of XRD in Catalyst Characterization   275
3. Static Versus Dynamic Analysis   284
4. Observations of Catalysts in Reactive Atmospheres   287
5. Cost of Powder Diffraction Under Catalytic Conditions   288
6. Aim and Scope of Powder Diffraction Experiments   289
7. Scattering Phenomena   289
8. Nanostructures and Diffraction   296
9. Phases in Working Catalysts   303
10. Applications of XRD to Working Catalysts   307
11. Combination of XRD Data with Auxiliary Information   309
12. Instrumentation   310
13. Case Studies   313
14. Summary and Conclusions   330
References   332

**6. Characterization of Catalysts in Reactive Atmospheres
    by X-ray Absorption Spectroscopy**   **339**

Simon R. Bare and Thorsten Ressler

1. Introduction   342
2. Importance of XAFS Spectroscopy to Catalyst Characterization
    in Reactive Atmospheres   349
3. Catalytic Reactors that Serve as Cells   369
4. XAFS Spectroscopy of Samples in Reactive Atmospheres
    Under Static Conditions   404
5. XAFS Spectra Measured Under Dynamic Conditions   428
6. Look to the Future   446
7. Concluding Remarks   456
Acknowledgment   456
References   456

Index   467

CONTRIBUTORS

Masakazu Anpo
Department of Applied Chemistry, Graduate School of Engineering, Osaka Prefecture University, Sakai, Osaka 599-8531, Japan.

Miguel A. Bañares
Instituto de Catálisis y Petroleoquímica, Catalytic Spectroscopy Laboratory, CSIC, E-28049-Madrid, Spain.

Simon R. Bare
UOP LLC, a Honeywell Company, Des Plaines, IL 60016, USA.

Alexei Barinov
Sincrotrone Trieste, Microscopy Section, 34012 Trieste, Italy.

Hendrik Bluhm
Lawrence Berkeley National Laboratory, Chemical Sciences Division, Berkeley, CA 94720, USA.

Raoul Blume
Fritz Haber-Institut der Max-Planck-Gesellschaft, Department of Inorganic Chemistry, 14195 Berlin, Germany.

Valerii I. Bukhtiyarov
Boreskov Institute of Catalysis SB RAS, Novosibirsk 630090, Russia.

Michel Che
Laboratoire de Réactivité de Surface, UMR 7609 — CNRS, Université Pierre et Marie Curie, 75252 Paris Cedex 05, France and Institut Universitaire de France.

Pavel Dudin
Sincrotrone Trieste, Microscopy Section, 34012 Trieste, Italy.

Stanislaw Dzwigaj
Laboratoire de Réactivité de Surface, UMR 7609 — CNRS, Université Pierre et Marie Curie, 75252 Paris Cedex 05, France.

Michael Hävecker
Fritz-Haber-Institut der Max-Planck-Gesellschaft, Department of Inorganic Chemistry, 14195 Berlin, Germany.

Friederike C. Jentoft
Fritz-Haber-Institut der Max-Planck-Gesellschaft, Department of Inorganic Chemistry, 14195 Berlin, Germany.

Vasily V. Kaichev
Boreskov Institute of Catalysis SB RAS, Novosibirsk 630090, Russia.

Maya Kiskinova
Sincrotrone Trieste, Microscopy Section, 34012 Trieste, Italy.

Evgueni Kleimenov
ETH Zürich, Laboratory of Physical Chemistry, HCI E 209, 8093 Zürich, Switzerland.

Axel Knop-Gericke
Fritz-Haber-Institut der Max-Planck-Gesellschaft, Department of Inorganic Chemistry, 14195 Berlin, Germany.

Gerhard Mestl
Süd-Chemie AG, Research and Development, Catalytic Technologies, D-83052 Bruckmühl—Heufeld, Germany.

Alexander I. Nizovskii
Boreskov Institute of Catalysis SB RAS, Novosibirsk 630090, Russia.

Igor P. Prosvirin
Boreskov Institute of Catalysis SB RAS, Novosibirsk 630090, Russia.

Thorsten Ressler
Technische Universität Berlin, Fachgruppe Anorganische und Analytische Chemie, Institut für Chemie, D-10623 Berlin, Germany.

Robert Schlögl
Fritz-Haber-Institut der Max-Planck-Gesellschaft, Department of Inorganic Chemistry, 14195 Berlin, Germany.

Detre Teschner
Fritz-Haber-Institut der Max-Planck-Gesellschaft, Department of Inorganic Chemistry, 14195 Berlin, Germany.

Spiros Zafeiratos
LMSPC-UMR 7515 du CNRS, 67087 Strasbourg Cedex 2, France.

We are pleased to report that Dr. Friederike C. Jentoft of the Fritz-Haber-Institut der Max-Planck-Gesellschaft, Berlin, Germany (and newly of the University of Oklahoma), has joined us as Associate Editor, beginning with this volume of *Advances in Catalysis*. We welcome her to the team.

Volumes 50 and 51 of the *Advances*, published in 2006 and 2007, respectively, were the first of a set of three focused on the physical characterization of solid catalysts in the functioning state. This volume completes the set. The six chapters presented here are largely focused on the determination of structures and electronic properties of components and surfaces of solid catalysts. The first chapter is devoted to photoluminescence spectroscopy; it is followed by chapters on Raman spectroscopy; ultraviolet-visible-near infrared (UV-vis-NIR) spectroscopy; X-ray photoelectron spectroscopy; X-ray diffraction; and X-ray absorption spectroscopy.

Anpo, Dzwigaj, and Che summarize the application of dynamic photoluminescence spectroscopy for characterization of surface catalytic sites, specifically of photocatalysts. Photoluminescence is useful for characterization of highly dispersed supported transition metal oxides with low loadings, as it is highly sensitive and nondestructive. The method allows monitoring of the interactions of reactant molecules with active sites as a result of dynamic changes of the photoluminescence intensities and lifetimes. Dynamic photoluminescence of bulk semiconductor oxide catalysts such as TiO_2 and ZnO can provide valuable information about the chemical nature of surface states and defect centers of the catalysts and their roles in photocatalytic reactions. There is, however, still a dearth of applications of this method for the investigation of working catalysts, and the authors provide a critical evaluation of the prospects for such applications.

Raman spectroscopy is one of the most versatile spectroscopies for the characterization of solid catalyst surfaces and of surface species under reaction conditions. Bañares and Mestl provide an in-depth description of catalytic reaction cells that allow recording of Raman spectra simultaneously with measurements of catalytic activities and selectivities. The authors discuss the advanced modern equipment and methodologies that permit the detection of Raman spectra at elevated pressures and temperatures (>1270 K) with good time resolution and spatial resolution (Raman microscopy). Measurements can be made during catalyst synthesis, during temperature-programmed processes, and during induction and deactivation periods. Furthermore, Raman spectroscopy is possible in the presence of solvents and supercritical media. An extensive literature overview with emphasis on bulk and supported oxide catalysts is included in the chapter.

Jentoft reports on UV–vis–NIR spectroscopy. Spectra of powder samples in the UV–vis and NIR regions can be measured either in the transmission mode or in the diffuse reflection mode, depending on the wavelength regime and on the particle size and morphology, which sensitively determine the sample's light scattering properties. The chapter begins with an account of these two spectroscopic methods and continues with a discussion of the theoretical aspects of diffuse reflectance spectroscopy, including the Kubelka-Munk formalism. Jentoft further reports on optical accessories such as mirror optics, fiber optics, and cells, especially those that permit measurements in a controlled gas atmosphere at elevated temperatures. The most desirable cell designs are those that facilitate the characterization of the same catalytic material with more than one physical technique (e.g., UV–vis spectroscopy and EPR spectroscopy). The chapter includes an overview of a number of experimental investigations of adsorption and of catalytic reactions under conditions of controlled pressure and temperature. The experiments comprise measurements of UV–vis spectra of catalytic materials after and during treatments under well-defined conditions. Of particular interest are investigations in which the spectroscopy and measurements of catalytic reaction rates or conversions and selectivities are carried out simultaneously.

Knop-Gericke *et al.* report methods and equipment for measurement of X-ray photoelectron spectra of solid catalysts in the presence of gases at pressures up to about a millibar. The unavoidable inelastic interactions of the photoelectrons as information carriers with the gas molecules are minimized by keeping the path length of the photoelectrons in the gas atmosphere between the sample and the detector as small as possible and by pumping the cell differentially. The practical pressure range is thus expanded by ~5 orders of magnitude relative to the conventional upper pressure limit of X-ray photoelectron spectroscopy (XPS) of $\sim 10^{-5}$ mbar. The authors demonstrate that the application of this technique under the conditions mentioned permits the determination of adsorbate structures and surface coverages. The potential of XPS in the presence of a gas phase at pressures up to ~1 mbar is exemplified by investigations of adsorption and catalysis, including the interaction of carbon monoxide with Pd(111), the dehydrogenation and oxidation of methanol on Pd(111), ethene epoxidation catalyzed by silver, methanol oxidation catalyzed by copper, and carbon monoxide oxidation on Ru(0001).

Schlögl summarizes the use of powder X-ray diffraction (XRD) for determination of the structures of bulk components in solid catalysts.

XRD is valuable because it provides unambiguous determinations of phases, their dynamics, and evidence of their structures at the nanoscale, even at high pressures and temperatures. This technique is still not frequently applied for characterization of functioning catalysts, but the advent of ultra-high-brilliance radiation sources such as free-electron lasers and improved synchrotron sources will open new possibilities for determination of time-resolved XRD patterns to establish details of the temporal evolution of structural dynamics and of the nanostructures of functioning catalysts. Examples in this chapter illustrate the value of the method for characterization of mixed oxide catalysts, such as those containing molybdenum oxide structures.

Bare and Ressler thoroughly summarize the applications of extended X-ray absorption fine structure (EXAFS) spectroscopy and X-ray absorption near edge structure (XANES) spectroscopy in catalyst characterization. EXAFS spectroscopy has developed into one of the standard methods for characterization of catalysts, providing information about the average local geometric structures of catalysts in the working state, but it has usually been applied at temperatures not much more than about 400 K. A wide range of reactors that serve as EXAFS cells are summarized in this chapter, along with a number of case studies illustrating how EXAFS spectroscopy determines local structural information characterizing small catalyst structures. The method has been used, for example, to determine the locations of Sn atoms in Sn-β zeolite, to determine average sizes and structural data characterizing small metal clusters on supports, and to observe the formation of metal sulfide structures in hydroprocessing catalysts. XANES provides information about oxidation states of metals in catalysts and is applicable at much higher temperatures than EXAFS spectroscopy. For example, the method was used to demonstrate the evolution of the average oxidation state of molybdenum in an activated oxidation catalyst, $H_5[PV_2Mo_{10}O_{40}] \cdot 13H_2O$, during changes in gas-phase composition (propene to propene + O_2). The X-ray absorption spectroscopy techniques for the most part require synchrotron radiation, and the authors report how improvements in synchrotrons are leading to improved characterization of catalysts.

The contents of this volume and the two that preceded it constitute the most comprehensive summary of methods for characterization of functioning solid catalysts and a thorough set of examples illustrating how much insight has emerged from the efforts of researchers to focus on the structures that matter most in surface catalysis—the working species rather than the spectators. The pace of the advances in these techniques has been amazingly rapid, reflecting both the development in experimental instruments and techniques and a sharpened appreciation by researchers of the importance of the functioning structures in catalysts. There is now a clear trend toward application of multiple techniques simultaneously to investigate working catalysts, such as XRD and EXAFS spectroscopy, as illustrated in the chapter by Schlögl, and UV–vis and NIR spectroscopy, as illustrated in the chapter by Jentoft.

B.C. GATES
H. KNÖZINGER

Applications of Photoluminescence Spectroscopy to the Investigation of Oxide-Containing Catalysts in the Working State

Masakazu Anpo,* Stanislaw Dzwigaj,‡ and Michel Che†,‡

Abstract

This chapter is an update of an earlier version (*Adv. Catal.* **44**, 119 (1999)) devoted to the applications of photoluminescence (PL) spectroscopy to the characterization of solid surfaces in relation to adsorption, catalysis, and photocatalysis. This chapter concerns highly dispersed or bulk oxide-containing catalysts, including a discussion of the effects of pressure and temperature on PL spectra of various materials. The aim of this chapter is to describe recent advances in the identification of catalytically active sites by PL spectroscopy and other spectroscopic techniques, such as IR, EPR, XAS (XANES and EXAFS), UV–vis, and NMR. One can thus obtain precise information about the nature and local environment of transition metal ions in zeolites and oxide and hydroxide ions on MgO surfaces. PL appears particularly useful in the case of zeolites and supported samples with low metal loadings (e.g., vanadium-containing Siβ zeolite). Indeed, PL spectroscopy in static and

* Department of Applied Chemistry, Graduate School of Engineering, Osaka Prefecture University, Sakai, Osaka 599-8531, Japan
† Institut Universitaire de France
‡ Laboratoire de Réactivité de Surface, UMR 7609 – CNRS, Université Pierre et Marie Curie, 75252 Paris Cedex 05, France

Advances in Catalysis, Volume 52
ISSN 0360-0564, DOI: 10.1016/S0360-0564(08)00001-1

dynamic modes allows one to distinguish, in VSiβ zeolite, three kinds of tetrahedral V(V) species, the relative amounts of which depend on vanadium content and calcination/rehydration treatments. PL, which is far more sensitive than UV–vis and ^{51}V NMR spectroscopies, gives electronic and vibrational (fine structure and energy) information about each type of vanadium. Moreover, dynamic PL allows investigation of the chemical nature and properties of catalytically active surface sites in the working state and their role in photocatalytic reactions. This chapter provides a description of the advantages of PL techniques to characterize acidic and basic sites by means of probe molecules, as well as various materials used in catalysis and electroluminescence, including a consideration of future directions of research involving PL spectroscopy.

Contents

1. Introduction   3
2. Characterization of Catalytically Active Sites by PL Spectroscopy   4
   2.1. Coordination Chemistry of Cations   4
   2.2. Coordination Chemistry of Anions: MgO as a Model System   19
3. Characterization of Acidic and Basic Sites by Means of Probe Molecules and Photoluminescence Spectroscopy   21
   3.1. Characterization of Acidic Sites   21
   3.2. Characterization of Basic Sites   23
4. Photoluminescence Investigations of Photocatalytic Processes Involving Bulk Semiconductor TiO$_2$-Containing Photocatalysts   23
5. Effect of Pressure on Photoluminescence Spectra   28
   5.1. Low-Pressure Domain: Physical versus Chemical Quenching   28
   5.2. High-Pressure Domain   29
6. Effect of Temperature on Photoluminescence Spectra   32
7. Use of Photoluminescence Spectroscopy in High-Throughput Experiments   34
8. Conclusions and Outlook   35
Acknowledgments   38
References   39

ABBREVIATIONS

CT	charge transfer
EPR	electron paramagnetic resonance
EXAFS	extended X-ray absorption fine structure

FT	Fourier transform
HT	high-throughput
I	photoluminescence intensity under gas pressure
I_0	photoluminescence intensity under vacuum
iPr	isopropyl
IR	infrared
k_0	quenching rate constant
LC	low coordination
LDH	layered double-hydroxide
λ	wavelength
λ_{exc}	wavelength of excitation
λ_{em}	wavelength of emission
MBOII	?-methylbut-3-yn-2-ol
NMR	nuclear magnetic resonance
O_{LC}^{2-}	oxide ion in position of low coordination
O_{3C}^{2-}	tricoordinated oxide ion
O_{4C}^{2}	tetracoordinated oxide ion
O_{5C}^{2-}	pentacoordinated oxide ion
P_g	partial pressure of the quenching gas
PL	photoluminescence
PPV	poly(phenylene–vinylene)
PVG	porous vycor glass
Q	phosphorescence yield in the presence of quencher molecules
Q_0	phosphorescence yield in the absence of quencher molecules
$[Q]$	concentration of the quencher
TP^+	2,4,6-triphenyl-pyrylium
TSL	thermally stimulated luminescence
τ	lifetime of the excited triplet state of species in the absence of quencher
UV	ultraviolet light ($200 < \lambda < 400$ nm)
UV–vis	ultraviolet–visible
VSEPR	valence shell electron pair repulsion
XANES	X-ray absorption near edge structure
XAS	X-ray absorption spectroscopy

1. INTRODUCTION

There is a strong driving force for the development of new catalysts and processes that are environmentally sustainable—clean, safe, stable, and efficient—to address today's urgent need to reduce global environmental pollution and to develop clean and renewable energy resources.

Photocatalytic processes driven by solar or visible light that use stable, nontoxic, inexpensive, and highly efficient photocatalysts offer some of the best prospects for addressing these needs.

In an earlier review of the applications of photoluminescence (PL) techniques to the characterization of adsorption, catalysis, and photocatalysis (Anpo and Che, 1999), we addressed the basic principles of PL and the importance of PL measurements for understanding of (photo)catalytic processes. This chapter describes more recent developments and focuses on investigations of catalysts in the working state, with an emphasis on the role of local structure on photocatalytic reactions determined with PL and related techniques.

Dynamic PL spectroscopy is important for the investigation of the chemical nature and properties of catalytically active surface sites of photocatalysts in the working state and their role in photocatalytic reactions (Anpo, 2002; Anpo and Che, 1999; Matsuoka and Anpo, 2005). PL is particularly useful for characterization of highly dispersed supported transition metal oxides with low loadings, because it is highly sensitive and nondestructive. Thus, it is possible to monitor the interactions of reactant molecules with active sites *via* dynamic changes of the PL parameters, especially intensities and lifetimes (Anpo et al., 1999; Dzwigaj et al., 1998; Higashimoto et al., 2001b; Matsuoka and Anpo, 2005; Zhang et al., 1998). Dynamic PL of bulk semiconductor oxide catalysts such as TiO_2 and ZnO can provide valuable information about the chemical nature of surface states and defect centers of the catalysts and their role in photocatalytic reactions (Anpo, 2004; Anpo and Moon, 1999; Nakajima et al., 2002; Nakamura and Nakato, 2004).

This chapter addresses recent results typical of highly dispersed and bulk oxide catalysts in the working state.

2. CHARACTERIZATION OF CATALYTICALLY ACTIVE SITES BY PL SPECTROSCOPY

2.1. Coordination Chemistry of Cations

2.1.1. Titanium-Containing Zeolites

The characterization of titanium in zeolites or immobilized on oxide surfaces in reactive atmospheres can be carried out using UV–vis, EPR, PL, and XAS (XANES and EXAFS) spectroscopies. The spectra can provide important insight into the local structure of catalytically active sites and their photocatalytic activities, particularly in the decomposition of NO_x into N_2 and O_2 and in the reduction of CO_2 with H_2O to give CH_3OH and CH_4 (Anpo et al., 1998, 2001, 2002; Hu et al., 2003a; Ikeue et al., 1999, 2001, 2002; Matsuoka and Anpo, 2003b; Ogawa et al., 2001; Zhang et al., 2000, 2001).

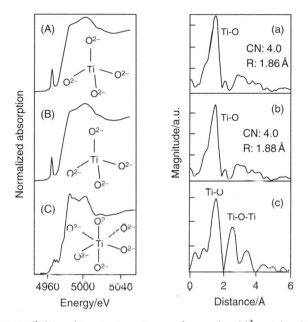

FIGURE 1   XANES (left) and Fourier transforms of normalized k^3-weighted EXAFS (right) without phase-shift correction recorded at room temperature of Ti–MCM-41 (A, a), Ti–MCM-48 (B, b), and powder TiO_2 (C, c) as reference sample. (Reproduced with permission from Anpo et al. (1998).)

Figure 1 shows the XANES and EXAFS spectra of titanium containing MCM-41 and -48 and TiO_2 (Anpo et al., 1998). A characteristic feature of XANES spectra of Ti–MCM-41 and Ti–MCM-48 is the single pre-edge peak which has a shape similar to that of the tetrahedral Ti(O-iPr)$_4$ complex, indicating that, in these solids, titanium is in a tetrahedral coordination, whereas bulk TiO_2 exhibits three pre-edge peaks of low intensity typical of octahedrally coordinated titanium, as in the rutile and anatase forms of TiO_2. However, the XANES pre-edge peak characterizing Ti–MCM-41 and Ti–MCM-48 constitutes ~30% of the edge jump, and this result suggests that a larger fraction of the titanium is in square pyramidal or octahedral coordination, probably because of the adventitious introduction of water before and/or during the XAS measurements. Moreover, the Fourier transforms of the EXAFS spectra of these catalysts exhibit only one peak, at ~1.6 Å, attributed to Ti–O contributions, indicating the presence of isolated titanium oxides. The data obtained from the fitting of the EXAFS data are summarized in Table 1 (Anpo et al., 2001).

Figure 2 shows the PL spectrum of Ti–MCM-48 (Anpo and Che, 1998), which together with its absorption spectrum (excitation spectrum), is in good agreement with those previously observed for highly dispersed

TABLE 1 Physical Parameters and Photocatalytic Properties of Titanium Oxide Species Occluded in Zeolites and on FSM-16 and SiO_2 Surfaces. (Reproduced with permission from Anpo et al. (2001).)

	TS-1	TS-2	Ti–MCM-41	Ti–MCM-48	Ti/FSM-16	Ti/Vycor glass	Bulk TiO_2
Coordination	Tetra	Tetra	Tetra	Tetra	Tetra	Tetra	Octa
Ti–O bond length (Å)	1.83	1.84	1.86	1.88	—	—	—
Photoluminescence wavelength (nm)	480	480	480	480	465	445	—
Selectivity for the CH_3OH formation (%)	20.6	26.5	30.8	28.8	41.2	35.0	$\cong 0$
Selectivity for the N_2 formation (%)	82.0	80.0	—	89.0	—	88.0	>18

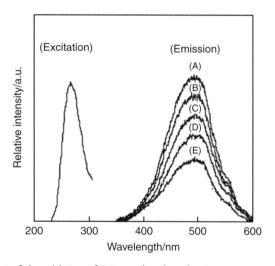

FIGURE 2 Effect of the addition of CO_2 on the photoluminescence spectra recorded at 77 K of Ti–MCM-48. Pressure of added CO_2: 0 (A) 0.1 (B) 0.5 (C) 2.0 (D) and (E) 10.0 Torr. (Reproduced with permission from Anpo et al. (1998).)

tetrahedral titanium species (Anpo and Che, 1999). These spectra originate from the absorption/emission (photoluminescence) of radiation
between the charge transfer (CT) excited triplet state and the singlet
ground state of highly dispersed titanium in tetrahedral coordination,
represented as follows:

$$(\text{Ti}^{4+}-\text{O}^{2-}) \underset{\text{Photoluminescence, } h\nu'}{\overset{\text{Absorption, } h\nu}{\rightleftharpoons}} (\text{Ti}^{3+}-\text{O}^{-})^{*} \tag{1}$$

Singlet ground state   Triplet excited state

The emission spectra in Figure 2 show that the addition of CO_2 (as well as of NO and H_2, results not shown) to the catalyst leads to an efficient quenching of the PL. A shortening of the lifetime of the CT excited state is also observed, and its extent depends on the amount of gas added. Such an efficient quenching of PL by molecules introduced from the gas phase indicates that they interact with tetrahedral titanium oxides in both their ground and excited states and that the titanium species are located at accessible positions (Anpo, 2000; Anpo et al., 1998, 2001, 2002; Hu et al., 2003a,b; Ikeue et al., 1999, 2001, 2002; Jung and Park, 2001; Matsuoka and Anpo, 2003b; Ogawa et al., 2001; Yamashita and Anpo, 2003; Yamashita et al., 2002; Zhang et al., 2000, 2001).

The selectivity towards N_2 in the photocatalytic decomposition of NO as a function of the coordination number of Ti obtained from EXAFS measurements for various titanium oxide photocatalysts is shown in Figure 3 (Yamashita and Anpo, 2004). The linear dependence observed demonstrates that the lower the coordination number of titanium, the higher the N_2 selectivity (Zhang et al., 2001). A similar dependence of CH_3OH selectivity on the coordination number of titanium is also observed in the photocatalytic reduction of CO_2 with H_2O (i.e., the

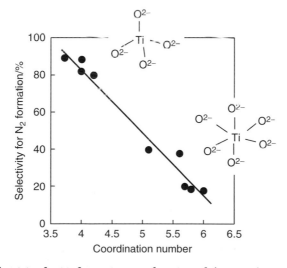

FIGURE 3 Selectivity for N_2 formation as a function of the coordination number of Ti determined by EXAFS in the photocatalytic decomposition of NO into N_2 and O_2 on various titanium oxide catalysts including highly dispersed, chemical mixture and bulk TiO_2 powder. (Reproduced with permission from Yamashita and Anpo (2004).)

lower the coordination number of titanium, the higher the CH_3OH selectivity) (Ikeue et al., 2002). These results show that efficient and selective photocatalytic decomposition of NO into N_2 and O_2 as well as reduction of CO_2 with H_2O into CH_3OH can be achieved by using photocatalysts with highly dispersed tetrahedral titanium-containing species, as catalytically active sites (Anpo, 2000; Hu et al., 2003a,b; Yamashita and Anpo, 2003; Yamashita et al., 2002).

2.1.2. Vanadium-Containing Zeolites

Spectroscopic techniques, particularly UV–vis, EPR, NMR, Raman, and IR, are often applied to characterize vanadium-containing species in various supported catalysts (Anpo, 2002; Anpo and Che, 1999; Anpo et al., 1999; Dzwigaj et al., 1998; Matsuoka and Anpo, 2005). The data obtained by IR, UV–vis, and ^{51}V NMR spectroscopies show that vanadium is present as tetrahedral V^{5+} ions. However, notwithstanding the combined development of these techniques, it is difficult to conclude whether in vanadium-containing β-zeolite there is a single or are several kinds of tetrahedral V^{5+} species. On the basis of PL spectra, one can distinguish three kinds of tetrahedral V^{5+} species in VSiβ zeolite containing 1.5 wt % V. The data show that their relative concentrations depend strongly on dehydration/rehydration treatments (Anpo et al., 2003; Dzwigaj, 2003; Dzwigaj et al., 2000, 2001; Higashimoto et al., 1999, 2000, 2001a,b; Hu et al., 2004; Matsuoka et al., 2000).

Figure 4 shows the ^{51}V wide-line NMR spectra of $V_{0.05}$Siβ and $V_{1.5}$Siβ zeolites with 0.05 and 1.5 wt % V, respectively, after calcination at 773 K (hence the prefix C in sample labeling) and further rehydration at 298 K (hence the prefix Hyd) (Dzwigaj et al., 2001). A single orthorhombic signal at -535 ppm is evident for $V_{0.05}$Siβ zeolite. By contrast, $V_{1.5}$Siβ zeolite is characterized by an orthorhombic signal at -605 ppm and another at -350 ppm. The former is likely indicative of V^{5+} in a distorted VO_4 tetrahedral environment, and the latter (quite different from that of V_2O_5) is indicative of V^{5+} in an axially distorted octahedral environment (Dzwigaj, 2003; Dzwigaj et al., 1998, 2000, 2001).

As shown in Figure 5, the diffuse reflectance UV–vis spectrum of C-Hyd-$V_{0.05}$Siβ includes two bands at 270 and 340 nm (Dzwigaj et al., 2001). Because of the absence of d–d transitions in the 600–800 nm range and of any V^{4+} EPR signal, we infer that these bands can only involve V^{5+} ions, and thus they are attributed to the CT transition from O^{2-} to tetrahedral V^{5+}, involving oxygen in bridging (V—O—Si) and terminal (V=O) positions, respectively. By contrast, the UV–Vis spectrum of C-Hyd-$V_{1.5}$Siβ is composed of two intense CT bands, at 265 and 370 nm, with a shoulder at 235 nm. The band at 375 nm can be attributed to CT from O^{2-} to octahedral V^{5+} with oxygen in a terminal position (V=O). The band at $\sim$235–270 nm can be attributed to CT from O^{2-} to tetrahedral

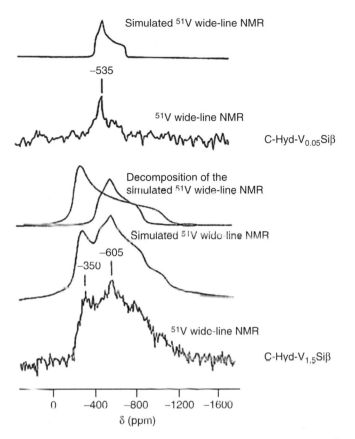

FIGURE 4 Experimental (recorded at 78.9 MHz and room temperature) and simulated ^{51}V wide-line NMR spectra of C-Hyd-$V_{1.5}$Siβ and C-Hyd-$V_{0.05}$Siβ samples. (Reproduced with permission from Dzwigaj et al. (2001).)

V^{5+} with oxygen in bridging position (V–O–Si). The shoulder at 235 and the band at 265 nm suggest the presence of two different kinds of tetrahedral V^{5+}, strongly distorted and less distorted, respectively.

C-Hyd-$V_{0.05}$Siβ and C-Hyd-$V_{1.5}$Siβ zeolites give PL spectra with maxima at $\sim$500 nm which exhibit complex vibrational fine structures (Dzwigaj et al., 2001), as shown in Figure 6. These spectra correspond to radiative transitions from the lowest vibrational energy level of the CT-excited triplet state $(V^{4+}\!\!-\!\!O^-)^*$ of tetrahedral $O\!\!=\!\!VO_3$ units to the various vibrational energy levels of its singlet ground state $(V^{5+}\!\!-\!\!O^{2-})$.

To facilitate the analysis of the complex spectra associated with the superimposition of various vibrational fine structures, the second-derivative presentation of spectra was used (Dzwigaj et al., 2001). Figure 7

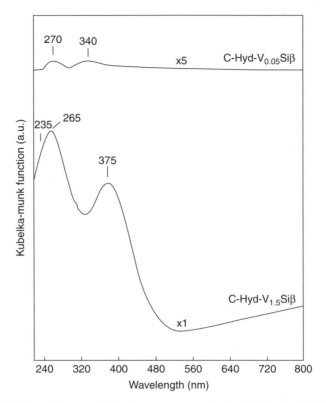

FIGURE 5 Diffuse reflectance UV–vis spectra recorded at 298 K of C-Hyd-V$_{1.5}$Siβ and C-Hyd-V$_{0.05}$Siβ samples with the parent V-free Siβ zeolite as reference. (Reproduced with permission from Dzwigaj et al. (2001).)

shows that C-Hyd-V$_{0.05}$Siβ exhibits three different types of vibrational fine structure, α, β, and γ, attributed to various kinds of tetrahedral V^{5+} species (Dzwigaj et al., 2001). The spectrum of C-Hyd-V$_{1.5}$Siβ is less resolved than that of C-Hyd-V$_{0.05}$Siβ and indicates mainly the α type species. The energy separation between the $(0 \rightarrow 0)$ and $(0 \rightarrow 1)$ vibrational transitions can thus be determined from the second derivative spectrum, and the vibrational energies are found to be equal to 1018 (species α), 1036 (species γ), and 1054 cm^{-1} (species β) (Dzwigaj et al., 1998) (Table 2). These values are in good agreement with the vibrational energy of V=O bonds determined by IR and Raman measurements for gaseous O=VF$_3$, liquid O=VCl$_3$, and O=VBr$_3$ compounds and various supported vanadium oxide catalysts (V$_2$O$_5$/SiO$_2$, V$_2$O$_5$/PVG (PVG stands for porous vycor glass), and V$_2$O$_5$/γ-Al$_2$O$_3$,) (Dzwigaj, 2003; Dzwigaj et al., 2000, 2001). The V=O bond length given in Table 2 (Dzwigaj et al., 1998) has been calculated on the basis of the correlation between the wave number of the V=O bond and

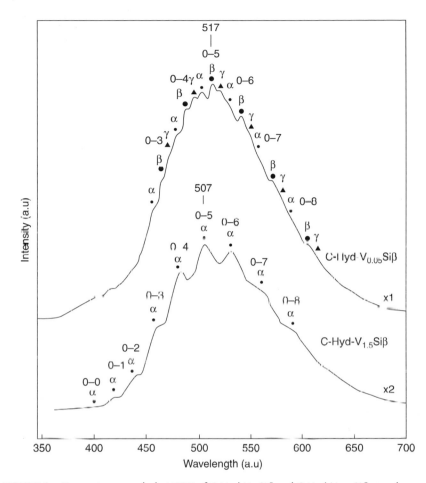

FIGURE 6   PL spectra recorded at 77 K of C-Hyd-$V_{1.5}$Siβ and C-Hyd-$V_{0.05}$Siβ samples outgassed at 473 K for 2 h (10^{-3} Pa). (Reproduced with permission from Dzwigaj et al. (2001).)

its length (Iwamoto et al., 1983). It decreases in the sequence $\alpha(1.58) > \gamma(1.56) > \beta(1.54)$. By comparison with the V–O single-bond length of 1.72–1.80 Å (Iwamoto et al., 1983), and on the basis of VSEPR arguments (Gillespie, 1972), the O—V–X angle, with X=O–(Si, H), increases in the sequence $\alpha < \gamma < \beta$. Thus, as the V=O double bond length increases toward the V—O single bond length, the symmetry moves closer to the ideal tetrahedral T_d symmetry. The increase of the lifetime of the excited triplet state in the sequence $\alpha < \gamma < \beta$ (Table 2) can be interpreted in terms of an increase in the distortion of the tetracoordinated V species and a decrease in the length of the V=O double bond (Dzwigaj et al., 1998).

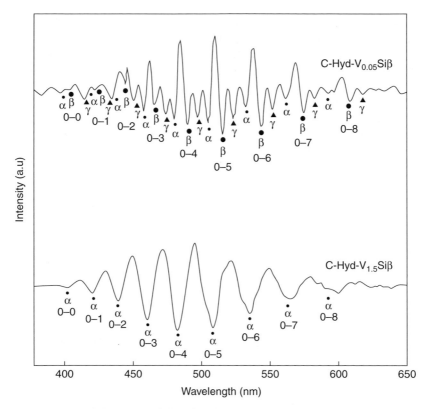

FIGURE 7   Second derivative of the photoluminescence spectra recorded at 77 K of C-Hyd-$V_{1.5}Si\beta$ and C-Hyd-$V_{0.05}Si\beta$ samples outgassed at 473 K for 2 h (10^{-3} Pa). (Reproduced with permission from Dzwigaj et al. (2001).)

TABLE 2   Physicochemical Parameters Related to Tetrahedral V Species in VSiβ zeolite. (Reproduced with permission from Dzwigaj et al. (1998).)

Type of tetracoordinated V species		α	γ	β
Vibrational energy (cm^{-1})		1018	1036	1054
V=O bond length (Å)		1.58	1.56	1.54
V–O single bonds lengths (Å)	$1.72–1.80^a$			
Lifetime of the excited triplet states (ms)		28	49	88

[a] Data for reference compounds containing isolated VO_4 tetrahedra taken from Nabavi et al. (1990).

The comparative investigation of vanadium species by PL and ^{51}V NMR spectroscopies allows one to suggest that there is a correlation between the vibrational energy of the V=O bond and the associated chemical shift of the ^{51}V NMR signal (Dzwigaj et al., 2001). The vibrational energy decreases in the order $\beta > \gamma > \alpha$, whereas the chemical shift increases from -535 ppm for the mixture of α, β, γ species to -605 ppm for the α species. VO_4^{3-} species with nearly perfect tetrahedral symmetry of V^{5+} are characterized by a chemical shift at approximately -740 ppm. These results characterizing tetrahedral V^{5+} species might suggest that, when the symmetry increases, the vibrational energy decreases whereas the chemical shift increases.

Vanadium silicalite (VS-1) prepared by hydrothermal synthesis was characterized by dynamic PL, XAS, UV–vis, and IR techniques, and its photocatalytic activity was investigated in the decomposition of NO in the absence and in the presence of C_3H_8, with reaction products determined by GC analysis (Anpo et al., 2003; Higashimoto et al., 2001a). The shape of the XANES spectrum of VS-1 was found to be similar to that of the axially distorted tetrahedral $O=V(O\text{-}iso\text{-}C_3H_7)_3$ complex, indicating that VS-1 consists of highly dispersed vanadium-containing species with tetrahedral coordination. Furthermore, in the curve-fitting analysis of the EXAFS spectrum determined at 298 K, the best fit was obtained with one short V=O bond of 1.68 Å and three long V–O bonds of 1.78 Å (Anpo et al., 2003; Higashimoto et al., 2001a).

VS-1 exhibits a phosphorescence spectrum attributed to the radiative transition from the CT excited triplet state of tetrahedral vanadium-containing species to its ground state. The phosphorescence was easily quenched by addition of NO or C_3H_8, showing that the CT excited state reacts with NO or C_3H_8. Indeed, UV-irradiation of VS-1 in the presence of NO leads to the photocatalytic decomposition of NO into N_2 and O_2. It was found that the reaction yield is dramatically enhanced in the presence of C_3H_8, with formation of N_2, C_3H_6, and oxygen-containing compounds such as CH_3COCH_3 and CO_2. The photocatalytic activity was found to be much higher than that of V/SiO_2 catalysts.

To understand the origin of such a difference in photocatalytic activity, the reaction rate constants were calculated for the quenching of the phosphorescence of VS-1 and V/SiO_2 catalysts by NO and C_3H_8 (Anpo, 2000; Anpo et al., 2003; Higashimoto et al., 2001a) by using the Stern–Volmer equation expressed as follows:

$$Q_0/Q = 1 + \tau \, k_0[Q] \tag{2}$$

where Q_0 and Q are the phosphorescence yields in the absence and in the presence, respectively, of quencher molecules such as NO, and τ, k_0, and Q are the lifetime of the CT excited triplet state of V species in the absence of quencher, the quenching (reaction) rate constant, and the concentration of the quencher present on the catalyst surface, respectively.

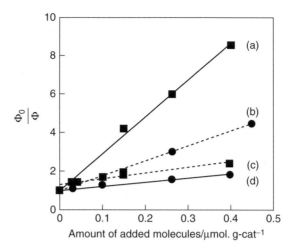

FIGURE 8 Stern–Volmer plots of Φ_0/Φ values versus [Q] for the yields of phosphorescence in the presence of NO or propane on the VS-1 and V/SiO$_2$ catalysts. VS-1, (a) propane: 2.4×10^9 g-cat/(mol s), (b) NO: 8.0×10^8 g-cat/(mol s), (c) V/SiO$_2$, propane: 4.8×10^8 g-cat/(mol s), (d) NO: 2.8×10^8 g-cat/(mol s). (Reproduced with permission from Higashimoto et al. (2001a).)

From an analysis of the Stern–Volmer plots (Higashimoto et al., 2001a) shown in Figure 8, the quenching rate constant (i.e., the reaction rate constant of C_3H_8) was determined to be 2.4×10^9 g of catalyst/(mol s) for the CT excited state of VS-1. This value is larger than that of NO, 8.0×10^8 g of catalyst/(mol s), indicating that the reactivity of C_3H_8 with the excited triplet state is greater than that of NO. Furthermore, these reaction rate constants of C_3H_8 and NO characterizing VS-1 are much larger than those of C_3H_8 (4.8×10^8 g of catalyst/(mol s)) and NO (2.8×10^8 g of catalyst/(mol s)) for V/SiO$_2$. These differences in the reaction rate constants for VS-1 and V/SiO$_2$ can be attributed to a condensation effect of NO/C_3H_8 quencher molecules in VS-1 cavities allowing an easier access to catalytically active sites and a better confinement of the reacting molecules, as compared with those in V/SiO$_2$.

2.1.3. Molybdenum-Containing Zeolites and Mo/SiO$_2$

As shown in Figure 9, UV-irradiation of Mo–MCM-41 in the presence of a mixture of NO and CO leads to the formation of N$_2$ and CO$_2$ with a linear dependence on UV-irradiation time, whereas the turnover frequency exceeds 1.0 after 2 h (Tsumura et al., 2000). After 3 h, NO conversion and selectivity towards N$_2$ are close to 100%, with only a small amount of N$_2$O formed. Figure 10 shows that there is a good correspondence between the yield of N$_2$ produced and the yield of PL of tetrahedral molybdenum species and the amount of Mo^{4+} ions generated upon

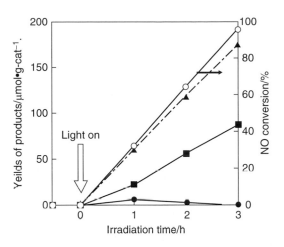

FIGURE 9 Reaction time profiles of the photocatalytic decomposition reaction of NO in the presence of CO on Mo–MCM-41 (1.0 wt % Mo): yields of CO_2 (▲), N_2 (■), N_2O (●), and conversion of NO (○). Amount of added NO or CO 180 mmol per gram of catalyst. (Reproduced with permission from Tsumura et al. (2000).)

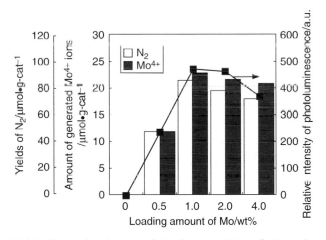

FIGURE 10 Yield of N_2 in the photocatalytic decomposition of NO in the presence of CO for 3 h (open bar), amount of generated Mo^{4+} ions in the photoreduction of Mo^{6+} with CO under UV-irradiation for 0.5 h (filled bars) and photoluminescence yield (solid line) on Mo–MCM-41 (with 0.5, 1.0, 2.0, and 4.0 wt % Mo). (Reproduced with permission from Tsumura et al. (2000).)

photoreduction of Mo^{6+} with CO and deduced from the number of photoformed CO_2 molecules (Tsumura et al., 2000). These results indicate that the CT excited triplet state of tetrahedral molybdenum species plays an important role in the photocatalytic decomposition reaction of NO in the presence of CO to form N_2 and CO_2 (Anpo and Higashimoto, 2001; Higashimoto et al., 2000; Tsumura et al., 2000).

UV-irradiation of Mo–MCM-41 in the presence of CO alone and subsequent evacuation at 293 K lead to efficient quenching of PL (Tsumura et al., 2000). Under these conditions, no EPR signal of Mo^{5+} ions can be detected, suggesting that the MO-CT in excited triplet state $(Mo^{5+}\!\!-\!\!O^-)^*$ reacts with CO to form Mo^{4+} ions and CO_2, in line with results reported by Shelimov et al. for the Mo^{6+}/SiO_2 catalysts (Lisachenko et al., 2002; Shelimov et al., 2003; Subbotina et al., 1999, 2001). It was also found that the Mo^{4+} ions thus formed readily react with NO and N_2O under dark conditions to produce N_2O and N_2, respectively, and also to reoxidize Mo^{4+} to Mo^{6+} ions, as indicated by the reappearance of the PL of Mo^{6+} ions after addition of NO and N_2O (Tsumura et al., 2000).

From these results, the catalytic cycle shown in Scheme 1 can be proposed for the photocatalytic decomposition of NO in the presence of CO for Mo–MCM-41 (Anpo and Higashimoto, 2001) and Mo^{6+}/SiO_2 (Lisachenko et al., 2002; Shelimov et al., 2003; Subbotina et al., 1999, 2001). This cycle is initiated by the formation of the CT excited triplet state of $(Mo^{5+}\!\!-\!\!O^-)^*$ upon UV-irradiation of the $(Mo^{6+}\!\!-\!\!O^{2-})$ surface ion-pair (Anpo and Higashimoto, 2001; Higashimoto et al., 2000; Lisachenko et al., 2002; Shelimov et al., 2003; Subbotina et al., 1999, 2001; Tsumura et al., 2000). Scheme 1 suggests that $(Mo^{6+}\cdots NO^{2-})$ is a key intermediate, which was observed indeed by EPR under conditions close to those of the photocatalytic reaction of NO + CO (Subbotina et al., 1999).

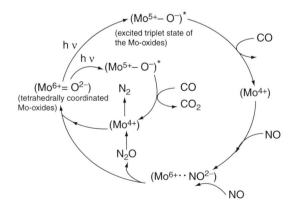

SCHEME 1 Photocatalytic decomposition of NO in presence of CO on Mo catalysts. (Reproduced with permission from Anpo and Higashimoto (2001).)

2.1.4. Silver-Containing Zeolites

PL experiments characterizing Ag^+-exchanged ZSM-5 zeolite (Ag(I)/ZSM-5) show that Ag(I) clusters (Ag_m^{n+}) are present in the cavities of ZSM-5, as shown by a band α at 380 nm (species A) upon excitation at 332 nm (Ju et al., 2003; Kanan et al., 2000; Matsuoka and Anpo, 2001; Matsuoka et al., 2003). UV-irradiation at 285 nm of Ag(I)/ZSM-5 leads to the transformation of Ag_m^{n+} clusters into reduced silver clusters ($Ag_m^{(n-1)+}$) which include a band β at 465 nm (species B) upon excitation at 315 nm. Species B is stable at 77 K, except when the system is further UV-irradiated at 285 nm. However, once Ag^+/ZSM-5 is heated to an ambient temperature, band β disappears and whereas band α is again observed with its initial intensity, that is, that observed before UV-irradiation. Moreover, these UV-induced changes of Ag^+/ZSM-5 are completely reversible under vacuum.

Figure 11 shows the intensity of the PL bands α and β as a function of the time of UV-irradiation at 285 nm (Matsuoka et al., 2003). Although, the intensity of band α (species Ag_m^{n+}) decreases, that of band β (species $Ag_m^{(n-1)+}$) follows the opposite trend, that is, it increases, suggesting that the former transforms into the latter. This transformation, which is reversible, suggests that the changes in PL are not related to the photoformation of silver clusters of different sizes, because such changes of cluster size cannot be easily restored by thermal annealing at temperatures close to 293 K (Anpo, 2000; Ju et al., 2003, 2004; Kanan et al., 2000, 2001a,b, 2003; Matsuoka and Anpo, 2001; Matsuoka et al., 2001; Matsuoka et al., 2003).

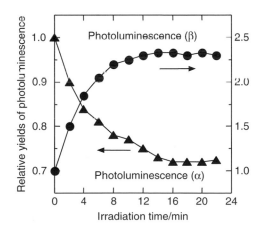

FIGURE 11   Time profiles of the intensities of the photoluminescence spectra observed at 380 nm (Exc = 332) (▲) and at 465 nm (Exc = 315 nm (●) under UV irradiation at 285 nm at 77 K of the Ag(I)/ZSM-5. (Photoluminescence spectra were measured at 77 K). (Reproduced with permission from Matsuoka et al. (2003))

2.1.5. Copper-Containing Zeolites

The relationship between the PL behavior of Cu(I) species and the photocatalytic activity of Cu(I)/SAPO-5 has been investigated as a function of the Si/Al ratio. The results show that the excited state of Cu(I) species plays a significant role in the photocatalytic decomposition of N_2O into N_2 and O_2.

As shown in Figure 12, Cu(I)/APO-5 (Si/Al = 0) exhibits only one PL band, at 460 nm, whereas Cu(I)SAPO-5 shows two main bands, at 490 and 570 nm with a shoulder at 460 nm (Chen et al., 2004a). On the basis of EPR measurements, which evidence only one type of Cu(II) species in APO-5 zeolite, the shoulder at 460 nm can be assigned to the radiative decay process of photoexcited Cu(I) ions $(3d^94s^1 \rightarrow 3d^{10})$ generated by reduction of Cu(II) ions, introduced in SAPO/APO-5 by exchange with the protons of Al(P)–OH groups. Cu(I)/APO-5 shows no activity in N_2O decomposition, despite the rather strong band at 460 nm, indicating that the exchanged Cu(II) ions do not significantly contribute to this reaction.

On the other hand, the bands at 490 and 570 nm can be attributed to the radiative decay process of two types of photoexcited Cu(I) ion, α and β, respectively, formed upon reduction of Cu(II) ions introduced by exchange of the protons of bridging hydroxyl groups (Si–OH–Al). Cu(I) ions of type β are assumed to be coordinated by –SiO(AlOPOAlO)$_n$Si– groups with relatively large values of n as compared to Cu(I) ions of type α.

In conclusion, Cu(I) α ions characterized by the band at 490 nm play an important role as stable active sites in the photocatalytic decomposition of N_2O. However, when Cu(I) ions are formed by reduction of Cu(II) ions exchanged with the hydroxyl groups of APO-SAPO, they contribute little to this reaction in spite of a fairly strong PL at 460 nm (Chen and Zink, 2000; Chen et al., 2004a,b; Matsuoka and Anpo, 2003a,b).

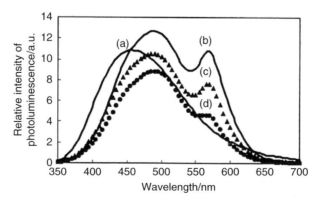

FIGURE 12 Photoluminescence spectra of (a) Cu(I)APO-5 and (b) Cu(I)SAPO-5 (Si/Al = 0.45), and the effect of the addition of (c) 0.3 Torr N_2O and (d) 0.6 Torr O_2 on the photoluminescence of Cu(I)SAPO-5. (Reproduced with permission from Chen et al. (2004a).)

2.1.6. Chromium-Containing Zeolites

Chromium-containing mesoporous silica molecular sieves (Cr-HMS) with tetrahedrally coordinated isolated chromium oxide (chromate) moieties can operate as efficient photocatalysts for the decomposition of NO and the partial oxidation of propane with molecular oxygen under visible light irradiation (Yamashita et al., 2001).

Cr-HMS exhibits PL bands at $\sim$550–570 nm, upon excitation of the absorption (excitation) bands at 250, 360, and 480 nm. These PL and absorption bands are similar to those obtained earlier with well-defined, highly dispersed chromium oxides grafted onto PVG or silica (Anpo et al., 1982). They can be assigned to CT processes in tetrahedral oxide moieties involving electron transfer from O^2 to Cr^{6+} and the reverse radiative decay, respectively.

2.2. Coordination Chemistry of Anions: MgO as a Model System

Because of its importance as model material and model catalyst, MgO has been much investigated. Thus, numerous bulk point defects such as F and V centers and their surface analogues have been identified and investigated by spectroscopy after irradiation of MgO by neutrons and γ rays (Che, 1985). These investigations, largely pioneered by Tench (Amphlett, 1985), have contributed much to improve our understanding of the resistance of oxides under intense irradiation of particles and rays emitted in nuclear reactors under working conditions. MgO has turned out to be an important catalyst for the oxidative coupling of methane, which is the principal component of natural gas and an abundant hydrocarbon resource. Lunsford and coworkers have been able to unravel the mechanism of the reaction catalyzed by MgO (Lunsford, 1993), by applying a matrix-isolation EPR technique (Driscoll et al., 1987). They pointed out the importance of oxygen species, the reactivity of which has been reviewed earlier (Che and Tench, 1982, 1983). Because of their ability to abstract hydrogen, mononuclear species were identified as catalytically active sites, the concentration of which could be increased by lithium-doping of MgO. More recently, MgO has gained a renewed interest in investigations of the basic properties of catalysts (Bailly et al., 2005a). These properties can be characterized by a test catalytic reaction such as the conversion of 2-methylbut-3-yn-2-ol (MBOH), which has been reviewed by Lauron-Pernot (2006).

PL is a useful technique to characterize the coordination not only of transition metal cations, as described above, but also of oxide anions. A typical coordination of oxide anions in the bulk lattices of oxides can be defined by the number of nearest neighbor cations; for octahedrally coordinated alkaline earth oxides, this number is 6. Ions of lower coordination are designated by the subscript LC to denote a low coordination. Earlier work (Anpo and Che, 1999) has shown that O_{LC}^{2-} surface oxide ions with different coordinations (O_{5C}^{2-}, pentacoordinated oxide ion on a MgO(100) plane; O_{4C}^{2-}, tetracoordinated oxide ion on an edge; O_{3C}^{2-}, tricoordinated oxide ion on a corner) could be discriminated by PL.

Recent improvement in the designs of PL cells and coupling of the cells to a dynamic vacuum system (Bailly et al., 2004) have made it possible to avoid quenching phenomena from gas-phase molecules. The emission spectra obtained with MgO are better resolved and more stable over time leading to an optimized PL yield of the more reactive surface species, namely O_{LC}^{2-} ions.

Outgassing of MgO samples at 1273 K leads to so-called "clean" surfaces, essentially free of carbonates and hydroxyl groups (Bailly et al., 2004, 2005b), thus allowing marked improvement the discrimination of O_{LC}^{2-} ions. Each type of O_{LC}^{2-} ions can be identified by means of a couple of wavelengths (λ_{exc}; λ_{em}) associated with the excitation and emission processes, respectively. The luminescence fingerprints (λ_{exc}; λ_{em}) of O_{5C}^{2-}, O_{4C}^{2-}, and O_{3C}^{2-} were shown to correspond to the (<220; <350), (230; 380), and (270; 460) couples, respectively (Bailly et al., 2004, 2005b).

Complementary to this approach based on the analysis of the position and shape of the PL signal, the lifetimes of photoluminescent emitting species on MgO nanocubes have been measured and modeled for the first time by means of a simple kinetic energy transfer model (Chizallet et al., 2006). Lifetime values ($\geq 10^{-6}$ s) suggest that phosphorescence takes place. The relative values of the lifetime of the species characterized by the (240; 380) and (280; 470) couples of wavelengths are consistent with higher delocalization of the exciton related to the former (Shluger et al., 1999; Sushko and Shluger, 1999; Sushko et al., 2002) and to their assignment to O_{4C}^{2-} and O_{3C}^{2-} ions, respectively. However, as illustrated in Figure 13, two types of O_{3C}^{2-} can be distinguished with comparable lifetimes: kinks and corners (Chizallet et al., 2006). Only corners are involved in energy transfer from O_{4C}^{2-}.

New developments are also proposed to evaluate the relative distribution of oxide ions of low coordination: values proportional to the O_{3C}^{2-}/O_{4C}^{2-} ratio can be obtained from the ratio of surface areas of the corresponding contributions of O_{LC}^{2-} (LC = 3C, 4C) extracted from excitation spectra. The shift of the emission band of O_{4C}^{2-} ions from 380 nm to lower wavelength depends on the O_{4C}^{2-}/O_{5C}^{2-} ratio: the lower the wavelength value, the lower the O_{4C}^{2-}/O_{5C}^{2-} ratio (Bailly et al., 2005a,b). Results are consistent with the morphology of samples obtained from TEM, XRD, and BET measurements and are dependent on the nature of the precursor (metallic magnesium or $Mg(OH)_2$ brucite phase) and on the morphology of the brucite phase, in relation with the synthesis method.

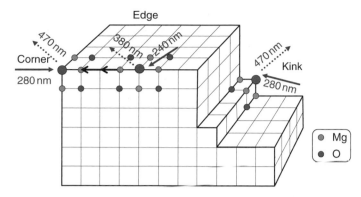

FIGURE 13 Schematic representation of excitation and emission processes occurring on a model MgO surface exhibiting O$_2$-LC ions at edges, corners, and kinks. Full and dotted arrows represent excitation and emission processes respectively, at the noted wavelength. The horizontal multiple arrow along the edge represents energy transfer phenomena. (Reproduced with permission from Chizallet et al. (2006).)

Besides the characterization of clean MgO surface, that is, free from carbonates and hydroxyls, the PL response of hydroxylated MgO surfaces is also specific, being sensitive to hydrogen bonding lateral interaction, between OH groups and/or molecularly adsorbed water molecules (Bailly et al., 2005b). Excitation of the hydroxylated MgO surface leads to three bands (250, 350, and 370 nm), whereas deexcitation occurring *via* energy transfer to hydroxyls in Mg–OH groups gives only two distinguishable contributions, at 410 and 470 nm. By comparing the relative stabilities of the three corresponding photoluminescent species, it has been shown that two of them, with the fingerprints (350–410) and (370–410), are involved in hydrogen bonding and/or interaction with water, whereas the thermally most stable one, with the fingerprint (250; 410), corresponds to isolated Mg–OH groups (Bailly et al., 2005b).

3. CHARACTERIZATION OF ACIDIC AND BASIC SITES BY MEANS OF PROBE MOLECULES AND PHOTOLUMINESCENCE SPECTROSCOPY

Materials can be characterized directly by PL if they exhibit intrinsic photoluminescent properties as described in Section 2 or, if this is not the case, *via* the use of probe molecules.

3.1. Characterization of Acidic Sites

Phosphorescence measurements of benzophenone adsorbed on Ti/Al binary oxides show the presence of its protonated form in addition to benzophenone hydrogen-bonded to surface OH groups (Nishiguchi

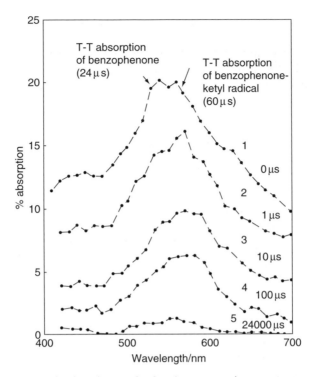

FIGURE 14 Time-resolved triplet–triplet (T–T) transient absorption spectra of benzophenone adsorbed on Al_2O_3 following excitation at 355 nm (pulse duration = 1.6 ms, pulse intensity = 10 mJ). 1, 0 ms; 2, 3 ms; 3, 5 ms; 4, 100 ms; 5, 24,000 ms after laser flash. (Reproduced with permission from (55).)

et al., 2001a,b). As shown in Figure 14, a triplet–triplet (T–T) absorption spectrum is observed at 570 nm, which is caused by benzophenone ketyl radicals (Nishiguchi et al., 2001b). Direct detection of time-resolved T–T transient absorption spectra of the adsorbed benzophenone ketyl radicals indicates that these radicals are formed on the surface of Al_2O_3 patches *via* hydrogen abstraction from acidic surface OH groups by the excited triplet state of benzophenone, in addition to the T–T absorption spectrum at 520–540 nm, because of the presence of adsorbed benzophenone (Nishiguchi et al., 2001b).

After adsorption of benzophenone on Ti/Al binary oxides, the systems were UV-irradiated, and benzhydrol and benzpinacol were detected as major products of photolysis. Their yields increased with UV-irradiation time, whereas the intensity of the phosphorescence of benzophenone decreased. Furthermore, the yield of benzhydrol depends strongly on the Ti/Al ratio. Benzpinacol is the only product observed in the photolysis of benzophenone adsorbed on SiO_2 and porous vycor glass

(Fujii et al., 1999; Li et al., 2005; Nishiguchi et al., 2001a,b; Zhang et al., 1999).

These results show that surface acidic sites of Ti/Al binary oxides play a key role in the photolysis of benzophenone. Indeed, protonation is favored when benzophenone is adsorbed, thus leading to the formation of benzhydrol.

3.2. Characterization of Basic Sites

The Brønsted basicity of a surface is related to its deprotonation ability, which can be probed by investigating the dissociative adsorption of protic molecules (Bailly et al., 2005a; Chizallet et al., 2006). The $O_{LC}^{2-} \leftrightarrow O_{LC}H^-$ transformation thus induced can be followed by PL, which is one of the few techniques able to simultaneously characterize oxide ions and their protonated forms. The same kind of equilibrium is also involved when a hydroxylated surface is undergoing thermal pretreatment (Section 2.1). PL is thus an interesting tool to evaluate the surface basic properties of alkaline earth oxides.

Adsorption of propyne and methanol on MgO has been followed by PL and IR spectroscopies to determine the Brønsted basicity (Chizallet et al., 2006). Gas in excess and physisorbed species were removed by evacuation at room temperature before PL spectra were recorded.

No conclusion could be reached from the global quenching of the signal after adsorption of methanol, but propyne can discriminate oxide ions of low coordination. Indeed, PL results indicate that O_{3C}^{2-} and O_{4C}^{2-} ions are involved in propyne dissociation. On the basis of PL and IR (semiquantitative evaluation of the number of molecules dissociated per square meter) characterizing dissociative adsorption of propyne and in line with theoretical results showing that acetylene does not dissociate on O_{5C}^{2-} (Nicholas et al., 1998), it is inferred that $Mg_{3C}-O_{4C}$ and $Mg_{4C}-O_{3C}$ ion pairs are involved in propyne dissociation (Chizallet et al., 2006).

4. PHOTOLUMINESCENCE INVESTIGATIONS OF PHOTOCATALYTIC PROCESSES INVOLVING BULK SEMICONDUCTOR TiO$_2$-CONTAINING PHOTOCATALYSTS

Anpo and coworkers (Anpo, 2004; Anpo and Che, 1999) investigated the PL behavior of TiO$_2$ photocatalysts in the presence of various kinds of reactants. The dependence of the PL intensity on the nature of the atmosphere was explained in terms of surface band bending of TiO$_2$ particles, its extent depending on the electronegativity or electroaffinity of the reactant molecules. Furthermore, such an effect on the PL intensity was found to be reversible after elimination of the reactant molecules by flowing N_2, allowing one to propose the idea of using such an effect for remote-controlled gas sensors.

Murabayashi coworkers (Nakajima et al., 2001, 2002) investigated the PL properties of rutile and anatase powders at room temperature in air in the absence or presence of reactants such as methanol or ethanol, with the goal of understanding the mechanisms of gas-phase photocatalytic reactions. In experiments with rutile, they observed that the PL intensity in the presence of methanol or ethanol increased linearly with the square root of UV-irradiation time, as shown in Figure 15. They also found a linear time dependence of the integrated amount of photodesorbed O_2 (Nakajima et al., 2002). The time dependence of the PL intensity and the effect of O_2 photodesorption were explained (as mentioned above) in terms of band bending of the powder. These results, however, were not observed for anatase powders. The authors explained that such a difference in PL behavior was related to the difference in photocatalytic activities of these alcohols.

It is important to understand the dynamics of photogenerated electrons and holes and their role in semiconductor systems to facilitate the design and development of efficient photocatalysts (Anpo, 2002; Anpo and Che, 1999; Matsuoka and Anpo, 2005). As reported earlier (Anpo and Che, 1999), the presence of a small number of platinum particles on TiO_2 is known to strongly increase the yield of photoreactions. This effect is attributed to a migration of the photoformed electrons from the conduction band of TiO_2 to the platinum particles, whereas the holes remain in the valence band, leading to effective charge separation of the

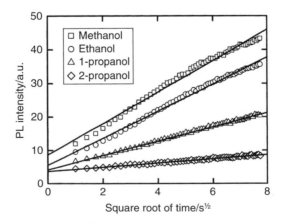

FIGURE 15 Photoluminescence intensity of rutile TiO_2 powder Kanto-R in air with C_1–C_3 alcohols (methanol, ethanol, 1-propanol, and 2-propanol) as a function of the square root of UV-irradiation time from 1 to 60 s. The squares of the regression coefficients are larger than 0.98. (Reproduced with permission from Nakajima et al. (2002).)

photogenerated carriers, in addition to the catalytic effect of platinum. However, the direct observation of the photoformed electron-hole charge separation process in TiO_2 photocatalysts has not been reported. Time-resolved pump-probe spectroscopy is a powerful method to investigate such dynamics of charge carriers with high time resolution.

Furube et al. have investigated the dynamics of charge separation of electrons and holes photogenerated in TiO_2 and platinum-loaded TiO_2 (Pt/TiO_2) photocatalysts by transient absorption measurements in the whole visible region by femtosecond diffuse reflectance spectroscopy (Furube et al., 1999, 2001a,b). Transient absorption spectra and intensity–time profiles for TiO_2 and Pt/TiO_2 (Furube et al., 2001a) are shown in Figure 16, for an excitation wavelength of 390 nm. A spectral shape with stronger absorption at longer wavelengths can be observed for Pt/TiO_2 for any delay time, especially at shorter delay times. A fast decay of 4–5 ps is observed in the absorption at 600 nm for Pt/TiO_2, whereas there is almost no absorption and decay for TiO_2 (Furube et al., 2001a). Similar transient absorption spectral shapes are observed for platinum loadings exceeding 0.2 wt %, as shown in Figure 16, whereas the decay curve depends on the amount of platinum.

Figure 17 shows that the decay is more rapid for increasing platinum loading, with a larger relative amplitude of the decay component. It was found that the inverse of the decay time is approximately proportional to platinum loading (Furube et al., 2001a), suggesting an electron migration from TiO_2 to platinum, its extent depending on the amount of platinum. Thus, the charge separation in Pt/TiO_2 was directly measured for the first time with a process taking place within a few picoseconds (Furube et al., 2001a).

Slower electron–hole recombination was observed in air, which was explained in terms of upward band bending near the surface because of adsorbed O_2. In catalysts with high photocatalytic activity, the observed trapped carriers mainly exist near the particle surface and do not undergo rapid recombination after photoexcitation, as illustrated by Figure 16 (Anpo et al., 1991; Furube et al., 2001a).

Furthermore, femtosecond diffuse reflectance spectroscopy with a white continuum probe pulse has been applied to detect the dynamics of hole transfer from photoexcited TiO_2 to adsorbed reactant molecules. As shown in Figure 18, at pH < 7 of the TiO_2 aqueous suspension with KSCN, ultrafast hole transfer takes place in less than 1 ps (Furube et al., 2001b). Subsequent structure stabilization of dimer anion radicals, $(SCN)_2{}^-$, within a few picoseconds and slow hole transfer with a time constant of a few hundred picoseconds are clearly observed (Furube et al., 2001b). Fast hole transfer is caused by a surface-trapped state interacting strongly with adsorbed molecules. Slow hole transfer observed at pH values >7 is caused by deep trapped states with a Boltzmann distribution

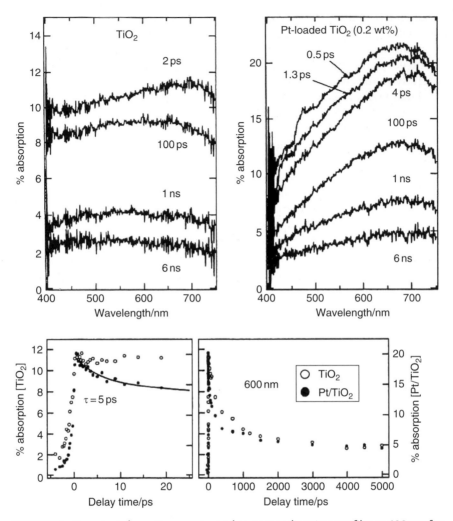

FIGURE 16 Transient absorption spectra and corresponding time profiles at 600 nm for TiO$_2$ and 0.2 wt % Pt-loaded TiO$_2$ powder with 390 nm excitation. (Reproduced with permission from (Furube et al. (2001a).)

of energy levels. Time-resolved spectroscopic techniques can easily be applied to opaque samples for heterogeneous photocatalytic reactions. It is thus expected that such direct analysis of interfacial CT for various kinds of photocatalyst such as surface-modified, size-controlled, ion-implanted and doped TiO$_2$, and hybrid multilayer systems such as SiO$_2$–TiO$_2$ will contribute to the design of efficient TiO$_2$ photocatalysts (Dutta and Kim, 2003; Fujihara et al., 2000; Miyashita et al., 2003; Yu et al., 2003).

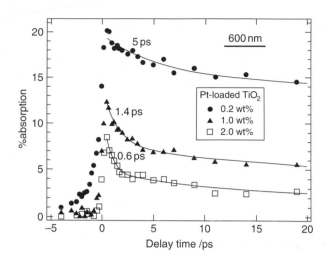

FIGURE 17 Dependence of transient absorption decay at 600 nm on platinum loading for Pt-loaded TiO_2 powder, with 390 nm excitation. (Reproduced with permission from Furube et al. (2001a).)

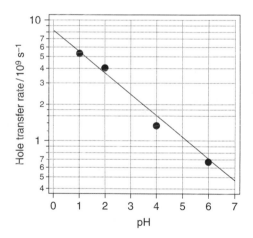

FIGURE 18 Rate of hole transfer from photoexcited TiO_2 to SCN- plotted as a function of the pH of KSCN solution. (Reproduced with permission from Furube et al. (2001b).)

In the photodecomposition of H_2O at the water–rutile TiO_2 interface, the primary intermediates of O_2 evolution can be investigated by multiple internal reflection IR absorption and PL. On the basis of spectroscopic data, Nakato et al. (Nakamura and Nakato, 2004) proposed a reaction mechanism in which O_2 photoevolution is initiated by nucleophilic attack of H_2O on a photogenerated hole at a surface lattice O site, and not by oxidation of a surface OH group by a hole. In the case of a photoetched rutile TiO_2 electrode, Nakato et al. (Nakamura and Nakato, 2004) observed a PL band at 820 nm that was more intense for the acidic (pH = 1.2) than for the alkaline (pH = 12.8) solution. These results were attributed to the photoformation of a hole at a surface lattice O site which upon nucleophilic attack of a H_2O molecule leads to oxygen evolution (Nakamura and Nakato, 2004).

5. EFFECT OF PRESSURE ON PHOTOLUMINESCENCE SPECTRA

There have been few investigations of the effect of pressure on PL spectra. In what follows, two pressure domains (0.003–66 mbar of gas, Section 5.1 (40d) and hydrostatic pressures of 1 bar to 7.6 GPa, Section 5.2 (Park et al., 2002; Sapelkin et al., 2000)) have been investigated for Mo^{6+}/SiO_2 photocatalysts and intercalation compounds, respectively.

5.1. Low-Pressure Domain: Physical versus Chemical Quenching

The quenching effect of gaseous NO, O_2, CO, and N_2O on the PL of $Mo^{6+}/$ SiO_2 photocatalysts has been investigated at room temperature either under vacuum or as a function of pressure (0.003–66 mbar) of the quenching gas (Shelimov et al., 2003). Nonlinear plots of the relative PL intensity I_0/I (I_0, initial intensity under vacuum) versus gas pressure were obtained.

The quenching effect is quantitatively well described in the entire pressure range investigated in the framework of a simple kinetic scheme, assuming that only adsorbed molecules efficiently quench the $(Mo^{5+}=O^-)^*$ excited triplet state. A Langmuir-type adsorption isotherm can be applied for calculating the fraction of adsorbed quencher molecules leading to an expression that correctly describes the $I_0/I = f(P_g)$ relationship in the entire pressure range. At low pressures, this expression reduces to the Stern–Volmer equation (Equation (1)) which predicts a linear dependence of I_0/I on the pressure.

Paramagnetic NO and O_2 molecules and reactive CO molecules efficiently quench the PL in the order NO > O_2 > CO, whereas N_2O only weakly affects the PL intensity. This work allowed confirmation of some important steps of the mechanism proposed earlier for the photocatalytic reduction of NO by carbon monoxide on Mo/SiO_2 (Subbotina et al., 1999). In particular, the NO photoreduction kinetics was consistently described by Subbotina et al. (1999) assuming first that the deactivation rate constant is much smaller than the quenching rate constant and second that the NO molecules efficiently quench the $(Mo^{5+}=O^-)^*$ excited triplet state without chemical interaction (i.e., "physical quenching"), in contrast to the "chemical quenching" by CO molecules to yield CO_2 molecules. The PL data demonstrate the correctness of those assumptions.

The ratio of the quenching rate constants for NO and CO calculated from the computer best fits of the experimental pressure dependence is in agreement with earlier data indicating the kinetics and mechanism of the photocatalytic reduction of NO by carbon monoxide on Mo^{6+}/SiO_2 photocatalysts reported earlier by Shelimov coworkers (Lisachenko et al., 2002; Subbotina et al., 1999, 2001).

5.2. High-Pressure Domain

In solution, the intensity of the emission plotted against the concentration of emitting species exhibits a maximum. This phenomenon is known as the concentration effect, leading to self-quenching and/or radiationless deactivation. It can be observed in the gas phase when emitting species are present under physically or chemically stressed environments (i.e., at high pressures or for sterically rigid states, respectively).

Intercalation compounds involving layered double-hydroxides (LDHs), illustrated by the structure shown in Figure 19 (Park et al., 2002), were synthesized to facilitate the measurement of the uniaxial stress (i.e., along the direction perpendicular to the layers), that host layers exert on intercalated species. Below a certain thickness of ~100 mm, the intercalated structures become more transparent to visible light, so that the information obtained by PL represents the properties of the bulk, not of the surface. Because it is sensitive to the deformation of the intercalants, the PL from samarium complexes was employed to measure the uniaxial stress, by using PL peak positions as a function of pressure (Park et al., 2002; Sapelkin et al., 2000).

Figures 20A and B show the PL spectra, recorded at 290 K, at 600 nm, and as a function of pressure, for $Cs_9(SmW_{10}O_{36})$ and $SmW_{10}O_{36}$–LDH, respectively (Park et al., 2002). For the sake of comparison, the line shapes are normalized and displaced along the vertical axis. In both cases, the peak position is red-shifted by 4–5 nm when the hydrostatic pressure increases from 1 bar to 61 kbar. It was shown that the red-shift from A to A* lies solely in the deformation of the samarium complexes by the uniaxial stress exerted by the host layers, whereas the shift from B to B* is also influenced by the change in the cation environment. Under the same conditions, B is not at the same position for the non-intercalated $(HN(n\text{-butyl})_3)_9(SmW_{10}O_{36})$ and $Cs_9(SmW_{10}O_{36})$ compounds (Park et al., 2002). Thus only peak A* is available to measure the unixial stress. This observation can be used to determine the uniaxial stress, when the external pressure is zero. For the $SmW_{10}O_{36}$—LDH system, the uniaxial stress varies significantly from 75 at 28 K to 140 kbar at 290 K (Park et al., 2002).

The strength of interaction between the electronic state of the complex and the environment (phonon) increases with the pressure,

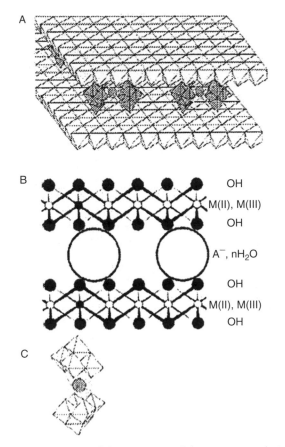

FIGURE 19   (A) Schematic view of the structure of the LDH intercalation compound. Shaded parts represent Sm ion complexes, the intercalants. (B) Lateral view of the LDH intercalation compound. The top and bottom layers are the host layers and the large circles in between them are negative ions (A^-) and water molecules (nH_2O). M(II) and M(III) represent Mg^{2+} and Al^{3+} ions, respectively. (C) Schematic drawing of the samarium ion complex $(SmW_{10}O_{36})^{9-}$. The central shaded circle is the Sm^{3+} ion; at the vertices of the small octahedra are oxygen atoms, whereas W atoms are located at the centers of the small octahedra. Five small octahedra compose a large lacunary octahedron (one small octahedron is vacant) and the two lacunary octahedra are in a twisted configuration with respect to each other.(Reproduced with permission from Park et al. (2002).)

leading generally to a reduction of the number of peaks and, as illustrated in Figure 20, to a broadening of the peaks.

Figure 21A shows the time evolution of the intensity of peak A at 290 K at several pressures. The PL decay rate increases with hydrostatic pressure. In Figure 21B, apart from the fast-decay components appearing as sharp spikes at 2.984 μs, the PL at 1 bar has almost a single exponential decay with a lifetime of 102.6 μs. This long-time component is still a single exponential at

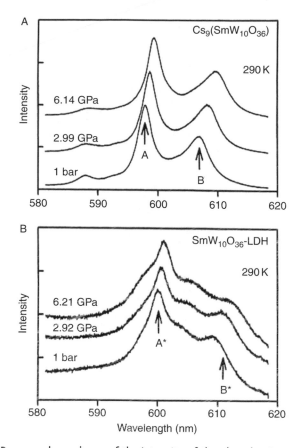

FIGURE 20 Pressure dependence of the intensity of the photoluminescence spectrum recorded at room temperature of (A) $Cs_9(SmW_{10}O_{36})$ and (B) $SmW_{10}O_{36}$ -LDH. For the sake of clarity, the line shapes were normalized and displaced vertically. In both cases (A) and (B) the peak positions are red shifted with increasing pressure. (Reproduced with permission from Park et al. (2002).)

30 kbar with a lifetime of 64.4 μs. The temperature-dependent variation of the uniaxial stress by such a large magnitude looks unusual at first sight. However, this phenomenon can be well understood in terms of a structural transformation, referred to as staging.

The staging phenomenon is characterized by a staging number n, which refers to the number of host layers separating two adjacent guest layers (Park et al., 2002). Thus, whereas a stage-1 compound has guest layers in every gallery space, not all the gallery spaces of stage-n compounds for $n > 1$ are filled with guest ions. Because of the finite rigidity of the host layers, they pucker at intercalating ions. The competition between

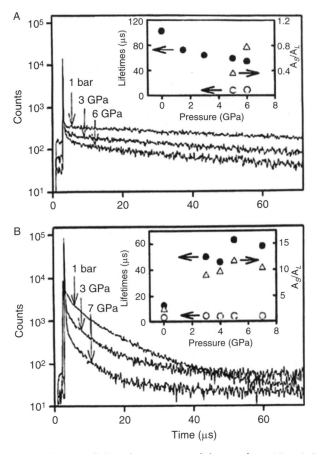

FIGURE 21   Time evolution of photoluminescence (A) at peak position A, indicated in Figure 19(A) for pre-intercalation compound ($Cs_9(SmW_{10}O_{36})$ salt), and (B) at peak position A*, indicated in Figure 19(B) for the LDH intercalation compound, under various pressures at room temperature. In the insets of (A) and (B) filled circles are long lifetimes T_L, empty circles short lifetimes T_S and triangles the corresponding ratio A_S/A_L. (Reproduced with permission from Park et al. (2002).)

the attractive interlayer elastic interaction and the repulsion interlayer elastic interaction is one of the main causes of the staging phenomenon.

6. EFFECT OF TEMPERATURE ON PHOTOLUMINESCENCE SPECTRA

The effect of temperature on the PL spectra is important as regards the internal conversion between the lowest excited triplet state and the ground state, because of the longer lifetimes of triplet states (Anpo and Che, 1999).

Therefore, the radiationless deactivation process for most emitting species in their excited triplet states is so efficient that phosphorescence spectra can be observed only when the system including the emitting species is frozen at liquid nitrogen temperature. Of course, the environmental aspects of radiationless deactivation do not determine all the whole properties. Pyridine molecules show neither fluorescence nor phosphorescence at room temperature or in the rigid glass state at liquid nitrogen temperature.

In general, the intensity of emission decreases when the temperature increases, because of the higher probability of other radiationless deactivation processes of excited molecules (Anpo and Che, 1999; Turro, 1978). Furthermore, a much better resolution of the vibrational fine structure of the emission (fluorescence and phosphorescence) can be observed at low temperature, as shown in Section 2.1.2, for the phosphorescence spectrum of highly dispersed tetrahedral vanadium species which exhibits a well resolved vibrational fine structure related to the $V=O$ double bond.

The other process, internal conversion, is strongly temperature-dependent: decreasing the temperature of the system causes a sharp increase in the emission, both for fluorescence and phosphorescence spectra.

In the field of solid materials, a number of mapping techniques have been proposed for defect profiling in SiC wafers, including PL, optical transmission, capacitive contactless resistivity, Raman imaging and, recently, thermally stimulated luminescence (TSL) (Amour et al., 1994; Bissiri et al., 2001; Korsunska et al., 2004). When the temperature increases from 4.2 to 150 K, the PL intensity is noticeably reduced, to a level below the sensitivity of the PL equipment (Korsunska et al , 2004). However, at temperatures above 165 K, the PL spectrum is dominated by a new broad "orange" luminescence with spectral maximum at 1.82 eV, assigned to the so-called O-band, which persists at temperatures above room temperature, as shown in Figure 22 (Korsunska et al., 2004).

The PL spectra were analyzed for Al- and B-doped p-type SiC single crystals in a broad spectral range (0.7–3.25 eV) at temperatures up to 573 K. The effect of thermal activation on the PL intensity is to strongly increase the visible luminescence at elevated temperatures. It is clear that the O-band is composed of at least two subbands with maxima at 1.86 (O_1) and 1.50 eV (O_2). The result of such a deconvolution is presented from 295 to 403 K, the intensity of the "orange" luminescence increasing exponentially by as much as 25 times. From an Arrhenius semilogarithmic plot of the PL intensity versus reciprocal absolute temperature a characteristic activation energy of 205 meV for the PL increase was obtained. The observed effect of PL activation with increasing temperature is attributed to an exponential rise of free-hole concentration caused by thermal ionization of the Al and B levels. The free-to-bound hole transition is identified as the recombination mechanism leading to the intense "orange" PL

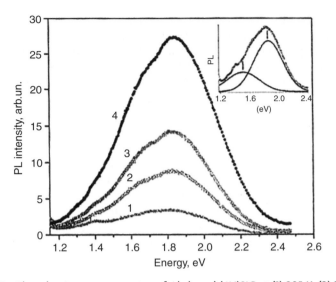

FIGURE 22 Photoluminescence spectra of Al-doped (4H)SiC at (1) 295 K, (2) 335 K, (3) 343 K and (4) 403 K. The inset shows a deconvolution of the 1.82-eV band in two individual Gaussian components. (Reproduced with permission from Korsunska et al. (2004).)

band at 1.8 eV and to the relevant thermally stimulated luminescence glow curves in p-type SiC semiconductors (Korsunska et al., 2004).

7. USE OF PHOTOLUMINESCENCE SPECTROSCOPY IN HIGH-THROUGHPUT EXPERIMENTS

In relation to the field of heterogeneous catalysis, an advantage of PL is that solid samples can be easily investigated. Other assets include its high sensitivity (often samples of less than a few tens of mg only are needed), the short analysis times required, and the different types of data, as discussed in the previous sections, which can be obtained from the investigation of the spectral and temporal profiles of both the excitation and emission processes (Anpo and Che, 1999).

Taking advantage of these characteristics, Corma and coworkers (Atienzar et al., 2004) have described and validated an automatic system for high-throughput (HT) characterization of large libraries of solid materials by PL in the range of 350–800 nm. The system proposed is able to provide time-resolved transient emission spectra in the microsecond range and can be employed to characterize materials, particularly in the fields of catalysis and electroluminescence.

The authors have first evaluated the HT system (Atienzar et al., 2004) by screening a 96-well sample library of $[Ru(bpy)_3]^{2+}$ encapsulated in zeolites. The potential of the technique was then investigated through the accelerated optimization of the synthesis of two host–guest materials with photoluminescent properties with potential application as active components in electroluminescent devices or as photocatalysts.

Two host–guest materials were selected and prepared: PPV oligomers and 2,4,6-triphenyl-pyrylium (TP^+) incorporated in large-pore zeolite hosts, hereafter referred to as PPV@zeolite and TP^+@zeolite. As an example of the potential value of the technique, the experimental preparation parameters investigated to prepare the PPV@zeolite system on a factorial design ($2 \times 3^2 \times 5$) were as follows: synthesis temperature ($3\times$), ranging from 473–573 K; zeolite structure including three large-pore zeolites, X, Y, and β; nature of the exchange ion ($5\times$) (H^+ and alkali-metal ions: Li^+, Na^+, K^+, and Cs^+); and two different monomer loadings ($2\times$), 4 and 6 wt %.

The HT luminescence characterization described above can also be applied to other materials, such as catalysts and sensor/probe molecules.

8. CONCLUSIONS AND OUTLOOK

The preceding review of PL which appeared in this series (Anpo and Che, 1999) gave the principles and applications of this technique to the investigation of solid surfaces in relation to adsorption, catalysis, and photocatalysis.

Since then, significant progress has been achieved in the characterization of solid materials and particularly of solid catalysts. A large part of the work published has dealt with the direct characterization of solids.

The work has largely focused on the coordination chemistry of transition metal ions (i.e., on the description of the nature and symmetry of their environments) (Section 2.1), in line with other spectroscopies, mainly optical (UV–vis), magnetic (EPR and NMR), which take advantage of partly filled d orbitals, and structural (EXAFS) (Sojka and Che, 2009). It has even become possible with PL *via* well-resolved fine structures to determine the extent of distortion of the environment of tetrahedral species (e.g., vanadium species in zeolites (Section 2.1.2)). It is likely that such information combined with that derived from other spectroscopies, vibrational on one hand, such as IR and Raman, and electronic on the other hand, such as EPR, will be applied by theoreticians to further improve the existing models and our understanding of the nature and role of surface species involved in catalytic processes.

By contrast, much less effort has been devoted to the coordination chemistry of anions (Section 2.2), a field in which PL, and to a lesser extent UV–vis spectroscopies (Anpo and Che, 1999; Che and Tench, 1982), have facilitated significant progress. One of the most outstanding results is that the coordination number of oxide ions of alkaline earth oxides, notably MgO, can now be determined. It is likely that, in future work, the same goal will be attained for hydroxyl groups. Their role becomes important, if not dominating, as soon as water is involved even in trace amounts either during thermal treatment or during catalysis, when water is one of the reaction products. Because of the basic character of anions of some materials, such as alkaline and alkaline earth oxides (Section 2.2), one can predict that PL will gain more importance in the field of catalysis involving basic catalysts. It is also likely that, not only oxide O_{LC}^{2-} ions, but other types of anions will be investigated more thoroughly in the future.

Beside its use for the direct characterization of solids, PL has been applied in combination with probe molecules, luminescent or not, to investigate the acidic and basic properties of oxide surfaces (Section 3). In relation to catalysis, the investigation of the formation of hydroxyl groups from hydrogen-containing reactants (such as hydrocarbons or alcohols) or reaction products (notably water) is an important step forward. Various types of hydroxyl group can be formed upon adsorption of such molecules. In the case of water adsorption on MgO, this can be illustrated schematically by the following reaction:

$$
\diagdown Mg^{2+} \quad O_{LC}^{2-} + H_2O \rightleftharpoons \left[\begin{array}{c} HO_{1C} \\ | \\ M \end{array} \right]^{+} \left[\begin{array}{c} H \\ | \\ O_{LC} \end{array} \right]^{-} \diagdown \qquad (3)
$$

which shows that several types of $O_{LC}H$ group can be formed. One denoted $O_{1C}H$, with its oxygen coordinated to only one Mg^{2+} ion, comes from the adsorption of the OH^- ion produced by heterosplitting of water. The others, denoted $O_{LC}H$, where the oxygen is coordinated to several Mg^{2+} ions, come from the adsorption of the H^+ ion produced by heterosplitting of water on O_{LC}^{2-} oxide ions. LC is equal to 5, 4, or 3, depending on whether the parent O^{2-} is on a face, edge, or corner, respectively. It can be expected that these different types of hydroxyl groups together with bare oxide O_{LC}^{2-} ions will exhibit different acid–base characteristics. It is probable that PL, together with other spectroscopies, will contribute to the improvement our understanding of acid–base catalysis.

Bulk semiconductor photocatalysts (Section 4), such as TiO_2 and TiO_2-containing materials (Sections 2.1 and 4), have continued to gain attention, because of their application to the electrochemical photolysis of water (Fujishima and Honda, 1972) and also because of their importance in photocatalysis (Anpo, 2005).

Because of the dependence of the PL intensity of TiO_2 on the nature of the gas-phase molecules introduced (alcohols) and its reversibility upon elimination of the molecules by flowing dinitrogen, there is hope that such an effect can be applied to gas sensors. With the combined use of several techniques (PL, time-resolved femtosecond diffuse reflectance spectroscopy, multiple internal reflection IR absorption), the dynamics and role of photogenerated electrons and holes in the absence or presence of metals (notably platinum) are now better understood, at both the gas–solid and liquid–solid interfaces. It is also likely that not only TiO_2, but other types of semiconductors will be more thoroughly investigated in the future.

Regarding the investigation of catalysts in the working state, the influence of pressure on PL spectra (Section 5) has been considered. For low pressures of gas (0.003–66 mbar) for the (NO, O_2, CO, N_2O)–Mo^{6+}/ SiO_2 system, the plot of PL intensity versus the gas pressure follows a Langmuir-type isotherm, pointing to the importance of desorbed molecules in the quenching process. A distinction was also made between "physical" (NO case) and "chemical" (CO case) quenching.

Because it is sensitive to the deformation of the intercalants, the PL from samarium complexes, such as $SmW_{10}O_{36}^{9-}$, was employed to measure the uniaxial stress exerted by the host such as LDHs on the intercalants. PL peak positions as a function of the hydrostatic pressure within the range of 1–75 kbar were measured (Park et al., 2002; Sapelkin et al., 2000).

Experimental results show that intense PL peaks can be observed, even at high pressure. Their positions, however, can be red-shifted because of the deformation of the samarium complexes by the uniaxial stress exerted by the LDH host layers. The shift may also have a chemical origin, that is, the change of the countercation of the samarium complex. By using counter cation-independent peaks, the uniaxial stress was measured to be 140 kbar at room temperature for $SmW_{10}O_{36}$–LDH. The shapes of the peaks are also affected. When the pressure increases, the strength of interaction between the electronic state and the environment (phonon) increases, leading generally to a reduction of the number of peaks and to a broadening of the peaks.

In line with such investigations, it is likely that in the future other spectroscopies will be used to investigate the effect of pressure on intercalants, particularly EPR, NMR, and EXAFS, which are useful for investigating the oxidation state (Dyrek and Che, 1997; Kuba et al., 2003; Mériaudeau et al., 1977) and environment (Carabineiro et al., 2006; Carrier et al., 1999; Villaneau et al., 2004) of tungsten, and for providing more insight into the physical versus chemical consequences of pressure changes. In relation to catalysis, hosts such as micro- and mesoporous solids require more investigation.

The influence of temperature on PL spectra is considered in section 6. In general, in agreement with the Boltzmann law, the intensity of PL increases when the temperature decreases. Furthermore, a much better resolution of vibrational fine structure of the emission (fluorescence and phosphorescence) can be observed at low temperature.

Besides this physical aspect, other effects can also be observed, as in the case of Al- and B-doped p-type SiC single crystals. The effect of thermal activation (in the range from room temperature to 573 K) is to lead to a new broad orange luminescence, the so-called "O-band." This is attributed to an exponential rise of free-hole concentration caused by thermal ionization of the Al and B levels. The free-to-bound hole transition is identified as the recombination mechanism leading to the intense "orange" PL band at 1.8 eV.

Because of the advantages of PL (Section 7 and Anpo and Che, 1999), the last section has been devoted to HT experiments, which give convincing evidence that PL is most adapted to characterization of a large number of samples in a short time.

This chapter is largely concerned with the characterization by PL of various types of solids under inert and reactive atmospheres, and at low temperatures and pressures (i.e., under "spectroscopic" conditions). In contrast, to the best of our knowledge, there has been no report so far of PL spectra of a catalyst under working conditions (i.e., in the presence of flowing reactants at pressures and temperatures typically involved in catalysis experiments in the laboratory).

To this end, we therefore have considered the data dealing with the effect of pressure and temperature on PL spectra. The analysis of these data suggests that pressure alone (i.e., exerted hydrostatically) does not appear to be a problem, whereas temperature appears to work against the measurement of intense spectra. This limitation can be, to some extent, compensated by the high sensitivity of the technique. The presence of gases able to quench the spectra is certainly an even more challenging problem which should be resolved in the near future.

The predictions concerning PL presented earlier (Anpo and Che, 1999) (applications to a broader range of systems, such as sulfides, (oxi)carbides, (oxi)nitrides; time-resolved equipment for in-depth investigation of excited states; increased attention to solar energy and photocatalysis related to environmental problems) are still largely valid.

ACKNOWLEDGMENTS

The authors gratefully acknowledge B. N. Shelimov (Zelinsky Institute of Organic Chemistry, Russian Academy of Sciences, Moscow, Russia) and G. Costentin, H. Lauron-Pernot and C. Chizallet (Laboratoire de Réactivité de Surface, UMR 7609-CNRS, Université Pierre et

Marie Curie-Paris 6) for most valuable discussions. S. Dzwigaj is grateful to CNRS and the Université Pierre et Marie Curie-Paris 6 for various appointments and financial support.

REFERENCES

Amphlett, C.B., *Stud. Surf. Sci. Catal.* **21**, 1 (1985).

Anpo, M. (Ed.) Special issue "Preparation, Characterization, and Reactivities of Titanium Oxide-based Photocatalysts, *Topics Catal.* **35**, 1 (2005).

Anpo, M., *Bull. Chem. Soc. Jpn.* **77**, 1427 (2004), and references therein.

Anpo, M. (Ed.), *Curr. Opin. Solid State Mater. Sci.* **6**, (2002) (Special issue on Photoluminescence).

Anpo, M., and Che, M., *Adv. Catal.* **44**, 119 (1999), and references therein.

Anpo, M., Chiba, K., Tomonari, M., Coluccia, S., Che, M., and Fox, M.A., *Bull. Chem. Soc. Jpn.*, **64**, 543 (1991).

Anpo, M., and Higashimoto, S., *Stud. Surf. Sci. Catal.* **135**, 123 (2001).

Anpo, M., Higashimoto, S., Matsuoka, M., Zhanpeisov, N., Shioya, Y., Dzwigaj, S., and Che, M., *Catal. Today* **78**, 211 (2003).

Anpo, M., Higashimoto, S., Shioya, Y., Ikeue, K., Harada, M., and Watanabe, M., *Stud. Surf. Sci. Catal.* **140**, 27 (2001).

Anpo, M., and Moon, S.C., *Res. Chem. Intermed.* **25**, 1 (1999).

Anpo, M., Tanahashi, T., and Kubokawa, Y., *J. Phys. Chem.* **86**, 1 (1982).

Anpo, M., Takeuchi, M., Ikeue, K., and Dohshi, S., *Curr. Opin. Solid State Mater. Sci.* **6**, 381 (2002).

Anpo, M., Yamashita, H., Ikeue, K., Fujii, Y., Zhang, S.G., Ichihashi, Y., Park, D.R., Suzuki, Y., Koyanao, K., and Tatsumi, T., *Catal. Today* **44**, 327 (1998).

Anpo, M., Zhang, S.G., Higashimoto, S., Matsuoka, M., and Yamashita, H., *J. Phys. Chem. B* **103**, 9295 (1999).

Atienzar, P., Corma, A., Garcia, H., and Serra, J.M., *Chem. Eur. J.* **10**, 6043 (2004).

Bailly, M.L., Chizallet, C., Costentin, G., Krafft, J.M., Lauron-Pernot, H., and *J. Catal.* **235**, 413 (2005a).

Bailly, M.L., Costentin, G., Krafft, J.M., and Che, M., *Catal. Lett.* **92**, 101 (2004).

Bailly, M.L., Costentin, G., Lauron-Pernot, H., Krafft, J.M., and Che, M., *J. Phys. Chem. B* **109**, 2404 (2005b).

Bissiri, M., Gaspari, V., Polimeni, A., Baldassarri, G., von Hogersthal, H., Capizzi, M., Frova, A., Fischer, M., Reinhardt, M., and Forchel, A., *Appl. Phys. Lett.* **79**, 2585 (2001).

Carabineiro, H., Villaneau, R., Carrier, X., Herson, P., Lemos, F., Ribeiro, F.R., Proust, A., and Che, M., *Inorg. Chem.* **45**, 1915 (2006).

Carrier, X., d'Espinose de la Caillerie, J.B., Lambert, J.F., and Che, M., *J. Am. Chem. Soc.* **121**, 3377 (1999).

Che, M., *Stud. Surf. Sci. Catal.* **21**, 11 (1985).

Che, M., and Tench, A.J., *Adv. Catal.* **31**, 77 (1982).

Che, M., and Tench, A.J., *Adv. Catal.* **32**, 1 (1983).

Chen, H.J., Matsuoka, M., Zhang, J.L., and Anpo, M., *Chem. Lett.* **33**, 1254 (2004a).

Chen, H.J., Matsuoka, M., Zhang, J.L., and Anpo, M., *J. Catal.* **228**, 75 (2004b).

Chen, J., and Zink, J.I., *Inorg. Chem.* **39**, 433 (2000).

Chizallet, C., Bailly, M.L., Costentin, G., Lauron-Pernot, H., Krafft, J.M., Bazin, P., Saussey, J., and Che, M., *Catal Today* **116**, 196 (2006).

Chizallet, C., Costentin, G., Krafft, J.M., Lauron-Pernot, H., and Che, M., *Chem. Phys. Chem.* **7**, 904 (2006).

Driscoll, D.J., Campbell, K.D., and Lunsford, J.H., *Adv. Catal.* **35**, 139 (1987).

Dutta, P.K., and Kim, Y.H., *Curr. Opin. Solid State Mater. Sci.* **7**, 483 (2003).

Dyrek, K., and Che, M., *Chem. Rev.* **97**, 305 (1997).

Dzwigaj, S., *Curr. Opin. Solid State Mater. Sci.* **7**, 461 (2003).

Dzwigaj, S., Matsuoka, M., Anpo, M., and Che, M., *Catal. Lett.* **72**, 211 (2001).

Dzwigaj, S., Matsuoka, M., Anpo, M., and Che, M., *J. Phys. Chem. B* **104**, 6012 (2000).

Dzwigaj, S., Matsuoka, M., Franck, R., Anpo, M., and Che, M., *J. Phys. Chem. B* **102**, 6309 (1998).

Fujihara, K., Izumi, S., Ohno, T., and Matsumura, M., *J. Photochem. Photobiol. A* **132**, 99 (2000).

Fujii, T., Tanaka, N., Kodaira, K., Kawai, Y., Yamashita, H., and Anpo, M., *J. Photochem. Photobiol. A.* **125**, 85 (1999).

Fujishima, A., and Honda, K., *Nature* **238**, 37 (1972).

Furube, A., Asahi, T., Masuhara, H., Yamashita, H., and Anpo, M., *Chem. Phys. Lett.* **336**, 424 (2001a).

Furube, A., Asahi, T., Masuhara, H., Yamashita, H., and Anpo, M., *J. Phys. Chem. B* **103**, 3120 (1999).

Furube, A., Asahi, T., Masuhara, H., Yamashita, H., and Anpo, M., *Res. Chem. Intermed.* **27**, 177 (2001b).

Gillespie, R.J., "Molecular Geometry". Van Nostrand-Reinhold, London, 1972.

Higashimoto, S., Matsuoka, M., Yamashita, H., and Anpo, M., *Jpn. J. Appl. Phys.* **38**, 47 (1999).

Higashimoto, S., Matsuoka, M., Yamashita, H., Anpo, M., Kitao, O., Hidaka, H., Che, M., and Giamello, E., *J. Phys. Chem. B* **104**, 10288 (2000).

Higashimoto, S., Matsuoka, M., Zhang, S.G., Yamashita, H., Kitao, O., Hidaka, H., and Anpo, M., *Micropor. Mesopor. Mater.* **48**, 329 (2001a).

Higashimoto, S., Tsumura, R., Matsuoka, M., Yamashita, H., Che, M., and Anpo, M., *Stud. Surf. Sci. Catal.* **140**, 315 (2001b).

Higashimoto, S., Tsumura, R., Zhang, S.G., Matsuoka, M., Yamashita, H., Louis, C., Che, M., and Anpo, M., *Chem. Lett.* **408** (2000).

Hu, Y., Higashimoto, S., Martra, G., Zhang, J.L., Matsuoka, M., Coluccia, S., and Anpo, M., *Catal. Lett.* **90**, 161 (2003a).

Hu, Y., Martra, G., Higashimoto, S., Zhang, J.L., Matsuoka, M., Coluccia, S., and Anpo, M., *Stud. Surf. Sci. Catal.* **146**, 593 (2003b).

Hu, Y., Wada, N., Matsuoka, M., and Anpo, M., *Catal. Lett.* **97**, 49 (2004).

Ikeue, K., Nozaki, S., Ogawa, M., and Anpo, M., *Catal. Today* **74**, 241 (2002).

Ikeue, K., Yamashita, H., and Anpo, M., *Chem. Lett.* 1135 (1999).

Ikeue, K., Yamashita, H., Anpo, M., and Takewaki, T., *J. Phys. Chem. B* **105**, 8350 (2001).

Iwamoto, M., Furukawa, H., Matsukami, K., Takenaka, T., and Kagawa, S., *J. Am. Chem. Soc.* **105**, 3719 (1983).

Jung, K.Y., and Park, S.B., *Korean J. Chem. Eng.* **18**, 879 (2001).

Ju, W.S., Matsuoka, M., and Anpo, M., *Int. J. Photoenergy* **5**, 17 (2003).

Ju, W.S., Matsuoka, M., Iino, K., Yamashita, H., and Anpo, M., *J. Phys. Chem. B* **108**, 2128 (2004).

Kanan, S.M., Kanan, M.C., Marsha, C., and Patterson, H.H., *J. Phys. Chem. B* **105**, 7508 (2001b).

Kanan, S.M., Kanan, M.C., and Patterson, H.H., *Curr. Opin. Solid State Mater. Sci.* **7**, 443 (2003).

Kanan, S.M., Omary, M.A., Patterson, H.H., Matsuoka, M., and Anpo, M., *J. Phys. Chem. B* **104**, 3507 (2000).

Kanan, S.M., Tripp, C.P., Austin, R.N., and Patterson, H.H., *J. Phys. Chem. B* **105**, 9441 (2001a).

Korsunska, N.E., Tarasov, I., Kushnirenko, V., and Ostapenko, S., *Semicond. Sci. Technol.* **19**, 833 (2004).

Kuba, S., Che, M., Grasselli, R.K., and Knözinger, H., *J. Phys. Chem. B* **107**, 3459 (2003).

Lauron-Pernot, H., *Catal. Rev.* **48**, 315 (2006).

Li, D.G., Zhang, J.L., and Anpo, M., *Opt. Mater.* **27**, 671 (2005).

Lisachenko, A.A., Chikhachev, K.S., Zhakarov, M.N., Basov, L.L., Shelimov, B.N., Subbotina, I.R., Che, M., and Coluccia, S., *Topics Catal.* **20**, 119 (2002).

Lunsford, J.H., *Stud. Surf. Sci. Catal.* **75A**, 103 (1993).

M. Anpo (Ed.), *"Photofunctional Zeolite—Synthesis, Characterization, Photocatalytic Reactions, Light Harvesting"*, Nova Science Publishers Inc., (2000).

Matsuoka, M., and Anpo, M., *Curr. Opin. Solid State Mater. Sci.* **7**, 451 (2003a).

Matsuoka, M., and Anpo, M., *in* "Catalysis by Unique Metal Ion Structures in Solid Matrices", (G. Centi, B. Wichterlova and A.T. Bell, Eds.), p. 249. Kluwer Academic Publishers, 2001.

Matsuoka, M., and Anpo, M., *J. Photochem. Photobiol. C.* **3**, 225 (2003b).

Matsuoka, M., and Anpo, M., *Shokubai (Catalysts & Catalysis)* **47**, 328 (2005).

Matsuoka, M., Higashimoto, S., Yamashita, H., and Anpo, M., *Res. Chem. Intermed.* **26**, 85 (2000).

Matsuoka, M., Ju, W.S., Chen, H.J., Sakatani, Y., and Anpo, M., *Res. Chem. Intermed.* **29**, 477 (2003).

Matsuoka, M., Ju, W.S., Yamashita, H., and Anpo, M., *J. Synchrotron Rad.* **8**, 613 (2001).

Mériaudeau, P., Boudeville, Y., de Montgolfier, P.h., and Che, M., *Phys. Rev. B* **16**, 30 (1977).

Miyashita, K., Kuroda, S., Tajima, S., Takehira, K., Tobita, S., and Kubota, H., *Chem. Phys. Lett.*, **369**, 225 (2003).

Nabavi, M., Taulelle, F., Sanchez, C., and Verdaguer, M., *J. Phys. Chem.* **51**, 1375 (1990).

Nakajima, H., Itoh, K., and Murabayashi, M., *Bull. Chem. Soc. Jpn.* **75**, 601 (2002).

Nakajima, H., Itoh, K., and Murabayashi, M., *Chem. Lett.* **304** (2001).

Nakamura, R., and Nakato, Y., *J. Am. Chem. Soc.* **126**, 1290 (2004).

Nicholas, J.B., Kheir, A.A., Xu, T., Krawietz, T.R., and Haw, J., *J. Am. Chem. Soc.* **120**, 10471 (1998).

Nishiguchi, H., Zhang, J.L., and Anpo, M., *Langmuir* **17**, 3958 (2001a).

Nishiguchi, H., Zhang, J.L., Anpo, M., and Masuhara, H., *J. Phys. Chem. B* **105**, 3218 (2001b).

Ogawa, M., Ikeue, K., and Anpo, M., *Chem. Mater.* **13**, 2900 (2001).

Park, T.R., Park, T.Y., Kim, H.G., and Min, P., *J. Phys. Condens. Mater.* **14**, 11687 (2002).

Sapelkin, A.V., Bayliss, S.C., Russell, D., Clark, S.M., and Dent, A.J., *J. Synchrotron Rad.* **7**, 257 (2000).

Shelimov, B., Dellarocca, V., Martra, G., Coluccia, S., and Che, M., *Catal. Lett.* **87**, 73 (2003).

Shluger, A.L., Sushko, P.V., and Kantorovich, L.N., *Phys. Rev. B* **59**, 2417 (1999).

Sojka, Z., and Che, M., *J. Chem. Educ.* (2009) submitted for publication.

St. Amour, A., Stum, J.C., Lacroix, Y., and Thewalt, M.L.W., *Appl. Phys. Lett.* **65**, 3344 (1994).

Subbotina, I.R., Shelimov, B.N., Che, M., and Coluccia, S., *Stud. Surf. Sci. Catal.* **140**, 421 (2001).

Subbotina, I.R., Shelimov, B.N., Kazansky, V.B., Lisachenko, A.A., Che, M., and Coluccia, S., *J. Catal.* **184**, 390 (1999).

Sushko, P.V., Gavartin, J.L., and Shluger, A.L., *J. Phys. Chem. B* **106**, 2269 (2002).

Sushko, P.V., and Shluger, A.L., *Surf. Sci.* **421**, L157 (1999).

Tsumura, R., Higashimoto, S., Matsuoka, M., Yamashita, H., Che, M., and Anpo, M., *Catal. Lett.* **68**, 101 (2000).

Turro, N.J., "Modern Molecular Photochemistry", Benjamin/Cummings Publ. Co Inc., Menlo Park, 1978.

Villaneau, R., Carabineiro, H., Carrier, X., Thouvenot, R., Herson, P., Lemos, F., Ribeiro, F.R., and Che, M., *J. Phys. Chem. B* **108**, 12465 (2004).

Yamashita, H., and Anpo, M., *Catal. Surv. Asia* **8**, 35 (2004).

Yamashita, H., and Anpo, M., *Curr. Opin. Solid State Mater. Sci.* **7**, 471 (2003).

Yamashita, H., Ikeue, K., Takewai, T., and Anpo, M., *Topics Catal.* **18**, 95 (2002).

Yamashita, H., Yoshizawa, K., Ariyuki, M., Higashimoto, S., Che, M., and Anpo, M., *Chem. Commun.* 435 (2001).

Yu, J.G., Yu, H.G., Cheng, B., Zhao, X.J., Yu, J.C., and Ho, W.K., *J. Phys. Chem. B.* **107**, 13871 (2003).

Zhang, J.L., Hu, Y., Matsuoka, M., Yamashita, H., Minagawa, M., Hidaka, H., and Anpo, M., *J. Phys. Chem. B* **105**, 8395 (2001).

Zhang, J.L., Matsuoka, M., Yamashita, H., and Anpo, M., *Langmuir* **15**, 77 (1999).

Zhang, J.L., Minagawa, M., Ayuzawa, T., Natarajan, S., Yamashita, H., Matsuoka, M., and Anpo, M., *J. Phys. Chem. B* **104**, 11501 (2000).

Zhang, S.G., Ariyuki, M., Mishima, H., Higashimoto, S., Yamashita, H., and Anpo, M., *Micropor. Mesopor. Mater.* **21**, 621 (1998).

CHAPTER 2

Structural Characterization of Operating Catalysts by Raman Spectroscopy

Miguel A. Bañares* and **Gerhard Mestl[†]**

Abstract

This chapter focuses on the application of Raman spectroscopy for the characterization of the reaction chemistry occurring during catalyst preparation and operation. Equipment for Raman experiments at various temperatures and in various environments is described, and developments in the methodology are compiled with examples demonstrating the versatility of the technique. Raman spectra of catalysts that have been recorded over wide temperature and pressure ranges, including spectra of catalysts in liquids and in supercritical fluids are presented. Raman spectroscopy is used with programmed variations in catalyst temperature or surrounding fluid composition and with catalysts under operating conditions, optimally when separate and simultaneous analysis of products is performed and the reaction kinetics data obtained in the spectroscopic cell approach those measured in an ideal reactor. The literature of such investigations is reviewed, with an emphasis on the analysis of bulk oxide and supported metal oxide catalysts.

Contents

1. Introduction   45
2. Advanced Raman Spectroscopy for Characterization
   of Samples in Reactive Environments   47
3. Raman Scattering   48

* Instituto de Catálisis y Petroleoquímica, Catalytic Spectroscopy Laboratory, CSIC, E-28049-Madrid, Spain
† Süd-Chemie AG, Research and Development, Catalytic Technologies, D-83052 Bruckmühl—Heufeld, Germany

Advances in Catalysis, Volume 52
ISSN 0360-0564, DOI: 10.1016/S0360-0564(08)00002-3

	3.1. Fundamentals of Raman Scattering	48
	3.2. Equipment for Raman Spectroscopy of Catalysts in Reactive Atmospheres	57
4.	Raman Investigations of Catalysts Under Controlled Environments	65
	4.1. Early Vacuum, Chemisorption, and Hydration/Dehydration Experiments	65
	4.2. Raman Investigations of Catalyst Preparation	72
	4.3. Raman Investigations of Catalysts under Conditions Relevant to Catalysis	76
	4.4. Raman Spectroscopy of Operating Catalysts (as Demonstrated by Simultaneous Activity Measurements)	92
	4.5. Raman Microspectroscopy	112
5.	Perspective and Conclusions	113
	Acknowledgments	115
	References	115

ABBREVIATIONS

$<Q02>$	the Boltzmann distribution
AFM	atomic force microscopy
AN	Acrylonitrile
BZC	Brillouin zone center
CCD	charge coupled device
CVT	chemical vapor transport
$DeNO_x$	removal of nitrogen oxides
$DeSO_x$	removal of sulfur oxides
DFT	density functional theory
DH	Dehydrogenation
DRIFT	diffuse reflectance IR Fourier transform
ε	dielectric constant
EPR	electron paramagnetic resonance
EXAFS	extended X-ray absorption fine structure
FT spectroscopy	Fourier-transform spectroscopy
GC	gas chromatography
HR-TEM	high-resolution transmission electron microscopy
IVCT	intervalence charge transfer (of an electron from one to another atom of the same or a different type)
K	wave vector
k_p	wave vector of the incident photons
LEED	low energy electron diffraction
LMCT	ligand-to-metal charge transfer

LRS	Laser Raman spectroscopy
MS	mass spectrometry
v	vibrational quantum number
NIR	near infrared
OCM	oxidative coupling of methane
ODH	oxidative dehydrogenation
PZC	point of zero charge
Q	the phonon normal coordinate
RR	resonance Raman
RRS	resonance Raman scattering
SCR	selective catalytic reduction
SERRS	surface enhanced resonance Raman scattering
SERS	surface enhanced Raman scattering
SFG	sum frequency generation
SNOM	scanning near-field optical microscopy
TOF	turnover frequency
TP	temperature programmed
TPO	temperature-programmed oxidation
TPR	temperature-programmed reduction
UV–vis DRS	UV–vis spectroscopy in diffuse reflectance
UV–vis	ultraviolet–visible absorption spectroscopy
VPO	vanadium phospho-oxides
ω	excitation frequency of incident light
XANES	X-ray absorption near edge structure
XRD	X-ray diffraction.

1. INTRODUCTION

The importance of catalysis in chemical technology provides a strong motivation for determination of relationships between catalytic activity and catalyst structure at the atomic scale. Spectroscopic techniques for characterization of catalysts in the working state are powerful, because they provide fundamental information about catalyst structures, including surface structures, under the appropriate conditions (Burch, 1991; Clausen et al., 1998; Dumesic and Topsøe, 1977; Hunger and Weitkamp, 2001; Niemantsverdriet, 1993; Somorjai, 1999; Thomas and Somorjai, 1999; Thomas, 1980; Topsøe, 2000; Weckhuysen, 2002). Such characterizations have permitted major advances in catalysis, as they can be the basis for the design or discovery of new catalysts.

Numerous spectroscopic techniques provide evidence of catalyst structures and surface states. Some are applied with model catalysts such as single crystals or well-defined clusters and under vacuum conditions. However, the complexity of surface structures in polycrystalline catalysts provides a wide variety of potential reaction sites, which are typically not present in a model (Somorjai and Rupprechter, 1999; Topsøe, 2000; Zaera, 2001, 2002). Furthermore, the state of a catalyst surface may be strongly affected by the temperature and by the chemical environment. For example, data obtained by infrared–visible sum frequency generation (SFG) indicate surface reconstruction induced by adsorbates (Somorjai and Rupprechter, 1999; Topsøe, 2000; Zaera, 2001, 2002); similar observations were made by LEED (Gauthier et al., 1991; Shih et al., 1981; Somorjai, 1996). Further, operation at relatively high pressures allows for the presence of weakly adsorbed intermediates in significant concentrations, which would not be observed under vacuum (Zaera, 2001, 2002).

For example, key intermediates of the catalytic CO oxidation on rhodium and on platinum at elevated pressure are not present under vacuum conditions (Somorjai and Rupprechter, 1999). These "gaps" in materials and pressure have to be bridged to establish quantitative structure–activity relationships (Topsøe, 2000; Wachs, 2003a, 2003b, 2003c). Raman spectroscopy is one of the most useful techniques for bridging these "gaps" and is the focus of this chapter.

A number of industrial catalysts containing transition metals undergo transient induction or deactivation processes, which are related to structural or electronic changes. In 1979, isotope labeling effects were investigated by Raman spectroscopy for the first time, and the changes that were observed in a bismuth molybdate catalyst during the transition from the fresh to the used state provided evidence about the nature of the specific active oxygen site in the lattice (Hoefs et al., 1979). Other oxides require long periods to show structural changes; for example, VPO catalysts undergo significant changes in performance for days during the start-up period, and investigations of the surfaces indicate extended reconstruction (Abdelouahab et al., 1992; Guliants et al., 1995, 2001; Hutchings et al., 1994; Volta et al., 1992). The reconstruction leads to an evident amorphitization of the outermost layer (Guliants et al., 1995), which interacts directly with the reactants, and thus this surface transformation was associated with the catalytic performance (Abdelouahab et al., 1992; Guliants et al., 1995, 2001; Hutchings et al., 1994; Koyano et al., 1997; Volta et al., 1992; Xue and Schrader, 1999).

These examples and many others have provided evidence of significant changes in catalyst structures resulting from changes in operating conditions. Techniques are thus necessary that can be applied to catalysts in the presence of probe molecules, in reactive environments (e.g., when catalysts undergo reduction, oxidation, etc.), and under catalytic reaction conditions. Moreover, the simultaneous determination of catalyst structure and activity or selectivity is needed to establish structure–activity or structure–selectivity relationships, which provide a basis for improvement and development of catalysts (Bañares, 2005; Thomas, 1980; Thomas, 1999; Topsøe, 2000; Wachs, 2005). The need for characterization of catalysts during operation has been emphasized and demonstrated by many authors (Bañares and Wachs, 2002; Bañares et al., 2002; Clausen et al., 1991; Dixit et al., 1986; García-Cortéz and Bañares, 2002; Guerrero-Pérez and Bañares, 2002, 2004; Knözinger, 1996; Kuba and Knözinger, 2002; Mestl, 2000; Mestl, 2002; Mestl et al., 1993; Rybaczyk et al., 2001; Thomas and Somorjai, 1999; Tibiletti et al., 2004; Topsøe, 2000; Wachs, 1996; Wachs and Hardcastle, 1993; Weaver, 2002; Weber, 2000; Weckhuysen, 2002, 2003), including those of the present volume and the preceding two volumes of *Advances in Catalysis*.

Raman spectroscopy is one of the most powerful tools for characterization of working catalysts. Raman experiments can be carried out at temperatures above 1000 °C and at elevated pressures, without interference from the gas phase. Current Raman spectrometers allow recording of the spectral range up to 4000 cm^{-1} in a single data acquisition. Time resolutions in the sub-second regime can be achieved with these spectrometers for materials with high Raman cross-sections. Thus, time-resolved transient temperature or pressure response experiments can be carried out (e.g., pulse experiments with isotopically labeled compounds), and reaction kinetics data can be measured directly and correlated with the spectroscopic data. Moreover, modern quartz fiber optics allows easy spectroscopic access to catalytic reactors. The Raman experiments can be carried out with either static or flowing mixtures of gases to mimic the conditions in a catalytic reactor. It is also possible to investigate reactions in the liquid phase or under supercritical conditions.

With respect to the information content of Raman spectra, it is appropriate to cite C. V. Raman's Nobel award lecture: ". . the character of the scattered radiation enables us to obtain an insight in the ultimate structure of the scattering substance." Given the capabilities of Raman spectroscopy listed in the preceding paragraph, application of this method to working catalysts appears extremely promising.

2. ADVANCED RAMAN SPECTROSCOPY FOR CHARACTERIZATION OF SAMPLES IN REACTIVE ENVIRONMENTS

A number of monographs and review articles on Raman spectroscopy in heterogeneous catalysis have been published (Bañares and Wachs, 2002; Bañares, 2004, 2005; Bartlett and Cooney, 1987; Chesters and Sheppard, 1981; Cooney et al., 1975; Delgass et al., 1979; Delhaye and Dhamelincourt, 1997; Dixit et al., 1986; Egerton and Hardin, 1975; Fleischmann et al., 1976; Garbowski and Coudurier, 1994; Grasselli et al., 1977; Kiefer and Bernstein, 1971; Knözinger and Mestl, 1999; Knözinger, 1991; Knözinger, 1996; Kuba and Knözinger, 2002; Mehicic and Grasselli, 1991; Mestl and Knözinger, 1997; Mestl, 2000, 2002; Mestl and Srinivasan, 1998; Mestl et al., 1993; Morrow, 1981; Nguyen, 1983; Payen and Kasztelan, 1994; Payen et al., 1978; Segawa and Wachs, 1992; Sheppard and Erkelens, 1984; Stair, 2001; Stencel, 1990; Takenaka, 1979; Tian et al., 2005; Wachs, 1986, 1996, 1999, 1999, 2000, 2002; Wachs and Hardcastle, 1993; Weaver, 2002; Weber, 2000a,b), and some of these also cover the physics of the scattering process. Therefore, the general physics of Raman spectroscopy is introduced only briefly here. The advantages and disadvantages of Raman spectroscopy (e.g., quantification problems) are also discussed.

This review does not deal with the general characterization of catalysts by Raman spectroscopy but is focused instead on the application to catalysts in reactive environments. Such experiments are often described as "*in situ,*" a term that is minimally used in this volume. The term "*in situ,*" Latin for "on site," implies that the sample is analyzed at the location where it has been treated or is being treated. Several levels of such experiments are described here:

(a) "*In situ*" spectroscopy: the spectra are recorded of a sample at the same location at which it has been or is being treated; the temperature or gas phase, however, may have been changed.
(b) Variable-conditions "*in situ*" spectroscopy: transformations occurring during the variation of a parameter, such as partial pressure of a component, temperature, etc. are monitored spectroscopically. Typical are temperature-programmed processes (Pieck et al., 2001; Waterhouse et al., 2003) such as TPR-Raman spectroscopy, in which Raman spectra characterize the reduction of a sample (Bañares et al., 2000a; Kanervo et al., 2003), TPO-Raman spectroscopy (Bañares et al., 2000a; Herrera and Resasco, 2003), or any temperature-programmed reaction with an adsorbate or a probe molecule (Burcham et al., 2000).
(c) "*In situ*" spectroscopy of the working catalyst. To demonstrate that the spectra correspond to an operating catalyst, quantitative analysis of the reaction progress (e.g., by gas chromatography) has to be performed, and in this case, structure and activity can be correlated. The term "*operando,*" which is Latin for "working," "operating," is sometimes used to emphasize the simultaneous evaluation of both structure and catalytic performance (Bañares and Wachs, 2002; Bañares et al., 2002; García-Cortéz and Bañares, 2002; Guerrero-Pérez and Bañares, 2002; Weckhuysen, 2002), including use of a cell that delivers reaction kinetics data that match those obtained in an ideal reactor.

3. RAMAN SCATTERING

3.1. Fundamentals of Raman Scattering

Atoms in crystalline solids move under well-defined, allowed phase relations, the vibrational modes of the crystal. Only movements of atoms are allowed that are parallel or perpendicular to the wave vector, termed acoustic or optical phonons, depending on the generation of dipole moments during the vibration. These vibrational modes also take part in inelastic light-scattering processes.

When electromagnetic radiation interacts with matter, scattering occurs together with absorption and reflection. Most of the scattered radiation is unchanged in frequency, and the process is called Rayleigh scattering (elastic scattering). A small part of the scattered radiation may have higher or lower energy than the incident radiation; this is the Raman scattering effect (inelastic scattering). The incoming photon excites the molecule or solid from its ground state into a virtual excited state from which it relaxes with the emission of a Raman-scattered photon of lower energy (Stokes scattering). If the molecule or solid is initially in its first vibrationally excited state, the scattering will lead to a Raman-scattered photon of higher energy (anti-Stokes scattering). The Raman scattering process can be understood as a series of three elementary steps:

Step 1: The incident photon is absorbed and a first virtual electron–hole pair is generated.
Step 2: This virtual electron–hole pair generates or annihilates a phonon and a second virtual electron–hole pair is formed.
Step 3: This second virtual electron–hole pair recombines and a Raman-scattered photon is emitted.

The laws of conservation of energy and momentum govern the Raman process. The intensities or line shapes of Raman bands are determined by the Raman scattering transition moment. The Raman scattering tensor relates the incoming and the scattered electromagnetic field for a particular vibrational mode and is always symmetric for normal Raman scattering. It is reduced to its irreducible components, which are identified with vibrations of well-defined group theoretical symmetries (Birman, 1974; Cardona, 1983; Loudon, 1964; Ovander, 1960).

The Raman scattering, developed as a Taylor series of the phonon normal coordinates, can be written as follows:

$$\varepsilon(Q,\ \omega) = \varepsilon(\omega) + \frac{d\varepsilon(\omega)}{dQ}Q + \frac{1}{22}\frac{d^2\varepsilon(\omega)}{dQ^2}Q^2 + \cdots, \tag{1}$$

where ε is the (frequency-dependent) dielectric constant, ω the frequency of the incident light, and Q the phonon normal coordinate. The series of derivatives describes the Raman tensor of first and second order and so on. The total intensity of the scattered radiation is described by the following proportionality:

$$I \propto \omega_L^4 \left|\frac{d\varepsilon}{dQ}\right|^2 \langle Q_0^2 \rangle, \tag{2}$$

where $\langle Q_0^2 \rangle$ is the Boltzmann distribution of the vibrational states and ω_L the excitation frequency. This is the ω^4-law: the higher the excitation frequency, the higher is the Raman scattered intensity. The Boltzmann term allows the calculation of sample temperatures from Stokes and anti-Stokes measurements, provided that the spectrometer function is accurately determined (Deckert and Kiefer, 1992; Knoll et al., 1990; Spielbauer, 1995).

Equation (2) also describes the relationship between the electronic states as described by the frequency-dependent dielectric constant, ε, and lattice vibrations, Q. The derivative $d\varepsilon/dQ$ becomes large through pronounced variations of ε, for example, at surfaces where ε shows a step change. Resonant Raman scattering, a result of resonant absorption of photons, can be understood in this context. Resonance experiments thus do not only provide information about the vibrational states but also about the electronic nature of the material under investigation, for example, about defect centers such as reduced transition metal ions. Resonance effects observed during Raman investigations are being reported with increasing frequency and are exemplified below. Resonance effects have a pronounced influence on the Raman spectra (Stair, 2007; Tian et al., 2005). Resonance effects are also of importance in UV-excited Raman spectroscopy, as recently discussed in detail by Stair (2007).

According to the spectroscopic exclusion rule, vibrations are Raman active if the polarizability of the oscillating bond changes during the vibration. This spectroscopic exclusion rule, however, strictly holds true only for centrosymmetric molecules or solids. The fundamental selection rule of vibrational Raman spectroscopy is the change of the vibrational quantum number v by ± 1. The conservation of the wave vector k is an important additional selection rule for crystalline materials (and in coherent Raman scattering) and has to be fulfilled for each of the above elementary steps. The length of the wave vector of the incident light is comparable to wave vectors at the Brillouin zone center (BZC). Only phonons with k_p vectors at the BZC interact with incident photons and lead to first-order Raman scattering. Raman scattering thus probes the symmetry of the crystal unit cell.

The rule of the conservation of momentum k does not apply strictly in the case of ill-defined crystallites, solids without translational symmetry (amorphous materials or small scattering volumes), and colored materials, in which both the incoming and scattered light waves are strongly attenuated (i.e., in the case of resonance with electronic transitions).

The electronic absorption spectrum of a catalyst shows which excitation energies will lead to resonance or far-from-resonance conditions (Clark and Dines, 1986; Nafie, 2001), the latter being common for colorless samples that have no electronic states in close proximity to the incident photon energy. If the incident photon energy is near the transition energy of an excited electronic state, the Raman scattering will change to resonance Raman scattering (RRS). It is important to realize that the resulting RRS is dominated by the properties of this resonant electronic state (Clark and Dines, 1986; Nafie, 2001). The RRS phenomenon accounts for variations in relative intensities observed in the Raman spectra of a given sample when different excitation energies are used and allows discriminating between different phases in both supported (Chua et al., 2001; Wu et al., 2005) and bulk oxides (Dieterle et al., 2002; Mestl, 2002; Ricchiardi et al., 2001). RR spectroscopy is applied in surface science mostly for the investigation of adsorbed probe molecules and thin solid films (Bartlett and Cooney, 1987; Geurts and Richter, 1987; Hicks et al., 1990). Thus, for instance, methyl red was used for RRS as a molecule to probe surface acidity of oxides such as SiO_2, Al_2O_3, TiO_2, and silica–aluminas (Bisset and Dines, 1995a,b).

In most cases though, Raman spectra of surface species cannot be obtained because of the very small molecular scattering cross-sections (Skinner and Nilsen, 1968) and their typically low concentrations, with the two exceptions of adsorbates on very high-surface-area powders and adsorbate layers on a series of particular metal surfaces, for which surface-enhanced Raman scattering (SERS) occurs (Arya and Zeyher, 1984; Otto, 1984). The SERS effect, first reported by Fleischmann and coworkers (Fleischmann et al., 1974) in 1974, is observed for the following metals: silver, gold, aluminum, copper, lithium, potassium, sodium, cadmium, platinum, mercury, nickel, and palladium, as well as silver–palladium alloys. It leads to an enhancement of the Raman scattering by a factor of about 10^4–10^7. Two different origins of the SERS effect are discussed: (i) the "physical", and (ii) the "chemical" SERS effect. The physical SERS effect arises from a resonant interaction of the Raman exciting electromagnetic wave and metal surface plasmons (Moskovits, 1985; Pettinger, 1992). For this interaction to occur, (i) the exciting laser has to have the right frequency, and (ii) the metal surfaces have to be rough on a scale smaller than the wavelength of the incident electromagnetic radiation, as may be the case, for example, for rough metal thin films, metal electrodes, or metal colloids (Liao, 1982; Vo-Dinh, 1998). On such surfaces, the adsorbate molecules experience a dramatically increased electromagnetic field strength, which in turn enhances the Raman scattering efficiency. The chemical SERS effect occurs only for atomically rough metal surfaces with metal adatoms, and arises from the resonant coupling of the exciting laser light into a "molecular" electronic transition between the electronic states of surface adatoms and the adsorbates (Avouris and Demuth, 1981; Campion and Kambhampati, 1998; Lombardi et al., 1986; Mrozek and Otto, 1990), and it leads to an enhancement by a factor of about 10^2.

Further enhancement can be observed if the exciting light additionally couples into an electronic transition of the adsorbate (surface-enhanced resonance Raman scattering, SERRS). In this case, enhancement factors of up to 10^{12} were reported (Pemberton, 1991). This "chemical" SERS effect is highly relevant to catalysis, because it directly probes the electronic structure of the adsorbate–metal adatom complex and thus provides information about its molecular structure.

Recently, the SERS effect has been used increasingly for highly sensitive sensor devices for biological and medical applications as well as in practical analytical chemistry (Alivisatos, 2004; Emory and Nie, 1998; Faulds et al., 2004; Ishikawa et al., 2002; Kneipp et al., 1995; Wang et al., 2003), and the field of catalysis certainly will see a comparable increase in SERS investigations of active metal catalysts (*vide infra*).

Summarizing, Raman spectra can be recorded of single crystals, powders, glasses, nanocrystalline, and amorphous materials, and of surface species such as transition metal compounds on high-surface-area oxide supports or surface adsorbates on some metals. Thus, Raman spectroscopy is a potentially valuable tool for the characterization of a broad range of catalytic materials and surfaces.

Raman spectroscopy is a bulk technique, although the depth of the analyzed volume is limited. The information depth and thus the spectra depend on the excitation frequency and the absorption coefficient and crystallinity of the sample (Cardona, 1983). To characterize catalyst surfaces and their interactions with reactants, the spectral contributions from the surface have to be discriminated from those of the catalyst bulk. This complication has to be considered when applying Raman spectroscopy to working catalysts (Bañares, 2005).

An advantage of Raman spectroscopy for the investigation of reacting catalysts arises from the usually negligible scattering of the gas phase. Raman spectra of operating catalysts can thus be recorded at elevated pressures without gas-phase interference. The Raman scattering of glass and quartz or sapphire is rather weak, allowing simple cell constructions. Typical high-surface-area catalyst supports, such as SiO_2 and Al_2O_3, are very weak Raman scatterers; therefore, their Raman spectra usually do not interfere with the Raman scattering of oxide moieties on these supports. Water is also a very weak scatterer, and Raman spectroscopy can be performed to characterize aqueous solutions or suspensions, and to monitor the adsorption of transition metal ions on supports, or crystallization processes such as the formation of zeolites.

3.1.1. Limitations of Raman spectroscopy

The major limitations of Raman spectroscopy are the following:

(i) *Laser heating* may alter temperature-sensitive samples and can result in loss of water of hydration (Payen et al., 1986, 1987), phase transitions, reduction, or even complete decomposition (Liu et al., 2000). In the event of such laser-induced changes, spectra are no longer representative of the catalyst under known conditions, and the data are not useful for their intended purpose. More deeply colored samples are more susceptible to laser heating effects. Temperature increases caused by laser heating of as much as 80 K have been recorded (Xie et al., 1999) for colorless samples, and still greater local temperature increases may occur, depending on the color of the sample and the power and wavelength of the excitation source (UV ≫ visible > IR). Laser heating can be reduced by applying low laser powers (<10 mW), facilitated by modern spectrometers equipped with holographic notch or edge filters and thermoelectrically cooled CCD detectors. Holographic notch and edge filters reject elastic scattering but do not reduce the Raman signal. Modern CCD detectors afford the necessary high signal-to-noise ratios. Alternatively, samples can be cooled with an inert gas of high thermal conductivity, or the laser energy can be distributed with a cylindrical lens focus (Freeman et al., 1981), by rotation of the sample (Chan and Bell, 1984; Cheng et al., 1980; Covington and Thain, 1975; Kiefer and Bernstein, 1971), or other focusing lens (Koningstein and Gachter, 1973; Snyder and Hill, 1991; Zimmerer and Kiefer, 1974) techniques.

Laser-induced damage is especially a serious concern in high-energy (10–300 mW) UV–Raman experiments with long collection times, during which the high photon energy may lead to serious sample degradation. For this reason, home-built UV–Raman instruments, typically constructed with far from optimum photon collection efficiency, always require rotating wafer samples or fluidized powder beds; otherwise, the results will be highly questionable (Stair, 2001; Xie et al., 1999). In the preceding few years, UV–Raman spectrometers with high photon collection efficiencies have become commercially available. These instruments feature single-stage monochromators with notch or edge filters and require only very low laser power, *ca.* 0.1 mW, and short collection times (of the order of seconds) (Tian et al., 2005).

(ii) *Fluorescence* can overwhelm the Raman spectrum. Typical causes of fluorescence are organic deposits, basic surface OH groups (Jeziorowski and Knözinger, 1977, 1979), proton superpolarizability, or reduced transition metal ions that can be resonantly excited. Fluorescence sometimes can be reduced by simply burning off the organic contaminants or dehydroxylating the surface. In other cases, changing the excitation frequency may lead to reduction of the fluorescence; for example, FT Raman instruments with NIR excitation (Chase, 1987; Hendra and Mould, 1988) or the UV–Raman technique (Stair, 2007) have been applied. Frequency modulation Raman spectroscopy (Brückner et al., 1984) can be an alternative approach, because the spectrum is recorded as the first derivative.

(iii) *Black-body* radiation can also overwhelm the Raman spectrum. The radiation emitted by a body depends on its temperature, whereby the emitted wavelength decreases with increasing temperature according to Planck's radiation law. For this reason, emitted infrared radiation typically prevents FT Raman investigations in the wave number range below $1000 \, cm^{-1}$ at temperatures above 200 °C. Visible Raman spectroscopy is limited to temperatures below *ca.* 400 °C for the 788-nm laser and approximately 800 °C for 514–532-nm excitation. Black-body radiation is not likely to affect Raman experiments with UV excitation lines, because the temperatures needed to emit UV radiation substantially exceed 1000 °C. Wachs and coworkers (Wang and Wachs, 2004) have successfully performed UV–Raman investigations (with excitation at 325 nm) during catalytic hydrocarbon reforming in the temperature range near 1000 °C. The quality of the UV–Raman spectra was found to be excellent at these high reaction temperatures. An interesting advantage of UV excitation is that the weak vibrations of gas phase molecules can also be detected under reaction conditions (especially those of O_2 and N_2 because of the high electron density in the bonds of these diatomic homonuclear molecules).

Further barriers to a broader use of Raman spectroscopy are problems with intensity calibration and benchmarking, lack of reference materials, and the expense of the equipment. The calibration equipment, which is typically supplied with the instruments by the manufacturers, varies; for instance neon or mercury lamps are delivered with dispersive instruments and HeNe lasers with FT spectrometers.

Accepted reference standards are still missing. Manufacturers and users report various methods of wavelength calibration and refer to the absorption bands of materials such as silicon, sulfur, polystyrene, or cyclohexane, or to the emission lines of neon or mercury lamps. This missing standardization in calibration and referencing, together with unavoidable differences in the investigated samples (origin, impurities, crystallinity, etc.), and variations in instrumental resolution and sensitivity may lead to inconsistencies when results of different research groups are compared. This dilemma, however, can be resolved by analyzing known crystalline reference compounds that contain multiple Raman bands (such as TiO_2, V_2O_5, MoO_3, and WO_3).

3.1.2. Advantages of Raman Spectroscopy

Infrared spectroscopy often achieves high signal-to-noise ratios, but the intense bands of typical oxide supports interfere with much of the interesting spectral region. Raman spectroscopy is an excellent alternative technique in such cases, because many supports exhibit only weak

Raman transitions. Moreover, there are no disturbances resulting from water absorption if visible excitation is used (Grasselli et al., 1981). Raman spectroscopy is easily conducted with the catalyst sample placed inside a cell or reactor, and the window materials and the typically weak absorptions of fluids allow for great flexibility regarding the pressure and temperature conditions.

3.1.3. Quantification of Raman Spectra

The inherently unknown Raman scattering cross-sections render quantitative Raman spectroscopy rather difficult. Even for pure components, the Raman scattering cross-sections may change as a function of temperature, pressure, or partial pressures of reactive gases. Raman cross-sections of surface species cannot be compared with those of pure reference compounds, because unknown electronic support effects cannot be ruled out, as in the case of the SERS effect (Pettinger, 1992). Furthermore, Raman spectra of supported transition metal oxides change as a function of treatment. Reduction of transition metals generally leads to colored samples, which have higher absorption coefficients, and as a consequence the Raman intensity can be strongly reduced by self-absorption (Kuba and Knözinger, 2002). The comparison of relative peak intensities can also lead to erroneous results, because relative Raman scattering efficiencies are usually not known and cannot be determined easily. This limitation is the most serious problem in determining quantitative structure–activity relationships from Raman spectra.

3.1.4. Raman Imaging

Raman microscopy in combination with automated xy-sample stages allows mapping larger areas or volumes of samples, and structural or compositional inhomogeneities can be characterized (Gardiner et al., 1988; Heimann and Urstadt, 1990; Watanabe and Ogawa, 1988). The xy-mapping components are typically complemented by an auto-focus (z position) to compensate for roughness of the sample surface. Depending on the goal of the investigation, two approaches are possible: mapping and global imaging. Mapping is a point-by-point analysis of the sample. It is possible to analyze random spots of the sample, a predetermined grid of spots, or along a line. The Raman mapping records complete Raman spectra in selected spots; it is thus time-consuming but rich in information.

Conventionally, wide-field Raman microprobes are applied for such mappings, but, recently, confocal microscope systems have also been used (Bridges et al., 2004; Puppels et al., 1990, 1991; Schlücker et al., 2003). Confocal microscopy originated from biological applications with the goal of analysis of the insides of cells without destruction of the cell membrane. Confocal microscopy selectively rejects any information from planes closer or further from the focal plane. Confocal microscopy is a common option in most of the commercial Raman microscopes; details of its application can be found elsewhere (Dhamelincourt et al., 1994; Lewis and Edwards, 2001).

However, in some cases, there is no need to record a complete Raman spectrum. In such cases, global imaging is the preferred option. Global imaging requires homogeneous illumination of an area of *ca.* 100 μm in diameter on the sample surface, which must be flat. The entire field is analyzed simultaneously, with the spectrometer used as a filter (Delhaye and Dhamelincourt, 1973), or by application of the Hadamard multiplexing technique (Bowden et al., 1990; Liu et al., 1991; Treado and Morris, 1990; Treado et al., 1990; Viers et al., 1990). These modes do not produce complete Raman spectra, but instead register only the intensities of selected Raman bands. Thus, spectra obtained by global imaging are more quickly acquired and more easily processed. In comparison to conventional Raman spectroscopy, global imaging is characterized by low spectral resolution (*ca.* 30 cm^{-1}). However, global imaging is not intended for molecular structure identification but instead for determination of the distributions of components across a surface. Raman mapping, on the other hand, provides complete spectroscopic information of the material and allows for more detailed investigations (Ferraro and Nakamoto, 1994; Lewis and Edwards, 2001). Imaging is applied for the analysis of semiconductors, microelectromechanical systems, fuel cells, polymers, pharmaceuticals, meteorites, and biomedical materials (Treado and Nelson, 2001). Raman mapping and global imaging in combination with statistical data evaluation will become easy-to-use, "online" characterization tools for quality control in catalyst preparation.

Raman microspectroscopy was recently combined with scanning near-field optical microscopy (SNOM) (Betzig and Trautman, 1992; Betzig et al., 1991, 1992; Fokas, 1999; Hoffmann et al., 1995; Jahnke et al., 1995, 1996; Münster et al., 1997; Noell et al., 1997; Pohl et al., 1984; Tanaka et al., 1998; Toledo-Crow et al., 1992; Valaskovic et al., 1995; Webster et al., 1998). In a SNOM experiment, a hollow, metal-plated light transmitting fiber tip is placed at such a distance to the surface that attractive van der Waals forces can be used to control the tip motion in the *z*-direction, as in noncontact mode atomic force microscopy (AFM), and, as in AFM, topographic information about the sample surface is obtained. The fiber tip has a diameter of about 50 nm; hence, the laser light transmitted through the tip excites the Raman (or fluorescence) spectrum of sample areas smaller than λ/10 nm. Thus, combined SNOM–Raman experiments simultaneously provide information about sample topography and structure. Because of the small Raman scattering cross-sections of adsorbed molecules, SNOM–Raman experiments are often performed with SERS–active metal surfaces (Deckert et al., 1998; Emory and Nie, 1998; Humbert et al., 2004; Pettinger et al., 2002, 2004; Vogel et al., 1998; Zeisel et al., 1998).

3.2. Equipment for Raman Spectroscopy of Catalysts in Reactive Atmospheres

In the preceding three decades, a number of cells that allow control of the conditions (pressure, temperature, reactants) have been designed for catalyst characterization. Even in 1977, Grasselli et al. (1977, 1981) described a heated cell, which was employed to follow transformations in molybdate catalysts. The design and use of cells was rather complex, because the instrument optics traditionally were aligned so that the scattered light was detected at 90° relative to the incident beam. Designing cells became much simpler with the introduction of Raman microscopes, which allow working in a backscattering mode (180°) (Bergin, 1990; Delhaye and Dhamelincourt, 1973; Messerschmidt and Chase, 1989; Rosasco, 1980; Schrader, 1990; Sommer and Katon, 1991; Treado and Morris, 1994). With microscope accessories and fiber optic probes, materials and processes under extreme pressure and temperature conditions can be investigated. For instance, the use of microreaction systems fitted to a Raman microscope permits analysis of geology-related processes at high pressures and temperatures (e.g., in diamond anvil cells; Adams et al., 1973; Gillet et al., 1998; Merkel et al., 2000; Sharma et al., 1985; Talyzin et al., 2002; Weinstein and Piermarini, 1975).

For example, Raman microscopy was applied to investigate the sequestration of CO_2 at 150 °C and at a partial pressure of CO_2 of 15 MPa in magnesite ($MgCO_3$) (Wolf et al., 2004). A cell with a Raman microprobe was used to monitor phase transformations of $CoMoO_4$ and $NiMoO_4$ upon heating in air and to follow the sulfidation of γ-alumina-supported $NiMoO_4$ catalysts at 320 °C in mixtures of H_2 and H_2S (Payen et al., 1980).

Commercially available Raman microscope cells often are not designed for the characterization of catalysts in reactive environments, and problems may arise from inhomogeneities in temperature and limited gas diffusion.

The investigation of liquids, gases, or solids in reactive atmospheres requires the use of vials or cells that have to be transparent to the radiation. Glass or quartz is suitable for the construction of cells, making Raman spectroscopy very versatile with respect to the media and conditions that can be used, particularly in comparison to IR spectroscopy. FT Raman spectroscopy uses IR radiation and shares the limitations inherent to IR radiation. UV–Raman spectroscopy uses UV radiation and shares the limitations inherent to UV radiation.

3.2.1. Reactor Cells for Solution Raman Spectroscopy

The future will see an increasing number of Raman investigations that aim at a better understanding of the processes occurring during catalyst preparation. In this respect, Raman cells or vials for the characterization of impregnation solutions, for instance, will become increasingly important. One example of an experimental design for solution Raman spectroscopy is shown in Figure 1. The respective precursor solutions containing various concentrations of transition metal ions are pumped through a cuvette with a plane parallel optical window. Both the incident beam and the Raman scattered light pass through this window. The pH value is monitored by a pH meter in the storage vessel, and a controller is used to adjust the solution temperature. Titroprocessors are used to control the addition of further precursor solutions or of the pH by automated acid or base addition (Dieterle, 2001). Such cells may also afford high temperature and high pressure in controlled environments and make possible experiments that provide insight into reactions demanding such conditions.

For example, in 1982 Woo and Hill (Woo and Hill, 1982, 1984) reported an investigation of $Co_2(CO)_8$ and $(RuCl(CO)_2)_2$ during propylene hydroformylation, with the complexes in solution or supported on silica or alumina. Figure 2A illustrates their Raman cell autoclave reactor, which was used in combination with online GC analysis of the effluent stream (Figure 2B). This successful attempt to use a spectroscopic cell as a reactor afforded the earliest Raman investigation of a working catalyst with simultaneous product analysis, in 1985 (Woo and Hill, 1985). Its design is essentially that of an autoclave, with a sapphire window to enable Raman spectroscopic measurements during reaction (Woo and Hill, 1985).

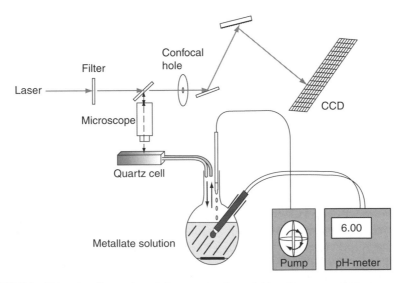

FIGURE 1 Example of experimental apparatus for solution Raman spectroscopy (Source G, Mestl).

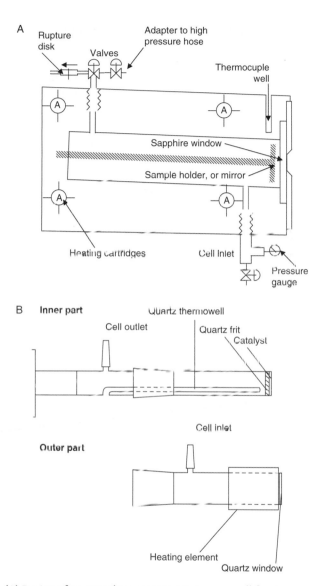

FIGURE 2 (A) Design of an autoclave reactor serving as a cell for Raman spectroscopic investigations during catalysis [Adapted from Woo S.I., and Hill, C.G., *J. Mol. Catal.*, **29**, 231 (1985) "*In situ* Raman-Spectroscopy Studies of the Hydroformylation of Propylene," copyright (1985), with permission from Elsevier); (B) Design of a flow reactor-like cell [Adapted from Wilson J.H., Hill C.G., and Dumesic J.A., *J. Mol. Catal.* **61**, 333 (1990) "Raman-Spectroscopy of iron Molybdate Catalyst Systems. 3. *In situ* Studies of Supported and Bulk Catalysts under Reaction and Redox Conditions," copyright (1990) (with permission from Elsevier) (*200*)].

3.2.2. Reactor Cells for Raman Spectroscopy of Solids in a Controlled Environment

Raman spectra of catalysts in a controlled environment may be acquired over a wide range of temperatures and pressures, including those in an autoclave or a reactor containing supercritical fluids. The upper temperature limit depends on the excitation wavelength, which determines the range of sensitivity of the detector. Thermal radiation in the same range can contribute significantly to the detector signal and can obscure spectral information (black-body radiation, *vide* Section 3.1.1).

Among the requirements that have to be considered in the design of spectroscopic cells for Raman experiments in controlled environments are the following:

(i) facile alignment and focusing of a laser spot for measurement;
(ii) environmental control (vacuum, high/low pressure, gas flow);
(iii) temperature control;
(iv) maximum optical transparence to minimize signal loss;
(v) minimization of local laser-induced heating at the sampling spot.

If catalytic activity measurements are to be performed simultaneously with the Raman spectroscopy, the cells have to also meet the specifications of catalytic reactors, such as the following:

(i) no mass or energy transfer limitations;
(ii) no bypassing, such as characterized by dead volume or preferential paths for gas flow.

Thus, activity and kinetics data and activation energy values measured with such a spectroscopic cell have to be consistent with those obtained with corresponding conventional catalytic reactors.

Raman spectroscopy has been used frequently to investigate the chemisorption of probe molecules (Cooney et al., 1975; Weber, 2000). Several groups reported variable Raman cells in which the temperature of the sample and the environment can be controlled so that catalytic reaction conditions can be simulated (Abdelouahab et al., 1992; Brown et al., 1977; Chan and Bell, 1984; Cheng et al., 1980; Lunsford et al., 1993; Mestl et al., 1997a; Vèdrine and Derouane, 2000). In these investigations, conversion and selectivity values were not measured simultaneously with the spectra. The developments of these Raman experiments have been reviewed elsewhere (Bañares, 2004; Knözinger and Mestl, 1999; Vèdrine and Derouane, 2000).

One of the first cells that allowed exposure of the catalyst sample to reaction conditions was developed by Schrader and coworkers (Cheng et al., 1980); key feature of the design is the placement of the catalyst, which is pressed into a wafer, on a rotor. The authors investigated the structural evolution of alumina-supported molybdenum oxide catalysts during calcination. Chan and Bell (1984) reported on the formation of

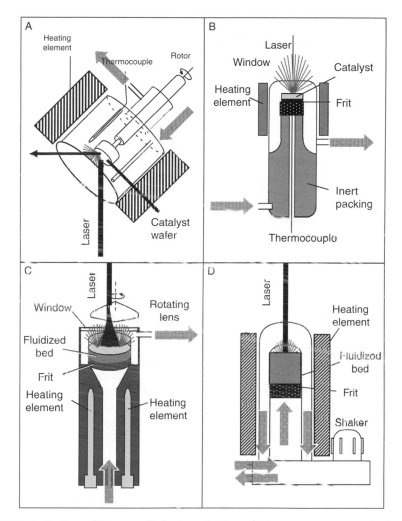

FIGURE 3 Designs of Raman cells for investigations during treatment or catalysis, (A) Schrader, (B) Lunsford, (C) Volta, (D) Stair (source, M.A. Bañares).

silica- and lanthana-supported palladium catalysts in this kind of reaction chamber. Cells using a rotating sample to prevent sample degradation in the beam have been described in detail (Brown et al., 1977; Chan and Bell, 1984; Cheng et al., 1980; Covington and Thain, 1975; Kiefer and Bernstein, 1971).

Figure 3 illustrates the concepts of Raman cells that can be used for experiments under reaction conditions. Several commercial cells are suitable for use in combination with Raman microscopy. The rotating sample design was modified by Wachs's group (Figure 3A, Bañares et al., 1994) and used to investigate supported oxides during selective alkane oxidation (Bañares et al., 2000C; Guliants et al., 1995; Sun et al., 1997) and various catalysts during oxidation of methanol (Bañares et al., 1994; Burcham et al., 2000). This cell was also employed to analyze the structures of catalysts in powder form when organic reactants underwent selective oxidation, or ethane or propane underwent oxidative dehydrogenation (ODH) (Bañares et al., 2000b,c; Santamaría-González et al., 1999). The Raman spectra afforded fundamental molecular structural information about the nature of the active sites by ^{18}O exchange (Bañares et al., 2000b,c) or the formation of surface carbonaceous species during reaction (Kuba and Knözinger, 2002; Mul et al., 2003).

Figure 3B shows the design of a Raman cell reported by Xie et al. (1999). The void volume upstream of the powder catalyst is minimized, and the gases flow through the bed. The cell was used for investigating catalysts for NO_x removal (Haller et al., 1996; Mestl et al., 1997a,b, 1998; Xie et al., 1997), including the determination of kinetics. To make possible the correlation of spectroscopic and catalyst performance data, the temperature of the analyzed spot has to be representative of the catalyst bed. Thus, local heating by the laser beam has to be minimized.

Figure 3C represents the cell design of Volta and coworkers (Abdelouahab et al., 1992) with a lens rotating off-axis, thus avoiding local heating at particular spots. Other lens systems distributing the laser light have also been described (Koningstein and Gachter, 1973; Snyder and Hill, 1991; Zimmerer and Kiefer, 1974). The Raman cell in Figure 3D was developed by Stair's group (Chua and Stair, 2000) for UV–Raman spectroscopy of a fluidized bed. To facilitate tumbling of the catalyst particles and thus to maintain the fluidization, the Raman cell was mounted on a shaker.

Experiments in which catalyst wafers are used may suffer from reactant mass transfer problems, which limit the validity of the data or complicate their analysis. To determine the reaction kinetics and activation energies, mass transfer effects have to be understood (Burcham and Wachs, 1999). These difficulties can be avoided if a conventional fixed-bed reactor is mimicked closely and the catalyst is used in powder form and the reactant gases flow through the bed. It is also important to prevent homogeneous gas-phase reactions by reducing the dead volume.

A criterion for the suitability of a spectroscopy cell for investigations of working catalysts can be formulated as follows: the activity or selectivity data and activation energy values have to be in agreement with the catalytic performance data measured with a conventional fixed-bed reactor. Table 1 is a comparison of the conversion and selectivity values characterizing an alumina-supported molybdenum–vanadium oxide catalyst during propane ODH obtained with a conventional fixed-bed reactor and with a spectroscopic cell that fulfills this requirement (Bañares and Khatib, 2004). Similar considerations have also been reported earlier for other methods, such as X-ray diffraction (Clausen et al., 1991).

Designs of Raman cells that seek to meet above-stated criterion are presented in Figure 4. Figure 4A shows the reactor cell used in an early

TABLE 1 Catalytic Activity of a Monolayer Catalyst of Molybdenum and Vanadium Oxides on Alumina with a Mo:V Atomic Ratio of 1:1 for Propane ODH (Bañares and Khatib, 2004), Measured in a Conventional Fixed-Bed Reactor and a Fixed-Bed Reactor Cell for Raman spectroscopy.

| Reactor | Temperature (°C) | Conversion (%) | Propene | Selectivity (%) | |
				CO	CO_2
Conventional	320	5.7	49.5	31.2	19.3
Conventional	320	5.6	45.6	35.6	18.8

Reaction conditions. catalyst mass, 300 mg; total flow rate, 90 ml/min; $C_3H_8/O_2/He$ — 1/12/8, molar.

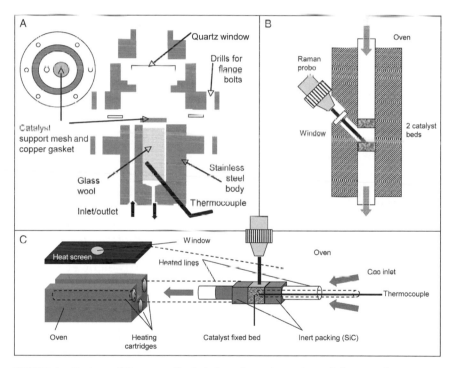

FIGURE 4 Designs of Raman cells that, for selected reactions, deliver catalyst performance data corresponding to those of an ideal reactor, (A) C.G. Hill [Adapted from Snyder T.P., and Hill C.G., *J. Catal.* **132**, 536 (1991) "Stability of Bismuth Molybdate Catalysts at Elevated-Temperatures in Air and under Reaction Conditions," copyright (1991), with permission from Elsevier (*311*)], (B) G. Mestl [source, M.A. Bañares], (C) M.A. Bañares [Source M.A. Bañares].

Raman investigation by Hill and coworkers (Wilson et al., 1990) of bulk and supported iron molybdate catalysts during methanol oxidation to formaldehyde. This Raman reaction cell was designed to minimize void volume and temperature gradients so that activity data would be representative of the catalytic process (Wilson et al., 1990). To avoid heating at the laser spot, a rotating lens was used. No significant structural transformations of the catalyst were observed during reaction at 275 °C. This result is consistent with those of other Raman investigations (Koningstein and Gachter, 1973; Snyder and Hill, 1991; Zimmerer and Kiefer, 1974). The authors reported a weakening of all the Raman bands, inferred to probably be a consequence of a partial reduction of the material, because the Raman intensity was restored upon reoxidation (Wilson et al., 1990). To minimize the (otherwise overwhelming) spectral contributions from the iron–molybdenum oxide bulk phases and to maximize the signal of surface species, the catalytic material was supported on silica. Experiments under reducing conditions (with methanol) and oxidizing conditions showed that segregated molybdena reoxidizes more readily than $Fe_2(MoO_4)_3$ (Wilson et al., 1990). More recent investigations emphasize how to take advantage of methanol-vapor induced changes of the dispersion of the active phase to prepare industrial catalysts (Wachs and Briand, 2002a, 2002b).

The Raman cell designed by Mestl and coworkers (Dieterle et al., 2001; Mestl et al., 2000; Ovsitser et al., 2002) includes a fiber optic Raman probe and a double bed configuration (Figure 4B). This is the only reported configuration with a tubular fixed-bed flow reactor that allows the investigation of the working catalyst at steady state in contact with the reactant and product mixture as a function of gas-phase residence time. The two-zone furnace has a small optical window that permits acquisition of spectra with a long-distance objective. Simultaneously with the Raman spectroscopy, online GC–MS analysis of the effluent stream was performed and provided activity and selectivity data. This configuration has been used to gain understanding of the function of bulk mixed metal oxides in selective partial oxidation reactions (Dieterle et al., 2001; Mestl et al., 2000; Ovsitser et al., 2002).

The apparatus designed by Bañares et al. (2000d) uses a Raman microscope system in combination with a fixed-bed microreactor and online GC and MS for analysis of the products and determination of the activity (Figure 4C). The microreactor walls have optical quality. Remarkable about this design is that no appreciable differences in activity and selectivity can be observed between the Raman reaction cell and a conventional fixed-bed microreactor (Bañares and Khatib, 2004). The furnace is designed with a small opening that allows spectra acquisition with a long-distance objective and does not lead to a local temperature decrease. This cell has been used to analyze the structural transformation of supported oxide catalysts during ethane oxidation (Bañares et al., 2000d), propane oxidation (García-Cortéz and Bañares, 2002, Bañares et al., 2002), and propane ammoxidation (Bañares et al., 2002; Guerrero-Pérez and Bañares, 2002, 2004).

Weaver and coworkers (Tolia et al., 1995, Williams et al., 1996) investigated NO oxidation with CO in a SERS experiment with MS analysis and reported simultaneously recorded Raman and activity data. The details of the cell design were not disclosed.

4. RAMAN INVESTIGATIONS OF CATALYSTS UNDER CONTROLLED ENVIRONMENTS

4.1. Early Vacuum, Chemisorption, and Hydration/Dehydration Experiments

Early Raman experiments characterizing catalysts (Brown et al., 1977; Chan and Bell, 1984; Cheng et al., 1980) already took advantage of the technology of sample rotation as an effective method to avoid sample damage and desorption of chemisorbed molecules (Covington and Thain, 1975; Kiefer and Bernstein, 1971). These early investigations concerned sulfidation and reoxidation treatments of molybdenum–cobalt oxide on silica–alumina and show the conversion of surface molybdenum oxides into sulfides. Partial reoxidation to oxysulfide could be avoided by applying vacuum, and the presence of MoS_2 on alumina was confirmed (Brown et al., 1977a,b). Cells with rotating samples were used to record many spectra of supported metal oxide catalysts after various treatments (Brown et al., 1977a,b; Stencel et al., 1984) and after use in catalytic reactions. For example, after coal liquefaction, the presence of carbon-containing deposits was evidenced by Raman features in the $1300–1700 \text{ cm}^{-1}$ range (Brown et al., 1977b). A cell designed to rotate the powder sample and to heat it to a temperature of $450\,^{\circ}C$, in vacuum or in a selected gas atmosphere, was used to investigate the chemisorption of pyridine on silica gel as a function of temperature (Schrader and Hill, 1975). A similar apparatus was used to monitor the effect of various treatments on various samples (Cheng et al., 1980). Investigations of chemisorption by Raman spectroscopy have been reviewed in detail elsewhere (Mestl and Knözinger, 1997; Weber, 2000).

One of the key discoveries during the first Raman experiments with cells that allowed control of the sample environment was a demonstration of the influence of moisture on the spectra of some catalysts. For example, supported metal oxide catalysts exhibit different Raman spectra depending on the loading of the metal oxide, such as molybdenum oxide. When the Raman spectra were recorded under ambient conditions, the spectra of the catalysts with low molybdenum loadings indicated the presence of monomolybdates, as characterized by the prominent band at about 900 cm^{-1}. In contrast, polymolybdates, indicated by Raman bands between 930 and 960 cm^{-1}, were detected at intermediate loadings, and crystalline MoO_3 nanoparticles were found in materials with high

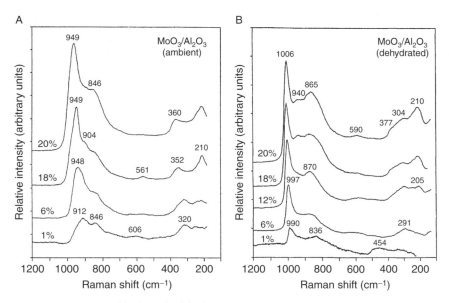

FIGURE 5   Spectra of hydrated–dehydrated–hydrated alumina-supported samples. (A) Raman spectra of MoO_3/Al_2O_3 catalysts with various molybdenum loadings recorded under ambient conditions (hydrated sample), displaying monomolybdates and polymolybdates. (B) Raman spectra of the same catalysts recorded after dehydration at 360 °C showing the presence of monooxo species [Reprinted from Hu, H., Wachs, I.E., and Bare, S.R., *J. Phys. Chem.* **99**, 10897 (1995), copyright (1995) American Chemical Society (231)].

loadings (Figure 5). This picture of the surface species, however, changed when the Raman spectra were recorded at high temperatures under controlled conditions (Hu et al., 1995); the noncrystalline surface-MoO_x species were then characterized by a Raman band at about 1000 cm^{-1}. The exact nature of the surface molybdenum species under conditions of low and high humidity was still under debate at the time, but the spectra recorded at high temperature (Figure 5B) show features that are otherwise not observed (Hu et al., 1995).

This dynamic behavior of the catalyst, depending on the degree of hydration, is typical of many supported transition metal oxides, as evidenced by the changes in their Raman modes (Brown et al., 1977, Brown et al., 1977), as discussed below. The transformations are usually reversible.

4.1.1. Hydrated Supported Metal Oxides

The initial Raman characterizations of supported metal oxides were conducted under ambient conditions. The first attempt to measure the Raman spectrum of MoO_3/Al_2O_3 was reported by Trifirò et al. in 1972

(Villa et al., 1974). The absence of detectable Raman bands led these investigators to conclude that the dispersed molybdate species must be Raman inactive. In 1977, however, Brown et al. (1977) repeated this Raman characterization experiment with supported MoO_3/Al_2O_3 and $MoO_3/SiO_2-Al_2O_3$ catalysts, as well as with NiO- and CoO-promoted catalysts, and reported that they successfully observed a new unique Raman band at *ca.* 950 cm^{-1} originating from the dispersed molybdate species. Furthermore, they also observed the presence of crystalline MoO_3 and $Al_2(MoO_4)_3$ in the Raman spectra, and the latter phase was not even detectable by XRD.

In 1978, Knözinger and Jeziorowski (1978) reported polymeric aggregates of octahedral molybdena on alumina-supported catalysts. This was a breakthrough in Raman characterization of supported metal oxide catalysts, which was rapidly followed by publications from other research groups reproducing the Raman spectrum of supported MoO_x/Al_2O_3 (Schrader and Cheng, 1983; Thomas et al., 1977; Wang and Hall, 1983) as well as other supported metal oxides: V_2O_5/Al_2O_3 (Roozeboom et al., 1978), Re_2O_7/Al_2O_3 (Kerkhof et al., 1979), CrO_3/Al_2O_3 (Iannibello et al., 1979), and NiO/Al_2O_3 (Payen et al., 1980). Over the past two decades, the ambient Raman spectra of numerous supported metal oxide catalysts have been reported (Wachs, 2002).

As more Raman spectra of supported metal oxide catalysts appeared in the literature, many contradictory models for the dispersed metal oxide structure were proposed. It was observed in 1983–1984 by Wang and Hall (1983), Chan et al. (1984), and Stencel et al. (1984) that supported Re_2O_7, MoO_3, and $WO_3-V_2O_5$ were in hydrated states during ambient Raman measurements. However, the molecular structures of the various hydrated dispersed metal oxide species on oxide supports were not fully understood at that time.

In subsequent years, various research teams realized that the Raman spectra of the hydrated dispersed metal oxide species present on oxide supports resemble those of the polyoxo anions of the respective metals in aqueous solution (Jeziorowski and Knözinger, 1979; Knözinger and Jeziorowski, 1978; Okamoto and Imanaka, 1988; Wang and Hall, 1980; Williams et al., 1991). Furthermore, comparable to the behavior of the polyoxo anions in aqueous solution, the molecular structures of the hydrated supported polyoxo anions varied with the isoelectric point (IEP) or the point of zero charge (PZC) (Hunter, 1988) of the oxide surface (Deo and Wachs, 1991). Figure 6 shows the dominant V_xO_y species in the liquid phase, depending on concentration and pH (Deo and Wachs, 1991). These observations allowed Deo and Wachs (1991) to predict the various hydrated molecular structures on surfaces. At low surface coverages of the hydrated dispersed metal oxide species, the net PZC is dominated by the PZC of the oxide support. At significant surface coverages by the hydrated dispersed metal oxide species, however, the PZC is a function of the PZC of the oxide support and the dispersed oxide component. At monolayer surface coverage, the PZC characterizing hydrated supported metal oxides appears to be approximately the mean of the values of the two components. The PZC theory was confirmed for many supported metal oxide materials under ambient conditions, for which both the Raman spectra and the pH at the PZC were determined: V_2O_5 (Deo and Wachs, 1991; Gil-Llambias et al., 1985), Re_2O_7 (Deo and Wachs, 1991; Hardcastle et al., 1988), MoO_3 (Hu et al., 1995; Roark et al., 1992), WO_3 (Horsley et al., 1987), CrO_3 (Weckhuysen et al., 1995), and Nb_2O_5 (Jehng and Wachs, 1992).

The Raman investigation of niobium species in aqueous solutions of niobium oxalate (Jehng and Wachs, 1991) nicely showed the dependence of their constitution on pH and concentration. The PZC theory was successfully applied to predict the hydrated, molecular structures of multicomponent supported metal oxide species, such as iron–molybdenum, iron–vanadium, molybdenum–vanadium, tungsten–vanadium, and sodium–vanadium oxide species (Vuurman et al., 1991; Wachs et al., 1993).

It is emphasized that the PZC theory for the prediction of the molecular structure of hydrated polyoxo anions holds true only under ambient conditions when the oxide surfaces are extensively hydrated. This condition is not satisfied when the supported metal oxide catalysts are heated

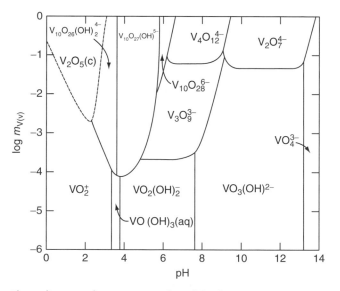

FIGURE 6 Phase diagram of aqueous vanadium (V) solution species, present in aqueous solutions at 25 °C, on the basis of data in Baes C.F., Jr., and Mesmer R.E., "The Hydrolysis of Cations," Wiley, New York, 1970 [Reprinted from Deo, G., and Wachs, I.E., *J. Phys. Chem.* **95**, 5889 (1991); copyright (1991) American Chemical Society (246)].

to elevated temperatures (>100 °C) and physisorbed water desorbs. Even in the presence of flowing steam, dispersed metal oxide species essentially become dehydrated at elevated temperatures (230–500 °C), because of the rapid desorption of water at such temperatures (Jehng et al., 1996). Recent Raman experiments carried out at temperatures up to 500 °C with samples in the presence of water confirmed this trend (Christodoulakis et al., 2004; Xie et al., 2000).

To summarize, the surface chemistry of adsorbed metallates in the fully hydrated state of the surface closely resembles their aqueous solution chemistry. A wealth of Raman spectra have been published characterizing pure molybdenum, vanadium, and tungsten polyoxometallates (Baes and Mesmer, 1976; Cheng and Schrader, 1979; Gonzales-Vildres and Griffith, 1972; Griffith, 1970; Griffith and Lesniak, 1969; Griffith and Wilkins, 1967; Himeno et al., 1997; Hunnius, 1975; Johanson et al., 1979; Ng and Gulari, 1984; Schrader et al., 1981; Tytko and Glemser, 1976; Tytko and Mehrnke, 1983; Tytko and Schönfeld, 1975; Tytko et al., 1975, 1980, 1983). Even the formation of heteropoly anions was reported from adsorbed metallates and dissolved support ions during the impregnation step (Carrier et al., 1997; Le Bihan et al., 1998) or after prolonged exposure to moisture (Bañares et al., 1994; Marcinkowska et al., 1986; Rocchiccioli-Deltcheff et al., 1990; Stencel et al., 1986).

4.1.2. Dehydrated Supported Metal Oxide Catalysts

As environmental Raman cells became more common, attention shifted to the molecular structures of dehydrated dispersed metal oxide species on supports. To the best of our knowledge, the first Raman spectra of dehydrated species concerned supported MoO_3/Al_2O_3 catalysts calcined in O_2 at 550 °C (Cheng et al., 1980). However, the Raman band positions of the dehydrated sample were not reported, and no attempt was made to identify the molecular structures. Subsequent Raman investigations of dehydrated supported Re_2O_7 (Wang and Hall, 1983), V_2O_5 (Chan et al., 1984), MoO_3 (Chan et al., 1984; Stencel et al., 1984), WO_3 (Chan et al., 1984; Stencel et al., 1984), CrO_3 (Hardcastle and Wachs, 1988), and Nb_2O_5 (Jehng and Wachs, 1991) demonstrated that the dehydrated surface metal oxide species had unique molecular structures that were dependent on both, the specific oxide support and the surface coverage of the dispersed metal oxide species. These results can be explained by the condensation reactions of the metallate ions with the support surface OH groups, which were confirmed by IR spectroscopy (Roark et al., 1992). The complete resolution of the structure required the combination of Raman spectra of the dehydrated dispersed metal oxides on supports with data obtained with complementary methods: XANES/EXAFS, solid-state NMR, IR, and diffuse reflectance UV–vis spectroscopies, and isotopic ^{18}O-exchange experiments. In particular, XANES and EXAFS data complement information from Raman spectra well.

On SiO$_2$, dehydrated surface metal oxide species were found to be generally present as isolated molecular structures (Wachs et al., 1993), because of the low density of the reactive anchoring surface Si–OH sites with which adsorbed metallate ions condense (releasing water). Polymeric species would exhibit Raman bands originating from bridging M–O–M bonds: δ(ca. 200–300 cm^{-1}), v_s (ca. 450–550 cm^{-1}), and v_{as} (ca. 650–750 cm^{-1}). These bands were absent from the spectra of dehydrated silica-supported metal oxide species. On other oxide supports, dehydrated surface metal oxide species were detected as both isolated and polymeric molecular structures (Bañares and Wachs, 2002). Isolated surface metal oxide species predominate at low surface coverages, and polymeric surface metal oxide species tend to be the major surface species at high surface coverages. The molecular structures of the group 5 and 6 supported metal oxides are characterized by the presence of a single oxo group and several bridging oxygen bonds to the support or other neighboring cations (Bañares and Wachs, 2002). The exact assignment of the vibrational modes is still a matter of debate. The band at wave numbers > 1000 cm^{-1} was attributed to the oxo mode of isolated and polymeric species and that near 910 cm^{-1} to vibrations of bridging oxygen atoms, on the basis of complementary IR and Raman data characterizing supported oxides and model catalysts, as well as DFT calculations and measurements of samples in reactive atmospheres (Deo et al., 1994; Freund, 2005; García-Cortéz and Bañares, 2002; Magg et al., 2004; Wu et al., 2005). The band at ca. 910 cm^{-1} appears to be associated to the bond between the supported oxide and the support, as was reported for zirconia-supported tungsten trioxide (Kuba and Knözinger, 2002). The Raman bands of dispersed oxides shift and weaken with temperature, which was attributed to thermal expansion and changes in the population of the vibrational energy levels, respectively (Xie et al., 2001).

Such variations with temperature have also led to preliminary conclusions regarding the exact structures of supported oxides (Gijzeman et al., 2004). A systematic investigation of supported oxides of the group 5, 6, and 7 elements provides some general trends (Weckhuysen et al., 2000): the group 5 (vanadium, niobium, and tantalum) surface oxide species consist of monoxo structures in the isolated and polymeric forms; the group 6 (chromium, molybdenum, and tungsten) surface oxides tend to assume dioxo structures in the isolated forms and monoxo in the polymeric forms. The group 7 (rhenium) surface oxide forms isolated ReO$_4$ species with a trioxo structure (Weckhuysen et al., 2000).

Combination of UV–vis DRS and Raman spectroscopy data has allowed for the quantitative determination of the monomer and polymer concentrations of the surface metal oxide species (Tian et al., 2006). The dispersed surface vanadia phase in V_2O_5/SiO_2 catalysts was found to consist predominantly of isolated surface VO_4 species. Isolated surface VO_4 units in supported V_2O_5/Al_2O_3 and V_2O_5/ZrO_2 catalysts, however, were present only at coverages below 20% of a monolayer (<1.5 V/nm^2). The fraction of polymeric surface vanadia increased more rapidly with coverage on alumina than on zirconia. At monolayer surface coverage, 100% of the vanadia species on alumina (and about 80% on zirconia) were polymeric.

Although UV–vis DR spectra of vanadia on other oxide supports (such as TiO_2, CeO_2, and Nb_2O_5) cannot be readily interpreted because of the overlap of their strong absorptions with those of vanadia, equivalent shifts of the Raman bands as a function of vanadia coverage suggest that the surface VO_4 species also polymerize on these supports.

In contrast to acidic surface metal oxides with cation oxidation states of $+5$ to $+7$, which are anchored to the support by surface hydroxyl groups, basic surface metal oxides with cation oxidation states of $+1$ to $+3$ are anchored at surface Lewis acid sites (Bredow et al., 1998; Cortéz et al., 2003; Diebold, 2003; Jehng and Wachs, 1992; Vuurman et al., 1996). Raman spectra demonstrated that supported basic metal oxides are, in contrast to acidic supported metal oxides, insensitive to moisture. The Raman spectra of basic surface metal oxide species do not show the bands at about 1000 cm^{-1} that would indicate terminal $M=O$ bonds. The spectra typically exhibit Raman bands in the wave number region of 500–700 cm^{-1}, characteristic of M—O bonds (Chan and Wachs, 1987; Tian et al., 2006; Vuurman et al., 1996); similar behavior was observed for TiO_x, ZrO_x, PtO_2, and other oxide surface species with cations in the $+4$ oxidation state.

Gao et al. (1999) performed combined Raman and UV–vis DRS analyses of TiO_2(anatase)/SiO_2 and found that the coordination of the surface TiO_x species varied from isolated TiO_4 to polymeric TiO_5 to TiO_2 (anatase) nanocrystals with increasing titania loading. The anatase nanocrystals formed only at coverages exceeding a monolayer (>3 Ti/nm^2), and their Raman bands did not depend on the crystallite size.

4.1.3. Hydrated and Dehydrated Bulk Metal Oxide Catalysts

In general, the Raman structural information characterizing crystalline bulk metal oxide phases is consistent with that provided by XRD. When the bulk metal oxide phase does not have long range order (i.e., is X-ray amorphous) with coherence lengths below ca. 4 nm, however, Raman spectroscopy can provide structural information (Cavani and Trifirò, 1994; Dutta and Shieh, 1986; Dutta et al., 1987; Guliants et al., 1995; Hutchings et al., 1994; Wachs et al., 1996). Bulk metal oxides in dry and moist environments typically produce identical Raman spectra, because water typically does not diffuse into the bulk of the lattice. Hydroxide-containing structures

(Jehng and Wachs, 1991) and heteropolyanions stabilized by water mole-cules (Bañares et al., 1994; Damyanova et al., 2000; Greenwood and Earnshaw, 1984; Tatibouët et al., 1988) were found to be an exception to this behavior.

The first Raman spectra of bulk metal oxide catalysts were reported in 1971 by Leroy et al. (1971), who characterized the mixed metal oxide $Fe_2(MoO_4)_3$. In subsequent years, the Raman spectra of numerous pure and mixed bulk metal oxides were reported; a summary in chronological order can be found in the 2002 review by Wachs (Wachs, 2002). Bulk metal oxide phases are readily observed by Raman spectroscopy, in both the unsupported and supported forms. Investigations of the effects of mois-ture on the molecular structures of supported transition metal oxides have provided insights into the structural dynamics of these catalysts. It is important to know the molecular states of a catalyst as they depend on the conditions, such as the reactive environment.

Beyond providing bulk structural information about 3-D metal oxide phases, Raman spectroscopy can also provide information about the terminating (and thus 2-D) surface layers of bulk metal oxides. For example, surface $Nb{=}O$, $V{=}O$, and $Mo{=}O$ functionalities were detected by Raman spectroscopy for bulk Nb_2O_5, and for vanadium–niobium, molybdenum–vanadium, molybdenum–niobium, and vanadium–antimony mixed oxide phases (Guerrero-Pérez and Bañares, 2004; Jehng and Wachs, 1991; Zhao et al., 2003).

4.2. Raman Investigations of Catalyst Preparation

There have been few Raman investigations of catalyst preparation (of oxides, zeolites, or metals). Such experiments deliver information about molecular structures, and the formation of crystalline phases is detected at earlier stages by Raman spectroscopy than by XRD. Moreover, cells that allow for variable conditions are easily constructed.

An example is the detailed report by Hill and Wilson (1990) on the effect of the following parameters on the formation of bulk iron molybdate: the concentrations of the parent solutions, temperature and pH of these solu-tions, the order of mixing, aging, and filtration and of evaporation, and washing; the influence of drying time and temperature, and calcination time and temperature were also determined. The spectra showed that iron and molybdenum precursors (namely, iron nitrate and ammonium hepta-molybdate) began to undergo transformations at 90 °C. New Raman bands characteristic of iron molybdate developed at temperatures of 105–120 °C. Segregated MoO_3 was not apparent for samples with excess of molybde-num before a temperature of 250 °C was reached; thus, iron molybdate appeared to form more readily than the segregated oxide phases.

Several reported Raman investigations concern the synthesis of zeolites (Angell, 1973; Barasnka et al., 1986; Dutta and Puri, 1987; Dutta and Shieh, 1986; Dutta et al., 1987; Roozeboom et al., 1983). Raman spectroscopy has provided information about the conformation of the organic template during the crystallization process (Brémard and Bougeard, 1995). For instance, the tetrapropylammonium ion template is initially trapped in an amorphous silica phase (Dutta and Puri, 1987). Raman spectra also showed that $Al(OH)_4^-$ polymerized into aluminosilicates that formed the nuclei for crystal growth (Roozeboom et al., 1983). However, $Al(OH)_4^-$ was shown to become detectable by Raman spectroscopy only at concentrations of 0.05–0.10 M (Dutta and Puri, 1987; Twu et al., 1991), and the Raman cross-section of this ion is higher than that of other anions (Roozeboom et al., 1983). A comprehensive review about the genesis of zeolites characterized by Raman spectroscopy is available (Mestl and Knözinger, 1997; Roozeboom et al., 1983).

TP-Raman spectroscopy was used to evaluate the thermal stability of titanium oxide nanotubes (Blume, 2001). Titania nanotubes are more stable in oxidizing than in inert atmospheres. Titania nanotubes were transformed into anatase at 250 °C in an inert atmosphere, whereas the nanotube structure remained stable up to a temperature of 400 °C in the presence of air. It was proposed that oxygen prevents segregation of hydroxyl and Ti^{4+} ions and the ensuing reduction to Ti^{3+}, which would otherwise lead to the generation of a nonstoichiometric anatase phase (Blume, 2001).

Mixed oxides represent an important class of selective partial oxidation catalysts. Typically, such catalysts contain at least three types of transition metal ions in optimized ratios and are industrially produced by spray-drying of highly concentrated precursor solutions. A wealth of Raman literature has been published characterizing the aqueous solution chemistry of pure molybdenum, vanadium, and tungsten polyoxometallates (Baes and Mesmer, 1976; Cheng and Schrader, 1979; Gonzales-Vildres and Griffith, 1972; Griffith and Lesniak, 1969; Griffith and Wilkins, 1967; Griffith, 1970; Himeno et al., 1997; Hunnius, 1975; Johanson et al., 1979; Ng and Gulari, 1984; Schrader et al., 1981; Tytko and Glemser, 1976; Tytko and Mehrnke, 1983; Tytko and Schönfeld, 1975; Tytko et al., 1975, 1980, 1983).

In the reported experiments, transition metal ion concentrations were typically low, and electrolytes were added, and, hence, the results are not necessarily pertinent to industrial catalyst preparation. However, Raman spectroscopy was also applied for the characterization of mixed metal ion solutions with compositions of industrial relevance (Dieterle, 2001). Figure 7A, for example, shows the Raman spectra of solutions containing molybdenum, tungsten, and vanadium ions as a function of pH. At high pH values, for example, between 9 and 6, Raman bands were observed that evidence MoO_4^{2-} and $V_{10}O_{28}^{6-}$ species. Raman bands of hepta- and octamolybdate, and decavanadate were observed upon acidification. At the pH

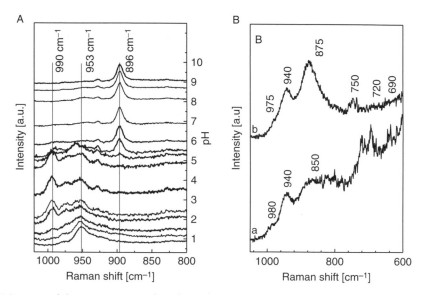

FIGURE 7   (A) Raman spectra of a solution containing molybdenum, tungsten, and vanadium as a function of pH; (B) spectra showing the main band of the wet mixed-metal polyoxo paste. Raman spectra of the wet paste obtained from Mo–V–W oxide solutions with a Mo:V:W ratio 6:2:1 during drying at 100 °C (a), and of the solid after drying at 100 °C (b).

value of 1, the bands of the pure polyoxo metallates disappeared, and the Raman spectrum indicated the formation of a mixed polyoxometallate species (Dieterle, 2001; Knobl et al., 2003; Mestl, 2002). This example shows the power of Raman spectroscopy for the characterization of the complex chemistry of mixed transition metal catalyst precursor solutions.

The next step in the preparation of typical multimetal mixed oxide catalysts after formation of the precursor solution is the removal of the solvent by drying. Various techniques are used including spray drying, freeze drying, vacuum evaporation, and evaporation at elevated temperatures. Raman spectroscopy was applied for the characterization of the structural changes during drying by heating (Knobl et al., 2003). Spectrum (a) of Figure 7B, recorded of a wet mixed molybdenum–vanadium–tungsten polyoxo paste at 87 °C, shows a main band at 940 cm^{-1}. The other broad bands were attributed to newly formed metal–oxygen bridges, which indicate structural changes of the mixed polyoxo metallate upon water evaporation. The similarity of the Raman spectrum (b) in Figure 7B of the dried precursor solid to those of nanocrystalline (MoVW)$_5$O$_{14}$-type mixed oxide catalysts (Dieterle et al., 2001; Mestl et al., 2000) suggests that structures similar to the ones in these oxides had already been formed at this stage of the preparation. Structural transformations during the synthesis of mixed oxide catalysts comprising molybdenum, vanadium, tellurium, and niobium were also investigated (Beato et al., 2006).

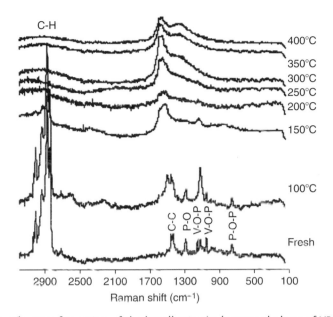

FIGURE 8   The transformation of the lamellar to the hexagonal phase of VPO precursors [Reprinted from Carreón, M.A., Guliants, V.V., Guerrero-Pérez, M.O., and Bañares, M.A., *Microporous Mesoporous. Mater.* **71**, 57 (2004) "Phase transformation in mesostructured VPO/surfactant composites," copyright (2004) with permission from Elsevier (*319*)].

VPO catalysts pass through various synthesis stages and undergo transformations during catalytic alkane oxidation reactions (Abdelouahab et al., 1992; Guliants et al., 1995, 2001; Hutchings et al., 1994; Volta et al., 1992). The phase transformations in mesostructured VPO phases during thermal treatment in N_2 were monitored by Raman spectroscopy (Carreón et al., 2004). The transformation of a lamellar into a hexagonal phase was evidenced by the spectra (Figure 8); Raman bands corresponding to symmetric P—O—P ($790\,cm^{-1}$), V—O—P (1070 and $1130\,cm^{-1}$) and P—O ($1300\,cm^{-1}$) stretching vibrations (Abdelouahab et al., 1992) were no longer visible at temperatures above 200 °C (Carreón et al., 2004). The characteristic bands of C—H stretching and C—C vibrations of the surfactant were initially present at 2900 and $1450\,cm^{-1}$; these bands were not observed at temperatures above 150 °C, a result that is indicative of the progressive decomposition of the surfactant (Carreón et al., 2004). Two new Raman bands, near 1590 and $1350\,cm^{-1}$, became evident as the treatment temperature increased. These bands are characteristic of graphite-like carbon species, which were evidently generated from occluded surfactant molecules. The absence of second-order Raman spectra of the carbonaceous deposits in the wave number range of $2700–3200\,cm^{-1}$ emphasizes the lack of long-range order in these aggregates (Angoni, 1998; Cuesta et al., 1994;

Vidano and Fischbach, 1978). Moreover, the relative intensities of the Raman bands at 1590 and 1350 cm^{-1} indicate that the graphite-like domains were roughly 60 Å in diameter (Tuinstra and Koenig, 1970). This size of the carbon aggregates confirmed the hypothesis of occluded surfactant molecules in the mesostructured phosphate.

Spatially resolved Raman spectroscopy has provided insights into the physicochemical processes that determine the distribution of the $H_2PMoO_{11}CoO_{40}{}^{5-}$ active phase in alumina pellets (Bergwerff et al., 2005). Molybdenum and cobalt complexes were found to diffuse through the pore structure of the alumina pellets at different rates: the transport of cobalt complexes was fast, whereas molybdenum complexes required several hours to reach an equilibrated distribution. Spatially resolved Raman monitoring provides information about how preparation conditions affect the distribution of molybdenum ions (Bergwerff et al., 2005).

An important stage of catalyst preparation is the formation of the actual active phase. Vanadium antimonate catalysts for propane ammoxidation are activated in the initial period on stream. This transient activity was reported to arise from a rearrangement into the active phase, which was not observed during calcination of the precursors. A $VSbO_4$ phase, which is believed to be essential for propane ammoxidation, formed from surface VO_x and SbO_x species on alumina (Guerrero-Pérez and Bañares, 2002). The catalytic cycle occurring on the vanadium and antimony sites can be understood by spectroscopic investigation of this catalyst during propane ammoxidation, as discussed below.

Few Raman investigations include reports of the preparation of catalysts containing dispersed metals, such as silica- or lanthana-supported palladium (Chan and Bell, 1984). The transformation of the $PdCl_2$ precursor into supported PdO was observed by Raman spectroscopy. During heat treatment, β-$PdCl_2$ on SiO_2 was converted to α-$PdCl_2$ and finally to PdO; on La_2O_3, however, the exact form of the palladium species could not be determined, and it was suggested that a lanthanum–palladium mixed oxide formed.

4.3. Raman Investigations of Catalysts under Conditions Relevant to Catalysis

The structures of catalysts may change significantly under reaction conditions. Most catalysts operate at elevated temperatures. Consequently, the structures of solid catalysts must be examined at working temperatures and in actual reaction environments. Even in 1977, Grasselli et al. (Grasselli et al., 1981) designed a heatable cell to follow transformations of molybdate catalysts. Since this early work, there have been many further developments in the understanding of transformations of catalysts under reaction conditions. Some of these investigations include the simultaneous quantitative kinetically relevant analysis of the product mixture.

4.3.1. Spreading, Segregation, and Phase Transformations

The dispersion and aggregation of oxides on supports has been investigated for a long time (Haber et al., 1985; Xie and Tang, 1999). Raman spectroscopy was applied early to follow spreading. For example, Knözinger and coworkers (Leyrer et al., 1990) reported on the spreading behavior of MoO_3 on alumina and on silica. The dispersion of metal oxides, which is relevant in the preparation of supported metal oxide catalysts, depends strongly on the nature of the support. This point was illustrated by a Raman microscopy investigation (Leyrer et al., 1990) in which a molybdenum trioxide wafer was in contact with an alumina wafer or a silica wafer. Upon thermal treatment, molybdenum oxide became dispersed on alumina but not on silica. Humidity promoted the dispersion of molybdenum oxide (Leyrer et al., 1990) and of other oxides, among them vanadium oxide (Spengler et al., 2001).

The driving force for spreading is the surface free energy, which is lower for surface-dispersed oxides than for the corresponding bulk oxides (Gonzalez-Elipe and Yubero, 2001; Leyrer et al., 1990). Surface species are stabilized through reaction with surface hydroxyl groups (Leyrer et al., 1990), as confirmed by IR spectra that indicate linear consumption of support hydroxyl groups with surface coverage (Roark et al., 1992).

High-temperature Raman spectroscopy was also used to follow the spreading process. The Raman spectrum of crystalline MoO_3 in a physical mixture of 9 wt% MoO_3 and Al_2O_3 completely lost any structure in the lattice mode regime at a temperature that was 250 K below its melting point (813 °C) (Mestl, 2000; Mestl and Knözinger, 1998) (Figure 9). This behavior was tentatively interpreted as spreading. The surface of crystalline MoO_3 seems to "melt" at approximately the Tammann temperature, as proposed in the unrolling carpet mechanism (Knözinger and Taglauer, 1993, 1997), and small $(MoO_3)_x$ oligomers spread across the Al_2O_3 surface.

Changes in the gas environment have a remarkable effect on the dispersion of supported metal oxides; for instance, spreading of metal oxides onto surfaces was induced during alcohol oxidation reactions (Cai et al., 1997; Wang et al., 1999). For example, when heating a physical mixture of 4% MoO_3 and 96% TiO_2 in dry air, spreading of MoO_3 occurs only at 500 °C, whereas during methanol oxidation catalysis it spreads at 230 °C (Wang et al., 1999). This trend was not observed for molybdena–silica, and this result emphasizes the low ability of silica to disperse oxides and is consistent with other aggregation phenomena in silica-supported oxides (Bañares et al., 2000a; Xie et al., 2000). Small alcohol molecules promote fast dispersion of metal oxides (Wang et al., 1999); the reported efficiency sequence is methanol ≫ ethanol > 2-butanol; water > oxygen.

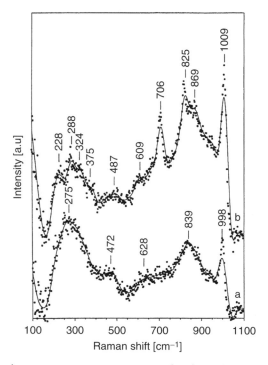

FIGURE 9 (A) High-temperature Raman spectra of a physical mixture of 9 wt% MoO_3 and Al_2O_3 recorded at 823 K in dry oxygen and (B) after quenching to room temperature. The Raman spectrum of crystalline MoO_3 in a physical mixture of 9 wt% MoO_3 and Al_2O_3 completely lost structure in the lattice mode regime at a temperature more than 250 °C lower than its melting point (813 °C); based on Mestl, G., *J. Mol. Catal.* A, 158, 45 (2000) "*In situ* Raman spectroscopy - A valuable tool to understand operating catalysts" and Mestl, G., and Knözinger, H., *Langmuir* 14, 3964 (1998) "Spreading of MoO_3 on γ-Al_2O_3 induced by mechanical activation".

The degree of dispersion of a metal oxide depends on the stabilization of the dispersed phase on the surface and varies with the nature of the support. The reaction-induced spreading of CrO_3, MoO_3, V_2O_5, Re_2O_7, and Cr_2O_3 during alcohol oxidation readily occurred on titania and tin oxide, but did not take place on silica (Wang et al., 1999), in line with previously observed trends (Leyrer et al., 1990). Reducing environments resulted in an induction period during the dispersion of bulk oxides (Haber et al., 1995).

Reaction-induced dispersion may be used as a substitute for conventional preparation methods for supported metal oxides (Wachs and Cai, 2001); it constitutes a particular case of solid—solid wetting, which is proposed to play an important role in catalyst preparation (Leyrer et al., 1990). Industrially relevant mixed metal oxide catalysts can be prepared by reaction-induced dispersion at temperatures that are significantly lower than those typically used in standard preparations. For example, the preparation of iron molybdate catalysts for methanol oxidation requires the coprecipitation of an aqueous solution of iron trichloride and ammonium heptamolybdate. Thus, a simple and efficient process to form iron molybdate from a physical mixture of molybdena and ferric oxide during methanol oxidation may be of interest for industrial application.

The fresh precursor mixture exhibited Raman bands characteristic of α-MoO_3. Iron oxide was not detected, because the Fe_2O_3 Raman cross-section is significantly lower than that of α-MoO_3. After exposure to the methanol oxidation environment, new Raman bands appeared that indicated $Fe_2(MoO_4)_3$. Thus, $Fe_2(MoO_4)_3$ readily formed from a mixture of the individual metal oxides during methanol oxidation (Wachs and Briand, 2002a, 2002b).

4.3.2. Surface Oxygen Species

Lunsford and coworkers (Haller et al., 1996; Mestl et al., 1997b, 1998; Xie et al., 1997) acquired Raman spectra during reaction to investigate peroxide species on barium-containing materials that are suitable for oxidative coupling of methane (OCM) and for NO_x storage. At 590 °C in a flow of N_2 containing 10% O_2, the BaO_2/MgO sample exhibited Raman bands near 829 cm^{-1} (crystalline BaO_2) and two shoulders at 820 and 810 cm^{-1}, characteristic of peroxide species (Haller et al., 1996). The intensities of the Raman bands of peroxides decreased with decreasing O_2 partial pressure during the OCM reaction (Mestl et al., 1998). A further decrease of the O_2 volume fraction from 0.75% to 0.50% resulted in the decomposition of the crystalline BaO_2 as indicated by the disappearance of the Raman band near 829 cm^{-1}. New Raman bands became visible at 123, 185, 900, and 925 cm^{-1} (Mestl et al., 1998). Thus, it appeared that noncrystalline peroxide species were still present during OCM even in the absence of crystalline BaO_2, and these could play an active role in the OCM reaction.

The effect of temperature on surface peroxide species on ceria catalysts was investigated by Pushkarev et al. (2004). Dioxygen was adsorbed on partially reduced CeO_2 to probe the various surface defect centers. Isotopic labeling was used in combination with temperature-programmed reaction of the adsorbed dioxygen species to identify peroxide and superoxide species adsorbed on one- or two-electron defect centers. These species were characterized by Raman bands in the wave number ranges of 831–877 and 1127–1135 cm^{-1}, respectively. The dynamic behavior of the different peroxide Raman bands at 831, 860, and 877 cm^{-1} during temperature-programmed experiments suggested the presence of three different defect species, namely, peroxides at isolated, linear, and triangular two-electron defects. However, these molecular oxygen species decompose or desorb under typical oxidation reaction conditions.

Li and Oyama (1997) investigated the oxidation of ethanol by ozone on manganese oxide catalysts. Transient Raman experiments characterizing ethanol chemisorbed on a manganese oxide catalyst at low temperatures (-196 to $-73\ ^\circ$C) showed a progressive removal of ethoxide species with increasing ozone concentration and the concurrent formation of peroxide species (874–$884\ cm^{-1}$), which originated from ozone decomposition. Zirconia-supported manganese oxide also formed surface peroxide species (Radhakrisnan and Oyama, 2001).

The assignment to peroxide species was confirmed by isotope labeling (Li et al., 1998). These results emphasize the higher reactivity of ozone as an oxidant, compared with molecular oxygen. The authors emphasized that the existence of peroxide species at temperatures below $-73\ ^\circ$C may not be relevant for the catalytic reaction, which occurs typically at temperatures above $27\ ^\circ$C (Li and Oyama, 1997).

4.3.3. Oxidation States

The oxidation states of the elements in a catalyst can in some cases be determined by Raman spectroscopy. During some reactions on catalysts containing elements such as platinum, palladium, rhodium, copper, and silver, Raman experiments were used to detect the presence of oxide species. These cases are discussed in the section on Raman investigations of metal catalysts during operation.

The determination of oxidation states of transition metals such as vanadium, chromium, or molybdenum in supported oxides is difficult, because many of the reduced phases are weak Raman scatterers. In such cases, the combination of UV—vis DRS and Raman spectroscopy is essential for identifying the reduced oxidation states and the extent of reduction. UV—vis DRS provides information about the changes in oxidation state (Weckhuysen, 2003) and the extent of reduction (Gao et al., 1999), and it permits determination of the specific wavelengths at which it is possible to resonantly enhance the Raman signal.

Weckhuysen (2003) reported on the simultaneous use of UV—vis DRS and Raman spectroscopy to obtain complementary information about the metal oxidation state and the molecular structure of alumina-supported chromia during propane dehydrogenation. The results suggested that Cr^{3+} ions were the active sites; Cr^{2+} species were not evident. In the presence of O_2, (i.e., during the ODH of propane), the metal oxidation state in the active phase appeared to be Cr^{6+} (Malleswara Rao et al., 2004). The observation of different chromium oxidation states during alkane dehydrogenation under oxidizing and nonoxidizing conditions is consistent with results obtained in complementary UV—vis and Raman experiments characterizing other supported oxides (Bañares et al., 2000c; Christodoulakis et al., 2004; Gao et al., 1999, 2002).

4.3.4. Resonance Raman Effects in Oxides

Evaluation of published IR and Raman spectra and of quantum chemical calculations characterizing TS-1 (titanium silicalite with Ti^{4+} incorporated into the silica framework) unraveled a resonance effect in these materials when the Raman spectra were excited in the UV range. The Raman bands observed at 960 and 1125 cm^{-1} were assigned to framework TiO_4 by comparison of LRS and RRS spectra, in line with previous suggestions (Deo et al., 1993; Knözinger and Ratnasamy, 1978; McMillan, 1986). Resonance enhancement was mainly observed for the Raman band at 1125 cm^{-1}, as a consequence of the resonant coupling of the UV laser light into the ligand-to-metal charge transfer (LMCT). This Raman band originates from the totally symmetric vibration of the TiO_4 tetrahedron in the silicalite framework, as shown by quantum chemical calculations. The resonance enhancement was not observed for the antisymmetric stretching mode at 960 cm^{-1}, but this band showed characteristic shifts depending on the local coordination. Interaction with H_2O or NH_3 quenched the RR effect through distortion of the TiO_4 units (Li et al., 1999a,b; Ricchiardi et al., 2001).

The research group of Zecchina (Bordiga et al., 2002, 2003; Tozzola et al., 1998) investigated the interaction of H_2O_2 with TS-1 to identify the nature of the active oxidizing agent. To achieve resonance enhancement for transitions of titanium peroxo complexes, which absorb in the visible range, a laser source at 442 nm was used. RRS demonstrated that upon peroxide adsorption the band at 960 cm^{-1} decreased in intensity and shifted to 976 cm^{-1}, and the band at 1125 cm^{-1} was quenched as a result of symmetry reduction. Resonant excitation led to the appearance of an additional band at 618 cm^{-1}, which was assigned to the symmetric breathing mode of a side-on titanium peroxo complex.

This species was proposed to play an important role in the catalytic activation of H_2O_2 on TS-1. IR spectroscopic investigations showed that a $Ti(\eta_2\text{-OOH})$ moiety oxidized small alkenes at room temperature in the dark (propene) or upon photoexcitation (ethene). These data appear to be the first direct detection of the active oxidation site in H_2O_2-loaded titanium silicalite (Lin and Frei, 2002).

Catalytic reaction conditions or the exposure to reducing environments may lead to the formation of reduced surface metal oxide species. It is generally difficult to obtain good Raman signals for reduced supported metal oxide species because of their low Raman cross-sections. On the other hand, many reduced transition metal ions have electronic absorption bands in the visible regime. Hence, the laser frequency may be tuned to these absorption bands, and resonantly enhanced Raman spectra should be obtained.

To unravel a possible resonance-enhanced Raman effect in oxides with partially reduced cations, single phase, well-crystallized binary, ternary, and quaternary mixed oxides were synthesized by chemical vapor

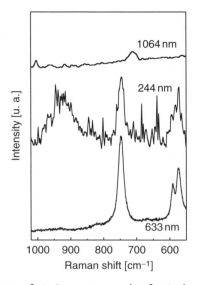

FIGURE 10   Raman spectra of MoO_2 as an example of a single-phase oxide as obtained by chemical vapor transport excited at three different frequencies, 1064, 244, and 632 nm (top to bottom); based on Blume, 2001, PhD Thesis, TU Berlin.

transport (CVT) or sintering to be used as reference compounds; these were investigated by Raman spectroscopy (Blume, 2001). For example, Figure 10 shows the Raman spectra recorded of MoO_2, excited at three different laser lines, namely 1064, 632, and 244 nm. Raman spectra were obtained for all excitation frequencies, but the comparison gave evidence of strong spectral differences as a function of the laser wavelength.

UV–vis spectroscopy of this molybdenum oxide demonstrated an absorption extending from below 580 to 1000 nm, which was attributed to an IVCT transition between fivefold coordinated Mo^{5+} centers and a neighboring sixfold coordinated Mo^{6+} center (Dieterle, 2001; Dieterle and Mestl, 2002; Payen et al., 1986). Hence, the variations in the Raman spectra were attributed to oxidation in the laser spot (to explain the new bands obtained during excitation at 244 nm), and also to resonance effects occurring in reduced molybdenum oxides (to explain the high signal-to-noise ratio obtained when the sample was excited at 632 nm).

The resonance Raman effect was corroborated by confocal Raman mapping of mixtures of MoO_3 and MoO_2 and of orthorhombic Mo_4O_{11}. Raman signals were recorded after dilution of the compounds 1:100 in BN, and statistical data analysis was performed (Dieterle et al., 2001, 2002). When the Raman spectrum was excited at 532 nm, (i.e., at a frequency close to the minimum absorption in the UV–vis spectrum) and the integration time was set to 200 s, only the characteristic bands of MoO_3 could be detected. Excitation at 632.8 nm produced, albeit at an integration time of only 30 s, distinct Raman bands of MoO_3, MoO_2 (Olson and Schrader, 1990; Spevack and McIntyre, 1997), and Mo_4O_{11} (Cross and Schrader, 1995; Dieterle and Mestl, 2002; Olson and Schrader, 1990). The very low signal intensities of the oxides with partially reduced molybdenum relative to those of the bands of BN indicated their low Raman cross-sections (as a result of strong absorption in the visible regime).

The resonance Raman effect was explained by a resonant excitation of the IVCT transition associated with the fivefold-coordinated Mo^{5+} centers. The Raman scattering intensity results then from the relative efficiencies of the resonant absorption of the incident photons and the resonant reabsorption of the Raman-scattered photons. The resonance enhancement was further demonstrated by investigation of a series of MoO_{3-x} samples with various degrees of reduction (Dieterle and Mestl, 2002; Dieterle et al., 2002).

In summary, resonance Raman enhancement occurred in reduced molybdenum oxides when the exciting frequency corresponded to the $Mo^{5+} \rightarrow Mo^{6+}$ IVCT transition at about 2 eV. This resonance enhancement was found to be crucial for catalyst characterization during operation. Excitation of the Raman spectra of such reduced oxides is of course also possible with other laser frequencies (*vide supra*). However, then the overall Raman scattering efficiency is much smaller, and small concentrations (e.g., in a catalytic reaction experiment) may not be detectable.

4.3.5. Reduction by H_2

Payen et al. (1986) investigated the reduction of alumina-supported molybdena and ascribed a Raman band at 760 cm^{-1} to reduced supported molybdenum oxide. The transformations could be reversed by reoxidization (Payen et al., 1986). Mestl and Srinivasan (1998) described some reduced phases formed from bulk molybdena, whereas reduced dispersed vanadia and chromia catalysts do not show Raman bands (Airaksinen et al., 2005; Bañares et al., 2000a; Gasior et al., 1988; Weckhuysen and Wachs, 1996).

No Raman bands characterizing reduced surface CrO_x species have been reported to date, but crystalline Cr_2O_3 (Cr^{3+}) gives rise to strong Raman signals (Hardcastle et al., 1988). Bañares et al. (2000a) observed structural transformations upon reduction and reoxidation of silica-supported vanadia. During TPR of V_2O_5/SiO_2 with a vanadia coverage of a third of a monolayer, the Raman intensity of the terminal $V{=}O$ bond of the surface vanadium oxide species at 1037 cm^{-1} monotonically decreased with H_2 consumption. The Raman spectrum of the fresh catalyst could be restored by reoxidation. However, at near monolayer coverage, the silica-supported surface vanadium oxide underwent a structural modification upon reduction in H_2 and reoxidation (Figure 11). At 200 °C, during TPR in H_2, new Raman bands assigned to crystalline V_2O_5 appeared at 994, 702,

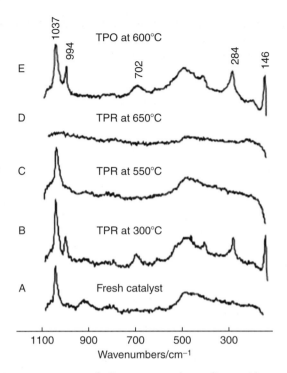

FIGURE 11   TPR-Raman spectra of silica-supported vanadium oxide near monolayer coverage (0.8 V atoms/nm^2) (source, M.A. Bañares).

284, and 146 cm^{-1}. The Raman bands of crystalline V_2O_5 reached a maximum intensity at 300 °C and then disappeared during further increases of the temperature. The Raman band of the isolated surface vanadium oxide species at 1037 cm^{-1} persisted up to a temperature of 550 °C. The TPO-Raman spectra demonstrated the formation of both crystalline V_2O_5 and isolated surface vanadium oxide species.

Therefore, it is inferred that near the maximum surface coverage, under reducing conditions, the surface vanadium oxide species compensate for the oxygen removal by sharing oxide ions. Any effect that promotes interactions among surface vanadia species must lead to the formation of crystalline V_2O_5.

Similar aggregation phenomena were confirmed for vanadia supported on several oxides through a combined Raman, XANES, and UV–vis spectroscopic investigation by Olthof et al. (2000); crystallization of V_2O_5 was induced even during repeated hydration–dehydration treatments (Xie et al., 2000). TPR profiles of supported oxides could be modeled by a linear combination of two phenomena, random nucleation and bidimensional nuclei growth; the latter became dominant for supported vanadia as the loading approached monolayer coverage (Harlin et al., 2000, 2001).

This description is consistent with the results of TPR/TPO-Raman investigations of supported vanadia on silica.

A similar comparison of results of TPR/TPO-Raman spectroscopy with those of quantitative TPR has also been made for alumina-supported vanadia (Kanervo et al., 2003). However, the Raman signal of V_2O_5 crystals is at least ten times more intense than that of surface VO_x species for excitation in the wavelength range of 514–532 nm, because of resonance enhancement (Xie et al., 2000). Thus, only a minor fraction of the surface VO_x species on alumina aggregated to form microcrystals during reduction and oxidation cycles.

Similar reduction experiments were performed with alumina-supported chromia (Kanervo and Krause, 2001, 2002), and several complementary techniques were employed, including DRIFT, Raman, and EXAFS spectroscopies (Airaksinen et al., 2003).

The redox properties of ceria–zirconia mixed oxides are interesting, because these materials find applications as electrolytes for solid oxide fuel cells, supports for catalysts for H_2 production, and components in three-way automobile exhaust conversion catalysts. The group of Kaspar and Fornasiero (Montini et al., 2004, 2005) used TPR/TPO-Raman spectroscopy to identify the structural features of more easily reducible zirconia–ceria oxides and the best method for their preparation by suitable treatments. TPR/TPO experiments and Raman spectra recorded during redox cycles demonstrated that a pyrochlore-type cation ordering in $Ce_2Zr_2O_8$ facilitates low temperature reduction.

Most of the supported oxides referred to above are reduced under the reducing conditions mentioned; however, contrasting behavior has been observed for ceria-supported vanadia. Ce^{4+} is reduced to Ce^{3+} while vanadium remains fully oxidized; it is assumed that, as a consequence of the strong interaction between vanadia surface species and the ceria support, reduction of ceria occurs at the interface (Martinez-Huerta et al., 2004). A TPR/TPO-Raman investigation by Martinez-Huerta et al. showed that the solid-state reaction between surface vanadia and the ceria support to form $CeVO_4$ is facilitated in a reducing atmosphere at approximately 200 °C (Martinez-Huerta et al., 2007).

4.3.6. Reduction by Oxygenates or Hydrocarbons

Supported dispersed metal oxides can also be reduced by alcohols. Hu and Wachs (1995) reported Raman bands of surface-reduced molybdena that was generated through contact with methanol. Zhao and Wachs (2006) recently investigated V_2O_5/Nb_2O_5 catalysts during propene oxidation to acrolein and detected a previously unknown Raman band at 978 cm^{-1}, which was tentatively assigned to a surface V^{4+} species.

An example of a material that is reducible by hydrocarbons is 1% V_2O_5/CeO_2, which in the dehydrated state is characterized by Raman

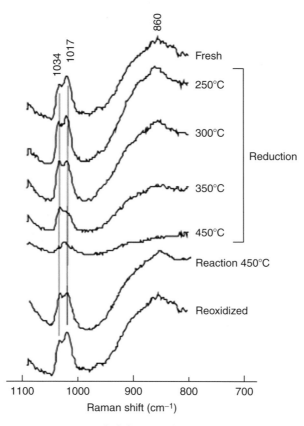

FIGURE 12 Raman spectra recorded during reduction of 1%V$_2$O$_5$/CeO$_2$ by ethane (gas feed C$_2$H$_6$:He = 1:8) [source, M. A. Bañares].

bands at 1034, 1017, and 860 cm^{-1}. The Raman bands at 1034 and 1017 cm^{-1} correspond to the stretching of the terminal V=O bond and that at 860 cm^{-1} to the V–O–V stretching mode of polymeric surface vanadia species (Figure 12) (Bañares et al., 2000b). The intensities of the Raman bands at 1017 and 860 cm^{-1} decreased faster in a reducing hydrocarbon atmosphere than that of the band at 1034 cm^{-1}. The polymeric surface vanadia species were more easily reduced than the isolated surface vanadium sites absorbing at 1037 cm^{-1}; thus, the Raman band at 1037 cm^{-1} corresponds to isolated surface vanadia species, and the two Raman bands at 1017 and 860 cm^{-1} correspond to polymeric surface vanadia species. Under reaction conditions (C$_2$H$_6$ + O$_2$ + He), only a small fraction of the surface vanadia sites was reduced, and the Raman spectrum at 450 °C was very similar to that of the fresh, dehydrated sample (Bañares et al., 2000b). Similarly, dehydrated zirconia-supported chromia exhibits Raman bands at 1030 cm^{-1}, characteristic of the terminal

$Cr=O$ bond of isolated surface chromate species, and bands at 1010 and 880 cm^{-1}, characteristic of the terminal $Cr=O$ and bridging Cr–O–Cr bonds of surface polychromate species. The assignments of the Raman modes of supported oxides are still a matter of debate (Deo et al., 1994; García-Cortéz and Bañares, 2002; Magg et al., 2004; Wu et al., 2005), as commented on in Section IV.A.2.

The preferential reduction of surface polymeric species was confirmed by UV–vis DR spectra that show an increase of the edge energy (Gao et al., 1999), which suggests a decrease in the degree of polymerization of the supported species. Furthermore, the vanadia–ceria system is different from other supported oxides because of the stabilization of Ce^{3+} sites at the interface between surface vanadia (V^{5+}) and the ceria (Ce^{4+}) support (Martinez-Huerta et al., 2004); a TPR-Raman investigation showed the formation of $CeVO_4$ (Martinez-Huerta et al., 2007). Recent work by Martínez Arias et al. (2006) further underlined the unique characteristics of the ceria support. They reported results of Raman and EPR spectroscopy experiments and XRD, which point toward a strong interaction between CeO_2 and CuO. Such an interaction was evidenced by a shift of the F_{2g} Raman mode of CeO_2 at ca. 462 cm^{-1} and the presence of an additional broad band near 600 cm^{-1} that is not observed for pure ceria. The shift of the F_{2g} Raman mode paralleled the formation of oxygen vacancies during incorporation of increasing amounts of copper into the fluorite ceria lattice.

This Raman band was used as an indicator to demonstrate a partial reoxidation of ceria at room temperature after a preceding reduction by CO at 473 K. According to the F_{2g} Raman mode, reoxidation of previously reduced CuO/CeO_2 restored its original, oxygen-vacancy-containing state (Martínez-Arias et al., 2006).

In general, the average oxidation state of a component during catalytic operation depends on the balance between the reduction by the hydrocarbon molecule and the reoxidation by gas-phase O_2. The support may also play a role; for example, UV–vis DR spectra did not indicate significant reduction of silica- and alumina-supported vanadia catalysts during ethane oxidation, and evidenced only moderate reduction of a 1% V_2O_5/ZrO_2 catalyst (Bañares et al., 2000c). UV–vis DRS investigations during ODH (Gao et al., 1999) and TPR/TPO experiments by López-Nieto et al. (1999) showed that the reoxidation of transition metal ions was in general faster than reduction. The oxidation state of a particular catalyst component during reaction will depend on its reducibility (which decreases in the order dispersed chromium > vanadium > niobium oxide species) and on the reducing power of the hydrocarbon molecule (which decreases in the order butanes > propane > ethane), as observed by UV–vis DRS during ODH (Gao et al., 1999). Thus, Raman spectra of supported metal oxide catalysts recorded during ongoing ODH do not show appreciable reduction under typical reaction conditions (Bañares et al., 2002;

Mul et al., 2003). More reducible oxide moieties (e.g., alumina-supported chromia) exhibited partial reduction during reaction (Puurunen and Weckhuysen, 2002; Puurunen et al., 2001). The average oxidation state of a particular catalyst will depend on the hydrocarbon-to-O_2 ratio in the feed (García-Cortéz and Bañares, 2002) and on the presence of other reactants. For example, a typical propane ammoxidation feed, which also contains ammonia, results in a more reduced catalyst than a propane ODH feed (Guerrero-Pérez and Bañares, 2002).

4.3.7. Carbonaceous Deposits

Carbon-containing deposits may accumulate on surfaces during reactions with hydrocarbons. Details of the formation and nature of such species have been described by Stair (2007). Raman spectroscopy is a tool that is well suited to the investigation of carbonaceous deposits. An early investigation of such carbon-containing deposits was reported by Brown et al. (Brown et al., 1977) in 1977. If sufficient oxygen is present in a hydrocarbon environment, carbon deposition will be minimized or will not occur, and the catalysts may remain fully oxidized.

The average oxidation state of a metal in a catalyst during reaction was found to be related to the presence of carbonaceous deposits on the surface. As the feed for propane ODH was depleted in O_2, the catalyst was readily reduced (Mul et al., 2003) and amorphous carbon-containing deposits formed. This behavior was corroborated by UV–vis DRS (Mul et al., 2003; Puurunen and Weckhuysen, 2002) and by combination of UV–vis DRS and Raman spectroscopy (Kuba and Knözinger, 2002; Nijhuis et al., 2003).

The nature of the carbon deposits depends strongly on the nature of the hydrocarbon. Deposits are more easily formed from more reactive molecules (Loridant et al., 2003). The redox and acid–base properties of the surface (Guisnet and Magnoux, 2001) and the reaction temperature are also significant. The relevance of the reactivity of the hydrocarbons and of the acid—base and redox properties was illustrated by Raman-spectroscopic investigations taking place during propane dehydrogenation (DH) on potassium-doped vanadia–alumina catalysts (Mul et al., 2003). Amorphous carbon-containing deposits formed after surface vanadium oxide species were reduced during reaction with a propane–helium mixture. The nature of the carbon-containing deposits changed with temperature; above a temperature of 450 °C graphite-like deposits were observed. Concomitant activity measurements evidenced production of propene simultaneously with the formation of graphite-like carbon (Mul et al., 2003).

Carbon-containing deposits darken the sample, which further complicates the already difficult quantification of Raman signals. However, it is possible to combine UV–vis DRS and Raman spectroscopy to correct the decreasing Raman signal by the decrease of reflectance in the same spot as determined by UV–vis DRS. This procedure was proposed by Kuba and Knözinger (2002) and further tested by Weckhuysen (Nijhuis et al., 2003, 2004).

The transition from amorphous carbon-containing deposits to graphite-like species and finally to graphitic carbon typically proceeds via polyaromatic heterocycles (Guisnet and Magnoux, 2001), which are not easily detected by conventional Raman spectroscopy because of fluorescence problems (Chua and Stair, 2003; Li and Stair, 1996). The use of UV excitation provides a powerful means to circumvent fluorescence problems and tackle the identification of the carbonaceous deposits (Chua and Stair, 2003). This subject was discussed in detail by Stair (2007). Polyaromatic deposits were burned off very quickly upon restoration of oxidizing conditions (Boulova et al., 2001; Mul et al., 2003; Puurunen and Weckhuysen, 2002; Puurunen et al., 2001).

Rapid changes in materials properties as a result of interactions with a gas or vapor can potentially be exploited for sensor applications provided the appropriate materials can be found. Raman spectroscopy and conductivity measurements were applied in combination to evaluate the effect of carbon-containing deposits (as detected by Raman spectroscopy) on nanoscaled tungsten oxide. The conductivity of this material changes by five orders of magnitude during repeated switching of the feed between NO_2 and CH_4 at temperatures of 100–150 °C (Boulova et al., 2001). The results of this investigation also emphasize the relevance of the nanodimensions of the material, because the dramatic effects were observed only for 2-nm WO_3 particles (and not for larger WO_3 particles (of 35 nm)).

The same group of researchers also combined Raman spectroscopy and resistance measurements for the detection of H_2S with nanocrystalline SnO_2 (Pangier et al., 1999) and SnO_2–CuO (Pangier et al., 2000) as sensing materials.

Xie and Bell (2000) used a Raman cell that was a reactor to identify the carbon-containing species on the surface of zirconia during the synthesis of dimethyl carbonate from CO_2 and methanol. The results showed that surface methoxides and surface carbonates formed from monomethyl carbonate species that yield dimethyl carbonate upon reaction with methanol.

4.3.8. Raman Investigations of NO_x-Trap Catalysts

To our knowledge, the first Raman investigations of the states of the barium phase during ongoing NO_x decomposition were reported by Lunsford and coworkers (Haller et al., 1996; Mestl et al., 1997a,b, 1998; Xie et al., 1997). They used a MgO-supported barium oxide model catalyst for their investigation of NO decomposition. It was found that the supported barium oxide phase stabilized peroxide species in the presence of molecular O_2 in the feed, which in turn reacted with gas-phase NO to form barium-nitro species. These barium-nitro species decomposed at slightly higher temperatures to give BaO again. The highest rates of NO reduction were observed by GC exactly under the conditions that favor decomposition of these barium-nitro complexes. Too high an O_2 partial pressure, on the other hand, appeared to transform the active barium-nitro species into inactive nitrates (Xie et al., 1997), but it also restored the peroxide species when NO_x-compounds were absent and was proposed to account for NO activation (Mestl et al., 1997a,b; Xie et al., 1997).

Uy et al. (2002) employed UV–Raman spectroscopy to compare the reactions of technological Pt/Ba/γ-Al$_2$O$_3$ catalysts and Pt/γ-Al$_2$O$_3$ in feeds containing NO, SO$_2$, and O$_2$. Under lean conditions, barium-containing materials stored large amounts of NO as barium nitrates and of SO$_2$ as barium sulfate, which were released at high temperature in flowing H$_2$. The NO$_x$ storage appeared to be faster under reaction conditions than the sulfate formation, but prolonged exposure to SO$_2$ deactivated the catalysts. In the presence of O$_2$ or NO$_2$ in the feed, the Raman spectrum indicated surface Pt–O vibrations (band at 570 cm^{-1}) together with the typical spectral features of nitrate. The spectral intensity of the Pt—O vibration correlated with the platinum coverage by oxygen during CO oxidation catalysis (Uy et al., 2002).

UV- and visible Raman spectroscopies were also used to investigate the effects of aging on Pt/Ba/γ-Al$_2$O$_3$ catalysts for NO$_x$ storage. Particle sintering, phase separation, and oxide formation were deduced from the observed spectral features of thermally aged catalysts (Uy et al., 2004). Aging led to a diminished ability to form atomic O—Pt species as indicated by the decreased intensity of the Raman band at about 600 cm^{-1}. The barium nitrate phase in aged catalysts was at least partially crystalline and behaved independently of the Pt/Al$_2$O$_3$.

Payen and coworkers (Le Bourdon et al., 2003) investigated Pd/γ-Al$_2$O$_3$ catalysts under DeNO$_x$ reaction conditions with Raman and IR spectroscopy. Their instrument allows the quasi-simultaneous recording of both IR and Raman spectra. Several adsorbed NO$_x$ species were identified (ranging from nitrates, nitrito, nitrate, and nitro species) by exploiting the high sensitivity of IR spectroscopy for adsorbate vibrations and the specificity of Raman spectroscopy for the vibrations of catalyst—adsorbate bonds.

Raman spectroscopy was used to detect Pd—N bonds in Pd/γ-Al$_2$O$_3$ catalysts in atmospheres containing NO and CO. The dissociation of NO, together with the formation of PdO, was proposed (Mamede et al., 2003).

4.3.9. DeNO$_x$ (SCR) and DeSO$_x$ Reactions

The molecular structure of supported metal oxides under selective catalytic reduction (SCR) conditions was reported to be the same as that under conditions leading to catalyst dehydration (Wachs et al., 1996). Raman spectra of 4% V_2O_5/ZrO_2 that was partially exchanged with ^{18}O were recorded at 350 °C during SCR, and the time required to exchange the terminal $V=^{18}O$ bond to $V=^{16}O$ in a feed of $^{16}O_2$, $N^{16}O$, and NH_3 was found to be approximately 10 reaction cycles (Wachs et al., 1996). Therefore, the terminal $V=O$ bond was considered not to be directly involved in the rate-determining step of the SCR. The turnover frequencies (TOF) for SCR, like the TOF values characteristic of alkane oxidation, were strongly affected by the support.

Thus, the TOF decreased by a factor of six at 200 °C in the series: $ZrO_2 > TiO_2 \gg Al_2O_3 > SiO_2$ (Wachs et al., 1996). This decrease parallels an increase in the electronegativity of the cation of the support oxide (Zr < Ti < Al < Si). Hence, it was suggested that the bridging $V-O-M_{support}$ bond (with M the metal cation of the support) represents the catalytically active site involved in the rate-determining step of SCR.

The structure–activity relationships characteristic of supported vanadium oxide catalysts have been evaluated for SO_2 oxidation to SO_3 by Wachs and coworkers (Dunn et al., 1998, Dunn et al., 1999). It was demonstrated for individual supported vanadium oxide catalysts that the oxygen of the bridging $V-O-M_{support}$ bond was involved in the rate-determining step of SO_2 oxidation (Dunn et al., 1998). The specific reaction rate (TOF) was found to be the same for both isolated and polymeric surface vanadia species, and depended only on the oxide support; changes in the support led to dramatic changes in the TOF (Dunn et al., 1998).

Giakomelou et al. (2002) used Raman spectroscopy to investigate SO_2 oxidation to SO_3 on supported molten salt catalysts. Their work describes the transformations of the $V_2O_5–Cs_2SO_4$ and $V_2O_5–Cs_2S_2O_7$ molten salts under oxidizing conditions, during SO_x oxidation, and under reducing conditions in SO_2. The complex $(VO)_2O(SO_4)_2^{4-}$ formed during SO_x oxidation and was reduced in SO_2.

4.3.10. Supercritical Conditions and Autoclave Reactors

Raman spectroscopy is characterized by lower sensitivity than IR spectroscopy, but in contrast to IR spectroscopy, Raman spectroscopy may be used to investigate catalysts under supercritical conditions of CO_2 or H_2O, because there are no strong absorptions by these molecules that interfere with the absorptions by the catalyst as is the case in IR spectroscopy. Grünwaldt et al. (2003) reviewed cell designs for spectroscopic experiments under supercritical conditions that either feature a window (lens) to focus the laser beam inside the cell, or fiber optics that are directly inserted into the cell (Howdle et al., 1994; Poliakoff et al., 1995). In some cases, several techniques may be combined (Addleman et al., 1998; Hoffmann et al., 2000). Such cells are designed with minimal void volume so that reliable kinetics and time-resolved analyses can be performed.

Kinetics data have been obtained under supercritical conditions (Masten et al., 1993; Rice et al., 1996; Steeper et al., 1996). The developments in Raman spectroscopy for monitoring of materials under supercritical reaction conditions (Grünwaldt et al., 2003) present an important opportunity for investigations of catalysts in such environments (Koda et al., 2001).

4.4. Raman Spectroscopy of Operating Catalysts (as Demonstrated by Simultaneous Activity Measurements)

In the applications described so far, catalytic data were not acquired along with the spectroscopic data, or the cells were unsuitable for correct measurements of the former. The determination of the catalyst structure and performance in a single experiment is not only of interest for catalysts but for any functional material. For instance, rather similar developments as in the field of catalysis have been reported in the fields of gas sensors and electrochemical devices. Many techniques allow for the simultaneous characterization of electrochemical materials and their performance (Luo and Weaver, 2001; Novák et al., 2000). Conductance cells provide a powerful approach to understanding of the structure–performance relationships at the molecular scale (Loridant et al., 1995).

To the best of our knowledge, the earliest cell designed with a fixed-bed configuration that afforded true catalytic data was reported by Hill and coworkers (Snyder and Hill, 1991). Their design of a catalytic Raman cell is characterized by the avoidance of dead volume, the minimum amount of catalyst to achieve measurable conversions, and rotation of the focusing lens to distribute the energy of the laser beam (Snyder and Hill, 1991). The design of this cell is presented in Figure 4A.

Raman spectra of a bismuth molybdate catalyst obtained with this apparatus are shown in Figure 13 (Snyder and Hill, 1991); they represent the starting material and the catalyst after 24 h of proven operation (by online GC analysis) in propene oxidation to acrolein at 400 °C. The conversion values in the cell ranged from 20% to 40%, with the selectivity to acrolein reaching values between 90% and 50%. Long-term experiments did not show significant variations within 26 days of operation at 400 °C (Snyder and Hill, 1991) if the catalysts were properly annealed beforehand.

Lunsford and coworkers (Xie et al., 1999) reported kinetics data characterizing the catalytic NO_x decomposition measured in their Raman cell. Other attempts were made by Volta and coworkers (Abdelouahab et al., 1992), but only low conversions were reached for alkane oxidation because of significant temperature gradients (Volta et al., 1992). The cell designed by Stair (2001) appears to be suitable for simultaneous determination of spectra and catalytic activity. Several groups have addressed the issue of obtaining accurate catalytic performance data when the reactor is a spectroscopic cell (Bañares and Khatib, 2004; Guerrero-Pérez and Bañares, 2002; Kerkhof et al., 1979; Mestl, 2002).

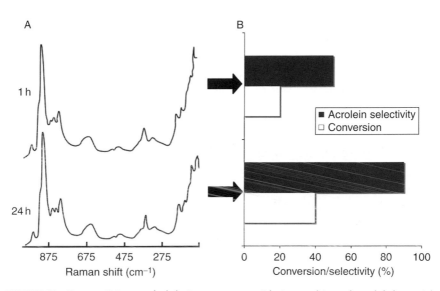

FIGURE 13   Raman-GC recorded during propene oxidation on bismuth molybdate with simultaneous activity measurement (A) Raman spectra (Reprinted from *J. Catal.* 132,536 (1991), Snyder T.P., Hill C.G., "Stability of bismuth molbydate catalysts at elevated temperatures in air and under reaction conditions," copyright (1991) with permission from Elsevier) (Snyder and Hill, 1991); (B) simultaneous conversion and selectivity (based on Snyder and Hill, 1991).

4.4.1. Raman Gas-Phase Characterization

Whereas most Raman investigations in catalysis evaluate the nature of the catalyst and surface intermediates, it is also possible to monitor the reaction in the gas or in the liquid phase with Raman spectroscopy. For example, during the combustion of methane at high pressure (Reinke et al., 2004), Raman spectroscopy was employed to detect the gas-phase signals. The Raman signal, however, was inherently weak, because the gas phase is several orders of magnitude less dense than solids or liquids. Nonetheless, Raman spectroscopy is used to measure gas-phase spectra, even in environments at pressures of a few millibars (Tejeda et al., 1996). High-pressure Raman spectra of gas phase species during the combustion of methane have provided information about the reaction mechanism and allowed evaluation of reaction models in the pressure range of 4–16 bar and the temperature range of 500–1000 °C (Reinke et al., 2004).

4.4.2. Raman Cell–Autoclave Reactors

A number of important reactions require the use of an autoclave. Even in 1982, Woo and Hill (Woo and Hill, 1982, 1984) reported Raman investigations of the states of organometallic catalysts in autoclave reactors. In 1985, they reported the first Raman spectra of a catalyst during proven operation in an autoclave (Woo and Hill, 1985). Their apparatus is described in Section 3.2.1. The hydroformylation of propene was investigated, and $Co_2(CO)_6(PPh_3)_2$ as a homogeneous catalyst was compared with $Co_2(CO)_6(PPh_3)_2$ supported on alumina or silica. The spectra demonstrated various surface species on the solid catalysts, and the combined catalytic and spectroscopic results provided information about the rate-determining step.

4.4.3. Bulk Mixed Metal Oxides for Alkane Oxidation Reactions

To the best of our knowledge, the first Raman investigation of operating bulk mixed-metal oxide catalysts was reported by Snyder and Hill in 1991 (Snyder and Hill, 1991). This bulk bismuth molybdate catalyst was also the first catalytic material that was analyzed by applying Raman spectroscopy and oxygen isotopic labeling to assess the reactivity of different oxygen sites (Hoefs et al., 1979). The work by Snyder and Hill demonstrated the importance of calcination treatments on the stability of the various bismuth molybdate phases during propene oxidation to acrolein.

Kovats and Hill (Kovats and Hill, 1986) in 1986 also reported Raman spectra of bulk iron and bismuth molybdates under reactive environments. Wilson et al. (1990) performed spectroscopy of iron molybdate catalysts during methanol oxidation and did not find appreciable changes in the bulk molecular structures of the catalyst. Ueda et al. (1982) and Glaeser et al. (1985) also employed ^{18}O labeling to better understand the functioning of bulk bismuth molybdate catalysts in the hydrocarbon oxidation to alkenes, oxygenates, and nitriles. It was shown that one type of bulk oxygen species participated in the production of alkenes (bridging Mo–O–Bi sites), whereas a second bulk oxygen was involved in the production of oxygenated or nitrogenated hydrocarbons (bridging Mo–O–Mo sites).

Ozkan and coworkers (Ozkan et al., 1990a,b) investigated bulk MoO_3 during alkane oxidation and demonstrated the participation of bulk lattice oxygen in this oxidation reaction by combining the use of O_2-rich and O_2-free feeds (Ozkan et al., 1990a) with the use of ^{18}O isotopic labeling (Ozkan et al., 1990b). Knözinger and coworkers (Mestl et al., 1994a–c) investigated physical mixtures of antimony and molybdenum oxides by Raman spectroscopy and ^{18}O labeling experiments with the aim of identifying an oxygen spillover step by the remote control mechanism in partial oxidation reactions.

However, the nature of the active sites in antimony–molybdenum mixed oxide catalysts for the selective oxidation of isobutene to methacrolein is not clear yet, notwithstanding significant research efforts in this area (Gaigneaux et al., 1998). There is evidence that the catalytically active site has to be part of the outermost layer of the catalyst (Gaigneaux et al., 1998), as expected. Recent catalytic experiments demonstrated that the TOF of the reaction on MoO_x is much higher than that on SbO_x, a result that suggests that surface MoO_x species constitute the catalytically active sites (Badlani and Wachs, 2001); however, it is necessary to analyze for amorphous surface mixed molybdenum–antimony oxide phases and to determine their reactivities.

The bulk structural features of molybdenum–vanadium–tungsten mixed oxide catalysts and their relevance for the oxidation of methanol, acrolein, and propene were examined by resonance Raman spectroscopy in combination with online GC-MS analysis of the products (Dieterle et al., 2001; Knobl et al., 2003; Mestl, 2002; Mestl et al., 2000; Ovsitser et al., 2002). The effect of addition of vanadium and tungsten on the formation of a Mo_5O_{14}-type structure can be assessed by taking advantage of the resonance Raman enhancement, which arises from the intervalence charge transfer between penta-coordinated Mo^{5+} and hexa-coordinated Mo^{6+} (Dieterle et al., 2002; Mestl, 2002). It is known from kinetics investigations on the industrial mixed metal oxide catalysts that there are two catalytically active "domains" present during acrylic acid production (Petzodt et al., 2001). Domain I catalyzes the partial oxidation of acrolein to acrylic acid, whereas domain II catalyzes total oxidation. Structural characterization demonstrated the presence of MoO_3-type and Mo_5O_{14}-type oxides in these catalysts (Dieterle, 2001; Dieterle et al., 2001; Mestl et al., 2000), and it was suggested that one phase was responsible for total oxidation, the other for partial oxidation. Catalytic tests of pure, reoxidized MoO_3 and of $(MoVW)_5O_{14}$ in the synthesis of acrylic acid from acrolein showed that stoichiometric MoO_3 was inactive for this reaction, whereas the mixed oxide was active (Blume, 2001). Therefore, a more demanding partial oxidation reaction, propene oxidation, was chosen for Raman investigations (Figure 14A).

Table 2 provides a comparison of the catalytic properties of the two oxides, MoO_{3-x} and $(MoVW)_5O_{14}$ (Dieterle, 2001; Mestl, 2002). MoO_{3-x} was additionally preconditioned in 5 vol% H_2/N_2 at 450 °C to induce partial reduction. At identical propene conversions and space velocities, MoO_{3-x} produced much more CO_2 (Figure 14B) than partial oxidation products compared to $(MoVW)_5O_{14}$, which showed a high selectivity to partial oxidation products in accordance with the two-domains model (Petzodt et al., 2001).

These observations were further corroborated by Raman spectra recorded during temperature-programmed and steady-state operation (Dieterle, 2001; Mestl, 2002). The highest propene conversion and the highest selectivity to partial oxidation products were found for the starting material directly after activation (i.e., for the most reduced MoO_{3-x}). With progressing reoxidation of MoO_{3-x} the propene conversion decreased, as did the selectivity to partial oxidation products (Figure 14C). These results indicate that the activity and selectivity of MoO_{3-x} catalysts are a function of the degree of reduction and imply that oxygen defects are necessary for catalytic activity (Dieterle, 2001; Mestl, 2002).

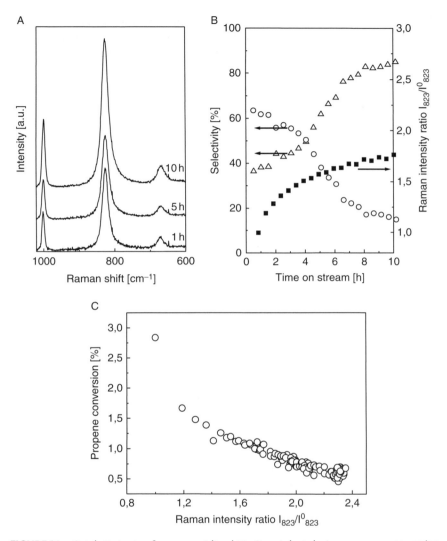

FIGURE 14 Catalytic tests of pure reoxidized MoO_3 catalyst during propene conversion to acrylic acid: (A) Raman spectra of MoO_{3-x} recorded during propene oxidation (500 mg MoO_3 in 5 vol% propene and 10% O_2 at the time on stream indicated. (B) Catalytic data (left axis) selectivity to acetic acid (open circles) and selectivity to CO_2 (open upward triangles), and the intensity I of the prominent Raman band of MoO_3 at 823 cm^{-1} referenced to its starting intensity I_0 (filled squares, right axis) as function of time on stream in the propene oxidation reaction. (C) Propene conversion during the *in situ* Raman experiment as function of the ratio of the intensity of the prominent MoO_3 band at 823 cm^{-1} and its initial intensity (i.e., with time on stream) (source, G. Mestl).

TABLE 2   Comparison of the Selectivity (%) of Oxygen-Deficient MoO_{3-x} and $(MoVW)_5O_{14}$ for a Propene Conversion of 20% (Dieterle, 2001; Mestl, 2002).

Catalyst	Selectivity (mol%)			
	CO_2	Acetic Acid	Acrolein	Acrylic Acid
MoO_{3-x}	60	40	0	0
$(MoVW)_5O_{14}$	36	16	33	15

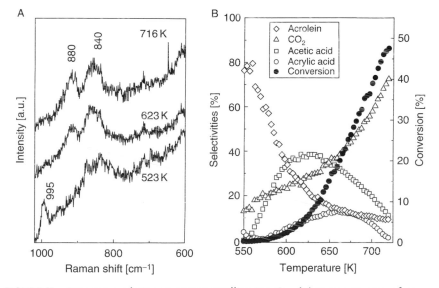

FIGURE 15   Propene oxidation on nanocrystalline Mo_5O_{14}. (A) Raman spectra of nano-crystalline $(Mo_6V_3W_1)_5O_{14}$, as obtained after decomposition of the ammonium precursors, recorded during propene oxidation (500 mg MoO_3 in 5 vol% propene, 10% O_2 and 85% N_2 at the temperatures indicated. (B) Catalytic performance of $(Mo_6V_3W_1)_5O_{14}$ obtained after the *in situ* decomposition of the ammonium precursors: propene conversion (filled circles), selectivity to acrolein (open diamonds), to CO_2 (open upward triangles), to acetic acid (open squares), and to acrylic acid (open circles) as function of reaction temperature (Source, G. Mestl).

A nanocrystalline Mo_5O_{14}-type mixed oxide was also subjected to temperature-programmed propene oxidation. The bands of $(MoV_3W_1)_5O_{14}$ became better resolved and increased in intensity upon crystallization of this oxide during partial oxidation (Figure 15A) (Dieterle, 2001; Mestl, 2002). Simultaneously, the conversion increased with increasing reaction temperature, as expected, and the selectivity to the primary product acrolein decreased, whereas that to CO_2 increased (Figure 15B). In comparison to MoO_{3-x}, the Mo_5O_{14}-type mixed oxide was a better catalyst for partial oxidation. The data obtained in these Raman measurements of the operating catalysts confirmed the results of the kinetics investigations (Petzodt et al., 2001) that had led to the model of two active domains in the industrial acrylic acid synthesis catalyst (*vide supra*) (Dieterle et al., 2001, 2002; Mestl et al., 2000). However, it was almost exclusively the bulk oxygen sites that were monitored in the experiments with the bulk oxides, and only little information was provided about the surface oxygen sites, which are the ones that were in contact with the reactants.

Hence, the challenge remains to fully analyze the surface layers of bulk oxide catalysts. Recently, Zhao et al. succeeded for the first time in detecting surface MoO_x and VO_x species on molybdenum–niobium and vanadium–niobium mixed metal oxides by Raman spectroscopy (Zhao et al., 2003).

Han and coworkers (Han et al., 2007) reported a Raman investigation of α-Mn_2O_3 during methane combustion, showing that during reaction, α-Mn_2O_3 was reduced to a Mn_3O_4-like phase.

4.4.4. Bulk Mixed-Metal Oxide Catalysts for Alkane Ammoxidation Reactions

Bulk mixed-metal oxides are also efficient catalysts for propane ammoxidation to acrylonitrile (AN). However, identification of the active site is seriously hampered, because strong contributions of the bulk phase dominate the spectra. Bulk antimony–vanadium mixed-metal oxides are typical catalysts for this reaction and have been investigated by a number of groups (Andersson et al., 2001; Barbaro et al., 2000; Brazdil and Bartek, 1998; Brazdil et al., 1999; Briand et al., 2000; Centi and Perathoner, 1995; Cortés-Corberán et al., 2000; Guttmann et al., 1988; Larrondo et al., 2003). Guerrero-Pérez et al. (2002) used alumina-supported model antimony–vanadium oxide catalysts with a coverage corresponding to a few monolayers, which leads to the formation of nanocrystals with a very high surface-to-volume ratio. This investigation showed that the formation of mixed antimony–vanadium oxide phases on alumina requires a coverage of at least one monolayer, corresponding to eight antimony or vanadium atoms per square nanometer of alumina support. Comparison of fresh and used catalysts showed that surface vanadium oxide and surface antimony oxide reacted to $VSbO_4$ phases. Vanadium antimonates exhibit a Raman (Xiong et al., 2005) and an IR (Landa-Cánovas et al., 1995) band at 880 cm^{-1} that can be assigned to the vibration of a twofold bridging oxygen next to a cation vacancy (Xiong et al., 2005). The intensity of this vibration associated to the cation vacancy was strongly affected by reducing or oxidizing conditions, which emphasized the extremely dynamic character of the $VSbO_4$ rutile phase.

The formation of $VSbO_4$ phases and their role in the ammoxidation reaction were investigated by Raman spectroscopy in combination with online GC (Guerrero-Pérez and Bañares, 2002). No appreciable

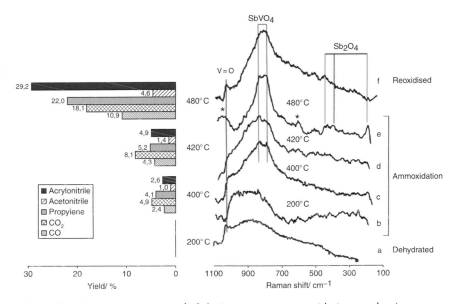

FIGURE 16   Raman spectra recorded during propane ammoxidation on alumina-supported nanocrystalline V-Sb-O system and simultaneous activity data determined by online gas chromatography (Guerrero-Pérez M.O., and Bañares, M.A. *Chem. Commun.* 1292 (2002), "*Operando* Raman study of alumina-supported Sb-V-O catalyst during propane ammoxidation to acrylonitrile with on line activity measurement," reproduced with permission of the Royal Society of Chemistry) (Guerrero and Bañares, 2002).

differences were found in conversion and selectivity in comparison to those recorded with a conventional fixed-bed microreactor.[1] In the fresh dehydrated state of the catalyst, the antimony and vanadium oxides were highly dispersed on the surface of the alumina support (Guerrero-Pérez and Bañares, 2002; Guerrero-Pérez et al., 2002). Surface-dispersed antimony oxide species on alumina do not appear to exhibit any strong Raman bands (Guerrero-Pérez et al., 2002); their weak Raman spectrum was recently reported (Guerrrero-Pérez and Bañares, 2004). The Raman mode of the surface vanadium oxide species near 1024 cm^{-1} tended to disappear during catalytic operation, and the shape of the broad Raman band centered at 900 cm^{-1} also changed (Figure 16). As propane ammoxidation became measurable at increasing temperatures, a broad Raman band became evident at about 800 cm^{-1} that is characteristic for the $VSbO_4$ phase (Guerrero-Pérez et al., 2002). At 480 °C, the catalyst became more selective for acrylonitrile formation. In contrast, CO_2 and propene were the main products at lower temperatures.

This dramatic change in product distribution with reaction temperature was not observed when the $VSbO_4$ phase was not formed, which is the case

---

[1] This investigation was the first in which the term "operando" was used.

for materials with less than a monolayer coverage of antimony–vanadium oxide species on alumina (Guerrero-Pérez et al., 2002). The broad Raman bands near 1060 and 620 cm^{-1} correspond to V—OC and VO—C vibrations of adsorbed alkoxy species, respectively (Burcham et al., 2000). The concomitant formation of $VSbO_4$ and Sb_2O_4 and a significant increase in acrylonitrile production indicated the transformation of the surface and the ensuing change in its selectivity from ODH to ammoxidation.

A further Raman investigation of the effect of the antimony-to-vanadium atomic ratio, gave evidence that the formation of $VSbO_4$ resulted in higher yields of acrylonitrile when surface vanadium oxide species were also present (Bañares et al., 2002; Guerrero-Pérez and Bañares, 2004). The presence of surface alkoxy species was not observed in the absence of dispersed surface vanadium oxide species.

A similar investigation of the $VNbO_4$ rutile phase under various environments showed that this phase formed and was stable in inert or reducing environments, but it segregated into the individual oxides in oxidizing atmospheres (Cavani et al., 2006).

4.4.5. Bulk VPO Catalyst for Maleic Anhydride Production

Raman investigations have provided information about bulk transformations of VPO phases in various environments. The precursor was converted to a disordered phase (Cavani and Trifirò, 1994; Guliants et al., 1995; Hutchings et al., 1994; Wachs et al., 1996) that slowly transformed into the active phase, depending on the environmental conditions (Hutchings et al., 1994). A comprehensive investigation of the Raman bands of VPO phases was reported (Abdelouahab et al., 1992). Volta and coworkers (Abdelouahab et al., 1992) monitored the phase transformations of VPO powders during n-butane oxidation in a Raman cell with online activity measurement. Simultaneously recorded conversion data were not reported because of their low values (Volta et al., 1992). The α_{II}-VOPO$_4$, β-VOPO$_4$, and γ-VOPO$_4$ phases did not undergo any structural transformations during catalytic operation up to a temperature of $ca.$ 440 °C (Abdelouahab et al., 1992). After prolonged exposure of the δ-VOPO$_4$ phase to reaction conditions (15 h at 440 °C), a new Raman band appeared at 994 cm^{-1}, which was assigned to formation of some α_{II}-VOPO$_4$. This interpretation is in agreement with a separate characterization of the catalyst in the absence of reactants by XRD and ^{31}P-NMR spectroscopy (Abdelouahab et al., 1992), but the band may also be attributed to the formation of crystalline V_2O_5. Xue and Schrader (1999) showed by Raman spectroscopy that the decomposition of VPO into V_2O_5 was reversible after cooling.

The nature and dynamics of the various VPO phases are still debated, mainly because of the overwhelming signal intensity from the bulk. HR–TEM has provided important complementary information, in that it showed that the disordered nanocrystalline $(VO)_2P_2O_7$ domains in the fresh catalyst (with a size near 15 nm) transformed during operation into well-crystallized $(VO)_2P_2O_7$ in the equilibrated catalyst (Guliants et al., 2001). Simultaneously, a *ca.* 2 nm-thick disordered layer that covered the surface (Fleischmann et al., 1974) planes of $(VO)_2P_2O_7$ disappeared (Guliants et al., 2001).

The use of nanoscaled materials can minimize the overwhelming Raman signal of the bulk compared to that of the surface. Guliants et al. (2001) used a nanocrystalline $VOHPO_4 \cdot 0.5H_2O$ precursor and followed its transformation to $(VO)_2P_2O_7$. During transformation in an *n*-butane air mixture, nanocrystalline oxidized δ-$VOPO_4$, invisible to XRD, was detected by Raman spectroscopy (Guliants et al., 2001). Similar $VOPO_4$ phases were formed on the surface of γ-titanium phosphate-supported vanadia catalysts during ethane oxidation (Santamaría-González et al., 1999).

VPO phases are extremely sensitive to temperature and feed composition, and rapid changes were evident during reaction (Conte et al., 2006). The evolution of the Raman signal can be used to monitor and also to control the transformation process (Bennici et al., 2007). Information about the reactivity of the various bulk lattice oxygen sites in VPO catalysts has been obtained for working catalysts. For example, Raman spectra of VPO catalysts recorded by Moser and Schrader (Moser and Schrader, 1984) showed that the terminal $V{=}O$ bond was not involved in the critical reaction step for the oxidation of *n*-butane. The bridging V–O–P or V–O–$M_{support}$ bonds appeared to be the active sites, as evidenced by two observations: during the oxidation of $(VO)_2P_2O_7$ to β-$VOPO_4$, ^{18}O was preferentially incorporated into bridging V–O–P bonds (Lashier and Schrader, 1991), and during reducing treatments defects were formed by elimination of the bridging V–O–P oxygen sites (Gai and Kourtakis, 1995; Gai et al., 1997). Breaking of the V–O–P bonds led to the formation of crystalline V_2O_5 (Xue and Schrader, 1999) and the migration of phosphorus species to the surface (Richter et al., 1998).

4.4.6. Heteropoly Acid Catalysts for Oxidations of Gas-Phase Reactants

Raman spectroscopy provides information about the stability of heteropoly acids. Bulk 12-molybdophosphoric acid ($H_3PMo_{12}O_{40}$), decomposed at 400 °C into β-MoO_3, as shown by the appearance of Raman bands at 851 and 775 cm^{-1} (Damyanova et al., 2000). At 500 °C, the α- and β-MoO_3 phases coexist; at 550 °C, α-MoO_3 dominates. Mestl et al. (2001) reported on the thermal stability of the Keggin anion $H_2PVMo_{11}O_{40}^{2-}$, and the effects of temperature, environment, and the presence of cesium were evaluated by a number of complementary techniques. The results evidenced a segregation of vanadium and molybdenum species and the progressive rearrangement of molybdena moieties into a defective bulk

MoO_3 phase in various environments. Hydrocarbon oxidation reaction conditions led to extensive disintegration of the bulk structure of the heteropoly acid.

Thus, heteropoly acids may not be stable under reaction conditions (Mestl et al., 2001). This statement is in line with the results of Raman investigations of supported β-silicomolybdic acid that unambiguously demonstrated its decomposition to surface molybdena during methane oxidation (Bañares et al., 1995). It was also shown that silica-supported silicomolybdic acid and silica-supported molybdenum surface oxide species with the same molybdenum loadings performed identically (Bañares et al., 1995). Thus, the presence of any heteropoly acid structure during high-temperature oxidation can be ruled out.

Notwithstanding this experimental evidence for the decomposition of heteropoly acid catalysts, their structural integrity at high temperatures is still assumed (Ueno, 2000).

4.4.7. Lanthana Catalysts for CCl_4 Destruction

Raman spectroscopy was used to elucidate the effect of the H_2O/CCl_4 ratio on the deactivation of La_2O_3 catalysts during CCl_4 destruction (Weckhuysen, 2003). Low H_2O/CCl_4 ratios led to the formation of inactive $LaCl_3$. Weckhuysen (2003) showed that $LaCl_3$ formation progressed via the oxychloride. With increasing H_2O/CCl_4 ratios, the transformation from active La_2O_3 into $LaCl_3$ was minimized and ultimately prevented.

4.4.8. Supported Metal Oxides

The group 5–7 supported transition metal oxides (of vanadium, niobium, tantalum, chromium, molybdenum, tungsten, and rhenium) are characterized by terminal oxo bonds ($M'=O$) and bridging oxygen atoms binding the supported oxide to the cation of the support ($M'-O-M_{support}$). The TOF values for ODH of butane or ethane on supported vanadia were found to depend strongly on the specific oxide support, varying by a factor of *ca.* 50 (titania > ceria > zirconia > niobia > alumina > silica).

This trend in the TOF values was found not to correspond with the variations in the strength of the terminal $V=O$ bond as measured by the respective Raman shifts (Bañares, 1999; Wachs et al., 1996). Potassium-doping of alumina-supported vanadia catalysts resulted in lower $V=O$ frequencies, which indicated a weakened terminal $V=O$ bond (Cortéz et al., 2003). However, the propane conversion and the catalyst reducibility decreased. Therefore, it was not considered to be likely that the terminal $V=O$ bond is the active site for alkane ODH on supported vanadia. The same effect was observed for titania-supported vanadia. DFT calculations described a close interaction of potassium ions with both the supported vanadia and the titania support (Si-Ahmed et al., 2007; Lewandowska et al., 2008). Such an interaction leads to an elongated $V=O$ bond with a

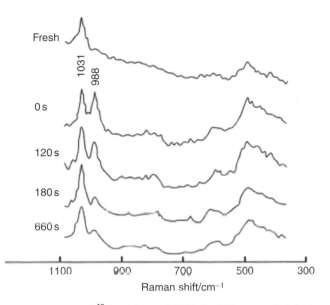

FIGURE 17   Raman spectra of ^{18}O-exchanged 2%V$_2$O$_5$/SiO$_2$ recorded during ethane oxidation reaction (Source, M.V. Martínez-Huerta, M.A. Bañares).

red-shifted vibrational mode, which was also observed experimentally (Si-Ahmed et al., 2007; Lewandowska et al., 2008).

Raman spectra of ^{18}O exchanged 2%V$_2$O$_5$/SiO$_2$ (Figure 17) were characterized by a band near 988 cm^{-1} (V$=$^{18}O vibration). Ethane ODH reaction was performed on this sample using gas-phase ^{16}O$_2$. The Raman intensities of the 988- and 1031-cm^{-1} bands of the 2% V$_2$O$_5$/SiO$_2$ catalyst changed with time on stream (Figure 17). The Raman band of the terminal V$=$^{18}O bond was still visible after 11 min on stream (near 8 catalytic cycles). The behavior of an exchanged 10% V$_2$O$_5$/Al$_2$O$_3$ catalyst was similar to that of the 2%V$_2$O$_5$/SiO$_2$ catalyst during ethane oxidation (Bañares et al., 2000b) and also during butane oxidation (Wachs et al., 1997). The same effect was observed for the oxidation of methane to formaldehyde (Cardoso, 1998).

The strong influence of the specific oxide support emphasizes the involvement of the bridging V–O–M$_{support}$ bonds in the rate-determining step. DFT calculations also suggested that the terminal V$=$O bond does not participate in the rate-determining step (Calatayud et al., 2004; Haber, 1998; Klisinska et al., 2004; Si-Ahmed et al., 2007; Lewandowska et al., 2008).

Vanadia species on supported on oxides were partially reduced during *n*-butane oxidation and during ethane oxidation, whereby the bridging V–O–V functionality was preferentially reduced relative to the terminal V$=$O bond (Bañares et al., 2000b; Wachs et al., 1996, 1997). The higher reducibility of the surface polymeric species had only a minor effect on the

TOF of ethane oxidation to ethene (4.4×10^{-3} s^{-1} for 5% V_2O_5/Al_2O_3 and 4.0×10^{-3} s^{-1} for 15% V_2O_5/Al_2O_3 at 530 °C) (Bañares et al., 2000c). An increasing degree of polymerization of the surface vanadia species with coverage did not significantly affect the TOF values for the conversion of propane to propene at 350 °C (4.4×10^{-3} s^{-1} for 1%V_2O_5/ZrO_2 and 4.0×10^{-3} s^{-1} for 2%V_2O_5/ZrO_2) (Segawa and Wachs, 1992, Gao et al., 2002). However, the TOF characterizing the reaction of propane on zirconia-supported vanadia was reported to increase with coverage (Argyle et al., 2002) or to remain constant (Christodoulakis et al., 2004).

These results led to questioning of the presumed similar performance of polymeric and isolated surface species. Raman spectroscopy and UV–vis DRS investigations of zirconia- and alumina-supported vanadia showed that the oxidation state of the surface vanadium oxide species was determined by the propane-to-O_2 ratio in the feed (García-Cortéz and Bañares, 2002). The combination of two spectroscopic techniques provided more detail about the structural state of the supported species during moderate reduction under reaction conditions (Gao et al., 2002).

A similar investigation combining UV–vis and Raman spectroscopy showed an equivalent effect of feed composition on the oxidation state of chromium in zirconia-supported chromia catalysts (Malleswara Rao et al., 2004).

These observations are consistent with those of other UV–vis experiments (Puurunen and Weckhuysen, 2002; Puurunen et al., 2001). Raman spectroscopy of a working alumina-supported vanadia catalyst, showed that the surface population ratio of polymeric-to-isolated vanadia species decreased during reduction (as indicated by a relative loss in intensity of the 1009-cm^{-1} band relative to that of the 1017-cm^{-1} band), whereas the total activity and selectivity in propane ODH essentially remained unaffected (García-Cortéz and Bañares, 2002). This result suggested that the active sites for ODH of propane and of ethane on alumina-supported vanadia should be isolated surface vanadia sites, whereas other arrangements of vanadium sites such as polymeric species did not seem to be crucial.

These conclusions are in line with a recent report that showed TOF values of propane ODH not to change with vanadia coverage on titania or zirconia (Christodoulakis et al., 2004). This Raman investigation corroborated the role of the bridging V–O–M$_{support}$ bond in the kinetically significant reaction steps (Christodoulakis et al., 2004).

The selective oxidation of propene to acrolein on supported vanadia catalysts was recently investigated by combined Raman, IR, and UV–vis DR spectroscopies (Zhao and Wachs, 2006). The surface vanadia species became more reduced under these reaction conditions as compared to those of alkane ODH (*vide supra*) because of the greater reducing power of alkenes relative to alkanes. Consequently, the reaction rates were dependent on the O_2 partial pressures, because the surface vanadia sites were deficient in oxygen. The terminal $V{=}O$ bond vibration at 1038 cm^{-1} characteristic of supported V_2O_5/Nb_2O_5 shifted to 978 cm^{-1} during propene oxidation. This band was tentatively assigned to a terminal $V{=}O$ bond of a surface V^{4+} species. The corresponding IR spectra exhibited the vibrations of surface allyl-type C_3H_5 intermediates that were suggested to be involved in the rate-determining step of this selective oxidation. The addition of steam shifted the selectivity from acrolein towards acrylic acid and increased the concentration of surface V^{5+} species.

The molecular structures of surface vanadium oxide species were found to be strongly affected during methanol oxidation catalysis (Burcham et al., 2000; Jehng, 1998; Jehng et al., 1996). The terminal $V{=}O$ and bridging V–O–V Raman bands weakened, as indicated by shifts to lower frequencies. UV–vis DR spectra evidenced a change in the coordination rather than a reduction of the surface V^{5+} species during methanol oxidation (Weckhuysen and Wachs, 1996). Thus, the shifts in vibrational frequency of the terminal $V{=}O$ Raman band were caused by a lengthening of the $V{=}O$ bond upon formation of chemisorbed methoxy species on vanadia.

Similar shifts resulting from coordination of methanol were observed for the $Mo{=}O$ mode in MoO_3/TiO_2 catalysts during methanol oxidation (Hu et al., 1995). Such shifts were reminiscent of those induced by adsorbed H_2O. Raman spectra showed the formation of various surface methoxides during methanol oxidation on metal oxides (specifically, oxides of vanadium, niobium, chromium, molybdenum, tungsten, and rhenium) supported on common oxides (Bañares et al., 1994, 1995; Burcham et al., 2001, 2001; Hu and Wachs, 1995; Jehng, 1998; Jehng et al., 1996; Weckhuysen and Wachs, 1996).

The V–OCH$_3$ species exhibited Raman bands near 1050–1060 cm^{-1} (C–O vibration in VO–CH$_3$) and near 600–660 cm^{-1} (V–O–CH$_3$ vibrations) (Figure 18A). The C—H vibrational modes of M—OCH$_3$ species were found to be sensitive to the nature of the metal ion. Thus, the vibrational bands also provide information on the surface metal oxide coverage of the support, for example, in the case of vanadia on silica. The intensity of the Raman bands attributed to V—OCH$_3$ methyl vibrations at 2930 and 2830 cm^{-1} increased with surface vanadium coverage relative to those of the Si—OCH$_3$ vibrations at 2960 and 2860 cm^{-1} (Figure 18B). Similar trends were found for supported V_2O_5/TiO_2 and V_2O_5/ZrO_2 (Busca et al., 1987, 1996). The surface vanadia monolayers on TiO_2, CeO_2, ZrO_2, and Al_2O_3 exhibited only V—OCH$_3$ bands. M—OCH$_3$ Raman bands were not evident for V_2O_5/TiO_2, Nb_2O_5/SiO_2, or WO_3/SiO_2 (Jehng et al., 1996). Fluorescence overwhelmed the region of the surface methoxy vibrations for V_2O_5/Al_2O_3 and Cr_2O_3/SiO_2 catalysts (Burcham et al., 2000; Jehng et al., 1996; Weckhuysen and Wachs, 1996).

The limited sensitivity of Raman spectroscopy for such adsorbates became apparent in these experiments, because surface methoxy vibrations

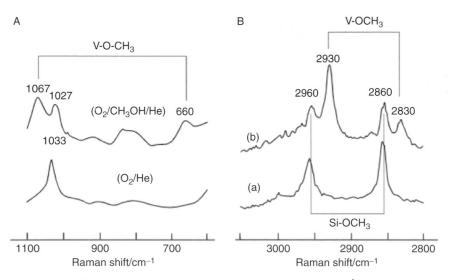

FIGURE 18 (A) The V-OCH$_3$ species exhibit Raman bands at 1067 cm^{-1} (C-O vibration in VO-CH$_3$) and 665 cm^{-1} (V-O-CH$_3$ vibrations). (B) The intensity of the Raman bands assigned to V-OCH$_3$ methyl vibrations at 2930 and 2830 cm^{-1} increase with respect to those of the Si-OCH$_3$ vibrations at 2960 and 2860 cm^{-1} with surface vanadium coverage. (Adapted from M.A. Bañares, I.E. Wachs, *J. Raman Spectrosc.* **33**, 359 (2002) "Molecular Structures of Supported Metal Oxide Catalysts Under Different Environments").

could be detected only on SiO$_2$, V$_2$O$_5$/SiO$_2$, TiO$_2$, and ZrO$_2$ surfaces (and not on V$_2$O$_5$/TiO$_2$ or V$_2$O$_5$/ZrO$_2$). The surface methoxy species were indeed present on those samples as evidenced by IR spectroscopy. However, the presence of surface alkoxy species on potassium-doped alumina-supported vanadia could be shown by Raman spectroscopy (Bañares et al., 2002).

Bronkema and Bell (2007) analyzed the Raman bands of surface methoxy species and of supported vanadia to elucidate the mechanism of methanol oxidation to formaldehyde. In their detailed investigation, insight from Raman spectroscopy was combined with information from EXAFS and XANES spectroscopies. The authors discussed the reaction pathways in the presence and absence of O$_2$, and identified the roles of various lattice oxygen sites. Formaldehyde was found to decompose to H$_2$ and CO in the absence of O$_2$ (Bronkema and Bell, 2007). Similar observations were reported by Korhonen et al. (2007) for methanol conversion on supported chromia catalysts.

Hu et al. (1995) reported extensive Raman experiments characterizing supported molybdenum oxide catalysts during methanol oxidation. In contrast to the stable oxidation state of vanadia, the valence of surface molybdenum species decreased during catalysis. The original band positions could not be fully restored by reoxidation at 230 °C for 1 h; a temperature of 500 °C was needed.

In a subsequent Raman experiment (Hu and Wachs, 1995), the behavior of the monooxo band of a 20 wt % MoO_3/Al_2O_3 catalyst during methanol oxidation was investigated as a function of the reaction temperature (230, 300, or 360 °C). With increasing temperature, the red shift of the band ceased, a result that was explained by a decrease in the methoxy concentration. The reduction of surface molybdena species during methanol reaction resulted in a new Raman band near 850 cm^{-1}, which grew at the expense of that near 850–870 cm^{-1}, characteristic of Mo–O–Mo bonds in surface polymeric molybdena (Brandhorst et al., 2006; Christodoulakis et al., 2006; Hu and Wachs, 1995); furthermore, a new band became apparent near 760 cm^{-1} (Christodoulakis et al., 2006). These features were assigned to partially reduced surface molybdenum oxide species.

The states of surface molybdenum species on silica during methanol oxidation catalysis depend on the molybdenum surface coverage. Highly dispersed surface molybdenum oxide species were detected in dehydrated MoO_3/SiO_2 catalysts at surface coverages below ca. 1 Mo/nm^2 (Bañares et al., 1994). As soon as the catalysts were exposed to the reaction gases (Figure 19A), the high-frequency band representative of the terminal Mo—O bond shifted to lower energies, and new bands appeared. A flow of pure O_2 restored the Mo=O band but did not affect the new signals.

The reversible Mo—O band shift was suggested to arise from the interaction of adsorbed methoxy species with the terminal Mo=O bond. The new bands at 842 and 768 cm^{-1} were assigned by McCarron (McCarron, 1986) to the so-called β-MoO$_3$, which under further O_2 treatment transformed to the stable α-MoO$_3$ phase. This Raman investigation demonstrated that the so-called β-MoO$_3$ is the stable surface molybdenum oxide phase at loadings exceeding 0.1 Mo/nm^2 under methanol oxidation conditions. In contrast, dispersed molybdenum oxide species were stable at elevated temperatures during methane oxidation (Bañares et al., 1994).

Ethanol oxidation on a supported molybdenum oxide catalyst was investigated by Zhang et al. (1995) by application of Raman spectroscopy. After activation in O_2, the Raman spectrum of a 1 wt % MoO_3/SiO_2 catalyst showed the presence of highly distorted octahedral monooxo species, as well as of small amounts of MoO$_3$. Introduction of ethanol at room temperature resulted in a drastic decrease in the intensity of the terminal Mo=O band, accompanied by a shift of this band to 960 cm^{-1} and a blue coloration of the sample. An increase in reaction temperature to 250 °C restored the white color and produced a shift of the Mo=O band to a frequency of 972 cm^{-1}, which is indicative of a dioxo site. Only oxidation at 500 °C for 2 h shifted the band to its original position.

These observations are in good agreement with the findings of Bañares et al. (1994) during methanol oxidation. Zhang et al. (1995)

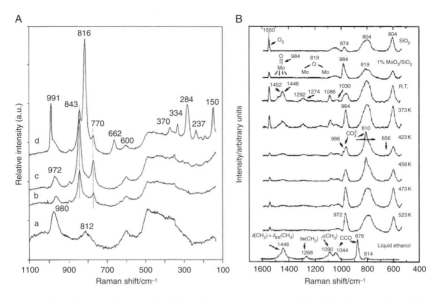

FIGURE 19 Changes during alcohol oxidation on supported molybdena catalysts (A) methanol oxidation (Reprinted from *Journal of Catalysis* **150**, 407 (1994), M.A. Bañares, H. Hu, I.E. Wachs, "Molybdena on Silica Catalysts - Role of Preparation Methods on the Structure Selectivity Properties for the Oxidation of Methanol," copyright (1994) with permission from Elsevier). (B) ethanol oxidation (Reprinted with permission from *Journal of Physical Chemistry*, **99**, 14468 (1995) by W. Zhang, A. Desikan, S.T. Oyama, "Effect of Support in Ethanol Oxidation on Molybdenum Oxide," copyright 1995, American Chemical Society).

spectroscopically detected two different adsorbed ethoxy species (Figure 19B). An interaction of the ethoxy intermediate with Mo=O groups was deduced from the reversion of the Mo=O band shift in parallel to the decreasing intensity of the band of the ethoxy species (2892 cm^{-1}). A remaining weak band at 819 cm^{-1} indicated the presence of some crystallites of α-MoO$_3$.

In addition to the formation of bulk crystalline phases from supported oxides, structural transformations occurring in supported oxides during reactions may lead to solid-state reactions that involve the support. In recent work, Martinez-Huerta et al. (2008) reported on the transformation of surface vanadium oxide species on ceria into CeVO$_4$ during ethane oxidative dehydrogenation catalysis. The catalyst undergoes deactivation at temperatures above 500 °C. A Raman spectroscopic investigation of the working catalyst showed that the incipient formation of CeVO$_4$ by reaction of surface vanadium oxide species with the ceria support did not lead to deactivation of the catalyst, an inference that is based on the observation that such a process was evident even at temperatures near 460 °C (Fig. 20A). The formation of CeVO$_4$ became increasingly evident as the

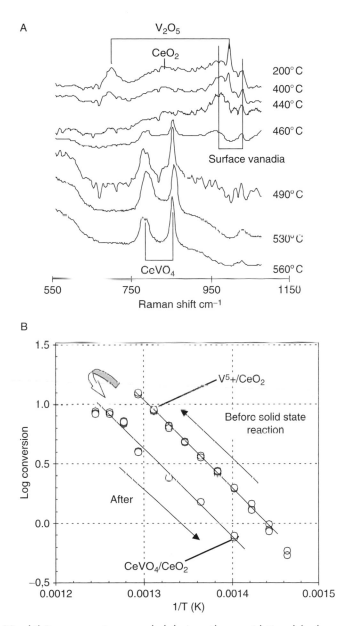

FIGURE 20   (A) Raman spectra recorded during ethane oxidative dehydrogenation on 5%V_2O_5/CeO_2 and (B) Arrhenius plot of fresh and aged catalyst measured online during Raman spectra acquisition. Reaction conditions, C_2H_6/O_2/He = 1/2/8; 50 mg; total flow rate of 100 cm^3/min; 140-400 mg of catalyst, and a particle size of 0.12–0.25 mm. Adapted from M.A. Bañares, M.V. Martínez-Huerta, G. Deo, I.E. Wachs, J.L.G. Fierro, *J. Phys. Chem. B* "Operando Raman-GC study on the structure-activity relationships in V^{5+}/CeO_2 catalyst for ethane oxidative dehydrogenation: The formation of $CeVO_4$") (Martínez-Huerta *et al.*, 2008).

reaction temperature increased, but the catalyst was deactivated at temperatures above 500 °C. The work shows that the Raman bands of $CeVO_4$ become sharper at temperatures above 500 °C. The Arrhenius plots measured with reaction taking place in the Raman cell (Fig. 20B) show that the apparent activation energy did not change significantly as the catalyst aged, that is, as the structure changed from surface vanadia on ceria to a material in which the surface vanadia had reacted with the ceria to form $CeVO_4$. Thus, it is inferred that the nature of the active site did not change; it is inferred that the V-O-Ce bonds present in both the fresh and aged catalysts were part of the active sites.

4.4.9. Metal catalysts

Silver is an industrial oxidation catalyst that has attracted much attention regarding the nature of the surface oxygen species. Bulk AgO exhibits major Raman bands at 425 and 215 cm^{-1} and a broad band centered at about 793 cm^{-1}. Bulk silver oxide decomposes to the metallic silver phase at elevated temperatures (Millar et al., 1997). Oxygen exists in metallic silver as surface and bulk species. The most common Raman bands observed upon oxidation of metallic silver at elevated temperatures are located at 970, 800, 620, 460, and 240 cm^{-1}, and they do not correspond to the characteristic Raman bands of bulk AgO.

The assignment of these Raman bands and the assessment of the reactivity of the corresponding surface species demonstrate the relevance of combining Raman spectroscopy with corresponding reactivity experiments employing isotopes (Wachs, 2002; Wang et al., 1999). Wachs and coworkers (Wang et al., 1999; Wachs, 2003c) showed by isotopic oxygen experiments that these vibrations originated from atomic oxygen species and that adsorbed molecular oxygen species were not present. Experiments with D_2O confirmed that none of the vibrations was associated with surface hydroxyl groups. The Raman bands were assigned as follows: the vibration at *ca.* 970 cm^{-1} to a terminal Ag=O bond, that at 800 cm^{-1} to bridging Ag–O–Ag bonds, and the vibrations at 600–700 cm^{-1} to atomic oxygen species dissolved in the bulk silver lattice. The Raman band at 460 cm^{-1} was attributed to an atomic oxygen species dissolved in the silver lattice. The Raman bands at 240, 760, and a shoulder near 790 cm^{-1} thus may originate from Ag–O–Ag vibrations associated with bending, symmetric stretching, and antisymmetric stretching, respectively.

Raman spectra demonstrated that the bridging Ag–O–Ag bond contained the active oxygen that participated in methanol oxidation and ethene epoxidation (Bao et al., 1993, 1995; Millar et al., 1995; Pettinger et al., 1994; Wang et al., 1997, 1999). The Raman band at 800 cm^{-1} was preferentially removed by reaction with methanol and restored in the presence of O_2/CH_3OH at 600 °C (Wang et al., 1997). This mechanism was demonstrated by Raman spectroscopy supported by isotopic oxygen labeling experiments and online MS analysis (Wang et al., 1999). The terminal Ag=O bond appeared to be unreactive, and the bulk atomic oxygen appeared to serve as a reservoir of atomic oxygen for surface Ag–O–Ag species (Michel et al., 1989; Miciukiewicz et al., 1995).

The nature of the working silver catalyst was different during methanol oxidation and ethene oxidation reactions as a result of variations in the reaction conditions (Wachs, 2002). In methanol oxidation (at 600 °C), H_2 was a major by-product, and Raman spectroscopy showed that the silver catalyst was essentially reduced and contained only trace amounts of atomic oxygen in the subsurface. During ethene oxidation (at *ca.* 230 °C), H_2 was not formed as a byproduct. The absence of H_2 and the lower reaction temperature during ethene oxidation result in a silver surface with atomic oxygen species.

Cao et al., 2006 recently used microfabricated reactors with online GC for a Raman investigation of silver catalysts during methanol oxidation catalysis. Their results are consistent with those described above.

The development of direct methanol fuel cells with platinum- or palladium-containing electrocatalysts requires fundamental understanding of the catalyst deactivation by CO. Weaver and coworkers (Chan et al., 1998; Williams et al., 1998) evaluated the decomposition (into $CO + H_2$) and the total oxidation of methanol on rhodium and palladium catalysts by SERS. During methanol decomposition in the presence of an O_2-free methanol feed, both rhodium and palladium produced CO and H_2. In a second run, carbon-containing deposits decreased the CO production (Chan et al., 1998; Williams et al., 1998). O_2 prevented the formation of carbon-containing deposits on both catalysts and the deactivation of rhodium but not of palladium (Chan et al., 1998; Williams et al., 1998). SER spectra gave evidence of the irreversible formation of PdO (700, 500 cm^{-1}) and the reversible formation of Rh_2O_3 (510 cm^{-1}) (Chan et al., 1998). Thus, O_2 removed carbon-containing deposits, but it also oxidized the catalyst, a process that was not reversible in the case of platinum.

The oxidation states of palladium on γ-alumina during methane combustion were investigated by Demoulin et al. (2003), who observed that, regardless of its initial state, palladium reached the same (equilibrium) state during reaction. Palladium was found to be partly oxidized during methane combustion.

In a Raman spectroscopic investigation of a rhodium catalyst during NO reduction with CO by SERS (Tolia et al., 1995; Williams et al., 1996), the CO_2 partial pressure was determined simultaneously by MS. The Raman band characteristic of the Rh–N mode (315 cm^{-1}) of adsorbed atomic nitrogen (Tolia et al., 1993) decreased sharply at 350 °C. A broad band at 530 cm^{-1} evidenced the formation of a surface rhodium oxide (Rh_2O_3) at 300 °C. These changes were thought to be responsible for a decrease in the activation energy of the catalytic reaction at higher temperatures. The variations in the surface adsorbate population and oxidation state affected the overall reaction rate, which was proposed to be limited by NO dissociation (Chin and Bell, 1983). The results of transient experiments suggested that the production of N_2O decreased with the ability of the catalyst to dissociate NO. Rhodium was found to be most effective in dissociating NO and did not produce N_2O (Tolia et al., 1995).

Uy et al. (2002) monitored the oxidation states of $Pt/\gamma\text{-}Al_2O_3$ catalysts in reactive gases, including O_2, CO, NO_2, and H_2. They showed that the formation of Pt–O sites depended on the reaction conditions; for example during reaction at 350 °C in $CO/O_2/N_2$ there were only half as many Pt–O sites as were observed with the sample in O_2. Raman spectroscopy was used to simultaneously follow the oxidation state of the supported platinum and the chemisorption of CO on metallic platinum in reducing environments: the intensity of the Pt–O vibrational mode decreased while the intensities of the Pt–CO and PtC–O vibrations increased (Uy et al., 2002). The frequency of the Pt–O vibrations varied with the nature of the oxide support. The Pt–O vibration was located at approximately $610\ \text{cm}^{-1}$ in the spectrum of PtO_x/Al_2O_3 and at approximately $690\ \text{cm}^{-1}$ in that of PtO_x/CeO_2. The different frequencies were thought to represent the different states of the PtO_x phases: PtO_x 3-D nanoparticles on Al_2O_3 and a 2-D surface phase on CeO_2. The PtO_x nanoparticles on Al_2O_3 were reduced under mild conditions, but the surface PtO_x phase on CeO_2 required high reduction temperatures (>400 °C) because of its strong interaction with the ceria surface.

4.5. Raman Microspectroscopy

Raman imaging is finding increasing application in catalysis, for example, in the investigation of the spreading of molybdenum oxide on silica and on alumina supports (Leyrer et al., 1990) and the distribution of cobalt and molybdenum species in alumina pellets (Bergwerff et al., 2005). Wachs and Briand (2002a, 2002b) recently employed Raman mapping to determine the distribution of crystalline MoO_3 in spent bulk commercial iron molybdate methanol oxidation catalysts in pellet form. The Raman mapping demonstrated that MoO_3 was absent from the outer regions of the pellets, but was present in the interior. This nonuniform distribution was suggested to be a result of the volatilization of Mo—OCH_3 species and of methanol transport limitations associated with its rapid oxidation to formaldehyde in the outer region of the pellet. This insight led to a novel method for regeneration of iron molybdate catalysts in commercial plants by methanol treatments for 15–30 min. In the absence of reactive O_2, methanol diffuses into the interior regions of the catalyst pellets and redistributes MoO_3 to the exterior via transport as Mo—OCH_3 species.

Recently, novel microreactors and membrane microreactors are becoming increasingly popular because of their unique mass and heat transfer properties (Au and Yeung, 2001; Wan et al., 2001; Yeung and Yao, 2004; Yeung et al., 2005). Catalysts in such microfabricated reactors may be investigated by Raman imaging methods (Cao et al., 2006). Furthermore, the combination of scanning near-field optical microscopy (SNOM) with Raman spectroscopy (Betzig and Trautman, 1992; Betzig et al., 1991, 1992; Fokas, 1999; Fokas and Deckert, 2002; Hoffmann et al., 1995; Houlne et al., 2002; Jahnke et al., 1995, 1996; Münster et al., 1997; Noell et al., 1997; Pohl et al., 1984; Tanaka et al., 1998; Toledo-Crow et al., 1992; Valaskovic et al., 1995; Webster et al., 1998) yields high spatial resolution and can be applied under the conditions of heterogeneous catalysis. The combined SNOM–Raman experiments simultaneously provide information about sample topography and structure.

5. PERSPECTIVE AND CONCLUSIONS

Raman spectroscopy can be used to investigate samples under a wide range of conditions, and it is thus one of the most powerful tools for characterization of working catalysts. Raman experiments can be carried out at any temperature or pressure used in catalytic processes and without any interference of gas or liquid (water) phases. It is thus possible to investigate reactions in liquid phases or under supercritical conditions. Quartz fiber optics allow easy spectroscopic access to catalytic reactors and the direct coupling of Raman spectrometers to large industrial reactors, thus offering the prospect of providing direct structural information regarding the state of the operating catalyst. Raman fiber probes are already used to control refinery columns and polymerization processes.

Catalytic molecular surface species may undergo drastic changes in their structure in the presence of reactants. For example, polymeric clusters may transform into highly distorted monomeric species. A crystalline phase may become mobile at its Tammann temperature, as shown by Raman spectroscopy, and it may spread over oxide supports driven by the reduction of the overall surface free energy. Reactive environments trigger many structural transformations, exemplified by particle sintering, dispersion of bulk phases, segregation of surface species into bulk phases, and solid-state reactions between supported oxides and supports.

Resonance Raman effects, which arise from resonant coupling of the exciting laser light to electronic transitions, can be exploited to overcome difficulties with detection limits and are of particular interest for the characterization of working catalysts that contain reduced transition metal ions.

Time resolution in the sub-second regime can be achieved with state-of-the-art spectrometers and when materials with high Raman cross-sections are investigated. Thus, time-resolved transient temperature or pressure response experiments can be carried out (e.g., pulse experiments with isotope labels), and reaction kinetics can be correlated directly with spectra. Raman spectroscopy is excellently suited to the characterization of structural changes occurring in catalytic materials during their action, because the spectra are sensitive to crystalline, amorphous, glassy, or molecular species. Phase transitions, decompositions, and solid-state reactions between different phases can be characterized. It is even possible to observe disintegration and phase separation in bulk mixed oxides, as shown by laterally resolved confocal Raman microspectroscopy of a molybdenum–vanadium–tungsten mixed oxide catalyst for selective partial oxidation.

Because of its broad applicability, Raman spectroscopy is expected to be used in the near future to characterize numerous catalytic materials in the functioning state, specifically, to unravel the nature of the catalytically active sites, to identify surface reaction intermediates, and to follow catalyst deactivation processes. Moreover, Raman spectroscopy is a powerful tool for the characterization of all synthesis and activation steps of catalysts. It can be used to investigate species formed in aqueous solution, depending on the pH, metal concentrations, or the presence of complexing agents. Such structural information is potentially valuable in laying the groundwork for the reproducible synthesis of industrial catalysts.

The main advantages of Raman spectroscopy as compared to other techniques are the following:

(a) ease of construction of reactors that simultaneously serve as Raman cells;
(b) possibility to work at high reaction temperatures;
(c) absence of significant spectral interference from the gas phase or the support.

These advantages outweigh by far the disadvantages, such as the occasionally occurring fluorescence and the inherently low sensitivity. Raman spectroscopy provides fundamental insight; however, complementary information from other spectroscopic techniques is extremely valuable. Many spectroscopic techniques are described in this volume and in the preceding two volumes of *Advances in Catalysis*, and less comprehensively in a review by Brückner (2003).

Several groups have reported combining several techniques for investigations of catalysts in reactive environments, such as Raman and EPR spectroscopy (Dai et al., 2000), Raman, IR, and NMR spectroscopy (Gao et al., 2004), Raman and EPR spectroscopy and XRD (Martin et al., 2003), and EPR, UV–vis and Raman spectroscopy (Brückner and Kondratenko, 2006). Weckhuysen and coworkers (Tinnemans et al., 2006) recently reviewed the combination of several techniques in one reaction cell. An interesting example for fundamental catalysis research and catalyst discovery at the same time is found in an automated high-throughput system with eight parallel reactors allowing spectroscopic characterization of the working catalysts (Bañares et al., 2007).

Combined spectroscopic monitoring of processes is expected to be of value during industrial operations to keep the catalysts in their best working state and to regenerate them at appropriate stages of deactivation.

For example, Bennici et al. (2007) reported a combined Raman and UV–vis spectroscopy monitoring system for a chromia/alumina catalyst that is used in propane dehydrogenation. With spectroscopic information at hand, coke burn-off can be controlled so that the temperature increase during regeneration can be reduced by a factor of three (25 vs. 75 K) (Bennici et al., 2007). Such an apparatus was also used to monitor and control VPO phase transformations during butane oxidation to maleic anhydride (Bennici et al., 2007). Of course in some cases, there will be significant hurdles to cross in the successful combination of several techniques (Brückner, 2003; Weckhuysen, 2003).

In a best-case scenario, several kinds of spectroscopic data will be collected simultaneously along with kinetics data characterizing the catalytic reaction. Such an approach will generate understanding of structure–activity relationships on a molecular scale, and provide a powerful process control tool. Bridging the pressure, temporal and materials gaps in catalysis is important for determination of structure–activity relationships that are pertinent to industrial operation. Raman spectroscopy, in combination with other spectroscopic techniques and theoretical modeling, can provide key information about reactants, surface species (spectators and intermediates), and catalyst bulk and surface structures present during catalyst operation. The results of such investigations will permit major advances in catalysis as they can provide the basis for the rational design and discovery of new and improved catalysts.

ACKNOWLEDGMENTS

MAB acknowledges the support from the Spanish Ministry under grant CTQ2005–02802/PPQ and from the European Science Foundation COST Action D36/006/06. GM acknowledges support from the European Coordination Action Network "CONCORDE," the European Science Foundation COST Action D36/006/06, and Süd-Chemie AG. The authors are indebted to H. Knözinger for helpful discussions and to I. E. Wachs for providing recent results.

REFERENCES

Abdelouahab, F.B., Olier, R., Guilhaume, N., Lefebvre, F., and Volta, J.C., *J. Catal.* **134**, 151 (1992).
Adams, D.M., Payne, S.J., and Martin, K., *Appl. Spectrosc.* **27**, 377 (1973).

Addleman, R.S., Hills, J.W., and Wai, C.M., *Rev. Sci. Instrum.* **39**, 3127 (1998).

Airaksinen, S.M.K., Bañares, M.A., and Krause, A.O.I., *J. Catal.* **230**, 507 (2005).

Airaksinen, S.M.K., Krause, A.O.I., Sainio, J., Latineen, J., Chao, K.-J., Guerrero-Pérez, M.O., and Bañares, M.A., *Phys. Chem. Chem. Phys.* **5**, 4371 (2003).

Alivisatos, P., *Nat. Biotechnol.* **22**, 47 (2004).

Andersson, A., Hansen, S., and Wickman, A., *Topics Catal.* **15**, 103 (2001).

Angell, C.L., *J. Phys. Chem.* **77**, 222 (1973).

Angoni, K., *J. Mater. Sci.* **33**, 3693 (1998).

Argyle, M.D., Chen, K., Bell, A.T., and Iglesia, E., *J. Catal.* **208**, 139 (2002).

Arya, K., and Zeyher, R., *in* "Topics in Applied Physics, Vol 54, Light Scattering in Solids IV", p. 419. Springer, Berlin, 1984.

Au, L.T.Y., and Yeung, K.L., *J. Membr. Sci.* **194**, 33 (2001).

Avouris, Ph., and Demuth, J.E., *J. Chem. Phys.* **75**, 4783 (1981).

Badlani, M., and Wachs, I.E., *Catal. Lett.* **75**, 137 (2001).

Baes, C., and Mesmer, R.E., *in* "Hydrolysis of Cations" (F.A. Cotton, Ed.) Wiley New York, p. 258 (1976).

Bañares, M.A., *Catal. Today* **100**, 71 (2005).

Bañares, M.A., *Catal. Today* **51**, 319 (1999).

Bañares, M.A., *in* "In situ Characterization of Catalytic Materials" (B.M. Weckhuysen, Ed.) American Scientific Publishers, 2004.

Bañares, M.A., Cardoso, J.H., Agulló-Rueda, F., Correa-Bueno, J.M., and Fierro, J.L.G., *Catal. Lett.* **64**, 191 (2000a).

Bañares, M.A., Guerrero-Pérez, M.O., García-Cortéz, G., and Fierro, J.L.G., *J. Mater. Chem.* **11**, 3337 (2002).

Bañares, M.A., Hu, H., and Wachs, I.E., *J. Catal.* **150**, 407 (1994).

Bañares, M.A., Hu, H., and Wachs, I.E., *J. Catal.* **155**, 249 (1995).

Bañares, M.A., and Khatib, S.J., *Catal. Today* **96**, 251 (2004).

Bañares, M.A., Martínez-Huerta, M.V., Gao, X., Fierro, J.L.G., and Wachs, I.E., *Catal. Today* **61**, 295 (2000c).

Bañares, M.A., Martínez-Huerta, M.V., Gao, X., Fierro, J.L.G., and Wachs, I.E., *in* "Metal oxide catalysts: active sites, intermediates and reaction mechanisms", Symposium, 220th ACS National Meeting, Washington, USA (2000d).

Bañares, M.A., Martínez-Huerta, M.V., Gao, X., Wachs, I.E., and Fierro, J.L.G., *Stud. Surf. Sci. Catal.* **130**, 3125 (2000b).

Bañares, M.A., Prieto, J., Goberna-Selma, C., Guerrero-Pérez, M.O., and García-Casado, M., *Prepr. Pap.-Am. Chem. Soc., Div. Petr. Chem.* **52**(2), 62 (2007).

Bañares, M.A., Spencer, N.D., Jones, M.D., and Wachs, I.E., *J. Catal.* **146**, 204 (1994).

Bañares, M.A., and Wachs, I.E., *J. Raman Spectrosc.* **33**, 359 (2002).

Bao, X., Muhler, M., Pettinger, B., Schlögl, R., and Ertl, G., *Catal. Lett.* **22**, 322 (1993).

Bao, X., Muhler, M., Pettinger, B., Uchida, Y., Lehmpfuhl, G., Schlögl, R., and Ertl, G., *Catal. Lett.* **32**, 171 (1995).

Bao, X., Pettinger, B., Ertl, G., and Schlögl, R., *Ber. Bunsen Ges J. Phys Chem.* **97**, 322 (1993).

Barasnka, H., Czerwinska, B., and Labudzinska, A., *J. Mol. Struct.* **143**, 485 (1986).

Barbaro, A., Larrondo, S., Duhalde, S., and Amadeo, N., *Appl. Catal A* **193**, 277 (2000).

Bartlett, J.R., and Cooney, R.P., *in* "Spectroscopy of Inorganic-Based Materials, Advances in Spectroscopy" (R.J.H. Clark and R.E. Hester, Eds.), Vol. 14, p. 187. Wiley, Chichester, 1987.

Beato, P., Blume, A., Girgsdies, F., Jentoft, R.E., Schlögl, R., Timpe, O., Trusnchke, A., Weinbert, G., Basher, Q., Hamid, F.A., Hamid, S.B.A., Omar, E., et al., *Appl. Catal. A* **307**, 137 (2006).

Bennici, Ç.S.M., Vogelarr, B.M., Nijhuis, T.A., and Weckhuysen, B.M., *Angew. Chem. Int. Ed.* **46**, 5412 (2007).

Bergin, F.J., *Spectrochim. Acta A* **46**, 153 (1990).

Bergwerff, J., van de Water, L.G.A., Visser, T., de Peinder, P., Liliveld, B.R.G., de Jong, K.P., and Weckhuysen, B.M., *Chem. Eur. J.* **11**, 4591 (2005).

Betzig, E., Finn, P.L., and Weiner, J.S., *Appl. Phys. Lett.* **60**, 2484 (1992).

Betzig, E., and Trautman, J., *Science* **257**, 189 (1992).

Betzig, E., Trautmann, J.K., Harris, T.D., Weiner, J.S., and Kostelak, R.L., *Science* **251**, 1468 (1991).

Birman, J.L., in "Encyclopedia of Physics" (S. Flügge and L. Genzel, Eds.), Vol. XXV/2b, p. 213. Springer, New York, (1974).

Bisset, A., and Dines, T.J., *J. Chem. Soc. Faraday Trans.* **91**, 499 (1995a).

Bisset, A., and Dines, T.J., *J. Raman Spectrosc.* **26**, 791 (1995b).

Blume, A., Ph.D. Thesis, TU Berlin, 2001.

Bordiga, S., Damin, A., Bonino, F., Ricchiardi, G., Lamberti, C., and Zecchina, A., *Angew. Chem. Intl. Ed.* **41**, 4734 (2002).

Bordiga, S., Damin, A., Bonino, F., Ricchiardi, G., Zecchina, A., Tagliapietra, R., and Lamberti, C., *Phys. Chem. Chem. Phys.* **5**, 4390 (2003).

Boulova, M., Gaskov, A., and Lucazeau, G., *Sens. Actuators B* **81**, 99 (2001).

Bowden, M., Gardiner, D.J., Rice, G., and Gerrard, D.L., *J. Raman Spectrosc.* **21**, 37 (1990).

Brandhorst, M., Cristol, S., Capron, M., Dujardin, C., VEzin, H., Le Bourdon, G., and Payen, E., *Catal. Today* **113**, 34 (2006).

Brazdil, J.F., and Bartok, J.P., *US Patent 5854172* (1998).

Brazdil, J.F., Kobarvkantei, F.A.P., and Padolwski, J.P., *JP Patent 11033399* (1999).

Bredow, T., Aprà, E., Catti, M., and Pacchioni, G., *Surf. Sci.* **418**, 150 (1998).

Brémard, C., and Bougeard, D., *Adv. Mater.* **7**, 10 (1995).

Briand, L.E., Farneth, W.E., and Wachs, I.E., *Catal. Today* **62**, 219 (2000).

Bridges, T.E., Houlne, M.P., and Harris, J.M., *Anal. Chem.* **76**, 576 (2004).

Bronkema, J.L., and Bell, A.T., *J. Phys. Chem. C* **111**, 420 (2007).

Brown, F.R., Makovsky, L.E., and Rhee, H.K., *Appl. Spectrosc.* **31**, 563 (1977).

Brown, F.R., Makovsky, L.E., and Rhee, K.H., *J. Catal.* **50**, 162 (1977a).

Brown, F.R., Makovsky, L.E., and Rhee, K.H., *J. Catal.* **50**, 385 (1977b).

Brückner, A., *Catal. Rev. Sci. Eng.* **45**, 97 (2003).

Brückner, A., and Kondratenko, E., *Catal. Today* **113**, 16 (2006).

Brückner, S., Jeziorowski, H., and Knözinger, H., *Chem. Phys. Lett.* **105**, 218 (1984).

Burcham, L.J., Briand, L.E., and Wachs, I.E., *Langmuir* **17**, 6164 (2001).

Burcham, L.J., Briand, L.E., and Wachs, I.E., *Langmuir* **17**, 6175 (2001).

Burcham, L.J., Deo, G., Gao, X., and Wachs, I.E., *Topics Catal.* **11/12**, 85 (2000).

Burcham, L.J., and Wachs, I.E., *Catal. Today* **49**, 467 (1999).

Burch, R. (Ed.), special issue "In situ mehtods in catalysis" *Catal. Today* **9**, (1/2) (1991).

Busca, G., *Catal. Today* **27**, 457 (1996).

Busca, G., Elmi, A.S., and Forzatti, P., *J. Phys. Chem.* **91**, 5263 (1987).

Cai, Y., Wang, C.-B., and Wachs, I.E., *Stud. Surf. Sci. Catal.* **110**, 255 (1997).

Calatayud, M., Mguig, B., and Minot, C., *Surf. Sci. Rep.* **55**, 169 (2004).

Campion, A., and Kambhampati, P., *Chem. Soc. Rev.* **27**, 241 (1998).

Cao, E., Firth, S., McMillan, P., Gavriilidis, A., *Catal. Today* **126**, 119 (2007).

Cardona, M., in "Topics in Applied Physics, Vol. 8, Light Scattering in Solids I" 2nd, (M. Cardona, Ed.), p. 254. Springer, Berlin, (1983).

Cardoso, J.H., PhD Dissertation, Universidade Federal de Sao Carlos, Sao Paulo, Brazil, 1998.

Carreón, M.A., Guliants, V.V., Guerrero-Pérez, M.O., and Bañares, M.A., *Microporous Mesoporous Mater.* **71**, 57 (2004).

Carrier, X., Lambert, J.F., and Che, M., *J. Am. Chem. Soc.* **119**, 10137 (1997).

Cavani, F., Ballarini, N., Cimini, M., Trifirò, F., Bañares, M.A., and Guerrero-Pérez, M.O., *Catal. Today* **112**, 12 (2006).

Cavani, F., and Trifirò, F., *CHEMTECH* **24**, 18 (1994).

Centi, G., and Perathoner, S., *Appl. Catal. A* **124**, 317 (1995).

Chan, H.Y.H., Williams, C.T., Weaver, M.J., and Takoudis, C.G., *J. Catal.* **174**, 191 (1998).

Chan, S.S., and Bell, A.T., *J. Catal.* **89**, 433 (1984).

Chan, S.S., and Wachs, I.E., *J. Catal.* **103**, 224 (1987).

Chan, S.S., Wachs, I.E., Murrell, L.L., Wang, L., and Hall, K., *J. Phys. Chem.* **88**, 5831 (1984).

Chase, B.C., *Anal. Chem.* **59**, 881A (1987).

Cheng, C.P., Ludowise, J.D., and Schrader, G.L., *Appl. Spectrosc.* **34**, 146 (1980).

Cheng, C.P., and Schrader, G.L., *J. Catal.* **60**, 276, (1979).

Chesters, M.A., and Sheppard, N., *Chem. Brit.* **17**, 521 (1981).

Chin, A.A., and Bell, A.T., *J. Phys. Chem.* **87**, 3700 (1983).

Christodoulakis, A., Heracleous, E., Lemonidou, A.A., and Boghosian, S., *J. Catal.* **242**, 16 (2006).

Christodoulakis, A., Machli, M., Lemonidou, A.A., and Boghosian, S., *J. Catal.* **222**, 293 (2004).

Chua, Y.T., and Stair, P.C., *J. Catal.* **196**, 66 (2000).

Chua, Y.T., and Stair, P.C., *J. Catal.* **213**, 39 (2003).

Chua, Y.T., Stair, P.C., and Wachs, I.E., *J. Phys. Chem. B* **105**, 8600 (2001).

Clark, R.J.H., and Dines, T.J., *Angew. Chem. Int. Ed. Engl.* **98**, 131 (1986).

Clausen, B.S., Steffensen, G., Fabius, B., Villadsen, J., Feidenhans'l, R., and Topsøe, H., *J. Catal.* **132**, 524 (1991).

Clausen, B.S., Topsøe, H., and Frahm, R., *Adv. Catal.* **42**, 315 (1998).

Conte, M., Budroni, G., Bartley, J.K., Taylor, S.H., Carley, A.F., Schmidt, A., Murphy, D.M., Girgsdies, F., Ressler, T., Schlögl, R., and Hutchings, G.J., *Science* **313**, 1270 (2006).

Cooney, R.P., Curthoys, G., and Nguyen, T.T., *Adv. Catal.* **24**, 293 (1975).

Cortés-Corberán, V., Savkin, V.V., Ruiz, P., and Vislovskii, V.P., *J. Mol. Catal. A* **158**, 271 (2000).

Cortéz, G.G., Fierro, J.L.G., and Bañares, M.A., *Catal. Today* **78**, 219 (2003).

Covington, A.K., and Thain, J.M., *Appl. Spectrosc.* **29**, 386 (1975).

Cross, J.C., and Schrader, G.L., *Thin Solid Films* **295**, 5 (1995).

Cuesta, A., Dhamelincourt, P., Laureyns, J., Martínez-Alonso, A., and Tascón, J.M.D., *Carbon* **32**, 1523 (1994).

Dai, H.Z., Ng, C.F., and Au, C.T., *Appl. Catal. A* **202**, 1 (2000).

Damyanova, S., Gomez-Sainero, L.M., Bañares, M.A., and Fierro, J.L.G., *Chem. Mater.* **12**, 501 (2000).

Deckert, V., and Kiefer, W., *Appl. Spectrosc.* **46**, 322 (1992).

Deckert, V., Zeisel, D., Zenobi, R., and Vo-Dinh, T., *Anal. Chem.* **70**, 2646 (1998).

Delgass, W.N., Haller, G.L., Kellerman, R., and Lunsford, J.H., *in* "Spectroscopy in Heterogeneous Catalysis" p. 55. Academic Press, New York (1979).

Delhaye, M., and Dhamelincourt, P., *J. Raman Spectrosc.* **3**, 33 (1973).

Delhaye, M., and Dhamelincourt, P., *J. Raman Spectrosc.* **3**, 33 (1997).

Demoulin, O., Navez, M., Gaigneaux, E.M., Ruiz, P., Mamede, A.S., Grange, P., and Payen, E., *Phys. Chem. Chem. Phys.* **5**, 4394 (2003).

Deo, G., Turek, A.M., Wachs, I.E., Huybrechts, D.R.C., and Jacobs, P., *Zeolites* **13**, 365 (1993).

Deo, G., and Wachs, I.E., *J. Phys. Chem.* **95**, 5889 (1991).

Deo, G., Wachs, I.E., and Haber, J., *Crit. Rev. Surf. Chem.* **4**, 141 (1994).

Dhamelincourt, P., Delhaye, M., and Da Silva, E., *J. Raman Spectrosc.* **25**, 3 (1994).

Diebold, U., *Surf. Sci. Rep.* **48**, 53 (2003).

Dieterle, M., PhD Thesis, Technical University, Berlin, 2001.

Dieterle, M., and Mestl, G., *Phys. Chem. Chem. Phys.* **4**, 822 (2002).

Dieterle, M., Mestl, G., Jäger, J., Uchida, Y., and Schlögl, R., *J. Mol. Catal. A* **174**, 169 (2001).

Dieterle, M., Weinberg, G., and Mestl, G., *Phys. Chem. Chem. Phys.* **4**, 812 (2002).

Dixit, L., Gerrard, D.L., and Bowley, H.J., *Appl. Spectrosc. Rev.* **22**, 189 (1986).

Dumesic, J.A., and Topsøe, H., *Adv. Catal.* **26**, 121 (1977).

Dunn, J.P., Jehng, J.M., Kim, D.S., Briand, L.E., Stenger, H.G., and Wach, I.E., *J. Phys. Chem. B.* **102**, 6212 (1998).

Dunn, J.P., Koppula, P.R., Stenger, H.G., and Wachs, I.E., *Appl. Catal. B Environ.* **19**, 103 (1998).

Dunn, J.P., Stenger, H.G., and Wachs, I.E., *Catal. Today* **51**, 301 (1999).

Dunn, J.P., Stenger, H.G., and Wachs, I.E., *Catal. Today* **53**, 543 (1999).

Dunn, J.P., Stenger, H.G., and Wachs, I.E., *J. Catal.* **181**, 233 (1999).

Dutta, P.K., and Puri, M., *J. Phys. Chem.* **91**, 4329 (1987).

Dutta, P.K., and Shieh, D.C., *J. Phys. Chem.* **90**, 2331 (1986).

Dutta, P.K., Shieh, D.C., and Puri, M., *J. Phys. Chem.* **91**, 2332 (1987).

Egerton, T.A., and Hardin, A.H., *Catal. Rev.–Sci. Eng.* **11**, 1 (1975).

Emory, S.R., and Nie, S., *J. Phys. Chem. B* **102**, 493 (1998).

Faulds, K., Smith, W.E., and Graham, D., *Anal. Chem.* **76**, 412 (2004).

Ferraro, J.R., and Nakamoto, K. "Introductory Raman Spectroscopy", Academic Press, New York, 1994.

Fleischmann, M., Hendra, P.J., and McQuillan, A.J., *Chem. Phys. Lett.* **26**, 163 (1974).

Fleischmann, M., Hendra, P.J., McQuillan, A.J., Paul, R.L., and Reid, E.S., *J. Raman Spectrosc.* **4**, 269 (1976).

Fokas, C., and Deckert, V., *Appl. Spectrosc.* **56**, 192 (2002).

Fokas, Ch.S., *Nachr. Chem. Tech. Lab.* **47**, 648 (1999).

Freeman, J.J., Heaviside, J., Hendra, P.J., Prior, J., and Reid, E.S., *Appl. Spectrosc.* **35**, 196 (1981).

Freund, H.J., *Catal. Today* **100**, 3 (2005).

Gaigneaux, E.M., Dieterle, M., Ruiz, P., Mestl, G., and Delmon, B., *J. Phys. Chem. B* **102**, 10542 (1998).

Gai, P.L., and Kourtakis, K., *Science* **267**, 661 (1995).

Gai, P.L., Kourtakis, K., Coulson, D.R., and Sonnichsen, G.C., *J. Phys. Chem. B* **101**, 9916 (1997).

Gao, J., Chen, Y., Han, B., Feng, Z., Li, C., Zhou, N., Gao, S., and Xi, Z., *J. Mol. Catal. A* **210**, 197 (2004).

Gao, X., Bañares, M.A., and Wachs, I.E., *J. Catal.* **188**, 325 (1999).

Gao, X., Bare, S.R., Fierro, J.L.G., and Wachs, I.E., *J. Phys. Chem. B.* **103**, 618 (1999).

Gao, X., Jehng, J.-M., and Wachs, I.E., *J. Catal.* **209**, 43 (2002).

Garbowski, E., and Coudurier, G., *in* "Catalyst Characterization. Physical Techniques for Solid Materials" (B. Imelik and J.C. Vedrine, Eds.), p. 45. Plenum Press, New York, 1994.

García-Cortéz, G., and Banares, M.A., *J. Catal.* **209**, 197 (2002).

Gardiner, D.J., Littleton, C.J., and Bowden, M., *Appl. Spectrosc.* **42**, 15 (1988).

Gasior, M., Haber, J., Machej, T., and Czeppe, T., *J. Mol. Catal.* **43**, 359 (1988).

Gauthier, Y., Baudoing-Savois, R., Heinz, K., and Landskron, H., *Surf. Sci.* **251**, 493 (1991).

Geurts, J., and Richter, W., *Springer Proc. Phys.* **22**, 328 (1987).

Giakomelou, I., Caraba, R.M., Parvulescu, V.I., and Boghosian, S., *Catal. Lett.* **78**, 209 (2002).

Gijzeman, O.L.J., van Lingen, J.N.J., van Lenthe, J.H., Tinnemans, S.J., Keller, D.E., and Weckhuysen, B.M., *Chem. Phys. Lett.* **397**, 277 (2004).

Gillet, P., Hemley, R.J., and McMillan, P.F., *Rev. Mineral.* **37**, 525 (1998).

Gil-Llambias, J., Escudey, A.M., Fierro, J.L.G., and López Agudo, A., *J. Catal.* **95**, 520 (1985).

Glaeser, L.C., Brazdil, J.F., Hazle, M.A., Mehicic, M., and Grasselli, R.K., *J. Chem. Soc. Faraday Trans. I* **81**, 2903 (1985).

Gonzales-Vildres, F., and Griffith, W.P., *J. Chem Soc. Dalton Trans.* **1416**, (1972).

Gonzalez-Elipe, A.R., and Yubero, F., "Handbook of Surfaces and Interfaces of Materials", p. 147, Academic Press, New York, 2001.

Grasselli, J.G., Hazle, M.A.S., and Wolfram, L.E., *in* "Molecular Spectroscopy" (A.R. West Ed.), p. 200. Heyden and Son, London, 1977.

Grasselli, J.G., Snavely, M.K., and Bulkin, J.J., "Chemical Applications of Raman Spectroscopy", Wiley Interscience (1981).

Greenwood, N.N., and Earnshaw, E., "Chemistry of the Elements", Pergamon Press, Oxford, New York, Toronto, Sidney, Paris, Frankfurt, 1984.

Griffith, W.P., *J. Chem. Soc. A* **286** (1970).

Griffith, W.P., and Lesniak, P.J.B., *J. Chem Soc. A* **1066**, (1969).

Griffith, W.P., and Wilkins, T.D., *J. Chem. Soc. A* **675**, (1967).

Grünwaldt, J.D., Wandeler, R., and Baiker, A., *Catal. Rev.* **45**, 1 (2003).

Guerrero-Pérez, M.O., and Bañares, M.A., *Catal. Today* **96**, 265 (2004).

Guerrero-Pérez, M.O., and Bañares, M.A., *Chem. Commun.* 1292 (2002).

Guerrero-Pérez, M.O., Fierro, J.L.G., Vicente, M.A., and Bañares, M.A., *J. Catal.* **206**, 339 (2002).

Guerrrero-Pérez, M.O., and Bañares, M.A., *Topics Catal.* (2004) Special issue on the Symposium on Catalysis by Metal Oxides, Philadelphia, August 2004.

Guisnet, M., and Magnoux, P., *Appl. Catal. A* **212**, 83 (2001).

Guliants, V.V., Benziger, J.B., Sundaresan, S., Yao, N., and Wachs, I.E., *Catal. Lett.* **32**, 379 (1995).

Guliants, V.V., Holmes, S., Benziger, J.B., Heaney, P., Yates, D., and Wachs, I.E., *J. Mol. Catal. A* **172**, 265 (2001).

Guttmann, A.T., Grasselli, R.K., and Brazdil, F.J., *US Patent*, 4746641, 4788173, and 4837233 (1988).

Haber, J., *in* "Catalytic Activation and Fuctionalisation of Light Alkanes. Advances and Challenges" (E.G. Derouane, J. Haber, F. Lemos, F.R. Ribeiro and M. Guisnet, Eds.), NATO ASI Series, Kluwer Academic Publishers, Dordrecht, 1998.

Haber, J., Machej, T., and Czeppe, T., *Surf. Sci.* **151**, 310 (1985).

Haber, J., Machej, T., Serwicka, E.M., and Wachs, I.E., *Catal. Lett.* **32**, 101 (1995).

Haller, J.H., Lunsford, and Laane, J., *J. Phys. Chem.* **100**, 551 (1996).

Han, Y.F., Ramesh, K., Chen, L., Widjaja, E., Chilukoti, S., and Chen, F., *J. Phys. Chem. C* **111**, 2830 (2007).

Hardcastle, F.D., and Wachs, I.E., *J. Phys. Chem.* **46**, 173 (1988).

Hardcastle, F.D., Wachs, I.E., and Horsley, J.A., *J. Mol. Catal.* **46**, 15 (1988).

Harlin, M.E., Niemi, V.M., and Krause, A.O.I., *J. Catal.* **195**, 67 (2000).

Harlin, M.E., Niemi, V.M., Krause, A.O.I., and Weckhuysen, B.M., *J. Catal.* **203**, 242 (2001).

Heimann, P.A., and Urstadt, R., *Appl. Opt.* **29**, 495 (1990).

Hendra, J., and Mould, H., *Int. Lab.* **34** (1988).

Herrera, J.E., and Resasco, D.E., *Chem. Phys. Lett.* **376**, 302 (2003).

Hicks, S.E., Fitzgerald, A.G., Baker, S.H., and Dines, T.J., *Phil. Mag. B* **62**, 193 (1990).

Hill, C.G., and Wilson, J.H., *J. Mol. Catal.* **63**, 65 (1990).

Himeno, S., Niiya, H., and Ueda, T., *Bull. Chem. Soc. Jpn.* **70**, 631 (1997).

Hoefs, E.V., Monnier, J.R., and Keulks, G.W., *J. Catal.* **57**, 331 (1979).

Hoffmann, M.M., Addleman, R.S., and Fulton, J.L., *Rev. Sci. Instrum.* **71**, 1552 (2000).

Hoffmann, P., Dutoit, B., and Salathe, R.-P., *Ultramicroscopy* **61**, 165 (1995).

Horsley, J.A., Wachs, I.E., Brown, J.M., Via, G.H., and Hardcastle, F.D., *J. Phys. Chem.* **91**, 4014 (1987).

Houlne, M.P., Sjostrom, C.M., Uibel, R.H., Keimeyer, J.A., and Harris, J.M., *Anal. Chem.* **74**, 4311 (2002).

Howdle, S.M., Stanley, K., Popov, V.K., and Bagratashviti, V.N., *Appl. Spectrosc.* **48**, 214 (1994).

Hu, H., and Wachs, I.E., *J. Phys. Chem.* **99**, 10911 (1995).

Hu, H., Wachs, I.E., and Bare, S.R., *J. Phys. Chem.* **99**, 10897 (1995).

Humbert, B., Grausem, J., and Courjon, D., *J. Phys. Chem. B* **108**, 15714 (2004).

Hunger, M., and Weitkamp, J., *Angew. Int. Ed.* **40**, 2954 (2001).

Hunnius, W.D., Z. *Naturforsch. B* **30**, 63, (1975).

Hunter, R.J., "Zetapotential in Colloid Science, Principle and Application", *in* "Colloid Science", (R.H. Ottewill and R.L. Rowell, Eds.) Academic Press, London, p. 233, 1988.

Hutchings, G.J., Desmartin-Chomel, A., Olier, R., and Volta, J.C., *Nature* **368**, 41 (1994).

Iannibello, A., Villa, P.L., and Marengo, S., *Gazz. Chim. Ital.* **109**, 521 (1979).

Ishikawa, M., Maruyama, Y., Ye, J.Y., and Futamata, M., *J. Biol. Phys.* **28**, 573 (2002).

Jahnke, C.L., Hallen, H.D., and Paesler, M.A., *J. Raman Spectrosc.* **27**, 579 (1996).

Jahnke, C.L., Paesler, M.A., and Hallen, H.D., *Appl. Phys. Lett.* **67**, 2483 (1995).

Jehng, J.-M., *J. Phys. Chem. B* **102**, 5816 (1998).

Jehng, J.-M., Deo, G., and Wachs, I.E., *J. Mol. Catal. A* **110**, 41 (1996).

Jehng, J.-M., Hu, H., Gao, X., and Wachs, I.E., *Catal. Today* **28**, 335 (1996).

Jehng, J.-M., and Wachs, I.E., *Chem. Mater.* **3**, 100 (1991).

Jehng, J.-M., and Wachs, I.E., *J. Mol. Catal.* **13**, 9 (1992).

Jehng, J.-M., and Wachs, I.E., *J. Mol. Catal.* **67**, 369 (1992).

Jehng, J.-M., and Wachs, I.E., *J. Phys. Chem.* **95**, 7373 (1991).

Jehng, J.-M., and Wachs, I.E., *J. Raman Spectrosc.* **22**, 83 (1991).

Jezlorowski, H., and Knözinger, H., *Chem. Phys. Lett.* **52**, 519 (1977).

Jeziorowski, H., and Knözinger, H., *J. Phys. Chem.* **83**, 1166 (1979).

Johanson, G., Petterson, L., and Ingri, N., *Acta. Chem. Scand. A* **33**, 305 (1979).

Kanervo, J.M., Harlin, M.E., Krause, A.O.I., and Bañares, M.A., *Catal. Today* **78**, 171 (2003).

Kanervo, J.M., and Krause, A.O.I., *J. Catal.* **207**, 57 (2002).

Kanervo, J.M., and Krause, A.O.I., *J. Phys. Chem. B* **105**, 9778 (2001).

Kerkhof, F.P.J.M., Moulijn, J.A., and Thomas, R.J., *J. Catal.* **56**, 279 (1979).

Kiefer, W., and Bernstein, H.J., *Appl. Spec.* **25**, 609 (1971).

Kiefer, W., and Bernstein, H.J., *Appl. Spectrosc.* **25**, 500 (1971).

Klisinska, A., Haras, A., Samson, K., Witko, M., and Grzyboska, B., *J. Mol. Catal. A* **210**, 87 (2004).

Kneipp, K., Wang, Y., Dasari, R.R., and Feld, M.S., *Appl. Spectrosc.* **49**, 780 (1995).

Knobl, S., Zenkovets, G.A., Kryukova, G.N., Ovsitser, O., Niemeyer, D., Schlögl, R., and Mestl, G., *J. Catal.* **215**, 177 (2003).

Knoll, P., Singer, R., and Kiefer, W. *Appl. Spectrosc.* **44**, 776 (1990).

Knozinger, H., *Catal. Today* **32**, 71 (1996).

Knözinger, H., *in* "Characterization of Heterogeneous Catalysts Studied by Particle Beams" (H.H. Brongersma and R.A. van Santen, Eds.), p. 167. Plenum Press, New York, (1991).

Knözinger, H., and Jeziorowski, H., *J. Phys. Chem.* **82**, 2002 (1978)

Knözinger, H., and Mestl, G., *Topics Catal.* **8**, 45 (1999).

Knözinger, H., and Ratnasamy, P., *Catal. Rev.–Sci. Eng.* **17**, 31 (1978).

Knözinger, H., and Taglauer, E. *in* "Handbook of Heterogeneous Catalysis" (G. Ertl, H. Knözinger and H. Weitkamp Eds.), Vol. 1, p. 216. Wiley-VCh, Weinheim, (1997).

Knözinger, H., and Taglauer, E., *in* "Catalysis", Vol. 10, p. 1. Royal Society of Chemistry, Cambridge, , 1993.

Koda, S., Kanno, N., and Fijiware, H., *Ind. Eng. Chem. Res.* **40**, 3861 (2001).

Koningstein, J.A., and Gachter, B.F., *J. Opt. Sci. Am.* **63**, 882 (1973).

Korhonen, S.T., Bañares, M.A., Fierro, J.L.G., *krause AOI Catal. Today* **126,** 235 (2007).

Kovats, W.D., and Hill, C.G. Jr., *Appl. Spectrosc.* **40**, 1215 (1986).

Koyano, G., Saito, T., and Misono, M., *Chem. Lett.* 415 (1997).

Kuba, S., and Knözinger, H., *J. Raman Spectrosc.* **33**, 325 (2002).

Landa-Cánovas, A., Nilsson, J., Hansen, S., Stahl, K., and Andersson, A., *J. Solid State Chem.* **116**, 369 (1995).

Larrondo, S., Irigoyen, B., Barenetti, G., and Amadeo, N., *Appl. Catal. A* **250**, 279 (2003).

Lashier, M.E., and Schrader, G.L., *J. Catal.* **128**, 113 (1991).

Le Bihan, L., Blanchard, P., Fournier, M., Grimblot, J., and Payen, E., *J. Chem. Soc. Faraday Trans.* **94**, 937 (1998).

Le Bourdon, G., Adar, F., Moreau, M., Morel, S., Reffner, J., Mamede, A.-S., Dujardin, C., and Payen, E., *Phys. Chem. Chem. Phys.* **5**, 4441 (2003).

Leroy, J.M., Peirs, S.L., and Tridot, G., *Comptes R. Acad. Sci.* **272**, 218 (1971).

Lewandowska, A.E., Calatayud, M., Lozano-Diz, E., Minot, C., Bañares, M.A., *Catalysis Today* **139**, 209 (2008).

Lewis, I.R., and Edwards, H.G.M. (Eds.) "Handbook of Raman Spectroscopy", Marcel Dekker Inc., New York, 2001.

Lewis, I.R., and Edwards, H.G.M., "Handbook of Raman Spectroscopy", Marcel Dekker, New York, 2001.

Leyrer, J., Mey, D., and Knözinger, H., *J. Catal.* **124**, 349 (1990).

Liao, P.F., *in* "Surface Enhanced Raman Scattering" (R.K. Chang and T.E. Furtak, Eds.), p. 379. Plenum Press, New York, 1982.

Li, C., and Stair, P.C., *Catal. Lett.* **36**, 119 (1996).

Li, C., Xiong, G., Xin, Q., Liu, J., Ying, P., Feng, Z., Li, J., Yang, W., Wang, Y., Wang, G., Liu, X., Lin, M., et al., *Angew. Chem.* **111**, 2358 (1999a).

Li, C., Xiong, G., Xin, Q., Liu, J., Ying, P., Feng, Z., Li, J., Yang, W., Wang, Y., Wang, G., Liu, X., Lin, M., et al., *Angew. Chem. Int. Ed.* **38**, 2220 (1999b).

Lin, W.Y., and Frei, H., *J. Am. Chem. Soc.* **124**, 9292 (2002).

Liu, H., Gaigneaux, E.M., Imoto, H., Shido, T., and Iwasawa, Y., *J. Phys. Chem. B* **104**, 2033 (2000).

Liu, K.-L.K., Cheng, L.-H., Sheng, R.-S., and Morris, M.D., *Appl. Spectrosc.* **45**, 1 (1991).

Li, W., Gibbs, G.V., and Oyama, S.T., *J. Am. Chem. Soc.* **120**, 9041 (1998).

Li, W., and Oyama, S.T., *Studs. Surf. Sci. Catal.* **110**, 873 (1997).

Lombardi, J.R., Birke, R.L., Lu, T., and Xu, J., *J. Chem. Phys.* **84**, 4174 (1986).

López-Nieto, J.M., Soler, J., Concepción, P., Herguido, J.H., Menéndez, M., and Santamaría, J., *J. Catal.* **185**, 324 (1999).

Loridant, S., Bonnat, M., and Siebert, E., *Appl. Spectrosc.* **49**, 1193 (1995).

Loridant, S., Marcu, I.C., Bergeret, G., and Millet, J.M.M., *Phys. Chem. Chem. Phys.* **5**, 4384 (2003).

Loudon, R., *Adv. Phys.* **13**, 423 (1964).

Lunsford, J.H., Yang, X., Haller, K., Laane, J., Mestl, G., and Knözinger, H., *J. Phys. Chem.* **97**, 13810 (1993).

Luo, H., and Weaver, M.J., *J. Electroanal. Chem.* **501**, 141 (2001).

Magg, N., Immaraporn, B., Giorgi, J.B., Schröder, T., Bäumer, M., Döbler, J., Wu, Z.L., Kondratenko, E., Cherian, M., Baerns, M., Stair, P.C., Sauer, J., et al., *J. Catal.* **226**, 88 (2004).

Malleswara Rao, T.V., Deo, G., Jehng, J.-M., and Wachs, I.E., *Langmuir* **20**, 7159 (2004).

Mamede, A.-S., Leclercq, G., Payen, E., Granger, P., and Grimblot, J., *J. Mol. Struct.* **651–653**, 353 (2003).

Marcinkowska, K., Adnot, A., Roberge, P.C., and Kaliaguine, S., *J. Phys.Chem.* **90**, 4773 (1986).

Martin, A., Narayana, V., and Lücke, B., *Catal. Today* **78**, 311 (2003).

Martínez-Arias, A., Gamarra, D., Fernández-García, M., Wang, X.Q., Hanson, J.C., and Rodríguez, J.A., *J. Catal.* **240**, 1 (2006).

Martínez-Huerta, M.V., Deo, G., Fierro, J.L.G., and Bañares, M.A., *J. Phys. Chem. C* **111**, 18708 (2007).

Martinez-Huerta, M.V., Coronado, J.M., Fernandez-García, M., Iglesias-Juez, A., Deo, G., Fierro, J.L.G., and Bañares, M.A., *J. Catal.* **225**, 240 (2004).

Martínez-Huerta, M.V., Deo, G., Fierro, J.L.G., and Bañares, M.A., *J. Phys. Chem. C* **112**, 11441–11447, (2008).

Masten, D.A., Foy, B.R., and Harradine, D.M., *J. Phys. Chem.* **97**, 8557 (1993).

McCarron, E.M. III, *J. Chem. Soc. Chem. Commun.* 336 (1986).

McMillan, P., *Am. Miner.* **69**, 622 (1986).
Mehicic, M., and Grasselli, J.G., *in* "Analytical Raman Spectroscopy" (J.G. Grasselli and B.J. Bulkin, Eds.), p. 325. Wiley, Chichester, 1991.
Merkel, S., Goncharov, A.F., Mao, H.-K., Gillet, P., and Hemley, R.J., *Science* **288**, 1626 (2000).
Messerschmidt, R.G., and Chase, D.B., *Appl. Spectrosc.* **43**, 11 (1989).
Mestl, G., *J. Mol. Catal. A* **158**, 45 (2000).
Mestl, G., *J. Raman Spectrosc.* **33**, 333 (2002).
Mestl, G., Ilkenhans, T., Spielbauer, D., Dieterle, M., Timpe, O., Kröhnert, J., Jentoft, F., Knözinger, H., and Schlögl, R., *Appl. Catal. A* **210**, 13 (2001).
Mestl, G., and Knözinger, H., *in* "Handbook of Heterogeneous Catalysis" (G. Ertl, H. Knözinger and J. Weitkamp, Eds.), Vol. 2, p. 539. Wiley-VCh, Weinheim, 1997.
Mestl, G., and Knözinger, H., *Langmuir* **14**, 3964 (1998).
Mestl, G., Knözinger, H., and Lunsford, J.H., *Ber. Bunsen Ges. Phys. Chem.* **97**, 319 (1993).
Mestl, G., Linsmeier, Ch., Gottschall, R., Dieterle, M., Find, J., Herein, D., Jäger, J., Uchida, Y., and Schlögl, R., *J. Mol. Catal. A* **27**, 455 (2000).
Mestl, G., Rosynek, M.P., and Lunsford, J.H., *J. Phys. Chem. B* **101**, 9329 (1997b).
Mestl, G., Rosynek, M.P., and Lunsford, J.H., *J. Phys. Chem. B* **102**, 154 (1998).
Mestl, G., Rosynek, M.P., Lunsford, J.J., *J. Phys. Chem.* **101**, 9321 (1997a).
Mestl, G., Ruiz, P., Delmon, B., and Knözinger, H., *J. Phys. Chem.* **98**, 11269 (1994b).
Mestl, G., Ruiz, P., Delmon, B., and Knözinger, H., *J. Phys. Chem.* **98**, 11276 (1994a).
Mestl, G., Ruiz, P., Delmon, B., and Knözinger, H., *J. Phys. Chem.* **98**, 11283 (1994c).
Mestl, G., and Srinivasan, T.K.K., *Catal. Rev.–Sci. Eng.* **40**, 451 (1998).
Michel, D., Van den Borre, M.T., and Ennaciri, A., *in* "Advances in Ceramics" (S. Somiya, N. Yamamoto and H. Yanagida, Eds.), Vol. 24A, p. 555. Am. Ceram. Soc., Columbus, OH 1989.
Miciukiewicz, J., Mang, T., and Knözinger, H., *Appl. Catal. A* **122**, 151 (1995).
Millar, G.J., Metson, J.B., Bowmaker, G.A., and Clooney, R., *J. Chem. Soc. Faraday Trans.* **91**, 4149 (1995).
Millar, G.J., Nelson, M.L., and Unwins, P.J.R., *Catal. Lett.* **43**, 97–105 (1997).
Montini, T., Bañares, M.A., Hickey, N., Di Monte, R., Fornaisero, P., Kaspar, J., and Graziani, M., *Phys. Chem. Chem. Phys.* **6**, 1 (2004).
Montini, T., Hickey, N., Fornasiero, P., Graziani, M., Bañares, M.A., Martinez-Huerta, M.V., Alessandri, I., and Depero, L.E., *Chem. Mater.* **17**, 1157 (2005).
Morrow, B.A., *in* "Vibrational Spectroscopy of Adsorbed Species, ACS. Symp. Series" (A.T. Bell and M.L. Hair, Eds.), Vol. 137, p. 119. ACS, Washington D.C., (1981).
Moser, T.P., and Schrader, G.L., *J. Catal.* **92**, 216 (1984).
Moskovits, M., *Rev. Mod. Phys.* **57**, 783 (1985).
Mrozek, I., and Otto, A., *J. Electron Spectrosc. Relat. Phenom.* **54**, 859 (1990).
Mul, G., Bañares, M.A., van der Linden, B., Khatib, S.J., and Moulijn, J.A., *Phys. Chem. Chem. Phys.* **5**, 4378 (2003).
Münster, S., Werner, S., Mihalcea, S., Scholz, C., and Oesterschulze, E., *J. Microsc.* **17**, 186 (1997).
Nafie, L.A., *in* "Theory of Raman Scattering", Handbook of Raman Spectroscopy" (I.R. Lewis, and H.G.M. Edwards, Eds.), Chap.1. Marcel Dekker, New York (2001).
Ng, K.Y., and Gulari, E., *Polyhedron* **3**, 1001 (1984).
Nguyen, T.T., *J. Singapore Natl. Acad. Sci.* **10–12**, 84 (1983).
Niemantsverdriet, J.W., "Spectroscopy in Catalysis", VCH, 1993.
Nijhuis, T.A., Tinnemans, S.J., Visser, T., and Weckhuysen, B.M., *Chem. Eng. Sci.* **59**, 5487 (2004).
Nijhuis, T.A., Tinnemans, S.J., Visser, T., and Weckhuysen, B.M., *Phys. Chem. Chem. Phys.* **5**, 4361 (2003).
Noell, W., Abraham, M., Mayr, K., and Ruf, A., *Appl. Phys. Lett.* **70**, 1236 (1997).

Novák, P., Panitz, J.-C., Joho, F., Lanz, M., Imhof, R., and Coluccia, M., *J. Power Sources* **90**, 52 (2000).

Okamoto, Y., and Imanaka, T., *J. Phys. Chem.* **92**, 7102 (1988).

Olson, U.A., and Schrader, G.L., *in* "Thin Film Structures and Phase Stability", (B.M. Clemens and W.L. Johnson, Eds.), Vol. 187, p. 167. Materials Research Society, Pittsburgh PA, 1990.

Olthof, B., Khodakov, A., Bell, A.T., and Iglesia, E., *J. Phys. Chem. B* **104**, 1516 (2000).

Otto, A., *in* "Topics in Applied Physics, Vol 54, Light Scattering in Solids IV", p. 289. Springer, Berlin, 1984.

Ovander, L.N., *Opt. Spectrosc.* **9**, 302 (1960).

Ovsitser, O., Uchida, Y., Mestl, G., Weinberg, G., Blume, A., Jäger, J., Dieterle, M., Hibst, H., and Schlögl, R., *J. Mol. Catal. A* **185**, 291, (2002).

Ozkan, U.S., Driscoll, S.A., and Ault, K., *J. Catal.* **124**, 183 (1990b).

Ozkan, U.S., Moctezuma, E., and Driscoll, S.A., *Appl. Catal.* **58**, 305 (1990a).

Pangier, T., Boulova, M., Galerie, A., Gaskov, A., and Lucazeau, G., *J. Solid State Chem.* **143**, 86 (1999).

Pangier, T., Boulova, M., Galerie, A., Gaskov, A., and Lucazeau, G., *Sens. Actuators B* **71**, 134 (2000).

Payen, E., Barbillat, J., Grimblot, J., and Bonnelle, J.P., *Spectrosc. Lett.* **11**, 997 (1978).

Payen, E., Dhamelincourt, M.C., Dhamelincourt, P., Grimblot, J., and Bonelle, J.P., *Appl. Spectrosc.* **36**, 30 (1980).

Payen, E., Grimblot, J., and Kasztelan, S., *J. Phys. Chem.* **91**, 6642 (1987).

Payen, E., and Kasztelan, S., *Trends Phys. Chem.* **4**, 363 (1994).

Payen, E., Kasztelan, S., Grimblot, J., and Bonnelle, J.P., *J. Mol. Struct.* **143**, 259 (1986).

Payen, E., Kasztelan, S., Grimblot, J., and Bonnelle, J.P., *J. Raman Spectrosc.* **17**, 233 (1986).

Pemberton, J.E. *in* "Electrochemical Interfaces: Modern Techniques for In Situ Interface Characterization" (H.D. Abruna, Ed.), p.193. VCh-Verlag, New York, (1991).

Pettinger, B., *in* "Adsorption of Molecules at Metal Electrodes" (J. Lipowski and P.N. Ross, Eds.), p. 285. VCh-Verlag, New York, 1992.

Pettinger, B., Bao, X., Wilcock, I., Muhler, M., Schlögl, R., and Ertl, G., *Angew. Chem. Int. Ed. Engl.* **33**, 171 (1994).

Pettinger, B., Picardi, G., Schuster, R., and Ertl, G., *Single Mol.* **3**, 285 (2002).

Pettinger, B., Ren, B., Picardi, G., Schuster, R. and Ertl, G., *Phys. Rev. Lett.* **92**, 096101–1 (2004).

Petzodt, J.C., Böhnke, H., Gaube, J.W., and Hibst, H., *in* "Proc. 4th World Congress on Oxidation Catalysis", Potsdam, Germany, Suppl. 16 (2001).

Pieck, C.L., Bañares, M.A., Vicente, M.A., and Fierro, J.L.G., *Chem. Mater.* **13**, 1174 (2001).

Pohl, D.W., Denk, W., and Lanz, M., *Appl. Phys. Lett.* **44**, 651 (1984).

Poliakoff, M., Howdle, S.M., and Kazarfian, S.G., *Angew. Chem. Int. Ed. Eng.* **34**, 1275 (1995).

Puppels, G.J., Colier, W., Olmikhof, J.H.F., Otto, C., de Mul, F.F.M., and Greve, J., *J. Raman Spectrosc.* **22**, 217 (1991).

Puppels, G.J., de Mul, F.F.M., Otto, C., Greve, J., Robert-Nicould, M., Arndt-Jovin, D.J., and Jovin, T., *Nature* **347**, 301 (1990).

Pushkarev, V.V., Kovalchuk, V.I., and D'Itri, J.L., *J. Phys. Chem. B* **108**, 5341 (2004).

Puurunen, R., Beheydt, B.G., and Weckhyuysen, B.M., *J. Catal.* **204**, 253 (2001).

Puurunen, R., and Weckhuysen, B.M., *J. Catal.* **210**, 418 (2002).

Radhakrisnan, R., and Oyama, S.T., *J. Catal.* **199**, 282 (2001).

Reinke, M., Mantzaras, J., Schaeren, R., Bombach, R., Inauen, A., and Schenker, S., *Combust. Flame* **136**, 217 (2004).

Ricchiardi, G., Damin, A., Bordiga, S., Lamberti, C., Spano, G., Rivetti, F., and Zecchina, A., *J. Am. Chem. Soc.* **123**, 11409 (2001).

Rice, S.F., Hunter, T.B., Ryden, A.C., and Hanush, R.G., *Ind. Eng. Chem. Res.* **35**, 2161 (1996).

Richter, F., Papp, H., Götze, Th., Wolf, G.U., and Kubias, B., *Surf. Interf. Anal.* **26**, 736 (1998).

Roark, R.D., Kohler, S.D., and Ekerdt, J.G., *Catal. Lett.* **16**, 71 (1992).

Roark, R.D., Kohler, S.D., Ekerdt, J.G., Kim, D.S., and Wachs, I.E., *Catal. Lett.* **16**, 77 (1992).

Rocchiccioli-Deltcheff, C., Amirouche, M., Che, M., Tatibouet, J.M., and Fournier, M., *J. Catal.* **125**, 2892 (1990).

Roozeboom, F., Medema, J., and Gellings, P.J., *Z. Phys. Chem.* **111**, 215 (1978).

Roozeboom, F., Robson, H., and Chan, S.S., *Zeolites* **3**, 321 (1983).

Rosasco, G.J., *in* "Advances in Infrared and Raman Spectroscopy" (R. Clark, and R.E. Hester, Eds.), Vol. 7, p. 223. Heyden and Son, London, 1980.

Rybaczyk, P., Berndt, H., Radnik, J., Pohl, M.M., Buyevskaya, O., Baerns, M., and Brückner, A., *J. Catal.* **202**, 45 (2001).

Santamaría-González, J., Martínez-Lara, M., Bañares, M.A., Martínez-Huerta, M.V., Rodríguez-Castellón, E., Fierro, J.L.G., and Jiménez-López, A., *J. Catal.* **181**, 280 (1999).

Schlücker, S., Schäberle, M.D., Huffman, S.W., and Levin, I.W., *Anal. Chem.* **75**, 4312 (2003).

Schrader, B., *Fresenius J. Anal. Chem.* **337**, 824 (1990).

Schrader, G.L., Batista, M.L., and Bergman, C.B., *Chem. Eng. Commun.* **12**, 121 (1981).

Schrader, G.L., and Cheng, C.P., *J. Catal.* **80**, 369 (1983).

Schrader, G.L., and Hill, C.G., *Rev. Sci. Instrum.* **46**, 1335 (1975).

Segawa, K., and Wachs, I.E., *in* "Characterization of Catalytic Materials" (I.E. Wachs, Ed.), Butterworth-Heinemann, Boston, (1992).

Sharma, S.K., Mao, H.-K., Bell, P.M., and Xu, J., *J. Raman Spectrosc.* **16**, 350 (1985).

Sheppard, N., and Erkelens, J., *Appl. Spectrosc.* **38**, 471 (1984).

Shih, H.E., Jonas, F., and Marcus, P.M., *Phys. Rev. Lett.* **46**, 731 (1981).

Si-Ahmed, H., Calatayud, M., Minot, C., Lozano-Diz, E., Lewandowska, A.E., Bañares, M.A., *Catal. Today* **126**, 96 (2007).

Skinner, J.G., and Nilsen, W.G., *J. Opt. Soc. Am.* **58**, 113 (1968).

Snyder, T.P., and Hill, C.G. Jr., *J. Catal.* **132**, 536 (1991).

Sommer, A.J., and Katon, J.E., *Appl. Spectrosc.* **45**, 527 (1991).

Somorjai, G.A., *Cattech* **3**, 84 (1999).

Somorjai, G.A., *Chem. Rev.* **96**, 1223 (1996).

Somorjai, G.A., and Rupprechter, G., *J. Phys. Chem. B* **103**, 1623 (1999).

Spengler, J., Anderle, F., Bosch, E., Grasselli, R.K., Pillep, B., Behrens, P., Lapina, O.B., Shubin, A.A., Eberle, H.-J., and Knözinger, H., *J. Phys. Chem. B* **105**, 10772 (2001).

Spevack, P.A., and McIntyre, N.S., *J. Phys. Chem.* **97**, 11020 (1997).

Spielbauer, D., *Appl. Spectrosc.* **49**, 650 (1995).

Stair, P., *Adv. Catal.* **51**, 75 (2007).

Stair, P.C., *Curr. Opin. Surf. Sci. Mat. Sci.* **5**, 365 (2001).

Steeper, R.R., Rice, S.F., Kennedy, I.M., and Aiken, J.D., *J. Phys. Chem.* **100**, 184 (1996).

Stencel, J.M., "Raman Spectroscopy for Catalysis", Van Nostrand, Reinhold, New York, (1990).

Stencel, J.M., Diehl, J.R., D'Este, J.R., Makowsky, L.E., Rodrigo, L., Marcinkowska, K., Adnot, A., Roberge, P.C., and Kaliaguine, S., *J. Phys. Chem.* **90**, 4739 (1986).

Stencel, J.M., Makovsky, L.E., Sarkus, T.A., De Vries, J., Thomas, R., and Moulijn, J.A., *J. Catal.* **90**, 314 (1984).

Sun, Q., Jehng, J.-M., Hu, H., Herman, R.G., Wachs, I.E., and Klier, K., *J. Catal.* **165**, 101 (1997).

Takenaka, T., *Adv. Colloid. Interface Sci.* **11**, 291 (1979).

Talyzin, A.V., Dubrovinsky, L.S., Le Bihan, T., and Jansson, U., *Phys. Rev. B* **65**, 245413 (2002).

Tanaka, Y., Fukuzawa, K., and Kuwano, H., *J. Appl. Phys.* **83**, 3547 (1998).

Tatibouët, J.M., Che, M., Amirouche, M., Fournier, M., and Rocchiccioli-Deltcheff, C., *J. Chem. Soc. Chem. Commun.* 1260 (1988).

Tejeda, G., Maté, B., Fenández-Sánchez, J.M., and Montero, S., *Phys. Rev. Lett.* **76**, 34 (1996).

Thomas, J.M., *Angew. Chem. Intl. Ed.* **38**, 3588 (1999).

Thomas, J.M., *in* "Characterization of Catalysts" (J.M. Thomas and R.M. Lambert, Eds.), Wiley, New York 1980.

Thomas, J.M., and Somorjai, G.A., *Topics Catal.* **8**, (1/2) (1999).

Thomas, R.J., Moulijn, J.A., and Kerkhof, F.P.J.M., *Recl. Trav.Chim. Pays-Bas* **96**, M114 (1977).

Tian, H., Ross, E.I., and Wachs, I.E., *J. Phys. Chem. B* **110**, 9593 (2006).

Tian, H., Wachs, I.E., and Briand, L.E., *J. Phys. Chem. B* **109**, 23491 (2005).

Tibiletti, D., Goguet, A., Meunier, F.C., Breen, J.P., and Burch, R., *Chem. Commun.* 1636 (2004).

Tinnemans, S.J., Mesu, J.G., Kervinen, K., Visser, T., Jijhuis, T.A., Beale, A.M., Keller, D.E., van der Eerden, A.M.J., and WEcjhysen, B.M., *Catal. Today* **113**, 3 (2006).

Toledo-Crow, R., Yang, P.C., Chen, Y., and Vaez-Iravani, M., *Appl. Phys. Lett.* **60**, 2957 (1992).

Tolia, A.A., Weaver, M.J., and Takoudis, C.G., *J. Vac. Sci. Technol. A* **11**, 2013 (1993).

Tolia, A.A., Williams, C.T., Takoudis, C.G., and Weaver, M.J., *J. Phys. Chem.* **99**, 4599 (1995).

Topsøe, H., *Stud. Surf. Sci. Catal.* **130**, 1–21 (2000).

Tozzola, G., Mantegazza, A., Ranghino, G., Petrini, G., Bordiga, S., Ricchiardi, G., Lamberti, C., Zulian, R., and Zecchina, A., *J. Catal.* **179**, 64 (1998).

Treado, P.J., Govil, A., Morris, M.D., Sternitzke, K.D., and McCreery, R.L., *Appl. Spectrosc.* **44**, 1270 (1990).

Treado, P.J., and Morris, M.D., *Appl. Spectrosc.* **44**, 1 (1990).

Treado, P.J., and Morris, M.D., *Appl. Spectrosc. Rev.* **29**, 1 (1994).

Treado, P.J., and Nelson, M.P., *in* "Handbook of Raman Spectroscopy" (I.R. Lewis and H.G. M. Edwards, Eds.) Chapter 5, Marcel Dekker Inc., New York, 2001.

Tuinstra, F., and Koenig, J.L., *J. Chem. Phys.* **33**, 1126 (1970).

Twu, J., Dutta, P.K., and Kresge, C.T., *J. Phys. Chem.* **95**, 5267 (1991).

Tytko, K.-H., Baethe, G., Hirschfeld, E.R., Mehrnke, J., and Stellhorn, D., *Z. Anorg. Allg. Chem.* **503**, 43 (1983).

Tytko, K.-H., and Glemser, O., *Adv. Inorg. Chem. Radiochem.* **19**, 239 (1976).

Tytko, K.-H., and Mehrnke, J., *Z. Anorg. Allg. Chem.* **503**, 67 (1983).

Tytko, K.-H., Petridis, G., and Schönfeld, B., *Z. Naturforsch. B* **35**, 45 (1980).

Tytko, K.-H., and Schönfeld, B., *Z. Naturforsch. B* **30**, 63, (1975).

Tytko, K.-H., Schönfeld, B., Cordis, V., and Glemser, O., *Z. Naturforsch. B* **30**, 834, (1975).

Ueda, W., Moro-oka, Y., and Ikawa, T., *J. Chem. Soc. Faraday Trans. I* **78**, 495 (1982).

Ueno, A., *in* "Catalysis" (J.J. Spivey, Ed.), Vol. 15, Cambridge, United Kingdom, 2000.

Uy, D., O'Neill, A.E., Li, J., and Watkins, W.L.H., *Catal. Lett.* **95**, 191 (2004).

Uy, D., O'Neill, A.E., and Weber, W.H., *Appl. Catal. B: Environ.* **35**, 219 (2002).

Uy, D., Wiegand, K.Q., O'Neill, A.E., Dearth, M.A., and Weber, W.H., *J. Phys. Chem. B* **106**, 387 (2002).

Valaskovic, G.A., Holton, M., and Morrison, G.H., *Appl. Opt.* **34**, 1215, (1995).

Vèdrine, J.C., and Derouane, E.G., *in* "Combinatorial Catalysis and High Throughput Catalyst Design and Testing", (E.G. Derouane, Ed.), Kluwer Academic Publishers, New York, (2000).

Vidano, N., and Fischbach, D.B., *J. Am. Ceram. Soc.* **61**, 13 (1978).

Viers, D.K., Ager, J.W., Loucks, E.T., and Rosenblatt, G.M., *Appl. Opt.* **29**, 4969 (1990).

Villa, P.L., Trifirò, F., and Pasquon, I., *React. Kinet. Catal. Lett.* **1**, 341 (1974).

Vo-Dinh, T., *Trends Anal. Chem.* **17**, 557 (1998).

Vogel, E., Kiefer, W., Deckert, V., and Zeisel, D., *J. Raman Spectrosc.* **29**, 693 (1998).

Volta, J.C., Bere, K., Zhang, Y.J., Olier, R., *ACS Symp. Ser.* **523**, 217 (1992).

Vuurman, M.A.,Stufkens, D.J., Oskam, A., Deo, G., and Wachs, I.E., *J. Chem. Soc. Faraday Trans.* **92**, 3259 (1996).

Vuurman, M., Wachs, I.E., and Hirt, A.M., *J. Phys. Chem.* **95**, 9928 (1991).

Wachs, I.E., *Catal. Commun.* **4**, 567 (2003a).

Wachs, I.E., *Catal. Today* **100**, 79 (2005).

Wachs, I.E., *Catal. Today* **27**, 437 (1996).

Wachs, I.E., *CATTECH.* **7**, 142 (2003b).

Wachs, I.E., *in* "Handbook of Raman Spectroscopy" (I.R. Lewis, Ed.), Marcel Dekker, (2002).

Wachs, I.E., *in* "Raman Scattering in Materials Science" (W.H. Weber and R. Merlin, Eds.), p. 271. Springer, Berlin, 2000.

Wachs, I.E., *Surf. Sci.* **544**, 1 (2003c).

Wachs, I.E., *Topics Catal.* **8**, 57 (1999).

Wachs, I.E., and Briand, L.E., *US Patent* 2002/0062048, 2002a.

Wachs, I.E., Briand, L.E., *WO Patent* WO 2002/022539, 2002b.

Wachs, I.E., and Cai, Y., *U.S. Patent*, US 6245708, 2001.

Wachs, I.E., Deo, G., Vuurman, M., Hu, H., Kim, D.S., and Jehng, J.-M., *J. Mol. Catal.* **82**, 443 (1993).

Wachs, I.E., Deo, G., Weckhuysen, B.M., Andreini, A., Vuurman, M.A., de Boer, M., and Amiridis, M.D., *J. Catal.* **161**, 211 (1996).

Wachs, I.E., and Hardcastle, F.D., *in* "Catalysis" Vol. 10, p. 102. The Royal Society of Chemistry, Cambridge, 1993.

Wachs, I.E., Hardcastle, F.D., and Chan, S.S., *Spectroscopy* **1**, 30 (1986).

Wachs, I.E., Jehng, J. M., Deo, G., Weckhuysen, B.M., Guliants, V., and Benziger, J.B., *Catal. Today* **32**, 47 (1996).

Wachs, I.E., Jehng, J.-M., Deo, G., Weckhuysen, B.M., Guliants, V.V., Benziger, J.B., and Sundaresan, S., *J. Catal.* **170**, 75 (1997).

Wang, C.-B., Cai, Y., and Wachs, I.E., *Langmuir* **15**, 1223 (1999).

Wang, C.-B., Deo, G., and Wachs, I.E., *J. Phys. Chem. B* **103**, 5645 (1999).

Wang, J., Xu, X., Deng, J., Liao, Y., and Hong, B., *Appl. Surf. Sci.* **120**, 99 (1997).

Wang, L., and Hall, K.W., *J. Catal.* **82**, 177 (1983).

Wang, L., and Hall, W.K., *J. Catal.* **66**, 251 (1980).

Wang, L., and Hall, W.K., *J. Catal.* **83**, 242 (1983).

Wang, X., and Wachs, I.E., ACS National Meeting, Book of Abstracts **229**, p. COLL-196 (2004).

Wang, Y., Li, Y.S., Zhang, Z.X., and An, D.Q., *Spectrochim Acta* **59**, 589 (2003).

Wan, Y.S.S., Chau, J.L.H., Yeung, K.L., and Gavriilidis, A., *Microporous Mesoporous Mater.* **42**, 157 (2001).

Watanabe, M., and Ogawa, T., *Jpn. J. Appl. Phys.* **27**, 1066 (1988).

Waterhouse, G.I.N., Bowmaker, G.A., and Metson, J.B., *Appl. Surf. Sci.* **214**, 36 (2003).

Weaver, M.J., *J. Raman Spectrosc.* **33**, 309 (2002).

Weber, W.H., "Raman Scattering in Materials Science", *Springer Ser. Mater. Sci.*, **42**, 233 (2000a).

Weber, W.H., *Mater. Sci.* **42**, 233 (2000b).

Webster, S., Smith, D.A., and Batchelder, D.N., *Vib. Spectrosc.* **18**, 51 (1998).

Weckhuysen, B.M., *Chem. Commun.* 97, (2002) (Feature Article).

Weckhuysen, B.M., *Phys. Chem. Chem. Phys.* **5**, 4351 (2003).

Weckhuysen, B.M., Jehng, J.-M., Wachs, I.E., *J. Phys. Chem B* **104**, 7382 (2000).

Weckhuysen, B.M., Schoonheydt, R.A., Jehng, J.-M., Wachs, I.E., Cho, S.J., Ryoo, R., Kijlstra, S., and Poels, E., *J. Chem. Soc. Faraday Trans.* **91**, 3245 (1995).

Weckhuysen, B.M., and Wachs, I.E., *J. Phys. Chem. B* **100**, 14437 (1996).

Weinstein, B.A., and Piermarini, G.J., *Phys. Rev. B* **12**, 1172 (1975).

Williams, C.C., Ekerdt, J.G., Jehng, J.-M., Hardcastle, F.D., and Wachs, I.E., *J. Phys. Chem.* **95**, 8791 (1991).

Williams, C.T., Takoudis, C.G., and Weaver, M.J., *J. Phys. Chem. B* **102**, 406 (1998).

Williams, C.T., Tolia, A.A., Weaver, M.J., and Takoudis, C.G., *Chem. Eng. Sci.* **51**, 1673 (1996).

Wilson, J.A., Hill, C.G., and Dumesic, J.A., *J. Mol. Catal. A* **61**, 333 (1990).

Wolf, G.H., Chizmeshaya, A.V.G., Diefenbacher, J., and McKelvy, M.J., *Env. Sci. Technol.* **38**, 932 (2004).

Woo, S.I., and Hill, C.G., *J. Mol. Catal.* **15**, 309 (1982).

Woo, S.I., and Hill, C.G., *J. Mol. Catal.* **24**, 165 (1984).

Woo, S.I., and Hill, C.G., *J. Mol. Catal.* **29**, 231 (1985).

Wu, Z., Kim, H.S., Stair, P.C., Rugmini, S., and Jackson, S.D., *J. Phys. Chem. B* **109**, 2793 (2005).

Xie, S., and Bell, A.T., *Catal. Lett.* **70**, 137 (2000).

Xie, S., Iglesia, E., and Bell, A.T., *J. Phys. Chem. B* **105**, 5144 (2001).

Xie, S., Iglesia, E., and Bell, A.T., *Langmuir* **16**, 7162 (2000).

Xie, S., Mestl, G., Rosynek, M.P., and Lunsford, J.H., *J. Am. Chem. Soc.* **119**, 10186 (1997).

Xie, S., Rosynek, M.P., and Lunsford, J.H., *Appl. Spectrosc.* **53**, 1183 (1999).

Xie, Y.C., and Tang, Y.Q., *Adv. Catal.* **37**, 1 (1999).

Xiong, G., Sullivan, V.S., Stair, P.C., Zajac, G.W., Trail, S.S., Kaduk, J.A., Golab, J.T., and Brazdil, J.F., *J. Catal.* **230**, 317 (2005).

Xue, Z.-Y., and Schrader, G.L., *J. Phys. Chem. B* **103**, 9459 (1999).

Yeung, K.L., and Yao, N., *J. Nanosci. Nanotechnol.* **4**, 1 (2004).

Yeung, K.L., Zhang, X.F., Lau, W.N., and Martin-Aranda, R., *Catal. Today* **110**, 26 (2005).

Zaera, F., *Prog. Surf. Sci.* **69**, 1 (2001).

Zaera, F., *Surf. Sci.* **55**, 947 (2002).

Zeisel, D., Deckert, V., Zenobi, R., and Vo-Dinh, T., *Chem. Phys. Lett.* **283**, 381 (1998).

Zhang, W., Desikan, A., and Oyama, S.T., *J. Phys. Chem.* **99**, 14468 (1995).

Zhao, C., and Wachs, I.E., *Catal. Today* **118**, 332 (2006)

Zhao, Z., Gao, X., and Wachs, I.E., *J. Phys. Chem. B* **107**, 6333 (2003).

Zimmerer, N., and Kiefer, W., *Appl. Spectrosc.* **28**, 279 (1974).

CHAPTER 3

Ultraviolet–Visible–Near Infrared Spectroscopy in Catalysis: Theory, Experiment, Analysis, and Application Under Reaction Conditions

Friederike C. Jentoft

Abstract

This review focuses on the application of ultraviolet–visible–near infrared (UV–vis–NIR) spectroscopy for the investigation of solid catalysts under operating conditions. The differences between measurements made in transmission mode and in reflection mode are summarized, and the corresponding equations for data analysis are introduced. The Kubelka–Munk function is derived, and its usage is discussed critically. The advantages of various optical accessories such as integrating spheres, mirror optics, and fiber optics are contrasted. Designs of cells for treatment of catalysts at various temperatures and in various environments are described. An overview of investigations of catalysts during and after various treatments is given, and investigations of catalysts under reaction conditions are emphasized. The advantages of using UV–vis spectroscopy in combination with other spectroscopic methods in a single apparatus are illustrated.

Contents

1. Introduction   132
   1.1. Ultraviolet–Visible–Near Infrared (UV–vis–NIR)
        Spectroscopy in Catalysis   132

Fritz-Haber-Institut der Max-Planck-Gesellschaft, Department of Inorganic Chemistry, 14195 Berlin, Germany
New address: University of Oklahoma, School of Chemical, Biological, and Materials Engineering, 100 East Boyd St, Norman, OK 73019-1004, USA, email: fcjentoft@ou.edu

Advances in Catalysis, Volume 52
ISSN 0360-0564, DOI: 10.1016/S0360-0564(08)00003-5

129

1.2. History and Previous Reviews 132
1.3. Scope and Outline 134
2. Theoretical Background 134
2.1. Transmission Versus Reflection Spectroscopy 134
2.2. Transmission and the Lambert–Beer Law 135
2.3. Reflection and the Kubelka–Munk Function 137
2.4. Use of the Kubelka–Munk Function 141
2.5. Referencing Diffuse Reflectance Spectra 145
2.6. One or Multiple Absorbing Species—
     Transmission Versus Diffuse Reflection 146
3. Experimental: Optical Accessories for UV–vis–NIR
   Spectroscopy in Transmission Mode 148
3.1. Optical Accessory Design for Transmission UV–
     vis–NIR Spectroscopy 148
3.2. Optical Accessories for Transmission UV–vis–
     NIR Spectroscopy 148
4. Experimental: Optical Accessories for UV–vis–NIR
   Spectroscopy in Reflection Mode 149
4.1. Optical Accessory Design for Diffuse Reflectance
     UV–vis–NIR Spectroscopy 149
4.2. Standard Materials for Diffuse Reflectance 150
4.3. Diffuse Reflectance Spectroscopy Using
     Integrating Spheres 153
4.4. Diffuse Reflectance Spectroscopy with Mirror
     Optics Attachments 155
4.5. Diffuse Reflectance Spectroscopy with
     Fiber Optics 158
4.6. Diffuse Reflectance Spectroscopy with
     Combined Optical Elements 159
5. Cells for Measurements in Controlled Environments 160
5.1. Design Parameters 160
5.2. Cells for Transmission Mode Spectroscopy 161
5.3. Cells for Use with Integrating Spheres 161
5.4. Cells for Use with Mirror Optics 163
5.5. Cells for Use with Fiber Optics 164
5.6. Cells for Combination of UV–vis Spectroscopy
     with Other Methods 165
5.7. Spectrometer Requirements 166
6. Data Acquisition and Analysis 167
6.1. Reference Measurement (Background
     Correction) 167
6.2. Sample Preparation 171
6.3. Handling and Representation of Spectra 172
6.4. Quantification by Independent Calibration 175
7. Examples 176
7.1. Room-Temperature Measurements after
     Catalyst Treatment 176
7.2. Spectroscopic Measurements during Treatments
     (In the Absence of Catalysis) 184

7.3. Spectroscopic Measurements under Catalytic
    Reaction Conditions   189
7.4. Simultaneous Acquisition of Spectroscopic
    and Catalytic Data   195
7.5. Correlation of UV–vis Spectral Features with
    Other Simultaneously Acquired Spectra   200
8. Conclusions   203
8.1. Experimental Limitations of Applying
    UV–vis–NIR Spectroscopy under Catalytic
    Reaction Conditions   203
8.2. Evaluation and Outlook   204
Acknowledgments   205
References   205

ABBREVIATIONS

A_e	napierian absorbance
A_{10}	decadic absorbance
ASTM	American Society for Testing and Materials
CIE	International Commission on Illumination
CT	charge transfer
d	particle diameter
DR	diffuse reflectance
ε	molar decadic absorption coefficient ($m^2\,mol^{-1}$)
f and F	selected areas of integrating sphere
$F(\rho)$	Kubelka–Munk or remission function
I, I_0	intensity
IVCT	inter valence charge transfer
k	absorption coefficient (cm^{-1} or m^{-1})
κ	molar napierian absorption coefficient ($m^2\,mol^{-1}$)
l	layer thickness (m)
λ	wavelength
LMCT	ligand-to-metal charge transfer
M	sphere multiplier
n_i	refractive index of species i
NIR	near infrared
r	particle radius
R_{reg}	reflectivity
ρ	reflectance or reflectance factor
σ	scattering coefficient (cm^{-1} or m^{-1})
τ	transmittance or transmission factor
UV	ultraviolet
vis	visible.

1. INTRODUCTION

1.1. Ultraviolet–Visible–Near Infrared (UV–vis–NIR) Spectroscopy in Catalysis

UV–vis–NIR spectroscopy is a valuable tool for characterization of solid catalysts on the basis of measurements of electronic and vibrational transitions. The following information may be obtained from the spectra:

- oxidation states and coordination of metals from metal-centered and charge transfer transitions;
- dispersions of supported oxide moieties from charge transfer transitions;
- particle sizes and shapes of dispersed (coinage) metals from surface plasmon resonance;[1]
- band gaps of semiconductors from analysis of the absorption edge;
- nature of surface functional groups from overtone and combination modes of their vibrations.

Adsorbates can also be detected: probes, reactants, products, and reaction intermediates—provided that they feature a chromophoric group or a vibrational pattern with pronounced overtone and combination modes.

The usefulness of UV–vis spectroscopy in catalysis research may be limited for various reasons. For example, the species of interest may not absorb sufficiently, or they may exhibit absorptions that are unspecific, or their absorption bands may overlap with other features.

Molar absorption coefficients can vary by several orders of magnitude; hence the sensitivity of the method varies strongly with the nature of the absorbing species. An estimate of the relative concentrations of multiple species cannot be made without knowledge of the ratio of the respective molar absorption coefficients. Quantification of absorbing species in or on a solid catalyst sample by diffuse reflectance spectroscopy is more difficult than quantification of components of nonscattering media by transmission spectroscopy. Nonetheless, a strong point of UV–vis–NIR spectroscopy in catalysis research is that spectra of a solid can be recorded in the presence of gas or liquid phases over a wide range of temperatures and with a time resolution of less than a second. The method is thus highly suitable for the investigation of catalysts as they work.

1.2. History and Previous Reviews

Optical spectroscopy of surface-coordinated species goes back to the work of de Boer (de Boer and Teves, 1930, 1931; de Boer and Broos, 1932; de Boer, 1932; de Boer and Dippel, 1933; de Boer and Custers, 1933, 1934; Custers

---

[1] Plasmons are quanta of plasma vibrations, which are collective excitations of electrons.

and de Boer, 1934; de Boer et al., 1934; Custers and de Boer, 1936; de Boer, 1938), who adsorbed alkali atoms, halogens, and aromatics on salt films. The general spectroscopic phenomena resulting from adsorption were discussed by Terenin (1964). This author addressed mainly the interactions of aromatic compounds with silica and silica–alumina surfaces, even at slightly elevated temperatures (373–423 K) and at low temperatures (70 K). Electron paramagnetic resonance spectroscopy was used as a second technique to corroborate the interpretations. An article published by Klier (1968) focused on fundamentals and application of the diffuse reflectance technique. Klier's examples concern the analysis of the electronic structures of solids and the nature of surface-coordinated species. The most comprehensive treatment of reflectance spectroscopy and its applications was compiled by Kortüm in his book (1969). The subject of diffuse reflectance spectroscopy was also integrated into a book on spectroscopy in catalysis (Kellermann, 1979); in one of the chapters, the essential parts of Kortüm's book are presented and the requirements for application of the Schuster–Kubelka–Munk theory are listed, supplemented by explanations. References to representative work in catalysis are given, and detailed discussions address the spectra of transition metal ions in zeolites, the nature and mobility of adsorbed water as deduced from NIR spectra, and the determination of the band gap. Stone (1983) advocated the use of UV–vis spectroscopy with a clear description of theory and experiment. A review by Schoonheydt (1984) focuses, after a brief presentation of theory and experiment, on the interpretation of spectra of transition metal ions with the aid of ligand field theory. A short introduction into NIR spectroscopy is also given. The spectral features of bulk and supported oxides of a number of elements are discussed. Garbowski and Praliaud (1994) presented a short version of the essentials and a few examples. The 1997 edition of the *Handbook of Heterogeneous Catalysis* includes a chapter on UV–vis–NIR and EPR spectroscopies by Che and Bozon-Verduraz (1997), which was revised by Sojka et al. (2008) for the second edition of the handbook in 2008. The types of transitions are described, followed by a number of examples concerning the various stages of catalyst preparation as well as the interaction with reactants. Weckhuysen and Schoonheydt (2000) briefly contrasted the properties of integrating spheres, mirror optics, and fiber optics with respect to their use for measurements under catalytic reaction conditions. The major part of the article is devoted to the crystal field, ligand field, and molecular orbital theories; and examples are presented concerning supported chromium, copper, and cobalt species.

Investigations under catalytic reaction conditions have moved into focus only in the past years. An article by Weckhuysen (2004) presents instrumental detail and several case studies involving supported chromium, copper, and cobalt species as catalysts.

1.3. Scope and Outline

Fundamentals of transitions in the UV–vis region are described in spectroscopy textbooks (Henderson and Imbusch, 1989) and have been addressed in previous articles in the catalysis literature (Section 1.2). The first part of the present article provides the theoretical background necessary to understand the various kinds of equipment available for measurement of reflection spectra, as well as how to record spectra and analyze data. As the focus of this volume is the characterization of catalysts in the working state, instrumentation and cells that allow such experiments are emphasized in this chapter, followed by a brief description of data acquisition and analysis. Examples start with treatments of materials in controlled gas atmospheres and at various temperatures and continue with the characterization of working catalysts. Finally, simultaneous applications of UV–vis spectroscopy and other methods are summarized.

2. THEORETICAL BACKGROUND

2.1. Transmission Versus Reflection Spectroscopy

It is usually considered more difficult to evaluate and quantify diffuse reflectance data than transmission data, because the reflectance is determined by two sample properties, namely, the scattering and the absorption coefficient, whereas the transmission is assumed to be determined only by the absorption coefficient. The absorbance is a linear function of the absorption coefficient, but its counterpart in reflection spectroscopy, the Kubelka–Munk function (sometimes also called remission[2] function), depends on both the scattering and the absorption coefficient. Often, researchers list a number of prerequisites for application of the Kubelka–Munk function, but, in contrast, transmittance is routinely converted without comment into absorbance.

To emphasize the similarities and differences between spectroscopy in transmission and reflection, both configurations are described in the following sections. The absorbance and the Kubelka–Munk functions are derived.

---

[2] From Latin "remittere", to send back.

2.2. Transmission and the Lambert—Beer Law

Figure 1 is a summary of the phenomena that can occur when a sample is illuminated (infrared range: irradiated). In a transmission measurement, the ratio of transmitted light (transmitted radiation) I to the incident light I_0 is recorded. This quantity is called τ, the transmittance or transmission factor. The desired quantity is the light absorbed by the sample, and it can be determined directly from the transmittance only in the absence of additional phenomena that reduce the transmitted light, such as luminescence, scattering, and regular reflection. Luminescence depends strongly on the chemical properties of the sample and occurs only in a fraction of compounds. Scattering is generally defined as the deflection of electromagnetic or corpuscular radiation from its original direction. Light scattering is negligible in molecularly dispersed media (solutions) but is significant in colloidal dispersions or in powders. The scattered intensity depends strongly on the relative dimensions of wavelength and particle size; and scattering can be reduced by embedding a powder in a material of similar refractive index (cf. the use of KBr wafers and Nujol immersion techniques in IR spectroscopy). Regular reflection occurs at all phase boundaries and amounts to only a few percent of the incident light, depending on the ratio of the refractive indices. The regular reflection can be eliminated mathematically after measurements are made of samples of various thicknesses; the amount of absorbed light varies, whereas the number of phase boundaries and thus the loss of light as a result of regular reflection remain constant. The regular reflection can also be eliminated experimentally by use of a reference specimen that has similar phase boundaries as the sample (e.g., a reference cuvette or a pure KBr wafer).

If luminescence, scattering, and regular reflection can be neglected or eliminated, then the absorption spectrum can be calculated from

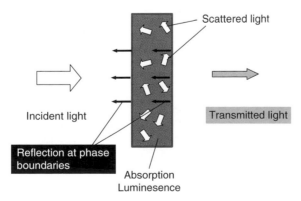

FIGURE 1   Fundamental phenomena resulting from the interaction of light with matter.

the transmittance. This simplification is not always possible; self-supporting wafers of fine catalyst powders, for example, often scatter significantly, so that absorption properties cannot be calculated directly from the transmittance.

To derive the relationship between transmittance and absorbance, the sample is considered to be sliced into infinitesimally thin layers, as shown in Figure 2. In each layer, the light intensity decreases, and the loss dI is proportional to the intensity of the incident light I, the layer thickness dl, and the absorption properties of the sample, which can be expressed by an absorption coefficient k.

$$dI = -Ik\,dl \tag{1}$$

Separation and integration from I_0 (intensity of incident light) to I (the intensity of transmitted light) and over the total sample thickness results in the following:

$$\int_{I_0}^{I} \frac{dI}{I} = \int_{0}^{l} k\,dl \tag{2}$$

The absorption coefficient can be expressed as the product of an absorber-specific, wavelength-dependent quantity, the molar absorption coefficient, and the concentration of this species. Integration gives the following:

$$\ln \frac{I}{I_0} = \ln \tau = -\kappa c l = A_e \tag{3}$$

with κ being the molar napierian absorption coefficient and c the molar concentration of the absorbing species. This relationship is the well-known Lambert–Beer law in the form using the natural logarithm and the napierian absorbance A_e. The common (decadic) logarithm is often the default in commercial spectroscopy software, yielding the decadic absorbance A_{10}, and making it necessary to use the decadic molar absorption coefficient ε:

$$\log \frac{I}{I_0} = \log \tau = -\varepsilon c l = A_{10} \tag{4}$$

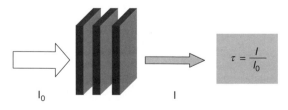

FIGURE 2   Model for the derivation of the Lambert–Beer law.

The absorbance can give an immediate indication of the quantitative composition because of the linear dependence on concentration. However, this is the case only when none of the above-mentioned phenomena interferes significantly and the product of κ (or ε) with c is not too large (this is the limiting case for weak absorption).

2.3. Reflection and the Kubelka–Munk Function

Because specimens cannot be infinitely thin, and dilution is not an option for reactive samples or atmospheres, it can become necessary to resort to reflection spectroscopy. It is possible to gain information about the absorption properties of a material by analyzing the reflected light. There are two limiting cases (Figure 3).

1. "Mirror-type" (polished) surfaces show ideal regular reflection (also known as specular reflection, mirror reflection, and surface reflection) according to the laws of geometric optics. Regularly reflected light is thus most intense in forward directions at an azimuth of 180° (Figure 3A and B). The reflecting power is called reflectivity. The reflectivity is high in the spectral range where absorption is high, as is shown by the Fresnel equations for absorbing media in the case of normal incidence (Klier, 1968; Wendlandt and Hecht, 1966):

$$R_{reg} = \frac{(n_1 - n_0)^2 + (n_1 \kappa_1)^2}{(n_1 - n_0)^2 + (n_1 \kappa_1)^2} \tag{5}$$

with n_0 and n_1 being the refractive indices of the surrounding medium and the sample and κ_1 the absorption coefficient of the sample. For spectral regions with high absorbance (κ_1 large), R_{reg} will approach unity.

2. "Mat" (dull, scattering) surfaces reflect the light diffusely. Ideal diffuse reflection is defined by an angular distribution of the reflected radiation that is independent of the angle of incidence (Figure 3C). A surface that is illuminated from one direction and appears equally bright at all observation angles is called a Lambertian surface.[3] The distribution of the reflected light is the result of multiple scattering. The causes of the redirection of the light are regular reflection, refraction, and diffraction "inside" the sample (e.g., through the many small particles of a powder). Under the conditions of ideal diffuse reflection, these phenomena are not separable. The reflecting power is called reflectance, and it is low in the spectral range where absorption is high.

In diffuse reflectance, the reflected light is analyzed, and attempts have been made to relate the reflectance to the true absorption coefficient.

Note that the term "isotropic" implies a spherical source and is thus not appropriate for a radiating surface.

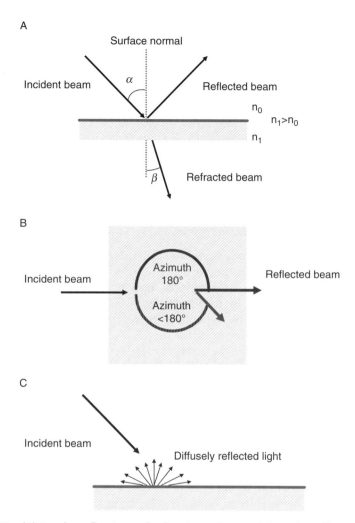

FIGURE 3   (A) Regular reflection and refraction, side view; (B) regular reflection, top view; and (C) diffuse reflection, side view.

Besides absorption in the sample, scattering needs to be considered. Scattering theories such as the Rayleigh or the Mie theory describe special cases requiring the wavelength of the light to be long versus the size of the scattering center, or the particles to be spherical. Scattering in a typical catalyst sample with a distribution of particle sizes and shapes is complex and eludes rigorous description. Most models treat the sample as parallel sheets of homogeneous layers ("continuum theories"); some statistical models address the particulate nature (Melamed, 1963).

The most frequently used description in catalysis research is the Schuster–Kubelka–Munk theory. To describe the transmission of light through aerosols ("Radiation through foggy atmospheres"), Schuster (1905) introduced a formalism involving two photon fluxes in opposite directions, and a scattering and an absorption coefficient. Kubelka and Munk (1931) described the optical properties of solids, namely coatings of paint. The authors treated the cases of finite and infinite thickness of the coating, and the cases of mere scattering (no absorption, ideal white paint) and mere absorption (no scattering, glossy coating). The deliberations indicated that the reflectance depends solely on the ratio of the absorption and scattering coefficients.

The Kubelka–Munk equation is the counterpart in diffuse reflectance spectroscopy of what the absorbance function represents in transmission spectroscopy. It can be derived in a similar manner as the Lambert–Beer law. The sample is thought to be sliced into infinitesimally thin layers with thickness dl (Figure 4) and illuminated by diffuse light; hence the average optical path corresponds to 2 dl. In each layer, part of the incident light is absorbed and part is reflected because of scattering processes, and the rest is transmitted. The photon flux in the forward direction is designated I, the flux in the reverse direction J. In each layer, light intensity is lost because of absorption (coefficient k) and scattering (coefficient s), but intensity is gained because light from the reverse flux is scattered. Both coefficients are functions of the wavelength. The two fluxes can be calculated as follows:

$$dI = -kI2\ dl - sI2\ dl + sJ2\ dl \tag{6}$$

$$dJ = -kJ2\ dl - sJ\ 2dl + sI\ 2dl \tag{7}$$

The equations are then divided by I or J, respectively, and added. Then the equations $K = 2k$ and $S = 2s$ are introduced, and the reflectance or reflectance factor ρ is the ratio:

$$\rho = \frac{J}{I} \tag{8}$$

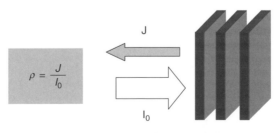

FIGURE 4   Model for the derivation of the Schuster-Kubelka–Munk equation.

Most authors use R instead of ρ; however, ρ is in accordance with IUPAC recommendations (Homann, 1996).

Separation of variables leads to

$$\int_{\rho_0}^{\rho_l} \frac{d\rho}{\rho^2 - 2\rho\left(1 + \dfrac{K}{S}\right) + 1} = S \int_0^l dl \qquad (9)$$

For an infinitely thin sample ($l \to 0$), no light will be reflected ($\rho_0 \to 0$). With increasing thickness ($l \to \infty$), the reflectance will approach a final value called ρ_∞. Integration is achieved via partial fraction expansion and the reflectance at infinite sample thickness is ρ_∞ (no transmission):

$$\rho_\infty = \frac{S}{K + S + \sqrt{K(K + 2S)}} \qquad (10)$$

The ratio of absorption and scattering coefficient can be expressed as

$$\frac{K}{S} = \frac{(1 - \rho_\infty)^2}{2\rho_\infty} = F(\rho_\infty) \qquad (11)$$

The function $F(\rho_\infty)$ is called the Kubelka–Munk or remission function; by analogy to the absorbance, it is dimensionless. It gives the correct values in the limiting cases: For a nonabsorbing sample ($K \to 0$), all light should be reflected, and indeed $\rho_\infty \to 1$. For a nonscattering sample ($S \to 0$), no light should be reflected, and indeed $\rho_\infty \to 0$. The Kubelka–Munk function depends on two quantities, and hence more than one measurement is necessary to obtain values of K or S. Typically, ρ_∞ is measured and additionally a value ρ_0 at a finite sample thickness with the specimen in front of a nonreflecting background (Perkampus, 1986).

The absorbance is a function of $\kappa(\lambda)$, c, and l. If the product of the molar absorption coefficient and the concentration $\kappa(\lambda)^*c$ is small (and l is constant), the Lambert–Beer law is valid and the absorbance is proportional to the concentration. At infinite sample thickness, the Kubelka–Munk function depends on $K(\lambda)$ and $S(\lambda)$. If the absorption coefficient K is small and S does not vary with c, the Kubelka–Munk function becomes proportional to the concentration:

$$F(\rho_\infty) = \frac{K}{S} \propto \frac{\kappa c}{S} \qquad (12)$$

If this proportionality is valid, and if c is constant, and S is constant over the considered wavelength range, then the logarithm of $F(\rho_\infty)$ equals the logarithm of the molar absorption coefficient plus a constant:

$$\log\left(F(\rho_\infty)\right) = \log \kappa + C \qquad (13)$$

The term $\log(F(\rho_\infty))$ plotted as a function of λ then represents a form of the true absorption spectrum of the sample, only shifted along the ordinate (Kortüm and Schreyer, 1955). This representation is also called the "typical color curve," and $\log(F(\rho_\infty))$ is frequently used, as it gives, under the conditions stated, the correct band positions. The term $\log(F(\rho_\infty)) - C$ equals the absorption spectrum as obtained from transmission spectroscopy only if the molar absorption coefficients κ are equal. This is the case when in both measurements the absorbing component is present in the same matrix. However, the same dye when in solution and measured in transmission, on the one hand, and when dispersed in a nonabsorbing solid and measured in reflectance, on the other, would lead to different values of $\kappa(\lambda)$. For mixed crystals of $KClO_4$ and $KMnO_4$ indeed $F(\rho)$ is strictly proportional to the absorption coefficient as obtained from transmission measurements (Kortüm, 1969, p. 186; Kortüm et al., 1963), meaning also that the scattering coefficient is independent of wavelength for this type of sample.

2.4. Use of the Kubelka–Munk Function

The motivation for transforming reflectance data into the Kubelka–Munk function is to obtain a representation of the absorption spectrum of the sample, which also allows one to relate intensities directly to concentration. It is sometimes debated as to whether the transformation should be performed, as described below.

As long as $0 < \rho < 1$ (deviations may occur in experiments), the conversion of reflectance into the Kubelka–Munk function can be performed, just as absorbance can be calculated from transmittance for $0 < \tau < 1$ (and it usually is without any further consideration). However, the proportionally to the concentration may not hold true. As applies for the Lambert–Beer law, the Kubelka–Munk function is a limiting case for the range of weak absorption. For low reflectance (transmittance), the Kubelka–Munk function (absorbance) will not be proportional to the concentration of the absorbing species, and quantification is not possible.

For an estimate of the error in $F(\rho)$, the dependence of the function on ρ needs to be considered. Figure 5 shows a comparison of $F(\rho)$ with $A_e(\tau)$. The Kubelka–Munk function is nearly flat at values of ρ close to 1 and steep at small values of ρ, whereas A_e exhibits less extreme slopes. Kortüm (1969, p. 250) determined the relative error of $F(\rho)$ with respect to ρ via differentiation. The favorable working range is $0.2 < \rho < 0.6$.

Reflectance data can be more deceiving with respect to the quantitative composition than transmittance data, as illustrated in Figure 6. Equivalent reflectance and transmittance spectra (Figure 6A) are converted into absorbance A_e and $F(\rho)$ (Figure 6B and C). The $F(\rho)$ representation demonstrates that absolute values and band areas differ much more than one would anticipate from the reflectance spectra.

Simmons (1975) compared various theories of diffuse reflectance. He introduced a modified remission function, which explains deviations from linearity when $F(\rho)$ is plotted versus k. He also concluded that the Kubelka–Munk function is proportional to the absorption coefficient k as obtained from transmission measurements for "weakly absorbing samples." Unfortunately, most literature is vague in that "weak" or "strong" absorption is not specified. One value given for "weak" is $F(\rho) < 1$ (Kellermann, 1979).

Klier (1972) deduced that for $0.6 < \rho < 1$, which corresponds to $0.13 > F(\rho) > 0$, the Kubelka–Munk absorption coefficient should be nearly proportional to the true absorption coefficient. Deviations from the proportionality up to a factor of two occurred for lower reflectance values. In the range $\rho > 0.6$, the Kubelka–Munk function should be nearly proportional to the absorber concentration. Through comparison with the radiative-transfer equation formulated by Chandrashekhar (1960), Klier related the phenomenological coefficients to the true absorption and scattering coefficients α and σ and obtained the following for $0.5 < \rho < 1$ (with 1.5% accuracy) (Klier, 1972):

$$\frac{K}{S} = \frac{8}{3}\frac{\alpha}{\sigma} \tag{14}$$

In the limit of zero absorption:

$$\alpha = \frac{1}{2}K \tag{15}$$

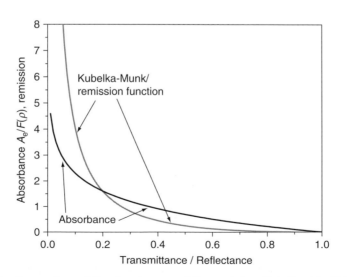

FIGURE 5   Comparison of the absorbance and the remission (Kubelka–Munk) function.

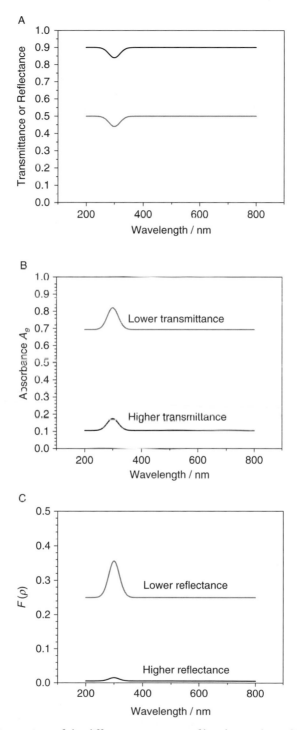

FIGURE 6 Comparison of the different appearance of band areas depending on representation. (A) reflectance or transmittance, (B) absorbance, and (C) Kubelka–Munk function.

$$\sigma = \frac{4}{3} S \tag{16}$$

For a typical catalyst, the scattering coefficient will be determined primarily by the matrix (support) and could be derived by measuring reflectance or transmittance of matrix specimens with different thicknesses. Absorption coefficients of surface species, however, are most likely not transferable from one catalyst (support) type to another. To make correlations between spectroscopic and catalytic data, it must be ensured that the analytical function, (i.e., $F(\rho)$) properly represents the concentration of an absorbing species; for an analysis of reaction kinetics, the actual surface concentration is needed. A calibration of the relationship between a spectral quantity and the concentration can be achieved by determining the concentration with a second independent method or by preparing calibration samples; such a calibration seems preferable to trying to determine actual scattering and absorption coefficients. If the scattering coefficient is wavelength-independent, the Kubelka–Munk function will reflect the absorption spectrum.

Kortüm et al. (1963) investigated the influence of particle size on the scattering coefficient. For particles large in comparison to the wavelength (i.e., for a mean particle size of $>2 \ \mu$m), the scattering coefficient σ was found to be practically independent of the wavelength. For smaller particles, σ was found to be inversely proportional to the square root of the mean particle size:

$$\sigma \propto \frac{1}{\sqrt{\overline{d^2}}} \tag{17}$$

It may not be appropriate to consider the scattering coefficient to be constant, and this point may become important in evaluation of trends monitored during a catalytic reaction experiment. Kortüm et al. (1963) pointed out that the scattering behavior depends on the ratio of refractive indices of the sample and the surrounding medium. As an example for a change in the refractive index of the sample, Kortüm described the adsorption of water, which reduces the scattering coefficient. Hence $F(\rho)$ will increase, which could erroneously be interpreted as an increase in absorption.

In practice, a simple prediction about the range of proportionality of $F(\rho)$ with respect to concentration of the absorbing species cannot be made because it is the product of concentration and molar absorption coefficient that is relevant. Furthermore, experiments show that linearity may extend over a larger or smaller range than those indicated above. Implying that the concentration of V^{4+} or V^{5+} species should be proportional to the vanadium loading, Catana et al. (1998) plotted the intensity of the respective bands versus the vanadium concentration, which ranged from 0 to 1 wt%. Deviations from linearity occurred at values of the Kubelka–Munk function of about 0.05 ($\rho \approx 0.73$) or 3.5 ($\rho \approx 0.11$).

Various concentrations of Pd ions in H-mordenite were investigated by Shimizu et al. (2000) after oxidative treatment. A d–d transition of Pd^{2+} ions at 460–480 nm was analyzed; $F(\rho)$ reached a maximum of 0.3 for this band ($\rho \approx 0.46$). The band area was linearly correlated with the palladium content, which ranged from 0 to 1.5 wt %. The point is that the proportionality needs to be verified experimentally for each individual system (as is necessary for the Lambert–Beer law).

To be able to quantify data by use of the Kubelka–Munk function, contributions from regular reflection must be excluded. From the Fresnel equations it follows that strongly absorbing materials will also feature a strong regular (surface or mirror) reflection, which will mix with the diffuse reflection. In order to derive the correct absorption spectrum, solely diffusely reflected light should reach the detector, whereas contributions from regular reflection should be suppressed or eliminated. This goal can be achieved in several ways: (1) the diffusely reflected light may be collected over a limited solid angle to exclude most of the regularly reflected light, which is most intense in the forward direction (azimuth of 180°, Figure 3); (2) powders can be finely ground, which leads to a lower fraction of regularly reflected light and a higher fraction of diffusely reflected light; (3) use of polarized light and placement of the sample between crossed polarizers will eliminate regular contributions because the diffusely reflected light will change polarization while the regularly reflected light will not; and (4) dilution. Kortüm and Schreyer (1955) reported that for "weakly absorbing" samples, grinding to a fine powder is sufficient to suppress the regular reflection contributions. For "strongly absorbing" samples, they introduced dilution as a method to eliminate regular reflection. Mixing was achieved by milling. The authors tested the influence of the mixing time and found that, in general, spectra would not change further after 12 h of milling. Such procedure may not be applicable for catalysts, which are metastable materials and may be altered through mechanical stress (Klose et al., 2003). Furthermore, such intense mixing led to adsorption of the sample on the diluent; in the cases of $KMnO_4$ or K_2CrO_4 mixed with $BaSO_4$ or MgO, the spectra were dependent on the nature of the diluent.

2.5. Referencing Diffuse Reflectance Spectra

In transmission spectroscopy, the light intensity transmitted through the sample is either simultaneously or consecutively compared with that transmitted by a reference, which is considered as 100% transmittance (I_0). The reference spectrum can either be obtained with an empty beam path (correction for changes in incident light), an empty cell (to correct additionally for window absorption) in the beam path, or a cell with only the gas phase or only the catalyst or only the solvent. The equivalent of the empty beam path in transmission mode is a reflecting material in diffuse reflection spectroscopy that ideally should diffusely reflect light to 100% over a wide wavelength range. There are no materials that diffusely reflect 100% over the entire UV–vis–NIR range, but some materials with close to ideal properties exist (Section IV.B). These materials are often referred to as "white standards." The spectrum of the sample is compared to the spectrum of such a reference material. Because the reflectance of the reference materials is <100%, we do not measure the absolute value of ρ_∞, but a slightly higher value, ρ'_∞:

$$\rho'_\infty = \frac{\rho_\infty}{\rho_{std}} \tag{18}$$

To obtain absolute values, the data have to be multiplied by the reflectance spectrum of the standard.

2.6. One or Multiple Absorbing Species—Transmission Versus Diffuse Reflection

The reflectance spectrum of a single absorbing species can be obtained from one measurement of the sample (catalyst) versus a standard, and with the relationship

$$\rho_{cat} = \frac{I_{cat}}{I_{std}} \tag{19}$$

and then application of the following equation:

$$F(\rho_{cat}) = \frac{(1-\rho)^2}{2\rho} = \frac{K_{cat}}{S_{cat}} \tag{20}$$

This type of measurement is performed in a manner similar to that of a transmission experiment:

$$\tau_{cat} = \frac{I_{cat}}{I_0} \tag{21}$$

and

$$A_{cat} = -\ln\frac{I_{cat}}{I_0} \tag{22}$$

There might be two absorbing species (e.g., a dye in a solvent or an adsorbate on a catalyst). In this case, the spectrum of the dye or adsorbate can be obtained as follows (shown here for an adsorbate):

$$A_{\text{cat} + \text{ads}} = A_{\text{cat}} + A_{\text{ads}} \tag{23}$$

$$A_{\text{ads}} = A_{\text{cat}+\text{ads}} - A_{\text{cat}} = -\ln \tau_{\text{cat}+\text{ads}} + \ln \tau_{\text{cat}} = -\ln \frac{I_{\text{cat}+\text{ads}}}{I_0} + \ln \frac{I_{\text{cat}}}{I_0} \tag{24}$$

and

$$A_{\text{ads}} = -\ln \frac{I_{\text{cat}+\text{ads}}}{I_{\text{cat}}} \tag{25}$$

A single measurement of the catalyst + adsorbate versus the catalyst is sufficient in a transmission experiment. For reflectance measurements, the scenario is different. Let us again consider two absorbing species, one of which (the majority species that forms the matrix) dominates the scattering properties of the sample. Examples of such systems are a catalyst and a diluent, an adsorbate (such as coke) on a catalyst, a supported species, or a fraction of reduced metal in a metal oxide. Again we choose the example of catalyst and adsorbate:

$$F_{\text{cat} + \text{ads}} = F_{\text{cat}} + F_{\text{ads}} = \frac{K_{\text{cat}}}{S_{\text{cat}}} + \frac{K_{\text{ads}}}{S_{\text{ads}}} \approx \frac{K_{\text{cat}} + K_{\text{ads}}}{S_{\text{cat}}} \tag{26}$$

The approximation is possible because the scattering properties of the sample are largely determined by those of the catalyst powder.

$$F_{\text{ads}} = F_{\text{cat} \mid \text{ads}} - F_{\text{cat}} = \frac{(1 - \rho_{\text{cat} \mid \text{ads}})^2}{2\rho_{\text{cat}+\text{ads}}} - \frac{(1 - \rho_{\text{cat}})^2}{2\rho_{\text{cat}}}$$

$$= \frac{\left(1 - \dfrac{I_{\text{cat}+\text{ads}}}{I_{\text{std}}}\right)^2}{2\dfrac{I_{\text{cat}+\text{ads}}}{I_{\text{std}}}} - \frac{\left(1 - \dfrac{I_{\text{cat}}}{I_{\text{std}}}\right)^2}{2\dfrac{I_{\text{cat}}}{I_{\text{std}}}} \tag{27}$$

It is important that a simplification as made above from Equations (19) and (20) is not possible, and

$$F_{\text{ads}} \neq \frac{\left(1 - \dfrac{I_{\text{cat}+\text{ads}}}{I_{\text{cat}}}\right)^2}{2\dfrac{I_{\text{cat}+\text{ads}}}{I_{\text{cat}}}} \tag{28}$$

Hence, F_{ads} cannot be derived from a simple measurement of catalyst + adsorbate (diluent + catalyst) versus the catalyst (the diluent). It is important to understand that other than in transmission spectroscopy, in the measurement of a supported species, the contribution of the support cannot be eliminated by using the support to record a reference (background correction). Exceptions can be made in regions where the support reflectance is near 1 (Section VI.A).

Two measurements are necessary, and the individual reflectance spectra have to be converted into $F(\rho)$, and then a subtraction can be made. There are two possibilities; either one measures $I_{cat + ads}/I_{std}$ and I_{cat}/I_{std} and F_{ads} is calculated as in Equation (27). Alternatively, $I_{cat + ads}/I_{cat}$ and I_{cat}/I_{std} can be measured and then $I_{cat + ads}/I_{std}$ is calculated as

$$\frac{I_{cat+ads}}{I_{std}} = \frac{I_{cat+ads}}{I_{cat}} \times \frac{I_{cat}}{I_{std}} \tag{29}$$

and then Equation (27) is applied. The latter possibility is preferable according to Kortüm (1969, p. 150); it is more practicable to once measure the catalyst versus the standard when using MgO because it ages fast.

3. EXPERIMENTAL: OPTICAL ACCESSORIES FOR UV–vis–NIR SPECTROSCOPY IN TRANSMISSION MODE

3.1. Optical Accessory Design for Transmission UV–vis–NIR Spectroscopy

Standard spectroscopic equipment can be used readily for investigations of homogeneous or enzymatic catalysis; for example, see Valentine et al. (1999). Transmission through self-supporting wafers of powders, however, is frequently low, and the strong scattering of fine catalyst powders leads to a wide angular distribution of the transmitted light. Hence either focusing or collecting elements are required, or the detector has to be close to the sample. Modern integrating spheres feature a sample position where the light enters the sphere.

3.2. Optical Accessories for Transmission UV–vis–NIR Spectroscopy

Förster et al. (1983) apparently used a simple quartz cuvette placed in a spectrometer without any further modifications; similarly, Pazé et al. (1999) referred to a quartz cell that was sealed after sample treatment and then inserted into the spectrometer. A design for transmission measurements was published by Melsheimer and Ziegler (1992), who referred to their device as a "scattered transmission accessory." The photomultiplier is placed only 28 mm behind the sample; attenuators of 1.7 or 10% were used in the reference beam.

4. EXPERIMENTAL: OPTICAL ACCESSORIES FOR UV–vis–NIR SPECTROSCOPY IN REFLECTION MODE

4.1. Optical Accessory Design for Diffuse Reflectance UV–vis–NIR Spectroscopy

The function of a diffuse reflectance accessory is to direct the light from the source onto the sample, collect the diffusely reflected light, and guide it to the detector. From a theoretical point of view, the sample should be illuminated by diffuse light. However, because of geometrical constraints, only light collection or illumination can be diffuse in the reflection mode. Upon regular illumination of a fine catalyst powder, the light is scattered in the surface layers, and the preferential direction from the original incidence will gradually be lost as the light penetrates the sample. For high scattering density, an isotropic distribution arises rapidly within the sample (Kortüm, 1969, p. 99). Indeed, the type of illumination of a sample, with regular or diffuse light, was found not to have any influence on the remission curves of fine powders (Kortüm and Schreyer, 1956). Klier reported similar experiences (Klier, 1968). Equations valid for regular illumination transform into the Kubelka–Munk equation for particles with sizes $2\pi r \ll \lambda$ (Kortüm, 1969, p. 130).

Types of apparatus with directed or diffuse irradiation and, correspondingly, diffuse or directed detection, have been compared, and the results were found to agree within the error limits of the method (Kortüm, 1969, p. 170). Many accessories feature regular illumination of the sample. There is a disadvantage to this configuration. There are no perfectly mat surfaces; for example, some crystals may align on the surface of a catalyst sample and form a "glossy" section. On such mirror-like patches, regular reflection will occur (Figure 7A and B). Regularly reflected light is

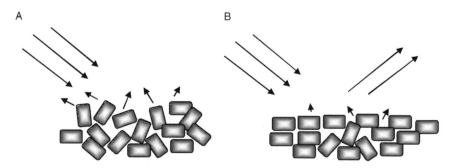

FIGURE 7 (A) Randomly oriented crystals and diffusely reflected light and (B) orientation of crystals (e.g., on flat window), formation of glossy sections, reflected light contains regular and diffuse contributions.

unwanted because it adds to the background on an already weak signal, and in spectral regions of strong sample absorbance the intensity of regularly reflected light is strong whereas that of the diffusely reflected light is weak; that is, in the collected light, information would be cancelled out. The intensity of regularly reflected light is high in the direction at the same angle to the surface normal as the incident beam, and in the forward direction (i.e., at an azimuth of 180°). Accessories are designed so that they collect a minimum of regularly and a maximum of diffusely reflected light. Integrating spheres operate at 0° or feature gloss traps; mirror accessories such as the Harrick Praying Mantis collect the light "off-axis" at an azimuth of 120° (Harrick, 2006), and in fiber-optic systems, illuminating and collecting fibers are placed at an angle to each other. Accessories may also be combined (i.e., mirror or fiber optics may be used together with an integrating sphere). For spectroscopy in catalysis, optical accessories should be evaluated in combination with the reaction cell. Various solutions may emerge as best depending on parameters such as the desired data quality, time resolution, and photometric accuracy.

4.2. Standard Materials for Diffuse Reflectance

Standard materials and their use have been reviewed by Springsteen (Springsteen, 1999, 2000a,b). CIE and ASTM both call for a "perfect reflecting diffuser" as a standard for diffuse reflectance measurements. A classical white standard formerly was MgO, which was favored by Kortüm. He summarized data from various groups showing that at wavelengths between 400 and 700 nm, freshly prepared MgO reflects more than 95% of the light (Kortüm, 1969, p. 146). A typical preparation method was burning of a high-purity Mg ribbon. Such MgO is reported not to be very stable, and it is not convenient to always prepare it freshly. MgO has recently been applied by researchers in the group of Iglesia; for example, see Barton et al. (1999).

Another possible reference material is $BaSO_4$, which reflects >95% of the light in the range 340–1400 nm (Figure 8A) and has been used recently by Weckhuysen et al. (1998) to record a baseline with the Harrick accessory. High-purity $BaSO_4$ that is specified for reflectance measurements is required. In principle, it is suitable for measurements reaching somewhat into the NIR range, but it easily takes up water, which generates two strong absorptions, at 1475 and 1950 nm, from OH vibrations. Sprayable $BaSO_4$ coatings are available (Optolon 2™, Optronics Laboratories).

Further standards tested by Kortüm were $MgSO_4$, NaF, NaCl, Li_2CO_3, Al_2O_3, Aerosil silica, and glucose (Kortüm et al., 1963).

In the mid-1970s, polytetrafluoroethylene (PTFE) standards were introduced. Packed PTFE shows almost perfect Lambertian behavior over the wavelength range of 190–2500 nm. These standards are available

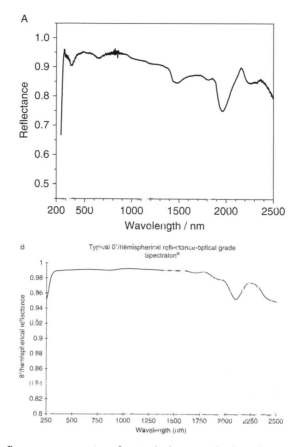

FIGURE 8 Reflectance properties of typical white standards in the UV–vis–NIR range. (A) Reflectance of barium sulfate. The measurement was carried out with a PerkinElmer Lambda 9 spectrometer equipped with an integrating sphere, and the reactor cell described in Ref. (Thiede and Melsheimer, 2002). For the background correction one piece of Spectralon® was placed at the reference port of the sphere, a second piece was placed inside the reactor cell at the sample port of the sphere. For measurement of the presented spectra, BasSO$_4$ was placed in the reactor, while the piece of Spectralon® at the reference port remained in position. (B) Reflectance of Spectralon® as provided by the manufacturer.

under trade names such as Spectralon® (Labsphere), the reflectance of which is shown as an example in Figure 8B, or Albrillon™ (Ancal Inc.). In publications, the PTFE trade names such as Teflon® (Du Pont) or Halon (former trademark of Allied Chemical Company) are sometimes used. The materials can be purchased as solid pieces, which are easier to handle than powders. Another advantage is that they can be machined, and reference specimens can be made to fit a reaction cell. However, when certified, these standards are expensive. Also, they attract dust and must be replaced or cleaned regularly. PTFE materials are employed frequently (Grubert et al., 1998; Thiede and Melsheimer, 2002; Melsheimer et al., 2003; Klose et al., 2006; Klokishner et al., 2002; Melsheimer et al., 2002; Jentoft et al., 2003).

Accuracy and linearity of the measured values of the reflectance ρ can be verified by measuring materials with known absolute reflectances of less than 100%. Examples of spectra of two such reference standards measured versus a 100% standard are shown in Figure 9. The reflectance of the two different standards is specified depending on the wavelength, as follows: 48.5% at 250 nm, 46.8 at 350 nm (minimum), and 56.6% at 2500 nm; 18.7% at 250 nm, 17.2% at 400 nm (minimum), and 26.3% at 2500 nm. It is evident from Figure 9 that the measured reflectance values do not correspond exactly to the specification. Furthermore, repeated measurements indicate less than perfect reproducibility. Analysis of the spectra in Figure 9 shows accuracy and precision of the measurements to be 2–5%. The deviations from the specified values could be caused by imperfect reference materials or could point towards a problem with the measurement apparatus (Thiede and Melsheimer, 2002).

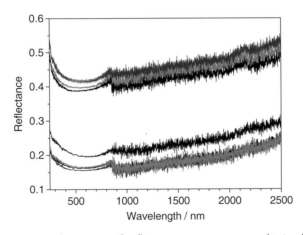

FIGURE 9   Precision and accuracy of reflectance measurements obtained with a PerkinElmer Lambda 9 spectrometer with a 60-mm integrating sphere (8 °). For the background correction, pieces of Spectralon® SRS-99–010 were placed at the reference and at the sample port behind quartz (Suprasil) windows. For measurement of the presented spectra, Spectralon® reflectance standards 50% and 20% (SRS-50–010 and SRS-20–10) were placed at the sample port behind a quartz window, while the piece of Spectralon® at the reference port remained in position. Multiple spectra represent repeated scans without substantial changes in the configuration (except for slight variations in sample position).

4.3. Diffuse Reflectance Spectroscopy Using Integrating Spheres

Integrating spheres (also called photometer spheres or Ulbricht spheres) are the classical optical devices used to collect diffusely reflected light. They cover a wide solid angle, which is advantageous, given the fact that the diffusely reflected is evenly distributed in all directions. Spheres are coated with materials that absorb as little light as possible and diffusely reflect it so that it can eventually reach the detector after multiple reflections. Materials suitable for coatings are the same as for reflectance standards (e.g., MgO, BaSO$_4$, or Spectralon®). At wavelengths from 800 nm into the MIR range, gold coatings may be used. Cost is a major factor here, in part because spheres may have to be recoated occasionally.

Spheres have ports to admit the sample and reference beams (double-beam spectrometers), to place the sample and reference, and to allow one or more detectors (e.g., a photomultiplier for the UV–vis range and a PbS detector for the NIR range). Samples are usually placed such that they virtually become part of the sphere wall. Although spheres collect almost all the light, the fraction of light reaching the detector is actually small, as the following considerations show (Figure 10):

The flux through the inlet port onto the sample is I_0; the flux reflected from the sample (with ρ_S being the sample reflectance) onto the entire sphere is then

$$I_{\text{sphere, 0}} = I_0\,\rho_S \tag{30}$$

The fraction of light falling onto the wall section that is projected onto the detector is

$$I_{\text{wallsection}} = I_{\text{sphere}}\frac{f_m}{F} \tag{31}$$

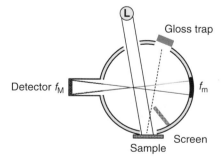

Detector f_M

Gloss trap

f_m

Screen

Sample

FIGURE 10   Sketch of an integrating sphere with an 8° arrangement including screen and gloss trap.

with f_m being the area of the wall section and F the total internal surface area of the sphere.

The wall section reflects diffusely in all directions, and only a fraction of the light coming from this area reaches the detector, giving the flux onto the detector as follows:

$$I_{detector} = I_{wallsection}\ \rho_W \frac{f_M}{F} = I_0 \rho_S \frac{f_m}{F} \rho_W \frac{f_M}{F} \tag{32}$$

where ρ_W is the wall reflectance and f_M the area of the detector.

The sphere contains the photons for a finite time; it is not only the flux generated by the incident light onto the sample that must be considered, but also the flux from reflections within the sphere. The fluxes from the first and second reflections are (on the entire sphere, the wall, and the detector):

$$I_{sphere,1st} = I_0\ \rho_S\ \rho_{ave} \tag{33}$$

$$I_{wallsection} = I_0\ \rho_S \frac{f_m}{F} \rho_{ave} \tag{34}$$

$$I_{detector} = I_0 \rho_S \frac{f_m}{F} \rho_W \frac{f_M}{F} \rho_{ave} \tag{35}$$

$$I_{sphere,2nd} = I_0\ \rho_S\ \rho_{ave}^2 \tag{36}$$

$$I_{wallsection} = I_0 \rho_S \frac{f_m}{F} \rho_{ave}^2 \tag{37}$$

$$I_{detector} = I_0 \rho_S \frac{f_m}{F} \rho_W \frac{f_M}{F} \rho_{ave}^2 \tag{38}$$

where ρ_{ave} is the average reflectance of the sphere, which consists of reflectance from the wall and reflectance from the ports:

$$\rho_{ave} = \rho_W \left(1 - \sum_{i=0}^{n} \frac{f_i}{F}\right) + \sum_{i=0}^{n} \rho_i \frac{f_i}{F} \tag{39}$$

with ρ_W being the reflectance of the sphere wall (coating), f_i the areas of the port openings, and ρ_i their reflectances. The second term of Equation (39), which represents the reflectance from the ports, is small and sometimes omitted.

For multiple reflections, the flux onto the detector is

$$I_{det} = I_0 \rho_S \frac{f_m}{F} \rho_W \frac{f_M}{F} \left(1 + \rho_{ave} + \rho_{ave}^2 + \rho_{ave}^3 + \cdots \right) \tag{40}$$

If there is a baffle (screen) in the sphere that prevents direct reflection of light from the sample onto the measurement area (a section of wall projected onto the detector), then the first term of the series disappears. The series can then be rewritten as follows:

$$I_{det} = I_0 \rho_S \frac{f_m}{F} \rho_W \frac{f_M}{F} \left(\frac{1}{1 - \rho_{ave}}\right) \tag{41}$$

The larger the sphere, the lower is the intensity of radiation incident onto the detector. However, the smaller the sphere, the larger is the fraction of the sphere devoted to ports, leading to smaller ρ_{ave} and to lower intensity on the detector. The ports cannot be infinitely small, but, on the other hand, the port fraction should not exceed 5%. As is obvious from Equation (32), the flux on the detector varies greatly with f_m and f_M. According to Sphereoptics (2004), the flux onto the detector corresponds to 1% of the incident flux or less. In diffuse reflectance spectroscopy with an integrating sphere, one generally operates with low intensities, increasing the demand on spectrometer performance. Sphere performance is sometimes characterized by the sphere multiplier (or sensitivity factor), which assumes high values while the light yield on the detector is effectively small. For an ideal integrating sphere (one without ports), the sphere multiplier is given as follows:

$$M = \frac{\rho}{1 - \rho} \tag{42}$$

For a real integrating sphere with ports:

$$M = \frac{\rho}{1 - \rho\left(1 - \sum_{i=0}^{n} \frac{f_i}{T}\right)} \tag{43}$$

M assumes typical values between 10 and 30.

Measurements may be performed either at an incident angle of $0°$ so that regularly reflected light will mostly be reflected towards the input port, or with an $8°$ angle and a gloss trap (a section that can be opened so that regularly reflected light escapes from the sphere and will not reach the detector, or closed so that the diffusely and regularly reflected light are collected). With the aid of the gloss trap, the contribution of regularly reflected light can be determined.

4.4. Diffuse Reflectance Spectroscopy with Mirror Optics Attachments

The designs of various commercially available mirror optics for diffuse reflectance in IR or UV–vis spectroscopy are, in principle, similar to each other. Each of these accessories is characterized by six mirrors, four flat and two curved mirrors. In this design, the incident beam and the light reflected from the sample are on the same optical axis, and the accessory can be placed easily into the sample chamber without the need for further physical modifications of the spectrometer. The accessories can be equipped with various base plates to allow them to fit into various spectrometers. Secure attachment to the spectrometer housing prevents difficulties with misalignment resulting from movement of the whole accessory.

An example of such an accessory that is marketed for application in the UV–vis regime is the Harrick Praying Mantis. A sketch and images are available at the Harrick webpage (Harrick, 2006) and have also been published by Weckhuysen (Weckhuysen and Schoonheydt, 1999, 2000; Weckhuysen, 2002, 2003, 2004) and other authors (Sojka et al., 2008).

It is advantageous to be able to change from a standard transmission configuration to one allowing diffuse reflectance measurements with the same spectrometer. Mirror optics are different from integrating spheres in that they need to be aligned. The solid angle over which diffusely reflected light is collected is smaller than for an integrating sphere. Intensity is lost at each of the mirrors. Designs with fewer mirrors are possible when the detector can be placed accordingly. Such a construction has been realized and is being used by Kazansky and coworkers for measurements in the IR regime (Kazansky et al., 1997).

However, in the MIR region, the throughput of mirror optics are good. For throughput measurements, a tilted mirror is placed in the sample position. The regularly reflected light from a flat mirror would be eliminated through the off-axis position, and a 30° angle versus horizontal is necessary to regularly reflect light towards the detector for the throughput measurement.

Figure 11A shows an original test throughput measurement delivered by Harrick with one of their accessories (DRP-P72) designed for a PerkinElmer Lambda 950 UV–vis–NIR spectrometer. The throughput varies between 72% and 80% over the MIR range 4000–400 cm^{-1}. The intense groups of bands with characteristic rotational fine structure are caused by air in the beam path and can be assigned to water and carbon dioxide.

Figure 11B shows the throughput of a Harrick DR accessory in the UV–vis–NIR range. The throughput is high in the NIR range, about 74% at 2500 nm (4000 cm^{-1}). However, the throughput decreases rapidly at wavelengths shorter than 1150 nm, to about 30% at 250 nm. There is a pronounced and broad minimum of about 30% at about 825 nm, which originates from the mirror properties. Overtone and combination modes of gaseous water from air in the beam path are visible as is an artifact created by the detector and grating change at 860 nm. In principle, the accessory throughput characteristics can be eliminated through the reference measurement. However, low throughput increases the demands on spectrometer performance, and the spectra can be noisier in such regions.

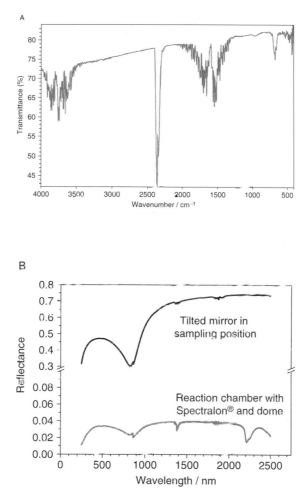

FIGURE 11   (A) MIR throughput of Harrick Praying Mantis DRP as provided by the manufacturer. (B) UV–vis–NIR throughput of Harrick Praying Mantis DRA-4-PE7 measured with a PerkinElmer Lambda 950 spectrometer. For the background correction the reference and sample beam were empty. For measurement of the spectra, the reference beam was left empty. The Praying Mantis mirror accessory was placed in the sample beam. The throughput of the accessory alone (upper curve) can be measured with a tilted mirror in the accessory's sample position; a horizontal mirror ideally gives a throughput of zero, as regularly reflected light should not be collected. The spectrum obtained with a Spectralon® white standard (lower curve) inside the reaction cell HVC-DR3 that was placed in the Praying Mantis accessory demonstrates the low intensity of the diffusely reflected light and further losses through the quartz window of the cell.

Spectral results obtained with mirror optics are sensitive to the sample position, which, in turn, affects the position of the collimated light spot on the detectors. The photomultiplier response is relatively homogeneous within the sensitive detector area, whereas PbS detectors (NIR range) exhibit poor spatial uniformity (Theocharous, 2006). If the sample is positioned differently than the reference or the position of the cell within the mirror optics accessory changes during the specimen exchange, the sample signal will not be truly comparable to the reference signal. The sample position may also vary if the sample volume changes during a treatment such as dehydration. Steps in the spectrum at the wavelength of the detector change or reflectance values exceeding 1 can be indications of variations in the sample position.

These problems can be remedied by use of an integrating sphere behind the mirror optics accessory; the light yield, however, is then very small. The advantages of mirror optics are (1) a good throughput over the entire UV–vis–NIR range, and (2) facile integration of a heatable reaction chamber.

4.5. Diffuse Reflectance Spectroscopy with Fiber Optics

Usually diffuse reflectance accessories are adapted to the design of a spectrometer. For example, the Praying Mantis optics can be placed into the regular sample compartment and, thus, into the normal beam path and integrating spheres can also be placed into the beam paths. Some designs have detectors mounted directly at the sphere. Fiber optics allow for a flexible redirection of the beam from its usual path from the source and spectrometer unit to the detector unit. Via an adapter, the beam is coupled out of its regular path, transmitted inside the fiber via total internal reflection, interacts with the sample at the head of the reflection probe, and returns through a different fiber and is finally coupled back into its regular path so that it arrives at the detector. The key parts are the adapters, the fiber itself, and the head.

Light reflected from the sample has to enter the fiber—that is, it has to pass from the optically less dense surrounding atmosphere of the catalyst into the optically more dense material of the fiber. Depending on the angle of incidence and the refractive indices, a fraction of the light will be reflected and a fraction will be refracted (as described by the Fresnel equations) at the phase boundary. Inside the fiber, light is guided via total internal reflection, which occurs for grazing incidence on the boundary of the fiber and the surrounding medium. The limiting angle (relative to the surface normal) is calculated from the refractive indices of the fiber (n_1) and of the surrounding medium (n_0):

$$\sin\alpha_g = \frac{n_1}{n_0} \qquad (44)$$

Fibers are usually made of silica. In the past, the lack of fibers with sufficient throughput in the NIR range has limited their application, but suitable fibers with low OH-group concentrations have now become available. To cover the entire spectral range from 200 to 2200 nm, two different fibers are usually necessary. Fiber bundles are used, with several (e.g., 6–18) illumination fibers surrounding a single read fiber. Combined bundles with two read fibers, UV–vis and vis–NIR, are also available. The length of the fiber bundle can be several meters. Contributions from specular reflection can be reduced if, at the probe head, the faces of the illumination fibers and the read fiber are not parallel; for example, the illumination fibers may be placed at a 30° angle relative to the read fiber (Hellma product information, 2006). Currently, probe heads are in operation that can withstand temperatures of 1023–1073 K (Weckhuysen, 2003; Fiberguide, 2006). The fibers can be positioned at will, and need not necessarily be placed outside the reactor in front of a window, but can instead be immersed into gas atmospheres and liquids. However, degradation of fibers over months in a high-temperature environment has been reported (Puurunen et al., 2001); formation of coke (Weckhuysen, 2003) or other deposits on the fiber can also be a problem.

Fiber optics can be integrated into a normal spectrometer, but complete solutions consisting of a combination of spectrometer and probe are commercially available (Oceanoptics, 2006; Avantes, 2006; Stellarnet, 2006). Some spectrometers are combined with CCD cameras for detectors, enabling fast data acquisition, one UV–vis spectrum in 6 ms (Weckhuysen, 2003). Besides the excellent time resolution, the main advantage of the fibers is that they can be integrated into typical reactors or added as a second or third method to a cell optimized for a different method.

4.6. Diffuse Reflectance Spectroscopy with Combined Optical Elements

Zou and Gonzalez (1992) published the diagram of a design that allows placement of an integrating sphere next to a quartz window in a reactor. Fiber optic cables were used as an interface between an integrating sphere and a PerkinElmer UV–vis spectrometer. The maximum temperature of the cell was stated to be 773 K, and spectra were presented for temperatures up to 473 K.

A commercially available flow cell (JASCO Co. Ltd), which was also operated in a configuration consisting of an integrating sphere and fiber optic cables, was used by Shimizu et al. (Shimizu et al., 2000). Spectra shown in their paper were recorded at temperatures up to 623 K.

5. CELLS FOR MEASUREMENTS IN CONTROLLED ENVIRONMENTS

5.1. Design Parameters

Reaction cells ideally should meet two requirements: they should allow for the collection of good catalytic data and good spectroscopic data. Often a compromise is made, and, depending on the catalyst and reaction conditions, one design may be superior to another. In general, the catalyst particles have to be in the volume that is analyzed spectroscopically; they may move within or in and out of this volume. Most designs place the catalyst in a fixed bed. Isothermal operation of the entire bed is often desired for treatments and catalysis. There should be fast and homogenous gas exchange throughout the entire bed to minimize gradients. This exchange is usually more easily achieved with beds of powder in flow reactors and more difficult with pelletized samples in batch reactors.

Specific challenges arise in UV–vis–NIR spectroscopy, depending on the optical arrangement that is used. The reactor with the catalyst has to be brought into the beam path. Mirror optics are designed for horizontally oriented fixed beds; fiber optics are flexible and can be oriented at will. However, most integrating spheres have their ports on the sides, requiring a reactor with a large enough flat window on a side. A typical port size of a 60-mm sphere is approximately 9×17 mm, requiring a relatively large bed. A large volume is advantageous to allow dilution of a strongly absorbing sample or use of enough catalyst with low activity to obtain a measurable conversion. A small volume is favored if the material is precious or cannot be easily produced in large quantities. A flexible volume is, thus, desirable.

The length of the beam path through the gas or liquid phase surrounding the catalyst in the reactor plays a role if the species in the reaction medium cause significant absorption (high concentration or high molar absorption coefficient). Identification of surface species may be hampered if gas- or liquid-phase contributions cannot be identified or eliminated from the spectra. For spectroscopy in the NIR range, thermal radiation from the sample or its heated surroundings onto the detector can be a problem. With increasing temperature, the noise in the NIR range can increase, and the reflectance may shift to higher values. NIR detectors can also be disturbed by pulsed heaters. There are limitations with respect to the pressure, because the cells need optically transparent windows. Fibers may be directly immersed into a reactive atmosphere, which, however, may lead to their rapid degradation.

For some catalytic applications, commercially available designs consisting of a cell and an optical arrangement may not always fulfil the requirements, and a number of cells have been designed by researchers, and commercial cells have been modified.

5.2. Cells for Transmission Mode Spectroscopy

Förster et al. (1983) published a schematic representation of their design, which consisted of a commercially available quartz cuvette connected to a vacuum and gas-dosage system. The samples were pressed into self-supporting wafers and mounted into a gold foil frame. Melsheimer and Ziegler (1992) designed a compact quartz cell to have a short beam path between sample wafer and detector. The cell, which is shown in a sketch in Melsheimer and Ziegler (1992), is heatable up to 573 K by means of a wire wrapped around its body. A cooling cell with circulating water thermally isolates the detector from the reaction cell. The volume of the cell is $\sim$45 cm^3, and small openings at each end allow for gas flow. The authors reported cases in which the small amount of sample (and maybe also mass transfer limitations in the wafer) led to immeasurably low conversions in the catalytic reaction.

5.3. Cells for Use with Integrating Spheres

For measurements with an integrating sphere, the sample is placed at the sampling port so that it fills the void opened by the port. This goal is easily met with a solid sample, and powders can be characterized in cuvettes, which are placed at the port. In principle, any container with a UV–vis transparent window of the right size and geometry can be used, and if designed properly, such an apparatus may allow for treatments of the sample under vacuum and in inert or reactive gases, in either flow or batch mode. Heating, however, must usually be effected only at a distance from the sphere, because of the need, first, to avoid damage of the sphere, and, second, to achieve homogeneous heating, including the window that is directed toward the "cold" sphere. One design solution is to relocate the powder from a section of a volume that is used for a treatment to a section that is used for spectroscopy. The only limitation of such an apparatus is that the measurement is at room temperature; in principle, catalytic reactions could be investigated in either batch or flow mode.

Sketches of designs that allow catalyst treatments and can be used with an integrating sphere have been published (Klier, 1968; Klier and Rálek, 1968; Kortüm, 1969, p. 242; Stone, 1983; Schoonheydt, 1984; Che and Bozon-Verduraz, 1997; Weckhuysen and Schoonheydt, 1999). Placement of the reference material and the sample next to each other, by using the same port of the sphere, yields exact position equivalency (Zecchina et al., 1975). Temperature gradients in the bed can be minimized by using a double set of windows separated by vacuum (Kellermann, 1979). Such a cell was described as usable at temperatures from 77 to 450 K, with the insulated windows permitting low-temperature operation because condensation is suppressed. Designs of flow-through cells have also been presented (Garbowski and Praliaud, 1994).

A set of heatable cells that can be used with an integrating sphere was reported by Melsheimer and Schlögl (1997). The design was further developed and has been described in detail by Thiede and Melsheimer (2002). The heated zone is set apart 12 mm from the sphere, and this gap is bridged by a tubular ceramic spacer with a high reflectivity. The reference port was equipped with an equivalent ceramic piece. The reactor consists of an inner tube (i.d., 15 mm; o.d., 20 mm) and an outer tube (i.d., 22 mm; o.d., 25 mm). An optical window closes one end of the outer tube and points towards the sphere. The powder bed is in the inner tube, and held in place between the window and a frit, which is located 28 mm into the inner tube. The sample can, thus, be considered "infinitely thick" with respect to the reflectance spectroscopy. The gas penetrates through the frit and escapes through crevices in the contact area between window and inner tube. Cooling blocks were added to the sides of the sphere and purged with room temperature water. With all the pieces made of quartz and the seals sufficiently distant from the heated zone, this reactor is suitable for high temperatures; measurements at temperatures up to 623 K have been reported (Thiede and Melsheimer, 2002).

A disadvantage of the design is the position of the bed at the end of the oven rather than in the isothermal zone. The volume of the cell requires gram-amounts of catalyst, and care must be taken to avoid shifting of the powder bed when the reactor is placed in the oven. At higher temperatures, thermal radiation disturbs the spectra in the NIR range (the thermal noise becomes more obvious at high gains).

The flow and diffusion paths in the bed were tested by flowing water vapor containing gas over a bed of silica gel with a color indicator. Depending on the flow rate, the bed was first wetted in the center or at the edges, indicating changing and, thus, suboptimal flow paths. An advantage from a spectroscopic point is that the intensities of sample and reference beam (with a white standard in the reference position) are similar. There are usually no problems with artifacts from changes of spectrometer components except for the detector and grating switch between the UV–vis and NIR ranges. A simple method to generate a background correction is to place a white standard in the reactor.

In a subsequent construction, Melsheimer et al. (2003) solved some of the above-mentioned problems by introducing a light conductor in the form of a tapered quartz rod, which guides the light by total reflection. This 120-mm long light conductor allows horizontal placement of the bed in the isothermal zone of the oven. The beam is focused so that the analyzed sample area is only 25% of its original size. The thermal noise is reduced accordingly, and smaller amounts of sample may be used. With this design, the sample and reference beam path become markedly different, so that a relatively elaborate correction procedure becomes necessary, which includes the measurement of a standard inside the reactor versus a standard in the reference position—and also the measurement of a soot sample in the reactor versus a standard in the reference position. This "dark spectrum" corrects for stray light that may enter through the rod. Spectra of good quality throughout the entire UV–vis–NIR range can be recorded at temperatures up to 723 K with a reactor similar to previous designs (Thiede and Melsheimer, 2002; Melsheimer and Schlögl, 1997).

By combining a Spectratech environmental chamber with an integrating sphere, Négrier et al. (2005) were able to record spectra UV–vis spectra at temperatures up to 503 K in controlled gas atmospheres.

A coolable cell design was presented by Bailes et al. (1996). The apparatus is made of silica and has a bulb-shaped section for pretreatment and a section for spectroscopy that can be cooled with liquid nitrogen.

A thick-walled (0.5-cm) quartz cell with Suprasil window was used by Weckhuysen (2002) and Weckhuysen et al. (2000) to monitor the formation of molecular sieves under hydrothermal conditions; data recorded at 448 K have been published.

5.4. Cells for Use with Mirror Optics

Reaction chambers fitting the Harrick Praying Mantis mirror optics are available commercially, and sketches or images are presented in the product description (Harrick, 2006), in the work of Weckhuysen and coworkers (Weckhuysen and Schoonheydt, 1999; Weckhuysen et al., 2000; Weckhuysen, 2002; Weckhuysen, 2003; Weckhuysen, 2004) and in a handbook article by Sojka et al. (2008). A low-pressure and a high-pressure version, suitable at pressures up to 202 303 kPa or 3.4 MPa (500 psi), are available; they are characterized by a dome with either three flat, circular windows or a dome with a single quartz half-sphere shaped quartz block with a small (also half-sphere shaped) volume above the catalyst. Evacuation to pressures less than 1.33×10^{-6} hPa and a maximum temperature of 873 K (under vacuum) are specified. A low-temperature version is specified for 123–873 K and up to 202–303 kPa. In the low-pressure versions, there are several centimeters of beam path through the gas phase, so that gas phase contributions are more likely to be observed than in experiments with cells holding the sample directly at the window (this depends on the gas phase concentrations and molar absorption coefficients).

Depending on the density of the sample, the Harrick reaction cell is filled with about 100–150 mg of catalyst. The catalyst rests on a wire mesh (various mesh sizes are available), allowing for flow through the bed. The cell has three ports, one of which leads to the space underneath the bed and can be used for gas removal in flow mode or via evacuation. Gas bypassing of the bed (channeling) has been suspected (Weckhuysen et al., 1998), and this suspicion led the group Iglesia to replace the wire mesh by a quartz frit (Argyle et al., 2004). Effluent stream analysis has been performed by gas chromatography (Weckhuysen and Schoonheydt, 1999; Klose et al., 2006).

The bed is heated by contact with the holder, which is heated from the bottom by a cartridge heater. Deviations between the actual surface temperature of the bed and the temperature given by the readout of the controller for various versions of the cell were reported by Venter and Vannice (1988), who used the unit for IR spectroscopy, and by Gao et al. (2002). The surface temperature was measured with an extra thermocouple placed into the powder bed. At a flow rate of 50 ml/min and a nominal temperature of 573 K, the surface temperature was found to be 96–125 K lower than the nominal value, depending on the gas in the cell (Gao et al., 2002). These deviations suggest that there may often be significant temperature gradients in the bed in the reaction chamber. Schulz-Ekloff et al. (1995) mounted wafers on a gold grid sample holder to minimize temperature gradients, and Venter and Vannice (1988) modified their cell (HVC-DRP) to minimize heat transfer losses and introduced an extra thermocouple in the sample bed. Brik et al. (2001) positioned a thermocouple "above the sample". In another publication, Brik et al. (2001) stated that they recorded all spectra at room temperature to avoid problems with light emission by the samples.

5.5. Cells for Use with Fiber Optics

Designs for use with fiber optics have been presented by a number of groups, many times in combination with other methods (Section 5.6). In 2001, the group of Weckhuysen reported placing an optical fiber probe a few millimeters above the catalyst bed inside a tubular reactor (Puurunen et al., 2001). Effluent gases were analyzed by online gas chromatography. Spectra were recorded at 853 K and atmospheric pressure. Although this placement of the fiber probe gave the best signal-to-noise ratio, the authors reported that the portion of the catalyst bed that was analyzed (the uppermost layer) was not always representative of the entire bed (Groothaert et al., 2003). Hence, Groothaert et al. (2003) described a slightly modified design with the fiber probe located outside the quartz reactor wall, and UV–vis spectra acquired at temperatures up to 773 K were published. UV–vis fiber probes have also been mounted above a movable sample holder to achieve spatial resolution (van de Water et al., 2005b).

A technique that is potentially of interest for heterogeneously catalyzed reactions with liquid-phase reactants was presented by Zimmermann and

Spange (2002). A diode array spectrometer with glass fiber optics was used to measure adsorbed species through a quartz window in the bottom of the reactor after sedimentation of the particles. Tromp et al. (2003) also mentioned an optical fiber probe for such an analysis.

5.6. Cells for Combination of UV–vis Spectroscopy with Other Methods

Weckhuysen and coworkers (Nijhuis et al., 2003) described equipment suitable for parallel Raman and UV–vis spectroscopic measurements. Openings on the opposite sides of a furnace allowed acquisition of Raman and UV–vis spectra through optical grade windows in a tubular quartz reactor. UV–vis spectra were recorded at 823 K. Gas-phase analysis was achieved with mass spectrometry and gas chromatography. A more advanced version of the design (Nijhuis et al., 2004) accommodates four optical fiber probes, placed at 10-mm vertical spacing along the tubular reactor. The temperature that the fibers can withstand is 973 K; the reported spectra characterize samples at 823 K.

Extending the equipment, the authors (Beale et al., 2005) recently added energy dispersive X-ray absorption spectroscopy (XAS). Raman and UV–vis spectra are recorded by illuminating opposite sides of a catalyst bed in a vertical tubular reactor and detecting the scattered and reflected light as described above. XAS is performed in the same horizontal plane but in transmission and with the beam orthogonal to the incident radiation of the other two methods. Example spectra were recorded for samples at 823 K. A combination of UV–vis (fiber optics) and XAFS spectroscopy for investigation of solids has also been described by Jentoft et al. (2004), who reported UV–vis measurements of samples at 773 K.

Weckhuysen and coworkers (Mesu et al., 2005; Tinnemans et al., 2006) also presented a combination of UV–vis (fiber optics) and XAFS (energy dispersive) spectroscopy for the analysis of homogeneous liquid-phase reactions. Both methods are performed in transmission, with the beams crossing in a cuvette (5-mm path length each) with quartz windows.

Brückner (2001) combined UV–vis with EPR spectroscopy, using online gas chromatography for product analysis. For many transition metal ions, EPR and optical spectra are complementary, in that some states are detectable or distinguishable with only one of the methods. The UV–vis facility was added to a previously described flow reactor system for EPR spectroscopy (Brückner et al., 1996). A fiber optical probe (Avantes, AVS-PC-2000 plug-in spectrometer) was inserted directly into the reactor via a Teflon®-sealed feedthrough and placed in the catalyst bed. UV–vis spectra were reported for temperatures up to 810 K. The design was later expanded to include a third method, Raman spectroscopy (Brückner, 2005; Brückner and Kondratenko, 2006). A hole in the

EPR cavity was used to focus the laser beam on the side of the catalyst bed; heating was achieved by hot flowing nitrogen. UV–vis spectra were measured at temperatures up to 673 K.

Hunger and Wang (2004) combined UV–vis with NMR spectroscopy. UV–vis fiber optics was integrated into the NMR equipment by attaching a fiber to the bottom of the stator. UV–vis spectra were recorded through a quartz window at the bottom of a 7-mm MAS NMR rotor (AvaSpec-2048 Fiber Optic spectrometer, Avantes glass fiber reflection probe). The achievable experimental conditions appear to be limited by the modified variable-temperature NMR probe (Bruker Biospin; Hunger and Horvath, 1995) rather than by the fiber optics system. Published data were recorded during operation of the system as a flow reactor at temperatures up to 673 K (Jiang et al., 2007).

Bürgi (2005) described equipment combining attenuated total reflection IR with reflectance UV–vis spectroscopy. A fiber optical probe was positioned parallel to the surface normal of the internal reflection element in front of a fused silica window in the reactor. The distance from the probe to the catalyst layer was about 4 mm. A second fiber optical arrangement allowed simultaneous acquisition of transmission UV–vis spectra of the effluent liquid from the reactor cell. All measurements were performed with the catalyst at 303 K.

Lezna and coworkers (Juanto et al., 1994; Lezna et al., 2003) combined diffuse reflectance UV–vis spectroscopy with cyclovoltammetry, using a specially designed spectrometer with an optical multichannel analyzer.

5.7. Spectrometer Requirements

The intensity of the UV–vis–NIR radiation that is transmitted or reflected by a powder sample can be very low. It is advantageous to work with a spectrometer with a large photometric range (e.g., the PerkinElmer Lambda 950 and the Varian Cary 5000 are specified to operate up to an absorbance A_{10} of 8). Monitoring of a catalyst during many hours on stream requires high instrument stability. Wavelength (x-axis) stability is usually not a problem, but slight, wavelength-dependent drifts with time in the intensity (y-axis) may occur, and the stability of the instrument should be tested to ensure that such drifts are not confused with changes in a catalyst.

Background correction and sample measurements are performed consecutively and, unless the original state of the sample is used for background correction, the specimen that is analyzed needs to be exchanged. Hence it is important that filling of material into any sample holder or reactor be reproducible. Specimen manipulation should not affect the relative positions of source, optics, sample holder and detector—these parts must remain in position, or their placement must be well defined. In general, instruments with integrating spheres are less sensitive than mirror optics with respect to deviations in sample positioning. A change in the beam path between background correction and sample measurement can lead to offsets in reflectance of sections of the spectrum, generating steps or spikes at wavelengths where optical elements are switched.

Typical artifacts arise from changes of (1) lamps (e.g., switches at about 300–350 nm from a deuterium lamp for the UV regime to a tungsten halogen lamp for the vis–NIR range); (2) monochromators (e.g., from UV–vis to NIR gratings between 700 and 900 nm); (3) detectors (e.g., from a photomultiplier to PbS detector, depending on the multiplier range); and (4) filters, several of which are usually used throughout the range. The wavelength positions of these changeovers are available from the instrument manual. Typically, lamp, monochromator, or detector changes can be shifted by some wave numbers in case there is interference with an important spectral feature.

In UV–vis–NIR spectrometers, the monochromator and detector are switched simultaneously. Step-like artifacts can be generated at this switch, and it is then questionable which part of the spectrum represents the correct absolute intensity. By nature, NIR detectors are susceptible to thermal radiation, and the step at the change-over to or from the NIR range and also the noise in the NIR range increase with temperature (Melsheimer et al., 2003). Sometimes authors present the UV–vis and NIR sections of the spectrum separately, disguising step-like artifacts at the transition.

When working with rapidly changing catalysts (e.g., during a start-up phase or deactivation), time resolution becomes an issue. The scan speed for modern spectrometers reaches more than 2000 nm/min (PerkinElmer Lambda 950); however, as the photon flux on the detector is small when specimens in a reactor are measured in diffuse reflection, it is usually not possible to obtain high-quality spectra at the highest speeds. Once sections of interest have been identified in the spectra, time resolution can be improved by restricting the spectral range or fixing the wavelength. For example, Argyle et al. (2005) reported measurements with a time resolution of 0.1–1.0 s at a wavelength of 1.86 eV (667 nm). Fiber optic probe spectrometers with array detectors require very short times to record a spectrum and thus offer a strong advantage over classical spectrometers. For example, Bürgi (2005) reported an integration time of 100 ms for one spectrum, with the wavelength range being ~300–650 nm.

6. DATA ACQUISITION AND ANALYSIS

6.1. Reference Measurement (Background Correction)

Because a measurement under a defined atmosphere requires that the sample be placed in a cell, corrections for both the optical components (of the spectrometer and the accessory) and the reflection and absorption by the window(s) become necessary.

In transmission spectroscopy, the cell serves as a reference to compensate for the window contributions. If there are no changes to the catalyst spectrum, and only the absorption of adsorbed species is of interest, then the activated catalyst sample in the cell may be taken as a reference. The diffuse reflectance of a sample of interest is always measured relative to a white standard (Section 4.2). The reference measurement, which by the instrument software may be referred to as "background correction" or "autozero", is thus performed with a white standard in the sample position. The use of the bare support as a reference material as in transmission spectroscopy is not *a priori* possible, according to the equations shown in Section 2.6. Nevertheless, such referencing is applied frequently (Weckhuysen and Schoonheydt, 1999), and in some scenarios it is justified. Essentially, the support can serve as a reference material for the background in those spectral ranges in which it acts like a white standard—at wavelengths at which its reflectance is close to 1. The deviation from $\rho = 1$ is often small for a white, fine powder, as many supports are, and not significantly greater than for typical white standards. A good example is the work by Argyle et al. (2003, 2004) characterizing VO_x/Al_2O_3 species. The preedge region was analyzed and does not show any absorption for the fully oxidized catalyst. Hence, the spectrum of the oxidized sample was used as a reference. The advantage here is that the change of specimen from reference material to catalyst is omitted, and slight deviations between background and catalyst measurement because of different fill levels of the cup or repositioning of the cell or optics as a consequence of the change are circumvented. Although a small error is introduced because the catalyst may not behave as a perfect white standard, errors induced by sample change are avoided and by instrument instabilities over time are minimized. In practice, the approach is convenient, as time-consuming sample change and cell cleaning are not necessary. On the other hand, one must be aware of the limitations: the only ranges of the final catalyst spectrum that are interpretable are those that do not show any absorption when the reference measurement is made. In ranges with absorption of the reference material—be it a standard, a pure support, or some state of the catalyst—the resulting spectrum of the sample recorded relative to the reference material will be distorted. Hence, in the investigation of supported species, the region of support absorption, which frequently consists of an edge-type feature, will be compromised.

Gao and Wachs (2000) investigated the influence of the reference material on the spectra using a diffuse reflectance attachment with an integrating sphere. VO_x species supported on alumina, zirconia, titania, niobia, ceria, and silica served as test compounds. Edge energies characterizing the supported species were affected by the choice of reference; they could not be extracted from the data when the support was used as a reference and its edge energy was less than or equal to that of the VO_x

species. The number of bands and their intensity varied, depending on the method of referencing, confirming the predictions stated in Section 2.6. Thus, whether the support can serve as a reference depends on its absorption characteristics and the wavelength range of interest. Further attempts to work with the support as a reference have been made by Rao et al. (2004) with a zirconia support, by Hartmann et al. (2001) with Al–MCM-41 as a standard when investigating ruthenium clusters supported on this material, and by Brosius et al. (2005) with Al_2O_3 as a reference for Ag/Al_2O_3.

An additional correction that can be made is the elimination of stray light contributions. A "dark spectrum", for example, of a black material, is recorded and its intensity subtracted from all measurements; the possibility of such a correction is available in modern spectroscopy software. Such a procedure is recommended with instruments that allow for stray light to enter the beam path. In many cases, the normal spectrometer cover cannot be used or closed, because then the cell could not be accommodated in the sample compartment. Feedthroughs for gases and heating also offer potential openings where light can enter. Shielding from light becomes more difficult as fibers are integrated into multimethod assemblies.

In some cases, researchers have recorded the reference spectra at the same temperature and in the same gas atmosphere as were applied for the respective measurements of the sample. Particularly in the NIR range, a temperature correction can become necessary. For example, Sels et al. (2001) used a spectrum of the catalyst matrix (layered double hydroxide) suspended in methanol and aqueous H_2O_2. Brik et al. (2001) recorded their $BaSO_4$ reference spectra at various temperatures. Lu et al. (2005) recorded the spectra of their catalyst support, $CaCO_3$, as references under the conditions of the catalytic reaction, that is, in a mixture of ethene or propene in O_2 and He at 473 K. Rao et al. (2004) heated their reference material to reaction temperature to generate the baseline.

Reference and sample measurements are performed consecutively, and the resultant (sample) spectrum is obtained as the ratio of the two photon fluxes onto the detector. In a single-beam spectrometer, there are no other options; in a double-beam spectrometer, the photon fluxes of the sample and reference beam path are compared. When an integrating sphere is used with two ports and a white standard in the reference position, the photon fluxes are comparable to each other, and no problems occur. Note that the ports are part of the sphere and that any material change in the reference or sample position will change the average sphere reflectance ρ_{ave}. The reference measurement should be conducted with exactly the same components (windows) as the sample measurement; otherwise, "substitution errors" may occur.

When mirror or fiber optics are used with a double-beam spectrometer, the difference in throughput between the sample and reference beam becomes large. An example is given in Figure 11B, which shows a through-put measurement with a mirror in the sampling position (upper curve) and a spectrum of Spectralon® inside the Harrick reaction cell in the Praying Mantis accessory (lower curve). The background correction was performed with two empty beam paths, and so the spectrum shows the cumulative attenuation through Spectralon®, windows, and mirrors. Spectral quality can be improved through attenuation of the reference side (Figure 12). Various attenuators were placed in the reference beam, a background correction was recorded, and then—without a change in configuration—the spectrum was recorded so that a flat line at $\rho = 1$ is expected over the entire wavelength range. The data show that in this case the 10% attenuator produces the best result.

An elaborate referencing procedure was suggested by Schulz-Ekloff et al. (1995), who could not fit their standard into the Praying Mantis reaction chamber. Groothaert et al. (2003) rescaled spectra to the most intense absorption bands because of instabilities of the fiber optics arrangement.

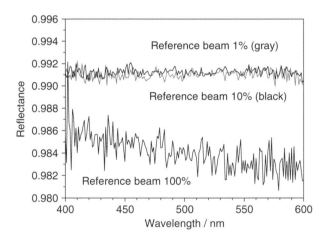

FIGURE 12   Effect of attenuation of reference beam. When working with a double beam spectrometer and a mirror optics accessory, the photon fluxes on the detector from the sample and the reference beams differ considerably (Figure 11). Better spectral quality can be obtained if the reference beam is attenuated to match the flux from the sample beam. The built-in attenuators of a PerkinElmer Lambda 950 spectrometer were used in this example. For the background correction, the reference beam was not attenuated, attenuated to 10%, or attenuated to 1% of its original intensity. In the sample beam was the Harrick Praying Mantis DRP-P72 accessory with a piece of Spectralon® inside the reaction cell HVC-VUV. The spectra were recorded without a change in configuration and hence the result should be $\rho = 1$ over the entire range. The accuracy is about equally good for the 1% and the 10% attenuator, but the signal-to-noise ratio is slightly better for the 10% attenuator.

Unfortunately, the exact procedure of referencing is omitted in many papers.

6.2. Sample Preparation

For transmission measurements, samples are typically pressed into thin wafers, whereas for measurements in diffuse reflection powders are used. To suppress regular reflection and particularly to work in a range in which $F(\rho)$ is proportional to the concentration, samples often have to be diluted. The diluent should be a nonabsorbing standard that does not interact with the sample, and the mixing procedure should be nondestructive (some samples are sensitive to mechanical stress (Section 2.4)).

The dilution method was investigated by Kortüm and coworkers (Kortüm and Schreyer, 1955, 1956, Kortüm et al., 1963), who used high dilutions of $1 \cdot 10^3$– $1:10^5$ (by molar fraction) Warnken et al. (2001) diluted SnO_2/NaY zeolite samples with the parent support when $F(\rho)$ exceeded 3. Rybarczyk et al. (2001) also diluted their vanadate catalysts; for nonreactive conditions they used $BaSO_4$, otherwise, they used $\alpha\text{-}Al_2O_3$ (calcined at 1473 K for 4 h). The ratio of catalyst/diluent or the desired $F(\rho)$ range was not specified, but in all the reported plots, $F(\rho)$ was <3. Gao et al. (2001) targeted $F(\rho) < 1$ when diluting their niobia catalysts with silica, which was the support material. Bentrup et al. (2004) diluted their VO_x/TiO_2 catalysts 1:10 with $\alpha\text{-}Al_2O_3$ that had been calcined at 1673 K for 4 h. The same material was used by Pérez-Ramírez et al. (2004) to dilute FeMFI zeolite samples in the ratio 1:3 (catalyst to diluent), and by Kumar et al. (2004) to dilute FeMFI zeolite samples in the ratio 1:10. Kondratenko et al. (2005) diluted $VO_x/MCM\text{-}41$ samples 1:10 with alumina.

Gao and Wachs (2000) diluted supported VO_x species to achieve the criterion $F(\rho) < 1$ and compared their spectra with those of the pure compounds. However, the diluent was also used as a reference, and so the effect of dilution alone was not evident. The band gap energy was found to be little affected by the diluent; resolution appeared improved. Furthermore, band location was reported to be "much better"; this statement presumably refers to more clearly identifiable maxima.

Melsheimer et al. (1999) diluted heteropoly acid catalysts with quartz in a ratio 1:13 by using an incipient-wetness impregnation from aqueous solution, the ratio being a compromise between measurable catalytic activity and spectral response. Lee et al. (2001) reported that dilutions of up to 1:100 in Suprasil® were necessary for heteropoly acids to obtain "well-resolved spectra."

For diffuse reflectance, the sample should also be "infinitely thick"— no light should be transmitted. Simmons (1975) applied the modified particle model theory and concluded that for most powdered samples,

1-mm-thick samples can be considered to be "infinitely thick"; his calculations show a rapid decay of the incident light intensity with increasing penetration into the sample. Other authors, in contrast, concluded that thicknesses of 2–3 mm and in some cases up to 5 mm are sufficient (Che and Bozon-Verduraz, 1997; Schoonheydt, 1984).

It is thus evident that, for investigations of catalytic reactions, the spectroscopically analyzed volume in most cases will represent only a fraction of the entire catalyst bed. The size of the sampled volume has been investigated as a function of the materials properties (particle size, refractive index) for MIR radiation (Moradi et al., 1999).

If catalytic performance is to be correlated with spectroscopic data, the analyzed volume must be representative of the working catalyst. Impurities from the feed that may accumulate in the first layers of a catalyst bed can be a problem (Groothaert et al., 2003). Furthermore, high conversions will lead to large composition gradients in the fluid phase, resulting in catalyst states and surface species that differ from position to position.

6.3. Handling and Representation of Spectra

A UV–vis measurement delivers a relative measurement of photon fluxes; depending on the positioning of the detector, the measured quantity is referred to either as transmittance τ or reflectance ρ, ranging from 0 to 1 (0–100%). Fine powder samples scatter so strongly that a conversion of transmittance into absorbance may not be justified. The data obtained in reflection mode measurements are presented in many ways. Logical choices include the "typical color curve", a representation of $\log F(\rho)$ as a function of the wavelength, which corresponds to the absorption spectrum of the sample shifted along the ordinate, and the Kubelka–Munk function $F(\rho)$, as discussed above. Sometimes $\log (1/\rho)$ is used, which is calculated from reflectance analogously as absorbance is calculated from transmittance. Spielbauer et al. (1996) and Kuba et al. (2003) used $1 - \rho_{\infty}$ as a measure of absorption by the sample. However, the equation $\alpha + \rho + \tau = 1$ is valid only in the absence of scattering (Homann, 1996); that is, $1 - \rho$ is simply a representation with absorption bands pointing in positive direction, equivalent to the conventions used for most other data and thus potentially more easily received. In many papers the terms "absorbance" (Brosius et al., 2005; Brückner, 2001) or "absorption" (Weckhuysen and Schoonheydt, 1999) are used for representing diffuse reflectance data, but it is not specified how these quantities have been calculated in this case. Sometimes, data are normalized for comparison; for example, Kumar et al. (2004) normalized their spectra of FeMFI zeolites to the absorption maximum.

It is sometimes desirable to extract information characterizing a particular species from a series of spectra. For example, a supported MO_x species may change during interaction with a feed gas. If the spectral features of the surface species overlap those of a support, then, provided the support does not change, difference spectra can be helpful. The only proper way to represent them is via Equation (27):

$$F_{ads} = F_{cat+ads} - F_{cat} \qquad (27)$$

In Figure 13, a series of spectra are presented that were obtained during isomerization of n-pentane catalyzed by sulfated zirconia. These spectra were measured with an integrating sphere, and a background correction was run with Spectralon® in the reference and in the sample position. Then the spectrum of the catalyst in the reactor was recorded with Spectralon® in the reference position. Figure 13A shows the spectra as measured in reflectance. With time on stream, a band grew in the range between 300 and 350 nm, which is assigned to unsaturated surface deposits.

Proper procedure according to Equation (27) involves conversion of all individual spectra into $\Gamma(\rho)$ and subtraction of the spectrum of the activated catalyst. The result is shown in Figure 13B. This method may not be applicable in regions of low reflectance, in which $F(\rho)$ will not deliver trustworthy results. Other analysis procedures have been applied previously (see Section 6.1 and this section) and are demonstrated here as a basis for estimating the errors associated with them. The first procedure is transferred from transmission spectroscopy, whereby the spectrum of the surface species can be obtained via measurement of support + surface species versus the support. Such a scenario was mimicked here mathematically by dividing ρ_{TOS} (the reflectance value of the catalyst with the surface species at certain time on stream) by ρ_{act} (the reflectance of the activated catalyst). The result of conversion into $F(\rho_{TOS}/\rho_{act})$ is presented in Figure 13C.

The second procedure is based on the simplified assumption of an "apparent absorbance" that is equal to $1-\rho$. The difference spectra calculated as the difference of the apparent absorbances of the ρ_{TOS} and ρ_{act} are depicted in Figure 13D.

It is evident that these "incorrect" procedures in the example lead to a shift of the band towards higher wave numbers, because the band characteristic of the deposit is located on a sloped and not on a flat section of the catalyst spectrum. Furthermore, the band shape is slightly changed, and the absolute intensity values are much smaller.

Values of $\rho > 1$ are occasionally obtained. In some cases, they may be experimental artifacts, arising from an imperfect background correction (e.g., an aged standard, differences in the specimen position for reference and sample measurement, or instrumental drifts). However, fluorescence can also contribute to an increased reflectance towards the high-energy end of the spectrum. Instruments usually have primary monochromators and, depending on the wavelength position of the fluorescence signal, the detector (photomultiplier) will respond. The photomultiplier does not

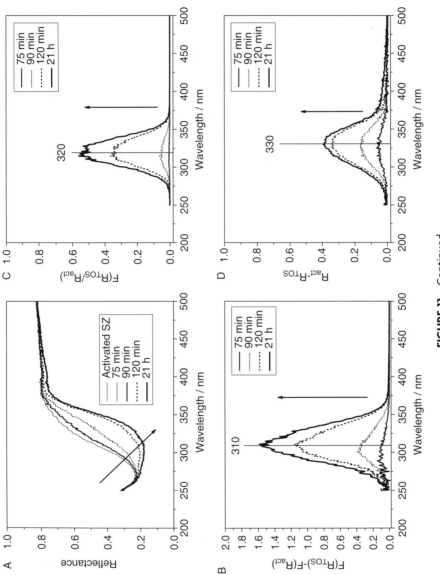

FIGURE 13   Continued

discriminate between different wavelengths but becomes more sensitive with increasing wavelength; hence the reflectance seemingly exceeds 1. Nelson et al. (1971) described this phenomenon for vacuum-activated MgO, attributing the fluorescence to excited surface states. Zecchina et al. (1975) confirmed these observations.

6.4. Quantification by Independent Calibration

Because the molar absorption coefficients of absorbing species in catalyst materials are usually not known, a calibration needs to be performed to relate intensity (at a certain wavelength or an integral over a range) to concentration. In a best-case scenario, the concentration is measured by a second independent method applied simultaneously. Published attempts include the use of EPR spectroscopy for the determination of the concentration of V^{4+} in VO_x/support materials; in this case, the EPR tube was a "side-arm" connected to the UV–vis cell, and samples could be transferred without exposure to air (Catana et al., 1998). EPR intensity and UV–vis intensity (KM function at a particular wavelength) were linearly correlated with each other.

However, some limitations apply to EPR spectroscopy, as only paramagnetic species can be detected. Brückner and Kondratenko (2006) determined the degree of reduction of vanadium in VO_x/TiO_2 by measuring H_2 consumption upon reduction by mass spectrometry and then related the change in "absorbance" in fiber optics UV–vis spectra to the O/V ratio. A nonlinear calibration curve was obtained. Argyle et al. (2004) analyzed the O_2 consumption upon reoxidation as measured by mass spectrometry to determine the degree of reduction of VO_x/Al_2O_3. Mirror optics UV–vis spectra were converted to the Kubelka–Munk function, and the intensity in the preedge range (1.49–1.86 eV) was integrated. A linear correlation of the

---

FIGURE 13   (A) Reflectance spectra of sulfated zirconia during catalysis of *n*-pentane isomerization. Data were obtained with a Perkin Elmer Lambda 9 spectrometer with 60 mm integrating sphere and reactor cell from Ref. (Thiede and Melsheimer, 2002). Background correction: reference position, Spectralon®; sample position, Spectralon® behind quartz window. Measurement: reference position, Spectralon®; sample position: 1.245 g of sulfated zirconia catalyst in reactor cell behind quartz window. Reaction temperature 300 K, 0.25 kPa partial pressure of *n*-pentane in He at a total flow rate of 50 ml/min. (B) Difference spectra after calculation of the Kubelka–Munk functions; spectrum of activated catalyst at 300 K in He subtracted from spectra taken during reaction. (C) Simulation of using the spectrum of the activated catalyst as a reference spectrum: calculation of the Kubelka–Munk function after division of the spectra taken during reaction by the spectrum of activated catalyst at 300 K in He. (D) Difference spectra after calculation of the apparent absorbance 1−ρ.

integrated intensity and the degree of reduction was obtained. There are several effects that could explain nonlinear relationships: different reduced vanadium species feature different molar absorption coefficients, and their ratio may change; the dispersion may change upon the reduction; and a reducible support may account for some H_2 or O_2 consumption.

7. EXAMPLES

This section is organized according to the increasing experimental difficulty and complexity of the measurements. Investigations that report only treatment in a defined atmosphere without the occurrence of catalysis are first presented in two brief sections for completeness and to provide an overview regarding which catalysts have been investigated by UV–vis spectroscopy (furthermore, important reference spectra are generated in this way).

The first section deals with spectra recorded at room temperature after a treatment and the second section with measurements in a defined gas atmosphere at a defined temperature higher or lower than room temperature. The succeeding sections address catalytic reaction experiments with simultaneous spectroscopy and gas-phase analysis.

7.1. Room-Temperature Measurements after Catalyst Treatment

Table 1 gives an overview on the types of catalyst that have been investigated by UV–vis spectroscopy at room temperature after treatments.

Several major areas of interest can be recognized, namely (1) the state and location of metal cations and clusters in zeolites, (2) the nature of carbocations and unsaturated hydrocarbon species (exclusively in zeolites), and (3) the nature of oxide-supported highly dispersed transition metal oxide species. The first and the third of these areas show that UV–vis spectroscopy is a useful laboratory method for analyzing the local environment of metal cations in cases where long-range order is absent.

Transmission geometry has been exclusively applied to the investigation of zeolites. Sendoda et al. (1975) compared transmission and reflection spectra and reported good agreement, except that the transmission mode gave "more intense bands". Förster et al. (1983) compared their transmission spectra of CoNaY to reflection spectra in the literature and claimed that the former are of at least equal quality.

Attempts have been made to relate spectroscopic quantities and catalytic behavior. Kim et al. (2007) correlated the performance of heteropoly acid catalysts in isobutyric acid oxidative dehydrogenation (ODH) to the position of the absorption edge after treatment at 603 K.

TABLE 1 Measurements of Catalyst Materials at Room Temperature after Treatment.

Catalyst	Treatment	Target	Equipment and further detail	Reference
Metal cations and clusters in microporous and mesoporous materials				
Ni-exchanged zeolite A	Dehydration and adsorption of N_2O, ethene, propene, cyclopropane	State of nickel, aquo complexes	DR/IS	Klier and Rálek, (1968)
Ni-exchanged zeolites A, X, and Y	Dehydration at temperatures up to 723 K	Location of nickel in cages depending on degree of hydration	DR/IS Crystal field theory	Garbowski et al. (1972)
NiX zeolite	Evacuation at up to 773 K, adsorption of ammonia, adsorption of pyridine followed by heating	Analysis of nickel complexes, location of nickel	TR, DR	Sendoda et al. (1975)
$Ru(NH_3)_6^{3+}$ exchanged zeolites X and Y	Evacuation at up to 573 K, flow of O_2 at temperatures up to 473 K, NO at temps. up to 423 K	$Ru(NH_3)_6^{3+}$ chemistry in zeolite cage	DR	Verdonck et al. (1981)
$Pt(NH_4)_4^{2+}$-exchanged HZSM-5	O_2-containing atmosphere, temperatures up to 723 K, moist air	State of platinum	DR/IS	van den Broek et al. (1997)

(continued)

TABLE 1 (*continued*)

Catalyst	Treatment	Target	Equipment and further detail	Reference
Co-doped ferrierites	Dehydration at 770 K	Location of Co^{2+}	DR/IS UV–vis–NIR	Kaucky et al. (1999)
CoZSM-5	Vacuum activation and NO adsorption and desorption	Symmetry of Co(II) species	DR/IS	Park et al. (2000)
NaY and Al-MCM-41 encapsulated rhenium complexes	Evacuation at 373 K, photoirradiation and CO_2 adsorption	Reactions of rhenium complexes	DR	Sung-Suh et al. (2000)
Al-MCM-41 supported ruthenium complexes	Dehydration in vacuum at up to 573 K	Ruthenium valence and environment	DR	Hartmann et al. (2001)
Rhenium complex encapsulated in NaY zeolite	Evacuation, photo-irradiation, and CO_2 adsorption	Reactions of rhenium complex	DR	Hwang et al. (2001)
Co-containing zeolites and CoO_x/Al_2O_3	Outgassing at 773 K in vacuum	State and location of cobalt	DR	Busca and coworkers (2003); Finocchio et al. (2003), Montanari et al. (2004), Resini et al. (2003)

	Treatment	Purpose	Technique	References
FeMFI	Air at 873 K, after use in selective catalytic reduction (SCR) of NO with isobutene	State of iron	DR/PM	Kumar et al. (2004)
Silver clusters in MFI zeolites	Treatment in O_2, H_2 or a mixture of NO and propene (for SCR)	Nature of silver clusters	DR	Shibata et al. (2004a,b)
Bulk oxides				
MgO	Outgassing at 1273 K, exposure to O_2 and γ-irradiation	Surface oxygen species (superoxide)	DR/IS	Nelson et al. (1971)
MgO, CaO	Outgassing at 1073 K, exposure to O_2, N_2O, CO_2, or H_2O	Surface excitons	DR	Zecchina et al. (1975)
MgO, CaO, and SrO	Outgassing at 1073 K, adsorption of NH_3, followed by O_2 or CO	NH_3 dissociation	DR	Garrone and Stone (1985)
MgO, CaO	Outgassing at 1073 K, adsorption of CO, coadsorption of CO and H_2	Adsorption and reaction products on surface	DR	Garrone and Stone (1987), Garrone et al. (1988)

(continued)

TABLE 1 (continued)

Catalyst	Treatment	Target	Equipment and further detail	Reference
Carbocations in microporous and mesoporous materials				
Zeolite Y with various cations (H^+, Pd^{2+}, Ni^{2+}, Co^{2+})	Adsorption of ethane, propene, butene and heating up to 573 K	Carbocations, formation of oligomers and aromatic compounds	DR	Garbowski and Praliaud (1979)
Zeolites	Outgassing at up to 870 K and exposure to hydrocarbons, alcohols, aldehydes	Spectral signature of carbocations	TR	Förster et al. (1983), Förster et al. (1986), Kiricsi (1987), Kiricsi and Förster (1988), Förster and Kiricsi (1988), Kiricsi et al. (1989a,b; 1990), for a review see Kiricsi et al. (1999)
HZSM-5	Evacuation at 773 K, adsorption of ethene or propene, desorption at up to 623 K	Analysis of surface intermediates of aromatization	DR/IS	Medin et al. (1989)

H-forms of various zeolites	Exposure to methylacetylene	Analysis of organic surface species	DR	Cox and Stucky (1991)
HZSM-5	Polymerization of acetylene on sample	Color changes of samples with polymer upon reaction with ammonia or propanamine	DR	Pereira et al. (1991)
H-mordenite	Outgassing at 673 K, propene adsorption	Unsaturated carbocationic species	DR	Geobaldo et al. (1997)
Ferrierite	Evacuation at 723 K, exposure to 1-butene, heating to 673 K	Cationic species and "coke"	DR UV-vis–NIR	Pazé et al. (1999)
Supported oxo-anions of transition metals				
VO_x supported on Al_2O_3	Dehydration	Nature of vanadium surface species	DR	Wu et al. (2005)
VO_x supported on silica, ceria, alumina, zirconia, niobia, titania-silica, zirconia-silica	Dehydration at 773 K in O_2/He, methanol adsorption	Band gap energies, nature of vanadium surface species	DR UV-vis–NIR	Gao et al. (1998, 1999b–c), Gao and Wachs (2000)
VO_x supported on alumina, silica	O_2/He at 773 K, adsorption of isopropanol	Edge energies, valence and dispersion of vanadium surface species	DR	Resini et al. (2005)

(continued)

TABLE 1 (continued)

Catalyst	Treatment	Target	Equipment and further detail	Reference
VO_x supported on silica	Preheated dry and wetted	Coordination of vanadium	DR/PM	Van Der Voort et al. (1997)
VO_x supported on MCM-41	Dehydration at 773 K in vacuum	Nature of vanadium species	DR/PM	Grubert et al. (1998)
VO_x supported on SBA-15	Dehydration in O_2/He at 573 K	Vanadium dispersion	DR/PM	Hess et al. (2004)
VO_x supported on SBA-15	Dehydration in dry air at 723 K	Valence and dispersion of vanadium	DR/PM	Liu et al. (2004)
$VO_x/MoO_x/Al_2O_3$ and $VO_x/CrO_x/Al_2O_3$	Dehydration in O_2/He at 723 K	State of chromium and vanadium, band gap energies	DR/PM	Yang et al. (2005)
Niobia dispersed on SiO_2 or MCM-41	Dehydration in O_2/He at 773 K	Edge energies, nature of niobium species	DR/IS UV–vis–NIR	Gao et al. (2001)
Cr-MCM-41	Dehydration	Chromium state	DR/IS	Zhu et al. (1999)
MoO_x/SiO_2	Reduction in ethanol or methanol vapor, "quenching" during ethanol or methanol oxidation	Molybdenum valence	DR/IS	Kikutani (1999a,b)
Titania-supported tungstate, EUROCAT SCR catalyst	dehydration, oxidation or rehydration	Valence and dispersion of vanadium and tungsten species	DR/IS	Busca et al. (2000)

Tungstated zirconia	O_2, H_2 at temperatures up to 673 K	Tungsten valence	DR	Kuba et al. (2003)
WO_x supported on Al_2O_3, Nb_2O_5, TiO_2, ZrO_2	Dehydrated in 10%O_2/He at 673 K	Edge energy	DR/PM	Kim et al. (2007)
Silica-supported V_2O_5, Nb_2O_5, Ta_2O_5, CrO_3, MoO_3, WO_3, Re_2O_7	Dehydrated in 10%O_2/He at 673 K, not clear if recorded at RT	LMCT bands and edge energies	DR/PM	Lee and Wachs (2007)

Miscellaneous bulk and supported catalysts

Ru/TiO_2	Reduction, use as catalyst in CO_2 methanation	Band gap energy	DR/IS	Melsheimer et al. (1991)
Na, K, or Cs-exchanged X- and Y-zeolites	Dehydration at 773 K and adsorption of iodine	Probing of basicity	DR/IS	Doskocil et al. (1999)
Si-, Ti-, and Al-MCM-41	Evacuation, adsorption of N-alkylphenothiazines, photoirradiation	Photoionization	DR	Krishna et al. (2000)
Titania-supported cobalt	Exposure to ethane and O_2 (for ODH) at 823 K	State of cobalt	DR/PM	Brik et al. (2001)
TiO_2/SiO_2	Used in photocatalytic degradation of dye	Surface intermediates	DR	Hu et al. (2003)
Heteropoly acid	Dehydrated at 603 K, not clear if reexposed to air	Edge energy	DR	Kim et al. (2006)

DR: Diffuse reflection geometry, TR: transmission geometry, IS: integrating sphere, PM: Praying Mantis

7.2. Spectroscopic Measurements during Treatments (In the Absence of Catalysis)

In Table 2, an overview of UV–vis spectroscopic measurements conducted during treatments is given. Novel experimental designs and observations of general importance are highlighted below.

Melsheimer et al. (1997) compared spectra measured in transmission and in reflection mode. Gas-phase contributions of H_2S were observable in the transmission spectra, which were believed to have their origin in the large volume (45 ml) of the cell. Hence, for measurements of catalysts in the presence of a fluid, the length of the beam path through the fluid, the concentration, and the absorption spectrum of the fluid have to be considered.

Van de Water et al. (2005a,b) developed a design for spatially resolved UV–vis spectroscopy. The fiber optical probe remains in a fixed position, while the specimen is moved by moving the platform with the cell. The spatial resolution is reported to be 100 μm. The apparatus was used to follow the impregnation of an alumina pellet with solutions containing nickel ions, chromate, or polyoxometallates.

Time-resolved measurements at fixed wavelength were used to investigate the kinetics of oxidation and reduction of bulk and supported oxides. Warnken et al. (2001) monitored the reduction of SnO_2 clusters; Bentrup et al. (2004) determined rate constants for the reduction and reoxidation of vanadium ions in VO_x/TiO_2 (V^{5+} to V^{4+}/V^{3+} and vice versa). A slightly different approach was taken by Barton et al. (1999), who investigated the kinetics of the reduction of WO_x/ZrO_2 by H_2 at 523 K. Integration of the Kubelka–Munk function over the section of the preedge region from 1.8 to 2.2 eV (690–563 nm) provided a measure of the degree of reduction, and the rate of formation of color centers was determined. Bensalem et al. (1997) investigated the reduction of CrO_x/Al_2O_3 by CO and used the intensity of a chromium charge transfer band in the spectra to follow the kinetics.

Tzolova-Müller et al. (unpublished data) followed the dehydration of $VO_x/SBA-15$, shown in Figure 14. The spectra provide a good example illustrating the usefulness of simultaneously acquired NIR data. The band at 1902 nm is typical for adsorbed water, and it can be seen by comparing the UV–vis and the NIR range that the vanadium environment changes the moment water desorbs (i.e., as the respective band vanishes).

Xie et al. (2001) measured UV–vis spectra of the bulk oxides niobia, molybdena, tungsten trioxide, and vanadia and determined the position of the band edge as a function of temperature. In a separate experiment, they investigated the samples by Raman spectroscopy. To understand the changes in the Raman intensities, they analyzed the intensity in the UV–vis spectra at the wavelength of the incident laser irradiation and at the

TABLE 2   Measurements of Catalyst Materials during Treatment.

Catalyst	Treatment or process	Equipment and further detail	Reference
Metal cations and clusters in microporous and mesoporous materials			
Platinum clusters in faujasites	Carbonylation of clusters at 363 K	DR/PM	Schulz-Ekloff et al. (1995)
Cobalt-exchanged zeolite Y	Mobility of Co^{2+}, temperature treatments	–	Weckhuysen and Schoonheydt (2000)
SnO_2 clusters—free and in zeolite NaY	Oxidation and reduction of SnO_2 clusters at 573 K	DR/PM Fixed wavelength mode	Warnken et al. (2001), Jitianu et al. (2003)
SnO_2, CdS, and rhodamine dyes hosted in MCM-41	Response of reflectance to CO, n-butane, and of fluorescence to SO_2 (use of materials as sensors)	DR/PM	Wark et al. (2003)
FeMFI	Reduction by H_2 or oxidation in air at 773 K, contact with N_2O or CO at 623 K	DR/PM	Pérez-Ramírez et al. (2004)
Bulk compounds			
HZSM-5	Adsorption of aromatic compounds, temperatures up to 573 K	TR	Melsheimer and Ziegler (1992)

(continued)

TABLE 2 (*continued*)

Catalyst	Treatment or process	Equipment and further detail	Reference
La_2O_3	Reaction of CO and O_2 on the surface at 90 K	DR	Bailes et al. (1996)
Various aluminas	Interaction with H_2S and SO_2, temperatures up to 400 K	DR, TR (Melsheimer and Ziegler, 1992; Melsheimer and Schlögl, 1997) comparison	Melsheimer et al. (1997), for reference spectra of heat-treated sulfur compounds see (Melsheimer and Schlögl, 1997)
SnO_2	O_2, temperatures up to 673 K	DR/PM	Popescu et al. (2001)
Bulk niobia, molybdena, tungsten trioxide, and vanadia	Temperatures up to 748 K	DR/PM	Xie et al. (2001)
Aluminophosphates (CoAPO)	Formation under hydrothermal conditions at temperatures up to 448 K	DR/IS	Weckhuysen et al. (2000), Weckhuysen (2004)
Heteropoly acids and their cesium salts	Redox behavior: He, O_2, and He/H_2O atmospheres, temperatures up to 663 K	DR (Thiede and Melsheimer, 2002) UV–vis–NIR	Melsheimer et al. (1999), Klokishner et al. (2002), Melsheimer et al. (2002), Jentoft et al. (2003),
Alumina	Impregnation with solutions containing nickel, chromium, or polyoxometallates	DR/FO Spatially resolved measurements of pellet	van de Water et al. (2005a,b)

Supported metal complexes and clusters

Material	Description	Technique	Reference
Ag/Al_2O_3	Reduction and reoxidation of silver at temperatures up to 823 K	DR/PM	Richter et al. (2004)
Ag/Al_2O_3	Formation of nitrates on the surface in mixtures of NO_2 (+NO), O_2, and H_2O at 423 K	DR/FO Quartz reactor (Groothaert et al., 2003)	Brosius et al. (2005)
Platinum, ruthenium, and mixed platinum–ruthenium species supported on silica	Dispersion of metallic species during treatments in O_2 and H_2 at temperatures up to 473 K	DR/IS at reactor, FO to spectrometer	Zou and Gonzalez (1992)
Various alumina-supported nickel complexes	Reaction of nickel complexes and interaction with support at temperatures up to 593 K	DR/IS Spectratech cell	Négrier et al. (2005)

Supported oxo-anions of transition metals

Material	Description	Technique	Reference
VO_x/TiO_2	Reduction in 1-butene and reoxidation in O_2 at 473 K	DR/PM	Bentrup et al. (2004)
VO_x/Al_2O_3	Oxidation in O_2 and reduction in 20% H_2 at 773 K	DR/PM	Steinfeldt et al. (2004)
VO_x/MCM-41	Reduction in 10% propane or 10% H_2 at 773 K	DR/PM	Kondratenko et al. (2005)

(continued)

TABLE 2 (*continued*)

Catalyst	Treatment or process	Equipment and further detail	Reference
VO_x/SBA-15	Dehydration at temperatures up to 533 K	DR/PM UV–vis–NIR	Tzolova-Müller et al. (unpublished data)
CrO_x/Al_2O_3	Reduction by CO at temperatures up to 773 K	DR/PM	Bensalem et al. (1997)
WO_x/ZrO_2	Reduction by H_2 at 523 K	DR/PM	Barton et al. (1999)
WO_x/ZrO_2	Reduction by H_2 or 2-butanol at temperatures up to 673 K	DR/PM	Baertsch et al. (2002)

FO: Fiber optics.

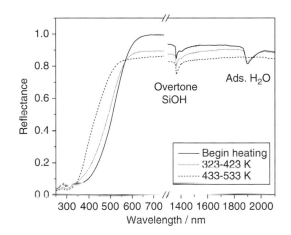

FIGURE 14   Diffuse reflectance UV–vis and NIR spectra of VO_x/SBA-15 recorded during heating. The experiment was carried out with Harrick Praying Mantis DRP-P72 with reaction cell HVC-VUV and a PerkinElmer Lambda 950 spectrometer. For the background correction a 10% attenuator was placed in the reference beam. In the sample beam was the Harrick Praying Mantis DRP-P72 with a piece of Spectralon® in the reaction cell. For the measurement of the presented spectra, the 10% attenuator was left in the reference beam and the piece of Spectralon® was exchanged with the catalyst. A heating rate of 10 K/min and a 20 ml/min flow (NTP) of $He:O_2 = 1:1$ were applied.

Raman scattering wavelength. The Raman intensities of the samples absorbing more light with increasing temperatures (up to 773 K) at these wavelengths were found to decline, a result that was ascribed to a decreased sampling depth as a result of absorption.

Correlations of catalytic and spectroscopic data have also been reported. Steinfeldt et al. (2004) investigated a series of VO_x/Al_2O_3 with varying vanadium loadings under oxidizing and reducing conditions. The activity in propane ODH was found to be related to the amount of tetrahedrally coordinated V^{5+}, and selectivity to propene increased with the dispersion of such species. Pérez-Ramírez et al. (2004) distinguished isolated Fe^{3+} ions, oligonuclear $Fe_x^{3+} O_y$ clusters, and Fe_2O_3 particles in FeMFI catalysts by their band positions in the UV–vis spectrum. The conversion of N_2O in the presence of CO was found to be linearly correlated to the "relative fraction of isolated Fe^{3+} sites" in the as-prepared catalysts.

7.3. Spectroscopic Measurements under Catalytic Reaction Conditions

If not otherwise specified, the experiments described in this section were performed with the Harrick Praying Mantis accessory and reaction cell.

Heteropoly acids containing molybdenum and vanadium were investigated by Lee et al. (2001) at temperatures up to 673 K in an inert atmosphere, in methanol, and in a methanol/O_2 mixture. Their equipment consisted of a quartz cell of in-house design combined with an integrating sphere (Melsheimer et al., 1997). The intensity of an inter valence charge transfer (IVCT) transition, indicative of reduced species, increased with temperature in two steps that coincided with the detection of water by mass spectrometry. The reaction products of methanol conversion in a temperature-programmed reduction (TPR) experiment were contrasted with these observations, and evolution of the methanol oxidation product formaldehyde was found in the plateau region between the first and second increase in the V^{4+}–O–Mo^{6+} IVCT band. The authors noted a 30 K higher reaction temperature in the UV–vis than in the TPR experiment, which was attributed to gas transport limitations. Under methanol oxidation conditions, the intensity of bands in the visible region and hence the degree of reduction was a function of the O_2:methanol ratio, with the reduction being partially irreversible.

Wachs and coworkers (Burcham et al., 2000; Gao et al., 1999a, 2002) focused on the characterization of VO_x-species on various supports. Gao et al. (1999a) followed the reduction of supported VO_x species in an O_2-containing or an O_2-free feed of ethane at 723 K (or butane at 503 K). Two different methods of analysis were applied to determine the degree of reduction, both of which used the spectra of the oxidized and the H_2-treated, and thus presumably completely reduced samples, as reference points. Either the area of the V(V) LMCT band or the intensity of the V^{4+}/V^{3+} d–d transitions at 17,000 cm^{-1} (all based on the $F(\rho)$ representation) were considered; the results agreed roughly with each other. The degree of reduction depended on the support and decreased in the following order: $ZrO_2 > Al_2O_3 > SiO_2$. Polymerized VO_x species were more easily reduced than monomeric species, and both were proposed to be catalytically active, implying that only a single V site is needed for ethane oxidation.

Edge energies characterizing VO_x supported on zirconia were determined from spectra measured at 573 K in a feed flow with various propane:O_2 ratios (Gao et al., 2002). The turnover frequency of the catalytic reaction, as measured in a separate reactor, was largely independent of the edge energy—and hence coverage of the surface with vanadia. The authors concluded that a single vanadia site suffices for propane oxidation. Selectivity was best for high O_2/propane ratios, and the results led the authors to propose that propene is produced on highly oxidized (+5) vanadium sites.

Burcham et al. (2000) presented measurements of VO_x supported on alumina, zirconia, or silica exposed to O_2/He and O_2/He/methanol atmospheres at 503 K. In the presence of methanol, d–d transitions ascribable to V^{4+} or V^{3+} appeared in the wavelength range 330–1000 nm

$(30,000–10,000 \text{ cm}^{-1})$ for the alumina- and the zirconia-supported VO_x, whereas the edge energies were almost unaffected. On the silica support, reduced vanadium could not be detected upon methanol exposure, but the edge energy increased by 0.57 eV. This strong shift was attributed to changes in the coordination of V^{5+} through adsorbed methoxy groups and water (from methanol dehydration). The difficulty in distinguishing reduction of V^{5+} from ligand effects was emphasized by the authors—the fraction of reduced vanadium cannot be deduced from changes in the position and intensity of LMCT bands alone.

Kondratenko and Baerns (2001) investigated alumina-supported vanadium oxide catalysts during treatments at 773 K. In air, V^{5+} species were present, which, depending on the loading, were either isolated or in the form of V_2O_5 clusters. Reduction with propane or H_2 produced V^{4+} and possibly also V^{3+}. Propane was also converted to coke-like species on the surfaces of catalysts with low vanadium contents (1 wt % VO_x). In a reaction mixture for ODH of propane, containing 30 vol % propane and either 15 vol.% O_2 or 30 vol.% N_2O, vanadium species were reduced, more so in the N_2O-containing than in the O_2-containing mixture. Previous data (Buyevskaya and Baerns, 1998) showing that the selectivity to propene increases with the degree of vanadium reduction were included in the interpretation and it was concluded that the active oxygen species should best be isolated.

The same reaction was investigated by Rybarczyk et al. (2001), who employed a catalyst containing the crystalline phases V_2O_5 and MgV_2O_6. Heating to 773 K in inert gas led to a decrease of the bands of V_2O_5 and to the formation of V^{4+}. In a 2:1 (molar) propane to O_2 mixture in N_2, "weakly condensed polyenyl species" absorbing at 300–350 nm developed with time on stream. Vanadium in a catalyst containing $Mg_2V_2O_7$ as a crystalline phase was hardly reduced. By comparison with catalytic data, the authors inferred that both V^{5+} and V^{4+} are active, with V^{4+} being less active but more selective.

Brückner (2000) exposed a $(VO)_2P_2O_7$ catalyst to a flow of 1% toluene in air at 623 K and observed increased absorption at wavelengths below 350 nm, which, on the basis of a comparison with reference spectra, was attributed to CT transitions of V^{4+}, formed by reduction of V^{5+} species.

Argyle et al. (2003, 2004) introduced a method to determine the average valence of Al_2O_3-supported VO_x species under the conditions of propane ODH catalysis. First, calibration measurements were made: the catalyst was reduced for various periods of time in H_2 at 603 K, and then the amount of O_2 required to fully restore the UV–vis spectra was measured by mass spectrometry. Spectra of the fully oxidized sample were recorded to generate the background. These relative reflectance spectra were converted by applying the Kubelka–Munk function and then the intensity in the range 1.5–1.9 eV was related to the extent of reduction as determined in the TPO experiment (the intensity was insignificant prior to reduction; thus the catalyst can be considered a white standard in this range when fully oxidized). A linear correlation was obtained between the integrated intensity in the pre-edge region and the degree of reduction (expressed as electrons per vanadium atom), which was used to calculate the average valence of vanadium in the sample in contact with a feed stream. The linear relationship suggests that only V^{4+} or V^{3+} was formed; or that the V^{4+}/V^{3+} ratio was constant; or that the molar absorption coefficients characterizing the respective d–d transitions are very similar to each other; or that the product of concentration and molar absorption coefficient is much larger for one species than the other.

Brückner and Kondratenko (2006) used a similar approach to characterize VO_x/TiO_2 catalysts. In a separate TPR experiment carried out with a quartz reactor equipped with a UV–vis fiber optical probe, the relationship between the "absorbance" at 800 nm and the degree of reduction as determined from H_2 consumption via mass spectrometry was established. The absorbance at 800 nm increased with increasing reduction of the vanadium, but not linearly. During the catalytic reaction experiment, the absorbance at 800 nm was then used to determine the average valence of vanadium. Because contributions of reduced titanium species in the analyzed spectral range could not be excluded, only a lower limit of the vanadium oxidation state could be determined, which was 4.86 at 523 K and $C_3H_8/O_2 = 1:1$.

In further experiments, Argyle et al. (2003, 2004) varied the propane to O_2 ratio and determined the degree of reduction as a function of feed composition. The fraction of rapidly and reversibly reducible and reoxidizable vanadium centers was identified by step-wise changes of the propane partial pressure in a C_3H_8/O_2 mixture and found to be 30–40%; another fraction of the vanadium reacted reversibly only on a time-scale of >1000 s and was thus deemed catalytically irrelevant; a third fraction (~10%) was irreversibly reduced (Argyle et al., 2004). The extent of reduction of the catalytically relevant vanadium sites depended solely on the C_3H_8/O_2 ratio and not on the individual partial pressures. The degree of reduction of the vanadium in the propane ODH feed was then compared with a propane conversion rate as determined separately with a conventional flow reactor (Argyle et al., 2002); the number of relevant reduced centers increased with increasing ODH rates. Additional work on this type of catalyst addressed the kinetics of reduction of VO_x/Al_2O_3 in C_3H_8/O_2 mixtures, which were monitored by using a fixed-wavelength mode with high time resolution (Argyle et al., 2005).

Alumina-supported chromia was monitored under propane dehydrogenation conditions at 853 K by the Weckhuysen group (Puurunen et al., 2001; Weckhuysen, 2002) using fiber optics. The authors observed reduction of Cr^{6+} to Cr^{3+} evidenced by characteristic bands and an overall increase of absorption interpreted as the formation of coke. The chromium was reoxidized and the coke removed by treatment in 5% O_2 at the same temperature. Brückner (2003) heated Cr–La/ZrO_2 in H_2 and observed a decrease in the intensities of bands at 290, 380, and 460 nm and an increase in those of bands at 460 and 700 nm; these results were interpreted as evidence of reduction of Cr^{6+} to Cr^{3+} ions. Exposure to 0.6 mol% octane produced a barely discernable band at 360 nm, which was ascribed to polyenyl coke precursors (Brückner, 2003). Cherian et al. (2002) heated alumina-supported CrO_x species to 773 K in a mixture of 30 vol.% propane in He, and a band at 370 nm, assigned to the Cr^{6+} LMCT band of chromate, disappeared, while a band at 280–290 nm was not affected. Zirconia-supported CrO_x species were investigated by Rao et al. (2004) at 573 K and varying propane/O_2 ratios. The relative extents of reduction were determined by analyzing band areas and applying a method developed for supported VO_x (Gao et al., 1999a); they increased with increasing propane concentration and decreasing chromium content. It was concluded that, similar to vanadium, polychromate is more easily reduced than monochromate. Both species are presumed to be active in propane dehydrogenation, because the turnover frequency (per total number of chromium atoms), obtained in a separate measurement, did not depend on the poly- to monochromate ratio.

Spectra of ZrO_2-supported WO_x species were recorded by Baertsch et al. (2002) after 1 h under 2-butanol dehydration conditions (0.5 kPa reactant, 323 K). The relative abundance of reduced centers was estimated from the Kubelka–Munk function in the range 1.5–3.2 eV (824–388 nm). Dehydration rates were obtained in a separate quartz reactor at 373 K. UV–vis band area and rate increased with the tungsten density up to a particular loading. Equivalent experiments with WO_x/Al_2O_3 were performed by Macht et al. (2004). A parallel increase of the initial dehydration rate at 373 K and the relative abundance of reduced centers at 423 K were pointed out.

The state of iron in FeZSM-5 was analyzed by Santhosh Kumar et al. (2006) during various treatments. Various samples were reduced in ammonia at 673 K and reoxidized in air at the same temperature. Changes of bands in the UV–vis spectra, representing isolated Fe^{3+} ions (241 nm, tetrahedral coordination, and 291 nm, octahedral) and oligonuclear Fe_xO_y clusters (350 nm), were used to deduce the pseudo-first-order rate constants for reduction and reoxidation. In an SCR feed with 0.1% each of NH_3 and NO, reduction of isolated Fe^{3+} was observed. The data were complemented by separately measured EPR and FTIR spectra. Groen et al. (2006) investigated FeZSM-5 samples for the decomposition of N_2O and analyzed the state of the iron after treatment in O_2, H_2, and N_2O, with the best catalyst exhibiting a high fraction of Fe^{2+} after reduction; some of these sites were oxidizable by N_2O but not by O_2.

Shimizu et al. (2000) investigated the state of palladium cations in H-mordenite at 623 K in various atmospheres (after oxidation, in 1000 ppm of NO, and in a mixture of NO, CH_4, and O_2). An integrating sphere was combined with fiber optics. First, dehydration of $Pd(H_2O)_n^{x+}$ complexes was observed; then, interaction of the resultant $Pd(O_{zeo})_n^{2+}$ complexes with NO was indicated by a slight broadening and a shift of the band at 480 nm to 450–470 nm. Only a slight further shift to lower wavelengths occurred in the reaction mixture, and it was concluded that Pd^{2+} ions were the principal species present during SCR of NO with CH_4.

A "homemade" quartz cell that was not described was used in combination with an integrating sphere by Sazama et al. (2005) to determine the state of silver supported on alumina under SCR conditions (523 K, NO, O_2, H_2O), with decane as a reducing agent; catalytic data obtained with a test reactor at the same reactant composition and GHSV were also presented. The authors deduced that Ag^+ centers are the active sites. In a follow-up paper, Wichterlová et al. (2005) tried to elucidate why H_2 has a positive influence on the reduction of NO by decane. UV–vis spectra indicated that silver clusters are formed in a reaction mixture containing either CO or H_2 as reducing agent; IR spectra showed a significantly higher number and variety of surface species in H_2 than in CO. Shimizu et al. (2007) also investigated the state of silver in Ag/Al_2O_3 catalysts under oxidizing, reducing, and C_3H_8-SCR conditions. The positive influence of H_2 on the SCR reaction was of particular interest, and the reduction of Ag^+ ions and the aggregation to form silver clusters was observed.

The state of silver supported on $CaCO_3$ was monitored by Lu et al. (2005) during exposure to a feed for ethene or propene oxidation at 473 K. Reduction of an initially present Ag^+ component was observed in $C_3H_6/O_2/He$, whereas in $C_2H_4/O_2/He$ some oxygen appeared to be retained by the silver. The authors considered a change of the silver particle shape under ethene oxidation conditions because of a shift of the surface plasmon resonance.

Spectra obtained during the decomposition of H_2O_2 catalyzed by MoO_4^{2-}-exchanged layered double hydroxides were presented by Sels et al. (2001). The experiments were performed at room temperature in a quartz cell with an integrating sphere. The catalyst was suspended in methanol. UV–vis and Raman spectra gave evidence of various peroxomolybdates, which are intermediates in the H_2O_2 decomposition reaction. Tromp et al. (2003) monitored the formation of dimeric and oligomeric palladium complexes during a homogeneous catalytic reaction, an allylic amination, using fiber optics. Time-resolved UV–vis spectra showed the growth of palladium particles, which ultimately precipitated in the cuvette. These last two examples illustrate how UV–vis spectroscopy can be applied for the investigation of catalysts in solution, colloidal dispersion, or suspension.

A considerable collection of data exists that describe the state of catalysts under reaction conditions. The cases presented here show that typical processes observed under reaction conditions include changes of the oxidation states (of transition metals), changes of particle size and shape (of metal clusters), or formation of coke. However, without the corresponding catalytic performance, relevant and spectator species cannot be distinguished.

7.4. Simultaneous Acquisition of Spectroscopic and Catalytic Data

Melsheimer et al. (2002) exposed heteropoly compounds ($H_4PVMo_{11}O_{40}$ and $Cs_2H_2PVMo_{11}O_{40}$) to various atmospheres at various temperatures; for a description of the equipment, see (Thiede and Melsheimer, 2002). Catalytic tests that were conducted in the spectroscopic cell with a reactive mixture containing 10% O_2 and 10% propene showed that oxidation products were formed only at 573 K (with the cesium salt of the heteropolyoxometallate) or at 613 K (with the heteropoly acid). Even at temperatures lower than these, the spectra were characterized by low signal-to-noise ratio, broadened features, and overlapping bands, so that band gap energies and band positions of transitions in the visible range could not be extracted. Thus, in this case, it was not possible to find a catalyst/diluent ratio and reaction conditions that would produce interpretable data.

Melsheimer and coworkers (Melsheimer and Schlögl, 1997; Melsheimer et al., 1997) investigated the oxidation of H_2S with O_2 or SO_2 on alumina catalysts. A transmission and a diffuse reflectance experiment were compared (Melsheimer et al., 1997). In the transmission experiments, the gas-phase spectra of the reactive gases had to be subtracted. A spectral response could always be observed, but conversions were difficult or impossible to detect because of the small mass of sample in a wafer. Spectra of adsorbed SO_2 obtained in the two experiments were similar to each other. During oxidation of H_2S with SO_2, a number of bands were recorded that were assigned to polysulfide species. The height of a band at approximately 435 nm was inferred to follow a parallel course to the H_2S conversion, and hence S_8^{2-} species were inferred to be "responsible for the steady-state conversion".

Subsequently, Melsheimer and Schlögl (1997) used various reflectance cells and tested pressed and unpressed powder samples, the spectra of which were characterized by different intensity distributions. From correlations of the integral intensity of individual components to the H_2S conversion with time on stream it was deduced that $S_2O_3^{2-}$ and S_2^{2-} species "affect the steady state" when SO_2 is used as an oxidant—and $S_2O_3^{2-}$ species when O_2 is the oxidant.

Melsheimer and Ziegler (1992) investigated the transformation of ethene on HZSM-5 wafers by transmission spectroscopy. Effluent gases were analyzed by online GC. Spectra were recorded during reactant flow at temperatures up to 573 K, showing maxima between 278 and 313 nm and additional bands or shoulders towards longer wavelengths. Cyclopentenyl and cyclohexenyl cations were identified as stable surface species. The authors stated explicitly that a correlation between the intensities of the UV–vis bands and ethene conversion was not found.

Spectra of $AgHf_2(PO_4)_3$ during dehydration and dehydrogenation of butan-2-ol were recorded by Brik et al. (2001), who used the Harrick accessory to determine the state of silver. The authors reported slightly lower conversions in the cell than in a U-shaped microreactor. The catalytic data corresponding to the spectra are not shown in the paper.

These five examples demonstrate several potential problems with the application of UV–vis spectroscopy to working catalysts: it may be difficult to obtain meaningful spectra; it may also be difficult to achieve a measurable conversion with a given sample size and form; and spectral results may vary depending on the sample preparation and measurement geometry. Online effluent gas analysis is compulsory, particularly if the cell and the form of the catalyst deviate strongly from those used in an ideal catalytic test experiment. However, even when good catalytic and spectroscopic data are collected, a correlation may not necessarily emerge immediately.

Melsheimer et al. (1999) recorded catalytic data and UV–vis spectra during the oxidation of methanol and ethanol on heteropoly compounds of the molybdophosphate type. A noncommercial cell was used in combination with an integrating sphere. Spectra in the feed stream recorded at temperatures up to 573 K were presented. A "structural parameter" consisting of a relative shift of the band gap energy was calculated for oxidation of methanol and of ethanol under aerobic and anaerobic conditions. Conversions obtained for each of these conditions were plotted versus the respective structural parameters, indicating a methanol conversion independent of the degree of reduction and a decreasing ethanol conversion with increasing degree of reduction of the catalyst.

Butane dehydrogenation was investigated by Weckhuysen and coworkers (Weckhuysen and Schoonheydt, 1999; Weckhuysen et al., 1998, 2000) with the Harrick equipment. For n-butane dehydrogenation catalyzed by silica–alumina-supported chromium, the conversion of butane at a particular time on stream observed with catalysts having various chromium loadings was correlated with the intensity of an absorption band ascribed to Cr^{3+} (Weckhuysen et al., 1998). The initial reduction of the Cr^{6+} in the fresh catalyst led to the formation of CO and CO_2. In the dehydrogenation of isobutane, a relationship between catalytic activity and the amount of reduced chromium (Cr^{2+} and Cr^{3+}, determined by using an advanced program for principle component analysis, named

SIMPLISMA) was established (Weckhuysen and Schoonheydt, 1999; Weckhuysen et al., 2000). Cr^{3+} ions were deemed the most active species for conversion of each butane (Weckhuysen et al., 1998, 2000).

Using fiber optics, Groothaert et al. (2003a) monitored the decomposition of NO and of N_2O on CuZSM-5. A bis(μ-oxo)dicopper species, $[Cu_2(\mu\text{-}O)_2]^{2+}$, which had been previously identified by XAFS and UV–vis analysis (Groothaert et al., 2003b), was proposed to be a key intermediate and O_2-releasing species. Although no quantitative correlation was established between decomposition rate and intensity of the absorption band of the copper species, the performance appeared to be related to the intensity of this band.

These three examples show that an elaborate analysis of the spectral information, including supporting methods to identify species, may be necessary to find correlations with the catalytic performance.

Weckhuysen and coworkers (Nijhuis et al., 2004; Tinnemans et al., 2005) correlated the catalytic performance of alumina-supported chromium catalysts in propane dehydrogenation at 823 K and the amount of coke on the surface. At the start of the reaction chromium was reduced but the band of Cr^{3+} was soon obscured because of darkening of the sample as a result of coke deposition. Spectra acquired with four fiber optical probes along the catalyst bed showed that coking (indicated by the absorbance at 1000 nm) was most severe towards the downstream end of the bed. Initially, coke was building up slowly, while the conversion was increasing; after about 2 hours, coke formation became more rapid, and the conversion declined. These results are a good example showing a correlation between catalyst performance and spectroscopic data; however, the typical difficulties become apparent in this work, such as overlapping features (of Cr^{3+} and coke) in the spectra.

Ahmad et al. (2003) employed a reactor placed near an integrating sphere as described in Thiede and Melsheimer (2002) to elucidate the causes of deactivation of sulfated zirconia during n-butane isomerization and n-pentane isomerization. Bands at 310 nm (observed with butane at 358 and 378 K) and at 330 nm (pentane, 298 and 308 K) were assigned to monoenic allylic cations, and bands at 370 and 430 nm (butane, at 523 K) were assigned to dienic and trienic allic cations. The performance of the sulfated zirconia catalyst changed drastically with time on stream; the conversion at first increased, then passed through a maximum, followed by rapid deactivation. The band at 310–330 nm evolved only as the conversion started to decrease, indicating that the allylic species are not reaction intermediates but side products, which are possibly formed in a competitive reaction to the isoalkane.

Kuba et al. (2003) monitored a WO_3/ZrO_2 catalysts with and without platinum during n-pentane isomerization and hydroisomerization at 523 K; their equipment consisted of a reactor placed next to an integrating sphere (Thiede and Melsheimer, 2002). Upon contacting of the WO_3/ZrO_2 with pentane, an absorption band developed rapidly at about 415 nm; it was assigned to polyalkenyl species. A redshift of this band with time on stream was attributed to its increasing chain length. The authors stated that no direct correlation between the growth of the UV–vis bands and the evolution of the catalytic activity could be observed. Similar spectra were recorded with a platinum-containing catalyst or in the presence of H_2 in the feed; only the combination of H_2 and platinum changed the observations. A broad absorption arose from metallic platinum, and the band at 415 nm was much weaker. Hence, it was inferred that the concentration of polyalkenyl species was so low that it had only a negligible effect on the catalysis, explaining the stable conversion to isopentane.

Discrete bands characteristic of surface hydrocarbon species have been detected during butane conversion catalyzed by zeolites, for example, in the reaction of n-butane catalyzed by H-mordenite at 573 K (Tzolova-Müller et al., unpublished). The spectra in Figure 15 include bands at 293, 330, 400, and about 455 nm.

These three examples concerning alkane isomerization demonstrate that UV–vis spectroscopy is a sensitive method to detect unsaturated hydrocarbon surface species on solid acid catalysts. Two facts make these combinations of catalyst and reaction more amenable to investigation by UV–vis spectroscopy than the dehydrogenation on supported chromium or vanadium catalysts, as follows: The reaction temperature is lower, leading to formation of defined species with discrete absorption bands rather than to multiple undefined species ("coke") that cause strong absorption over a wide wavelength range. Further, the contribution of catalyst absorption to the spectrum is simpler and less dynamic than in case of catalysts containing redox-active transition metal ions.

Changes in a Cr–Al–CM-41 catalyst during exposure to ethene at 373 K were monitored by UV–vis–NIR spectroscopy by Weckhuysen et al. (2000), who used the Harrick equipment. Reduction of chromium (initially Cr^{6+}, Cr^{5+}, and Cr^{3+}) was observed, and C–H vibrations of methylene groups that were detected in the NIR region indicated polymerization.

Kervinen et al. (2005) investigated a homogeneous catalytic reaction, namely, the oxidation of veratryl alcohol to its aldehyde in the liquid phase. In this case, UV–vis spectroscopy, performed by immersing a fiber probe in the reacting medium, allowed detection of changes in the Co (salen) catalyst as well as monitoring of product formation.

These two examples illustrate that NIR or UV–vis spectroscopy can be used not only for catalyst and adsorbate analysis but also for simultaneous product analysis.

Diffuse reflectance UV–vis spectroscopy was applied in electrocatalysis by El Mouahid et al. (1998), who followed the electropolymerization of a cobalt porphyrin complex on a vitreous carbon electrode. The thin polymer

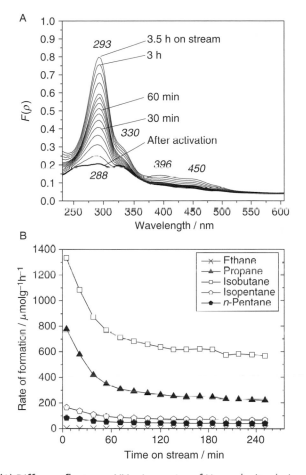

FIGURE 15   (A) Diffuse reflectance UV–vis spectra of H-mordenite during *n*-butane isomerization catalysis. The experiment was carried out with Harrick Praying Mantis DRP-P72 with reaction cell HVC-VUV and a PerkinElmer Lambda 950 spectrometer. For the background correction, a 10% attenuator was placed in the reference beam. In the sample beam was the Harrick Praying Mantis DRP-P72 with a piece of Spectralon® in the reaction cell. For the measurement of the presented spectra, the 10% attenuator was left in the reference beam and the piece of Spectralon® was exchanged with the catalyst. The reaction conditions were as follows: mass of catalyst 0.058 g, reaction temperature approximately 573 K (nominal cell temperature 643 K), 9 kPa *n*-butane in N_2, total flow 22 ml/min (NTP). (B) Corresponding catalytic data as recorded online with Varian 3800 GC and flame ionization detection.

film was then tested for electroreduction of O_2, and Co(II) was identified as the active species.

Because many solvents, including water, are largely transparent to UV–vis radiation, reactions in the liquid phase can be investigated, as well as reactions on surfaces in the presence of liquids.

7.5. Correlation of UV–vis Spectral Features with Other Simultaneously Acquired Spectra

Brückner (2001) presented UV–vis and EPR spectra and catalytic data measured during propane dehydrogenation in the presence of Cr–La/ Al_2O_3 and of Cr–La/ZrO_2 at various temperatures. Both catalysts initially contained chromate(VI), detected by UV–vis spectroscopy, and traces of Cr^{5+}, detected by EPR spectroscopy; the alumina-supported material also contained some Cr^{3+}, indicated by the UV–vis spectra. Upon heating of the sample in a 23% propane feed, the bands indicative of Cr^{6+} lost intensity, and absorption in the visible range increased, results that were interpreted as reduction of Cr^{6+} (and according to the EPR spectra, also of Cr^{5+}) to Cr^{3+} (which was detected by EPR) even before dehydrogenation set in. Increased absorption at 360 nm (sample with zirconia support) or at 470 nm (alumina) with time on stream was interpreted as evidence of formation of coke species; with the alumina support, this happens prior to measurable propane dehydrogenation. This result demonstrates that online analysis is important to enable correlations between catalytic activity and spectroscopic data. Activity-versus-time profiles may be different in a spectroscopic cell, and data from a separate reactor may not be related to the spectra.

Tinnemans et al. (2005, 2006) investigated the deactivation of CrO_x/ Al_2O_3 as a result of coking during propane dehydrogenation catalysis. Catalytic data, UV–vis, and Raman spectra were acquired simultaneously. The UV–vis spectra were used to correct the intensity of the Raman spectra for quantitative analysis. The Raman intensity is affected if species are present that absorb in the relevant wavelength range of the incoming laser beam or the much weaker, frequency-shifted Raman-scattered light (typically in the UV–vis range). Hence, changes in the concentrations of species absorbing in the range of laser wavelength + Raman shift (converted to nm) during the course of a reaction will be reflected in the intensities of the vibrational Raman spectrum. In this work, an excitation wavelength of 532 nm was used, and coke formation was monitored via the Raman band at $1580 \, \text{cm}^{-1}$ (equivalent to scattered light at approximately 580 nm). In previous work, the authors had shown parallel behavior of "coke" indicators in Raman and UV–vis spectra (absorbance at 850 nm; Nijhuis et al., 2003), but later reported a mismatch between UV–vis (absorbance at 1000 nm) and Raman spectroscopic determination of the coke amount, which was attributed to darkening of the sample and weakening of the Raman signal (Nijhuis et al., 2004). Hence, the Raman spectra were corrected by a factor derived from the reflectance at 580 nm, and the relative coke concentration was determined as a function of time on stream. The coke accumulation profile derived from the corrected Raman spectra approximately matched that recorded with a balance. Correction by this method gave results comparable to those obtained via correction with an internal standard (BN), but with the advantage that no foreign material had to be mixed into the catalyst bed. This example shows how UV–vis spectroscopy can be employed as a supporting method to extract more valuable information from data gained with other techniques.

The additional insight from UV–vis spectra also proved useful in an investigation of the homogeneous oxidation of benzyl alcohol to benzaldehyde in which copper complexes were employed as catalysts (Mesu et al., 2005; Tinnemans et al., 2006). It was shown that the synchrotron radiation used for XAFS spectroscopy affected the reacting solutions; besides a thermal effect, reduction of copper was induced. This effect was investigated in detail for a number of ligands (Mesu et al., 2006).

Similarly, Tinnemans et al. found that the onset temperature for the reduction of vanadium in VO_x/SiO_2 by H_2 to be lower according to Raman than to UV–vis spectra and concluded that the 70-mW laser beam caused local heating (Tinnemans et al., 2006), a problem that had been described earlier (Chan and Bell, 1984; Wovchko et al., 1998). In a further investigation, the local heating effect was found to become negligible at temperatures above 773 K (Tinnemans et al., 2006). Hence, as a solution to this problem, it was suggested to raise the reactor temperature.

In some applications, such as homogeneous oxidations catalyzed by Co(salen) complexes, Raman, UV–vis and ATR spectroscopies could not be applied simultaneously because the cobalt concentrations needed to be optimized individually for each technique (Kervinen et al., 2005; Tinnemans et al., 2006).

Beale et al. (2005) presented catalytic data, Raman, UV–vis, and XAFS spectra that were acquired simultaneously during ODH of propane catalyzed by alumina- or silica-supported molybdenum. Reduction of Mo^{6+} to Mo^{4+} in Mo/SiO_2 during reaction was demonstrated by the XANES; Raman spectra indicated the formation of coke and larger MoO_3 clusters, and the UV–vis spectra were unspecific. It was concluded that Mo^{4+} may be necessary for catalytic activity, that coke and solid-state reactions lead to temporary and permanent catalyst deactivation, respectively, and that MoO_x/Al_2O_3 is more stable than MoO_x/SiO_2 during successive dehydrogenation–regeneration cycles (Tinnemans et al., 2006).

Hunger and Wang (2004, 2006) published ^{13}C NMR and UV–vis spectra that were simultaneously recorded while methanol flowed over weakly dealuminated HZSM-5 at 413 K. The NMR spectra indicated methanol, dimethyl ether, and alkanes or alkylated cyclic compounds. Bands in the UV–vis spectra were interpreted as evidence of neutral aromatic compounds (275 nm) and mono- and di-enylic carbenium ions (315 and 375 nm). These species were not detected on non-dealuminated HZSM-5, and Lewis-acidic extra-framework-aluminum species were deemed responsible for hydrocarbon formation at these low temperatures. Spectra recorded after contact with ethene, a suspected intermediate in the methanol-to-olefins reactions, were also reported.

In further work (Jiang et al., 2007), the conversion of methanol on HSAPO-34 was investigated at temperatures up to 673 K, with simultaneous NMR and UV–vis spectroscopy and online GC analysis. In this experiment, the differing sensitivities of the two spectroscopies become apparent; dienes are indicated by the UV–vis spectra at 473 K but are not detected with NMR spectroscopy.

Hunger (2008) also monitored the H/D exchange of ethyl-d_5-benzene with zeolite HY at temperatures up to 523 K. The combination of NMR and UV–vis spectroscopy was found to be useful in this case; the progress of the exchange at the methyl groups of the side chain was evident from the NMR spectra, and carbenium ion intermediates were detected in the UV–vis spectra.

Brückner and coworkers (Brückner, 2005; Brückner and Kondratenko, 2006) presented simultaneously acquired EPR, UV–vis, and Raman spectroscopic data characterizing the state of VO_x/TiO_2 during propane ODH at temperatures up to 673 K. Upon contacting of the catalyst with a propane/O_2/N_2 mixture at 293 K, a band at 1034 cm^{-1} disappeared from the Raman spectrum; the absorbance at wave numbers exceeding 500 nm increased in the UV–vis spectrum; and one EPR signal ($g_{\parallel} = 1.940$, $A_{\parallel} = 176.3$ G) increased, and a new EPR signal ($g_{\parallel} = 1.925$, $A_{\parallel} = 199.2$ G) appeared. These changes were interpreted as consistently indicating reduction of isolated VO^{3+} into VO^{2+} species, more specifically into two types of VO^{2+} species, one bound to TiO_2 itself and one to sulfate present in the anatase sample. Absorption in the wavelength range >500 nm continued to increase with increasing temperature and this change was ascribed to further reduction of vanadium on the basis of analysis of the normalized EPR intensity, whereas coke formation was considered negligible. Simultaneously recorded catalytic data show the onset of propane conversion at 523 K and constant conversion at temperatures above 623 K when dispersed monomeric and polymeric VO_x species and a fraction of the V_2O_5 crystals were in a reduced state. Selectivity increased with increasing temperature up to 623 K, as did conversion and the degree of vanadium reduction. VO^{3+} species, which are reduced at room temperature, are considered to be active but not necessarily selective. The V^{4+} species are considered to be more selective for propene formation than

V^{5+} species because of their lower strength as oxidizing agents, and the reactive oxygen is proposed to originate from a V–O–Ti moiety.

Bürgi (2005) simultaneously recorded ATR (IR) and UV–vis spectra of the liquid–catalyst interface during ethanol oxidation catalyzed by Pd/Al_2O_3. In some experiments, the dissolved gas was switched from O_2 to H_2, which resulted in rapid changes in the spectra. The UV–vis absorbance increased in the presence of O_2 and decreased in the presence of H_2, a result that was attributed to slight oxidation and reduction of palladium. The interpretation was confirmed by the ATR spectra, which indicated a corresponding effect on the reflectivity of the catalyst. A slow increase in the UV–vis absorbance during repeated H_2 and O_2 treatments was attributed to dissolution and redeposition of palladium. Catalyst performance was not significantly affected by this change in the structure of the palladium particles, and it was concluded that concentration gradients within the catalyst grains generated oxygen-covered palladium particles, which are prone to dissolution, but do not contribute to the catalytic activity.

8. CONCLUSIONS

8.1. Experimental Limitations of Applying UV–vis–NIR Spectroscopy under Catalytic Reaction Conditions

The application of UV–vis–NIR spectroscopy is less routine than that of other methods such as, for example, FTIR spectroscopy in transmission or diffuse reflectance modes, and there are more difficulties in the former with artifacts and background corrections. The use of UV–vis–NIR spectroscopy to observe working catalysts is limited by the reaction conditions that are possible with the available equipment. The variations in instrumentation are large, and to solve a specific scientific problem, one specific design will be most suitable. Published work covers measurements over a temperature range from 77 to about 723–853 K; the high temperatures can be reached with diffuse reflectance cells used with integrating spheres, mirror optics, or fiber optics. For catalytic applications, the highest attainable temperature is important, and so are the gradients in the catalytic bed, which should be avoided. Fiber optics can be integrated into isothermal reactors, and gradient-free reactors with integrating spheres have been reported (Melsheimer et al., 2003), but the commercially available cells for mirror optics feature large deviations (as much as 125 K) between the surface temperature of the catalyst bed and the nominal cell temperature (Gao et al., 2002; Venter and Vannice, 1988); thus, the observations indicate that the temperature may vary considerably throughout the catalyst bed, particularly at high operating temperatures.

Many cells are made of quartz; hence, the pressure is limited. For use with an integrating sphere, a thick-walled cell for hydrothermal synthesis has been presented (Weckhuysen et al., 2000), and for mirror optics, a cell is available for pressures up to 3.4 MPa. Some reflectance fibers are specified for pressures up to 3.4 MPa (Stellarnet, 2006).

For the most part, gas–solid reactions have been investigated, but in a few cases, the gas–liquid interface has been analyzed by UV–vis spectroscopy (Bürgi, 2005; Sels et al., 2001).

The reported transmission cells exhibit severe problems in the measurement of catalytic activity because of the limited amount of catalyst in the cell, large dead volumes, and mass transfer limitations in wafers (Melsheimer and Schlögl, 1997). Cells used with mirror optics permit flow through a catalyst bed and feature small dead volumes, but allow little variation of the catalyst mass. Fiber optics requires almost no adaptation of a normal reactor for spectroscopic needs and offers the best solution from a *catalytic* viewpoint.

From a *spectroscopic* viewpoint, spectral resolution and wavelength accuracy are usually not an issue, because many features are broad and ill-defined. The best signal-to-noise ratios, obtained under typical operating conditions, seem to be obtained with mirror optics. The best accuracy in reflectance values is achieved with integrating spheres, which are the least sensitive to specimen positioning. With mirror or fiber optics, reflectance offsets are easily introduced. Fiber optics in combination with diode array detectors clearly exhibit the best time resolution—with, however, limited spectral quality. The best long-term stability for experiments with changes over hours or days is provided by high-end classical UV–vis spectrometers.

Developments in the preceding few years have occurred mainly in the fiber optics area, and the fibers have become popular because of their easy integration into reactor systems—also in combination with other methods—and the wide range of operating conditions. The optimal approach might be to use an integrating sphere or mirror optics accessory to obtain high-quality reference spectra and to use fibers solely to monitor rapid processes during start-up, deactivation, or step-changes in reaction conditions during catalysis.

8.2. Evaluation and Outlook

Analysis of the literature shows that in principle many scientific questions about catalysts can be addressed successfully with UV–vis spectroscopy. The quality of the data depends strongly on the instrumentation, and in some cases UV–vis spectroscopy has become an add-on supporting method. The variety of catalysts investigated decreases dramatically with increasing degree of complexity of the experiment; only few groups in the world combine a reaction cell with online analysis, and therefore only a few catalysts have been investigated during conditions of proven operation.

These are not necessarily the catalysts and reactions that are most suitable for analysis by UV–vis spectroscopy. Consequently, the power of the method has not been fully exploited, and one would hope for a broader range of applications in the future. A significant advantage of UV–vis spectroscopy is that many solvents are transparent in the respective wavelength ranges, and the method is suitable for investigating the preparation of catalysts and catalysis at solid–liquid interfaces.

Quantification of spectral intensity has been achieved via calibration with a second, independent analytical technique, and in this way properties such as the degree of reduction of a metal in an oxidation catalyst have been determined and related to the feed composition. However, notwithstanding the hopes expressed in 1984 (Schoonheydt, 1984), quantitative investigations are still scarce.

One inherent problem in the method is that spectral features are broad and may not be specific; changes observed during catalytic reactions may have multiple causes, such as changes in valence of metal ions, changes in dispersion of supported species, or deposition of feed components and side products on the catalyst. In this respect, input from theoretical chemistry for interpretation of the spectra (Garbowski et al., 1972; Klokishner et al., 2002; Melsheimer et al., 2002; Jentoft et al., 2003) should be considered more frequently.

ACKNOWLEDGMENTS

Special thanks to Genka Tzolova-Müller and Annette Trunschke, who performed numerous tests on different types of equipment. The Max Planck Society is acknowledged for providing a stipend for GTM. AT was supported by the Deutsche Forschungsgemeinschaft, project JE 267/2–1.

REFERENCES

Ahmad, R., Melsheimer, J., Jentoft, F.C., and Schlögl, R., *J. Catal.* **218**, 365 (2003).
Argyle, M.D., Chen, K., Bell, A.T., and Iglesia, E., *J. Catal.* **208**, 139 (2002).
Argyle, M.D., Chen, K., Iglesia, E., and Bell, A.T., *J. Phys. Chem. B* **109**, 2414 (2005).
Argyle, M.D., Chen, K., Resini, C., Krebs, C., Bell, A.T., and Iglesia, E., *Chem. Commun.*, 2082 (2003).
Argyle, M.D., Chen, K., Resini, C., Krebs, C., Bell, A.T., and Iglesia, E., *J. Phys. Chem. B* **108**, 2345 (2004).
Avantes. www.avantes.com, accessed Sept. 29, 2006.
Baertsch, C.D., Komola, K.T., Chua, Y.-H., and Iglesia, E., *J. Catal.* **205**, 44 (2002).
Bailes, M., Bordiga, S., Stone, F.S., and Zecchina, A., *J. Chem. Soc. Faraday Trans.* **92**, 4675 (1996).

Barton, D.G., Shtein, M., Wilson, R.D., Soled, S.L., and Iglesia, E., *J. Phys. Chem. B.* **103**, 630 (1999).

Beale, A.M., van der Eerden, A.M.J., Kervinen, K., Newton, M.A., and Weckhuysen, B.M., *Chem. Commun.*, 3015 (2005).

Bensalem, A., Weckhuysen, B.M., and Schoonheydt, R.A., *J. Phys. Chem. B* **101**, 2824 (1997).

Bentrup, U., Brückner, A., Rüdinger, C., and Eberle, H.-J., *Appl. Catal. A Gen.* **269**, 237 (2004).

Brik, Y., Kacimi, M., Bozon-Verduraz, F., and Ziyad, M., *Microp. Mesop. Mater.* **43**, 103 (2001).

Brik, Y., Kacimi, M., Ziyad, M., and Bozon-Verduraz, F., *J. Catal.* **202**, 118 (2001).

Brosius, R., Arve, K., Groothaert, M.H., and Martens, J.H., *J. Catal.* **231**, 344 (2005).

Brückner, A., *Appl. Catal. A: Gen.* **200**, 287 (2000).

Brückner, A., *Catal. Rev.—Sci. Eng.* **45**, 97 (2003).

Brückner, A., *Chem. Commun.*, 1761 (2005).

Brückner, A., *Chem. Commun.*, 2122 (2001).

Brückner, A., and Kondratenko, E., *Catal. Today* **113**, 16 (2006).

Brückner, A., Kubias, B., and Lücke, B., *Catal. Today* **32**, 215 (1996).

Burcham, L.J., Deo, G., Gao, X., and Wachs, I.E., *Top. Catal.* **11/12**, 85 (2000).

Bürgi, T., *J. Catal.* **229**, 55 (2005).

Busca, G., Martra, G., and Zecchina, A., *Catal. Today* **56**, 361 (2000).

Buyevskaya, O.V., and Baerns, M., *Catal. Today* **42**, 315 (1998).

Catana, G., Rao, R.R., Weckhuysen, B.M., Van Der Voort, P., Vansant, E., and Schoonheydt, R.A., *J. Phys. Chem. B* **102**, 8005 (1998).

Chandrashekhar, S., "Radiative Transfer." p. 9. Dover, New York, 1960.

Chan, S., and Bell, A.T., *J. Catal.* **89**, 433 (1984).

Che, M., and Bozon-Verduraz, F., *in* "Handbook of Heterogeneous Catalysis" (G. Ertl, H. Knözinger and J. Weitkamp, Eds.), Vol. 2, p. 641. Verlag Chemie, Weinheim, 1997.

Cherian, M., Rao, M.S., Hirt, A.M., Wachs, I.E., and Deo, G., *J. Catal.* **211**, 482 (2002).

Cox, S.D., and Stucky, G.D., *J. Phys. Chem.* **95**, 710 (1991).

Custers, J.F.H., and de Boer, J.H., *Physica (The Hague)* **1**, 265 (1934).

Custers, J.F.H., and de Boer, J.H., *Physica (The Hague)* **3**, 407 (1936).

de Boer, J.H., *Nederlands Tijdschrift voor Natuurkunde* **4**, 276 (1938).

de Boer, J.H., *Z. physikal. Chem.* **B16**, 397 (1932).

de Boer, J.H., *Z. physikal. Chem.* **B18**, 49 (1932).

de Boer, J.H., and Broos, J., *Z. physikal. Chem.* **B15**, 281 (1932).

de Boer, J.H., and Custers, J.F.H., *Z. physikal. Chem.* **B21**, 208 (1933).

de Boer, J.H., and Custers, J.F.H., *Z. physikal. Chem.* **B25**, 238 (1934).

de Boer, J.H., Custers, J.F.H., and Dippel, C.J., *Physica (The Hague)* **1**, 935 (1934).

de Boer, J.H., and Dippel, C.J., *Naturwissenschaften* **21**, 204 (1933).

de Boer, J.H., and Teves, M.C., *Zeitschr. f. Phys.* **65**, 489 (1930).

de Boer, J.H., and Teves, M.C., *Zeitschr. f. Phys.* **73**, 192 (1931).

Doskocil, E.J., Bordawekar, S.V., Kaye, B.G., and Davis, R.J., *J. Phys. Chem. B.* **103**, 6277 (1999).

El Mouahid, O., Rakotondrainibe, A., Crouigneau, P., Léger, J.M., and Lamy, C., *J. Electroanal. Chem.* **455**, 209 (1998).

Fiberguide. www.fiberguide.com, accessed Sept. 29, 2006.

Finocchio, E., Montanari, T., Resini, C., and Busca, G., *J. Mol. Catal. A: Chem.* **204–205**, 535 (2003).

Förster, H., Franke, S., and Seebode, J., *J. Chem. Soc. Faraday Trans. I* **79**, 373 (1983).

Förster, H., and Kiricsi, I., *Catal. Today* **3**, 65 (1988).

Förster, H., and Kiricsi, I., *Zeolites* **7**, 508 (1987).

Förster, H., Seebode, J., Fejes, P., and Kiricsi, I., *J. Chem. Soc. Faraday Trans. I* **83**, 1109 (1986).

Gao, X., Bare, S.R., Weckhuysen, B.M., and Wachs, I.E., *J. Phys. Chem. B* **102**, 10842 (1998).

Gao, X., Bañares, M.A., and Wachs, I.E., *J. Catal.* **188**, 325 (1999a).

Gao, X., Bare, S.R., Fierro, J.L.G., and Wachs, I.E., *J. Phys. Chem. B* **103**, 618 (1999b).

Gao, X., Fierro, J.L.G., and Wachs, I.E., *Langmuir* **15**, 3169 (1999c).

Gao, X., Jehng, J.-M., and Wachs, I.E., *J. Catal.* **209**, 43 (2002).

Gao, X., and Wachs, I.E., *J. Phys. Chem.* **104**, 1261 (2000).

Gao, X., Wachs, I.E., Wong, M.S., and Ying, J.Y., *J. Catal.* **203**, 18 (2001).

Garbowski, E.D., and Praliaud, H., *J. Chim. Phys.* **76**, 687 (1979).

Garbowski, É., Kodratoff, Y., Mathieu, M.V., and Imelik, B., *J. Phys. Chim.* **69**, 1386 (1972).

Garbowski, E., and Praliaud, H., in "Catalyst Characterization—Physical Techniques for Solid Materials" (B. Imelik and J.C. Védrine, Eds.), p. 61. Plenum Press, New York and London, 1994.

Garrone, E., and Stone, F.S., in "Adsorption and Catalysis on Oxide Surfaces" (M. Che and G.C. Bond, Eds.), p. 97. Elsevier, Amsterdam, 1985.

Garrone, E., and Stone, F.S., *J. Chem. Soc., Faraday Trans. I* **83**, 1237 (1987).

Garrone, E., Zecchina, A., and Stone, F.S., *J. Chem. Soc., Faraday Trans. I* **84**, 2843 (1988).

Geobaldo, F., Spoto, G., Bordiga, S., Lamberti, C., and Zecchina, A., *J. Chem. Soc. Faraday Trans.* **93**, 1243 (1997).

Groen, J.C., Bruckner, A., Berrier, E., Maldonado, L., Moulijn, J.A., and Pérez-Ramírez, P., *J. Catal.* **243**, 212 (2006).

Groothaert, M.H., Lievens, K., Leeman, H., Weckhuysen, B.M., and Schoonheydt, R.A., *J. Catal.* **220**, 500 (2003a).

Groothaert, M.H., van Bokhoven, J.A., Battiston, A.A., Weckhuysen, B.M., and Schoonheydt, R.A., *J. Am. Chem. Soc.* **125**, 7629 (2003b).

Grubert, G., Rathousský, J., Schulz-Ekloff, G., Wark, M., and Zukal, A., *Microp. Mesop. Mat.* **22**, 225 (1998).

Harrick, "The Praying Mantis," product description, from www.harricksci.com, accessed Sept. 29, 2006.

Hartmann, M., Bischof, C., Lan, Z., and Kevan, L., *Microp. Mesop. Mater.* **44–45**, 385 (2001).

Hellma product information, Datenblatt 131/06-D-1, retrieved from www.hellma-worldwide.com on July 7, 2006.

Henderson, B., and Imbusch, G.F., in "Optical Spectroscopy of Inorganic Solids", Monographs in Physics and Chemistry of Materials, (H. Frohlich, A.J. Heeger, P.B. Hirsch, N. F. Mott and R. Brook, Eds.), Vol. 44. Clarendon Press, Oxford, 1989.

Hess, C., Hoefelmeyer, J.D., and Tilley, T.D., *J. Phys. Chem. B* **108**, 9703 (2004).

Homann, K.-H. (Ed.) "International Union of Pure and Applied Chemistry (IUPAC)—Größen, Einheiten und Symbole in der Physikalischen Chemie." VCh Weinheim, 1996.

Hu, C., Tang, Y., Yu, J.C., and Wong, P.K., *Appl. Catal. B: Environ.* **40**, 131 (2003).

Hunger, M., *Prog. Nucl. Magn. Res.Spectrosc.*, **53**, 105 (2008).

Hunger, M., and Horvath, T., *J. Chem. Soc. Chem. Commun.*, 1423 (1995).

Hunger, M. and Wang, W., *Adv. Catal.* **50**, 150 (2006).

Hunger, M., and Wang, W., *Chem. Commun.*, 584 (2004).

Hwang, J.S., Kim, D.S., Lee, C.W., and Park, S.E., *Korean J. Chem. Eng.* **18**, 919 (2001).

Jentoft, F.C., Klokishner, S., Kröhnert, J., Melsheimer, J., Ressler, T., Timpe, O., Wienold, J., and Schlögl, R., *Appl. Catal. A: Gen.* **256**, 291 (2003).

Jentoft, R.E., Kniep, B.-L., Swoboda, M., and Ressler, T. in "Hasylab Annual Report 2004" (P. Gürtler, J.R. Schneider and E. Welter, Eds.), Hamburger Synchrotronstrahlungslabor HASYLAB at Deutsches Elektronen-Synchrotron DESY, a member of the Helmholtz Association, p. 875, 2004, http://www-hasylab.desy.de/science/annual_reports/2004_report/main.htm.

Jiang, Y., Huang, J., Marthala, V.R.R., Ooi, Y.S., Weitkamp, J., and Hunger, M., *Microp. Mesop. Mater.* **105**, 132 (2007).

Jitianu, A., Altindag, Y., Zaharescu, M., and Wark, M., *J. Sol–Gel Sci. Technol.* **26**, 483 (2003).

Juanto, S., Lezna, R.O., and Arvia, A.J., *Electrochim. Acta* **39**, 81 (1994).

Kaucky, D., Dedecek, J.I., and Wichterlova, B., *Microp. Mesop. Mater.* **31**, 75–87 (1999).

Kazansky, V.B., Borovkov, V.Yu., and Karge, H.G., *J. Chem. Soc. Faraday Trans.* **93**, 1843 (1997).
Kellermann, R., *in* "Spectroscopy in Heterogeneous Catalysis" (W.N. Delgass, G.L. Haller, R. Kellermann and J.H. Lunsford, Eds.), p. 86. Academic Press, New York, 1979.
Kervinen, K., Korpi, H., Mesu, J.G., Soulimani, F., Repo, T., Rieger, B., Leskelä, M., and Weckhuysen, B.M., *Eur. J. Inorg. Chem.*, 2591 (2005).
Kikutani, Y., *J. Molec. Catal. A* **142**, 247 (1999a).
Kikutani, Y., *J. Molec. Catal. A* **142**, 265 (1999b).
Kim, H., Youn, M.H., Jung, J.C., and Song, I.K., *J. Mol. Catal. A: Chem.* **252**, 252 (2006).
Kim, T., Burrows, A., Kiely, C.J., and Wachs, I.E., *J. Catal.* **246**, 370 (2007).
Kiricsi, I., and Förster, H., *J. Chem. Soc. Faraday Trans. I* **84**, 491 (1988).
Kiricsi, I., Förster, H., and Tasi, G., *Stud. Surf. Sci. Catal.* **46**, 355 (1989a).
Kiricsi, I., Förster, H., Tasi, G., and Nagy, J.B., *Chem. Rev.* **99**, 2085 (1999); Additions and Corrections in *Chem. Rev.* **99**, 3367 (1999).
Kiricsi, I., Förster, H., Tasi, Gy., and Fejes, P., *J. Catal.* **115**, 597 (1989b).
Kiricsi, I., Tasi, Gy., Molnar, A., and Förster, H., *J. Mol. Struct.* **239**, 185 (1990).
Klier, K., *Catal. Rev.—Sci. Eng.* **1**, 207 (1968).
Klier, K., *J. Optic. Soc. Am.* **62**, 882 (1972).
Klier, K., and Rálek, M., *J. Phys. Chem. Solids* **29**, 951 (1968).
Klokishner, S., Melsheimer, J., Ahmad, R., Jentoft, F.C., Mestl, G., and Schlögl, R., *Spectrochimica Acta Part A* **58**, 1 (2002).
Klose, B.S., Jentoft, F.C., Joshi, P., Trunschke, A., Schlögl, R., Subbotina, I.R., and Kazansky, V.B., *Catal. Today* **116**, 121 (2006).
Klose, B.S., Jentoft, R.E., Hahn, A., Ressler, T., Kröhnert, J., Wrabetz, S., Yang, X., and Jentoft, F.C., *J. Catal.* **217**, 487 (2003).
Kondratenko, E.V., and Baerns, M., *Appl. Catal. A: Gen.* **222**, 133 (2001).
Kondratenko, E.V., Cherian, M., Baerns, M., Su, D., Schlögl, R., Wang, X., and Wachs, I.E., *J. Catal.* **234**, 131 (2005).
Kortüm, G., "Reflectance Spectroscopy." Springer Verlag, Berlin, Heidelberg, New York, 1969.
Kortüm, G., Braun, W., and Herzog, G., *Angew. Chem.* **75**, 653 (1963); *Angew. Chem. Int. Ed.* **2**, 233 (1963).
Kortüm, G., and Schreyer, G., *Angew. Chem.* **67**, 694 (1955).
Kortüm, G., and Schreyer, G., *Z. Naturforschg.* **11a**, 1018 (1956).
Krishna, R.M., Prakash, A.M., and Kevan, L., *J. Phys. Chem. B* **104**, 1796 (2000).
Kuba, S., Che, M., Grasselli, R.K., and Knözinger, H., *J. Phys. Chem. B.* **107**, 3459 (2003).
Kuba, S., Lukinskas, P., Ahmad, R., Jentoft, F.C., Grasselli, R.K., Gates, B.C., and Knözinger, H., *J. Catal.* **219**, 376 (2003).
Kubelka, P., and Munk, F., *Z. Tech. Physik* **12**, 593 (1931).
Kumar, M.S., Schwidder, M., Grünert, W., and Brückner, A., *J. Catal.* **227**, 384 (2004).
Lee, E.L., and Wachs, I.E., *J. Phys. Chem. C.* **111**, 14410 (2007).
Lee, J.K., Melsheimer, J., Berndt, S., Mestl, G., Schlögl, R., and Köhler, K., *Appl. Catal. A: Gen.* **214**, 125 (2001).
Lezna, R.O., Romagnoli, R., de Tacconi, N.R., and Rajeshwar, K., *J. Electroanal. Chem.* **544**, 101 (2003).
Liu, Y.-M., Cao, Y., Yi, N., Feng, W.-L., Dai, W.-L., Yan, S.-R., He, H.-Y., and Fan, K.-N., *J. Catal.* **224**, 417 (2004).
Lu, J., Bravo-Suárez, J.J., Takahashi, A., Haruta, M., and Oyama, S.T., *J. Catal.* **323**, 85 (2005).
Macht, J., Baertsch, C.D., May-Lozano, M., Soled, S.L., Wang, Y., and Iglesia, E., *J. Catal.* **227**, 479 (2004).
Medin, A.S., Borovkov, V.Yu., and Kazanskii, V.B., *Kinet. Catal.* **30**, 152 (1989).
Melamed, N.T., *J. Appl. Phys.* **34**, 560 (1963).
Melsheimer, J., Böhm, M.C., Lee, J.K., and Schlögl, R., *Ber. Bunsenges. Phys. Chem.* **101**, 726 (1997).
Melsheimer, J., Guo, W., Ziegler, D., Wesemann, M., and Schlögl, R., *Catal. Lett.* **11**, 157 (1991).

Melsheimer, J., Kröhnert, J., Ahmad, R., Klokishner, S., Jentoft, F.C., Mestl, G., and Schlögl, R., *Phys. Chem. Chem. Phys.* **4**, 2398 (2002).

Melsheimer, J., Mahmoud, S.S., Mestl, G., and Schlögl, R., *Catal. Lett.* **60**, 103 (1999).

Melsheimer, J., and Schlögl, R., *Ber. Bunsenges. Phys. Chem.* **101**, 733 (1997).

Melsheimer, J., and Schlögl, R., *Fresenius J. Anal. Chem.* **357**, 397 (1997).

Melsheimer, J., Thiede, M., Ahmad, R., Tzolova-Müller, G., and Jentoft, F.C., *Phys. Chem. Chem. Phys.* **5**, 4366 (2003).

Melsheimer, J., and Ziegler, D., *J. Chem. Soc. Faraday* **88**, 2101 (1992).

Mesu, J.G., Beale, A.M., de Groot, F.M.F., and Weckhuysen, B.M., *J. Phys. Chem. B* **110**, 17671 (2006).

Mesu, J.G., van der Eerden, A.M.J., de Groot, F.M.F., and Weckhuysen, B.M., *J. Phys. Chem. B* **109**, 4042 (2005).

Montanari, T., Bevilacqua, M., Resini, C., and Busca, G., *J. Phys. Chem. B* **108**, 2120 (2004).

Moradi, K., Depecker, C., Barbillat, J., and Corset, J., *Spectrochim. Acta Part A* **55**, 43 (1999).

Négrier, F., Marceau, E., Che, M., Giraudon, J.-M., Gegembre, L., and Löfberg, A., *J. Phys. Chem. B* **109**, 2836 (2005).

Nelson, R.L., Hale, J.W., and Harmsworth, B.J., *Trans. Faraday Soc.* **67**, 1164 (1971).

Nijhuis, T.A., Tinnemans, S.J., Visser, T., and Weckhuysen, B.M., *Chem. Eng. Sci.* **59**, 5487 (2004).

Nijhuis, T.A., Tinnemans, S.J., Visser, T., and Weckhuysen, B.M., *Phys. Chem. Chem. Phys.* **5**, 4361 (2003).

Oceanoptics www.oceanoptics.com, accessed Sept. 29, 2006.

Park, S.K., Kurshev, V., Lee, C.W., and Kevan, L., *Appl. Magn. Res.* **19**, 21 (2000).

Pazé, C., Sazak, B., Zecchina, A., and Dwyer, J., *J. Phys. Chem. B* **103**, 9978 (1999).

Pereira, C., Kokotailo, G.T., and Gorte, R.J., *J. Phys. Chem.* **95**, 705 (1991).

Pérez-Ramírez, J., Kumar, M.S., and Brückner, A., *J. Catal.* **223**, 13 (2004).

Perkampus, H.-H., "UV–vis Spektroskopie und ihre Anwendungen." Springer Verlag, Berlin, 1986.

Popescu, D.A., Herrmann, J.M., Ensuque, A., and Bozon-Verduraz, F., *Phys. Chem. Chem. Phys.* **3**, 2522 (2001).

Puurunen, R.L., Beheydt, B.G., and Weckhuysen, B.M., *J. Catal.* **204**, 253 (2001).

Rao, T.V.M., Deo, G., Jehng, J.-M., and Wachs, I.E., *Langmuir* **20**, 7159 (2004).

Resini, C., Montanari, T., Busca, G., Jehng, J.M., and Wachs, I.E., *Catal. Today* **99**, 105 (2005).

Resini, C., Montanari, T., Nappi, L., Bagnasco, G., Turco, M., Busca, G., Bregani, F., Notaro, M., and Rocchini, G., *J. Catal.* **214**, 179 (2003).

Richter, M., Abramova, A., Bentrup, U., and Fricke, R., *J. Appl. Spectrosc.* **71**, 400 (2004).

Rybarczyk, P., Berndt, H., Radnik, J., Pohl, M.-M., Buyevskaya, O., Baerns, M., and Brückner, A., *J. Catal.* **202**, 45 (2001).

Santhosh Kumar, M., Schwidder, M., Grünert, W., Bentrup, U., and Brückner, A., *J. Catal.* **239**, 173 (2006).

Sazama, P., Čapek, L., Drobná, H., Sobalík, Z., Dědeček, J., Arve, K., and Wichterlová, B., *J. Catal.* **232**, 302 (2005).

Schoonheydt, R.A., *in* "Characterization of Heterogeneous Catalysts" (F. Delannay, Ed.), p. 125. Marcel Dekker, New York and Basel, 1984.

Schulz-Ekloff, G., Lipski, R.J., Jaeger, N.I., Hülstede, P., and Kubelkova, L., *Catal. Lett.* **30**, 65 (1995).

Schuster, A., *Astrophys. J.* **21**, 1 (1905).

Sels, B., De Vos, D.E., Grobet, P.J., and Jacobs, P.A., *Chem. Eur. J.* **7**, 2547 (2001).

Sendoda, Y., Ono, Y., and Keii, T., *J. Catal.* **39**, 357 (1975).

Shibata, J., Shimizu, K.-i., Takada, Y., Shichi, A., Yoshida, H., Satokawa, S., Satsuma, A., and Hattori, T., *J. Catal.* **227**, 367 (2004a).

Shibata, J., Takada, Y., Shichi, A., Satokawa, S., Satsuma, A., and Hattori, T., *J. Catal.* **222**, 368 (2004b).

Shimizu, K.-i., Okada, F., Nakamura, Y., Satsuma, A., and Hattori, T., *J. Catal.* **195**, 151 (2000).

Shimizu, K.-i., Tsuzuki, M., Kato, K., Yokota, S., Okumura, K., and Satsuma, A., *J. Phys. Chem. C* **111**, 950 (2007).

Simmons, E.L., *Appl. Optics* **14**(6), 1380 (1975).

Sojka, Z., Bozon-Verduraz, and Che, M., *in* "Handbook of Heterogeneous Catalysis" (G. Ertl, H. Knözinger, F. Schüth and J. Weitkamp, Eds.), Vol. 2, p. 1039. WILEY-VCH, Weinheim, 2008.

Sphereoptics. "Integrating Sphere Design and Applications—Technical Information", 2004.

Spielbauer, D., Mekhemer, G.A.H., Bosch, E., and Knözinger, H., *Catal. Lett.* **36**, 59 (1996).

Springsteen, A., *Anal. Chim. Act.* **380**, 379 (1999).

Springsteen, A., *Spectroscopy* **15**(5), 20 (2000a).

Springsteen, A., *Spectroscopy* **15**(6), 22 (2000b).

Steinfeldt, N., Müller, D., and Baerns, M., *Appl. Catal. A: Gen.* **272**, 201 (2004).

Stellarnet. www.stellarnet-inc.com, accessed Oct. 06, 2006.

Stone, F.S., *in* "Surface Properties and Catalysis by Non-Metals" (J.P. Bonnelle, B. Delmon and E.G. Derouane, Eds.), p. 237. D. Reidel Publishing Company, Boston, 1983.

Sung-Suh, H.M., Kim, D.S., Lee, C.W., and Park, S.E., *Appl. Organomet. Chem.* **14**, 826 (2000).

Terenin, A., *Adv. Catal.* **15**, 227 (1964).

Theocharous, E., *Appl. Optics* **45**, 2381 (2006).

Thiede, M., and Melsheimer, J., *Rev. Sci. Inst.* **73**, 394 (2002).

Tinnemans, S.J., Kox, M.F.H., Nijhuis, T.A., Visser, T., and Weckhuysen, B.M., *Phys. Chem. Chem. Phys.* **7**, 211 (2005).

Tinnemans, S.J., Kox, M.H.F., Sletering, M.W., Nijhuis, T.A., Visser, T., and Weckhuysen, B.M., *Phys. Chem. Chem. Phys.* **8**, 2413 (2006).

Tinnemans, S.T., Mesu, J.G., Kervinen, K., Visser, T., Nijhuis, T.A., Beale, A.M., Keller, D.E., van der Eerden, A.M.J., and Weckhuysen, B.M., *Catal. Today* **113**, 3 (2006).

Tromp, M., Sietsma, J.R.A., van Bokhoven, J.A., van Strijdonck, G.P.F., van Haaren, R.J., van der Eerden, A.M.J., van Leeuwen, P.W.N.M., and Koningsberger, D.C., *Chem. Commun.*, 128 (2003).

Tzolova-Müller, G., Hess, C., and Jentoft, F.C., unpublished data.

Tzolova-Müller, G., Jentoft, F.C., Villegas, J.I., and Murzin, D.Yu., unpublished results.

Valentine, A.M., Stahl, S.S., and Lippard, S.J., *J. Am. Chem. Soc.* **212**, 3876 (1999).

van den Broek, A.C.M., van Grondelle, J., and van Santen, R.A., *J. Catal.* **167**, 417 (1997).

Van Der Voort, P., White, M.G., Mitchell, M.B., Verbeckmoes, A.A., and Vansant, E.F., *Spectrochim. Acta Part A* **53**, 2181 (1997).

van de Water, L.G.A., Bergwerff, J.A., Leliveld, R.G., Weckhuysen, B.M., and de Jong, K.P., *J. Phys. Chem. B* **109**, 14513 (2005a).

van de Water, L.G.A., Bergwerff, J.A., Nijhuis, T.A., de Jong, K.P., and Weckhuysen, B.M., *J. Am. Chem. Soc.* **127**, 5024 (2005b).

Venter, J.J., and Vannice, M.A., *Appl. Spectrosc.* **42**, 1096 (1988).

Verdonck, J.J., Schoonheydt, R.A., and Jacobs, P.A., *J. Phys. Chem.* **85**, 2393 (1981).

Wark, M., Rohlfing, Y., Altindag, Y., and Wellmann, H., *Phys. Chem. Chem. Phys.* **5**, 5188 (2003).

Warnken, M., Lazar, K., and Wark, M., *Phys. Chem. Chem. Phys.* **3**, 1870 (2001).

Weckhuysen, B.M., *Chem. Commun.*, 97 (2002).

Weckhuysen, B.M., *in* "In Situ Spectroscopy of Catalysts" (B.M. Weckhuysen, Ed.), p. 260. American Scientific Publishers, Stevenson Ranch, CA, 2004.

Weckhuysen, B.M., *Phys. Chem. Chem. Phys.* **5**, 4351 (2003).

Weckhuysen, B.M., Baetens, D., and Schoonheydt, R.A., *Angew. Chem. Int. Ed.* **39**, 3419 (2000).

Weckhuysen, B.M., Bensalem, A., and Schoonheydt, R.A., *J. Chem. Soc. Faraday Trans.* **94**, 2011 (1998).

Weckhuysen, B.M., Rao, R.R., Pelgrims, J., Schoonheydt, R.A., Bodart, P., Debras, G., Collart, O., Van Der Voort, P., and Vansant, E.F., *Chem. Eur. J.* **6**, 2960 (2000).

Weckhuysen, B.M., and Schoonheydt, R.A., *Catal. Today* **49**, 441 (1999).

Weckhuysen, B.M., and Schoonheydt, R.A., *in* "Spectroscopy of Transition Metal Ions on Surfaces" (B.M. Weckhuysen, P. Van Der Voort and G. Cantana, Eds.), p. 228. Leuven University Press, Leuven, Belgium, 2000.

Weckhuysen, B.M., Verberckmoes, A.A., Debaere, J., Ooms, K., Langhans, I., and Schoonheydt, R.A., *J. Mol. Catal. A: Chem.* **151**, 115 (2000).

Wendlandt, W.W., and Hecht, H.G., "Reflectance Spectroscopy.", p. 19. Interscience Publishers (a division of John Wiley & Sons), New York, 1966.

Wichterlová, B., Sazama, P., Breen, J.P., Burch, R., Hill, C.J., Čapek, L., and Sobalík, Z., *J. Catal.* **235**, 195 (2005).

Wovchko, E.A., Yates, J.T., Jr., Scheithauer, M., and Knözinger, H., *Langmuir* **14**, 552 (1998).

Wu, Z.L., Kim, H.S., Stair, P.C., Rugmini, S., and Jackson, S.D., *J. Phys. Chem. B* **109**, 2793 (2005).

Xie, S., Iglesia, E., and Bell, A.T., *J. Phys. Chem. B* **105**, 5144 (2001).

Yang, S., Iglesia, E., and Bell, A.T., *J. Phys. Chem. B* **109**, 8987 (2005).

Zecchina, A., Lofthouse, M.G., and Stone, F.S., *J. Chem. Soc., Faraday Trans. I* **71**, 1476 (1975).

Zhu, Z.D., Chang, Z.X., and Kevan, L., *J. Phys. Chem. B* **103**, 2680 (1999).

Zimmermann, Y., and Spange, S., *J. Phys. Chem. B* **106**, 12524 (2002).

Zou, W., and Gonzalez, R.D., *J. Catal.* **133**, 202 (1992).

CHAPTER 4

X-Ray Photoelectron Spectroscopy for Investigation of Heterogeneous Catalytic Processes

**Axel Knop-Gericke,* Evgueni Kleimenov,[†]
Michael Hävecker,* Raoul Blume,* Detre Teschner,*
Spiros Zafeiratos,[‡] Robert Schlögl,*
Valerii I. Bukhtiyarov,[§] Vasily V. Kaichev,[§]
Igor P. Prosvirin,[§] Alexander I. Nizovskii,[§]
Hendrik Bluhm,[¶] Alexei Barinov,[∥] Pavel Dudin,[∥]
and Maya Kiskinova[∥]**

Abstract

X-ray photoelectron spectroscopy (XPS) is commonly applied for the characterization of surfaces in ultrahigh vacuum apparatus, but the application of XPS at elevated pressures has been known for more than 35 years. This chapter is a description of the development of XPS as a novel method to characterize surfaces of catalysts under reaction conditions. This technique offers opportunities for determination of correlations between the electronic surface structures of active catalysts and the catalytic activity, which can be characterized simultaneously by analysis of gas-phase products. Apparatus used for XPS investigations of samples in reactive atmospheres is described here; the application of

* Fritz-Haber-Institut der Max-Planck-Gesellschaft, Department of Inorganic Chemistry, 14195 Berlin, Germany
† ETH Zürich, Laboratory of Physical Chemistry, HCI E 209, 8093 Zürich, Switzerland
‡ LMSPC-UMR 7515 du CNRS, 67087 Strasbourg Cedex 2, France
§ Boreskov Institute of Catalysis SB RAS, Novosibirsk 630090, Russia
¶ Lawrence Berkeley National Laboratory, Chemical Sciences Division, Berkeley, CA 94720, USA
∥ Sincrotrone Trieste, Microscopy Section, 34012 Trieste, Italy

Advances in Catalysis, Volume 52
ISSN 0360-0564, DOI: 10.1016/S0360-0564(08)00004-7

synchrotron radiation allows the determination of depth profiles in the catalyst, made possible by changes in the photon energy. The methods are illustrated with examples including methanol oxidation on copper and ethene epoxidation on silver. Correlations between the abundance of surface oxygen species and yields of selective oxidation products are presented in detail. Further examples include CO adsorption and methanol decomposition on palladium and CO oxidation on ruthenium.

Contents

1. Introduction   215
2. History of XPS Applied at Substantial Pressures   218
3. Description of XPS Apparatus   221
    3.1. VG ESCALAB Photoelectron Spectrometer   221
    3.2. XPS Equipment Operated with an Electrostatic
         Lens System   224
4. Interaction of CO with Pd(111)   229
5. Dehydrogenation and Oxidation
   of Methanol on Pd(111)   234
6. Ethene Epoxidation Catalyzed by Silver   240
7. Methanol Oxidation Catalyzed by Copper   247
8. CO Oxidation Catalyzed by Ru(0001)   256
9. Outlook   266
Acknowledgments   267
References   267

ABBREVIATIONS

A	aperture
ALS	Advanced Light Source
BE	binding energy
BESSY	Berliner Elektronenspeicherringgesellschaft für Synchrotronstrahlung
BP	backing pump
DFT	density functional theory
E	energy
ESCA	electron spectroscopy for chemical analysis
FWHM	full width at half maximum
$h\nu$	photon energy
HP	high pressure
HREELS	high-resolution electron energy loss spectroscopy
HT	high temperature

IR	infrared
IRAS	infrared absorption spectroscopy
KE	kinetic energy
LT	low temperature
ML	monolayer
MS	mass spectrometer
PGM	plane grating monochromator
q	charge
QMS	quadrupole mass spectrometer
LEED	low energy electron diffraction
LV	leak valve
MFC	mass flow controller
NEXAFS	near edge X-ray absorption fine structure
P	pressure
q	elementary charge
R	radius of the first aperture of the electrostatic lens system
R	universal gas constant
PTRMS	proton transfer reaction mass spectrometer
RT	room temperature
SC	sample chamber
SFG	sum frequency generation
SIMS	secondary ion mass spectrometry
STM	scanning tunneling microscopy
SXRD	surface X-ray diffraction
T	temperature
t	time on stream
TDS	thermal desorption spectroscopy
TP	turbo molecular pump
TPRS	temperature programmed reaction spectroscopy
TPD	temperature programmed desorption
UHV	ultra high vacuum
UPS	ultraviolet photoelectron spectroscopy
U49/2	undulator U49/2 @ BESSY
XANES	X-ray absorption near edge spectroscopy
XAS	X-ray absorption spectroscopy
XPS	X-ray photoelectron spectroscopy
z	distance between the sample surface and the first aperture of the electrostatic lens system (measured in units of R, therefore dimensionless)
Θ	coverage
Φ	work function

1. INTRODUCTION

The surface science approach to heterogeneous catalysis takes advantage of simplified model versions of catalysts and reactions. Model reactions have been investigated under well-defined ultrahigh-vacuum (UHV) conditions with powerful methods that have allowed chemical and structural characterization down to the atomic level. Some investigations, carried out with complementary spectroscopic and structural surface science techniques have yielded detailed information about (1) elementary steps of surface reactions, such as molecular dissociation, adsorption, desorption, and interactions between the species present on surfaces, (2) responses of surface geometric and electronic structures to the presence of atoms and molecules that are participants in catalytic reactions, (3) effects of structural imperfections and controlled amounts of adsorbed atoms such as those used often as additives in the preparation of real catalysts, and (4) surface dynamics, such as propagation of reaction fronts and the related formation of stationary patterns, which may act as local chemical microreactors, etc. Such information provides basic knowledge about the elementary steps of surface reactions used in modeling reactivity and mechanisms of catalytic processes. Theoretical predictions of elementary steps, reaction barriers, the nature of transition states, and dynamics of surface processes based on DFT electronic structure calculations or Monte-Carlo simulations have been used to guide the experiments.

The major drawbacks of this "bottom ups" approach are the so-called "materials gap," and "pressure gap." Technological catalysts are morphologically complex multicomponent materials, often with microscopic dimensions, whereas the fundamental surface-science investigations focus on the catalytic behavior of well-structured and, most often, single-crystal metal, alloy, or oxide samples. Some attempts have already been made to bridge the materials gap by preparing microstructured and nanostructured model supported catalysts, made by using lithography or dispersing catalyst nanoparticles or microparticles on supports (Bäumer and Freund, 1999; Goodman, 1995; Graham et al., 1994; Günther et al., 2004; Schütz et al., 1999). The size dependence of the geometric and electronic structures of the nanomaterials implies the need for quantitative understanding the effects of the size of these structures on surface reactivity, an important issue for tailoring the catalyst properties that is still in infancy (Aballe et al., 2004; Valden et al., 1998).

Realistic operating pressures in catalytic reactions are orders of magnitude higher than those used in most surface-science experiments, and the chemical potential of the gas, usually neglected in the UHV experiments, becomes a significant contribution to the free energy of the surface layer. This pressure difference implies that the structures monitored under unrealistic UHV conditions do not necessarily involve the structures that play important roles in catalytic reactions. Significant changes in the morphology of the catalysts can occur as a consequence of dynamic structural rearrangements induced by the molecules participating in the reaction. The compositional and structural changes of the catalyst surface may exert dramatic effects not only on the catalyst performance, but also on the reaction mechanism (Stampfl et al., 2002).

The first attempts to bridge the pressure gap were *ex situ* experiments, allowing direct transfer of samples between a UHV station and a high-pressure reaction cell; thus, UHV surface-science techniques were coupled with reaction experiments carried out under realistic conditions (Rodriguez and Goodman, 1991). This strategy permits determination of the structure and composition of a model catalyst (usually a clean single crystal metal, alloy, or oxide, which may or may not be modified by controlled amounts of adsorbates) before and after the catalytic reaction to be correlated with the catalytic activity measured in the cell at a relatively high pressure.

An example illustrating the value of this approach is the work demonstrating the pressure gap effect in CO oxidation on Ru(0001). By use of *ex situ* TPD, STM, LEED, and XPS characterization, it was shown that the Ru(0001) surface, which appears to be inactive in surface-science investigations, is in reality a more efficient catalyst than platinum, because the active oxygen-rich phase cannot form under UHV conditions (Böttcher et al., 1997; Over et al., 2000, 2001).

The limitation of the *ex situ* approach is that the composition and structure of the catalyst surface are not investigated under reaction conditions. This limitation prevents the post-mortem analysis of the catalyst under UHV to determine an assessment of the surface restructuring, intermediate species, segregation of specific components, etc., that are characteristic of reaction conditions.

Thus, there is a strong motivation to implement techniques to characterize catalytic surfaces during the reaction and to elucidate their active phases. The first such reaction investigations were carried out with optical absorption, diffraction, and structural techniques that work at atmospheric pressures; the techniques included are IR spectroscopy (Beitel et al., 1996; Szanyi et al., 1994), SFG spectroscopy (Dellwig et al., 2000; Su et al., 1996), XAS (Knop-Gericke et al., 1998), SXRD (Peters et al., 2001), and STM (Hendriksen and Frenken, 2002; McIntyre et al., 1994).

X-ray photoelectron spectroscopy (XPS) has been recognized as one of the best analytical methods for probing composition and electronic structure of solid surfaces and interfaces. However, it is based on monitoring emitted photoelectrons, which imposes a limitation on the operating pressures; consequently, substantial time has passed before this technique was applied in reaction experiments in the near-atmospheric pressure range.

The section that follows is a review of the development of equipment for carrying out XPS at relatively high pressures. Although more than 35 years have passed since the first such XPS experiments were performed, only a few XPS investigations of catalysts under reaction conditions have yet been published. Subsequent sections provide experimental details and examples of XPS investigations of catalysts in the working state. Examples concerned with the interaction of CO and methanol with the Pd(111) surface demonstrate how XPS is useful for investigation of the pressure dependence of the structure of adsorption sites. The example of ethene epoxidation on silver illustrates a correlation of the catalytic activity with the abundance of oxygen species on the surface. XPS showed that in methanol oxidation catalyzed by copper a subsurface oxygen species is involved in the selective formation of formaldehyde.

2. HISTORY OF XPS APPLIED AT SUBSTANTIAL PRESSURES

The basic elements of the equipment for XPS with samples in reactive atmospheres are shown in Figure 1. The X-ray source (1) can be a conventional X-ray tube or a synchrotron radiation facility. The thin X-ray window (2) separates the volume of the X-ray source from the sample cell (4). X-rays from the source pass through the X-ray window, hit the sample (3), and induce the emission of photoelectrons. After traveling through the sample cell, a fraction of the photoelectrons reach the entrance aperture of the differential pumping stage(s) (5) and passes through it to the electron energy analyzer (6). Therefore, the application of differential pumping allows minimization of the path of photoelectrons in the gas phase. Gases can be introduced into the sample cell (in contrast to the UHV conditions of conventional XPS). The tolerated pressure in the sample cell is limited by the scattering of photoelectrons by gas-phase molecules, which leads to a decrease of the photoelectron signal. The maximum allowable pressure depends, for example, on the distance between the sample and the first aperture, the intensity of the X-ray source, the photoelectron

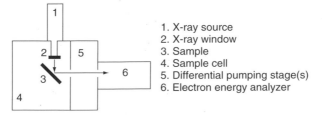

1. X-ray source
2. X-ray window
3. Sample
4. Sample cell
5. Differential pumping stage(s)
6. Electron energy analyzer

FIGURE 1   Schematic diagram of apparatus for the measurement of XP spectra of catalysts in reactive atmospheres.

collection and detection efficiencies, the kinetic energy of photoelectrons, and the composition of the gas. Electrostatic lenses introduced into the differential pumping stage (or stages) increase significantly the collection efficiency of photoelectrons (Figure 2).

The concept of differential pumping for XPS was first applied by K. Siegbahn and colleagues in 1969 for the investigation of gases at pressures up to a few tenths of a Torr (1 Torr = 1.33 mbar = 133 Pa) (Siegbahn et al., 1969). The system was based on a magnetic-type electron energy analyzer with one differential pumping stage. Four years later, K. Siegbahn and H. Siegbahn reported the first XPS experiment characterizing liquids (Siegbahn and Siegbahn, 1973). The apparatus allowed investigations of a beam of liquids having a vapor pressure of less than ~1 mbar and was based on a magnetic-type electron energy analyzer with one differential pumping stage. Other modifications of the spectrometer for liquid investigations were reported later by the same group (Fellner-Feldegg et al., 1975; Siegbahn et al., 1981).

In 1979, the construction of an XPS system for the investigation of solids in gas atmospheres at pressures of up to ~1 mbar was reported by Joyner and Roberts (1979a); this was later commercialized. One differential pumping stage around the high-pressure sample cell was used in combination with the commercial hemispherical electron energy analyzer ESCALAB of V.G. Scientific Ltd. Two "high-pressure" spectrometers of this type were supplied to the University of Wales (Cardiff, UK) and the Boreskov Institute of Catalysis (Novosibirsk, Russia).

The next design of a high-pressure XPS apparatus was reported by Ruppender et al. in 1990 (Ruppender et al., 1990). The system included three differential pumping stages and allowed the performance of experiments at pressures up to 1 mbar. In 2000, the first results obtained with the comparable XPS system designed by the group of Salmeron were reported. The spectrometer included a two-stage differential pumping system combined with an electrostatic lens system for the collection of

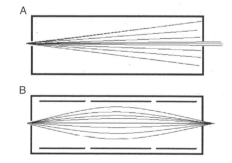

FIGURE 2   Collection of photoelectrons (A) without and (B) with electrostatic lenses.

photoelectrons. The equipment was used with a synchrotron X-ray source (Advanced Light Source (ALS) in Berkeley) and allowed investigations at pressures up to 7 mbar. This equipment was applied for the investigation of solid catalysts under reaction conditions with simultaneous monitoring of reaction products (Hävecker et al., 2003; Ogletree et al., 2002) and for the investigation of the process of ice premelting (Bluhm et al., 2002; Ogletree et al., 2002).

In 2001, Kelly et al. (Kelly et al., 2001) reported the construction of a high-pressure X-ray photoelectron spectrometer for monitoring the synthesis of thin films. A hemispherical electron energy analyzer with a specially constructed electrostatic lens system and a one-stage differential pumping system made possible measurements at pressures up to 0.03 mbar. The electrostatic focusing elements introduced into the differential pumping stage of this apparatus allowed collection of photoelectrons from a cone with a half angle of $15°$ and from a surface with an area of about $1–2 \text{ mm}^2$. This high collection efficiency was achieved because the electrostatic lens elements were mounted close to the sample. Nevertheless, the distance between the sample and the inlet aperture of the differential pumping equipment was about 40–50 mm (compared with 1–2 mm in all other reported designs), which limits the pressure in the sample cell to quite low values.

In March, 2004, the company Gammadata Scienta announced a commercial XPS system based on a spectrometer designated SES-100. Pressure drop in the four-stage differential pumping system was specified to be better than six orders of magnitude, which should correspond to a pressure limit in the sample cell of about 0.1 mbar (assuming a pressure in the analyzer not higher than 10^{-7} mbar).

In 2002, the group of Steinrück reported an apparatus for measurement of photoelectron spectra in the pressure range of up to 10^{-5} mbar (Denecke et al., 2002). The apparatus incorporated a hemispherical electron analyzer and a supersonic molecular beam. The authors observed a pressure dependence in the adsorption of CO on Pt(111). Kinetics parameters characterizing CO adsorption on Pt(111) were obtained in a further XPS investigation by the same group (Kinne et al., 2002, 2004). The adsorption of CO on and the desorption of CO from Pt(355) was also investigated (Tränkenschuh et al., 2006), as was the coadsorption of D_2O and CO on Pt(111) (Denecke, 2005; Kinne et al., 2004) and the adsorption of NO on Pt(111) as a function of oxygen precoverage (Zhu et al., 2003a,b). The adsorption of more complex molecules such as methane, ethene, and acetylene on Pt(111) and on Ni(100) were investigated by the Steinrück group (Fuhrmann et al., 2004, 2005; Neubauer et al., 2003), who reported equipment for the measurement of photoelectron spectra at pressures from 10^{-10} to 1 mbar (Pantförder et al., 2005). The apparatus included a modified hemispherical electron energy analyzer and several differential pumping stages. A modified twin anode was used as the X-ray source. The reaction gases were introduced by background dosing or as a directed gas beam from a small tube. It was not possible to measure the pressure close to the sample surface, but the authors observed gas-phase signals, an indication, that the partial pressure of the adsorbed gases exceeded 10^{-2} mbar (Pantförder et al., 2005).

The groups of Bukhtiyarov and Schlögl were also involved in the development of spectrometers to operate at substantial pressures. Researchers from the Boreskov Institute of Catalysis applied the Vacuum generators (VG)-based spectrometer described earlier, after modifying it for catalytic reaction experiments. The group reported its first results in this field in 2001 (Bukhtiyarov and Prosvirin, 2001), having used a system designed in collaboration with the group of Salmeron and representing the second version of the apparatus operated in Berkeley, as described earlier; the differential pumping and photoelectron collection systems were improved. Two almost identical versions of the new apparatus were produced, one used by the group of Salmeron at the ALS in Berkeley and the second operated at BESSY in Berlin. The latter equipment has worked since 2002, when it was tested with a copper and palladium catalyst (Bluhm et al., 2004; Teschner et al., 2005) The new apparatus was used to investigate the oxidation of n-butane to maleic anhydride on vanadium phosphorus oxides (Kleimenov et al., 2005). The oxidation of $CuGaSe_2$ thin films, which are used as absorber materials in solar cells, was investigated with the apparatus at BESSY (Würz et al., 2005). Recently an investigation of Pd/CeO_2 and Pt/CeO_2 catalysts for the oxidation of CO in hydrogen (PROX) was published (Pozdnyakova et al., 2006).

The apparatus in Berkeley has been used to investigate water on metallic and oxidic surfaces (Ghosal et al., 2005). The oxidation and reduction of Pd(111) were investigated in Berkeley (Ketteler et al., 2005). The facilities of the Boreskov Institute and of the Fritz Haber Institute are described in detail in the following section.

3. DESCRIPTION OF XPS APPARATUS

3.1. VG ESCALAB Photoelectron Spectrometer

The group working at the Boreskov Institute used a VG ESCALAB photoelectron spectrometer (Boronin et al., 1988; Joyner and Roberts, 1979a,b), shown in Figure 3; it consists of three main chambers (analyzer and two preparation chambers), each pumped by a separate diffusion pump providing a background pressure lower than 5×10^{-10} mbar. A high pressure is created in a gas cell, which is inserted inside the analyzer chamber of the spectrometer through the left preparation chamber same preparation chamber is used for movement of a sample to the cell and for separate pumping of an energy analyzer during exper The pressure in the gas cell is measured with an advanced Pira

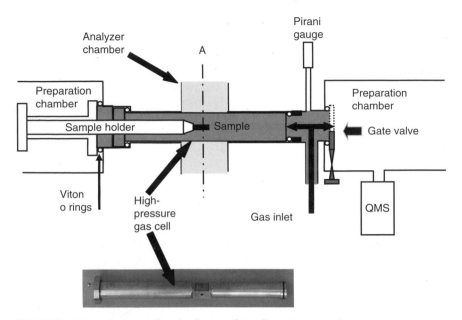

FIGURE 3   View of a gas cell and scheme of its allocation in a VG ESCALAB HP photoelectron spectrometer.

the composition of the gas phase is determined with a mass spectrometer (Figure 3, QMS) located in the right preparation chamber. Gas flow from the gas cell to the preparation chamber is regulated by a gate valve. The gas cell is sealed from other parts of the spectrometer with Viton O-rings. The pressure difference between the gas cell and other parts of the spectrometer is provided by small apertures (<4 mm) for the entry of X-rays and the exit of photoelectrons and for two-step differential pumping. Diffusion pumps are used to evacuate the analyzer chamber, the X-ray source recipient and the energy analyzer. Consequently, the pressures attained in the analyzer chamber and in the X-ray tube/electron energy analyzer zones are 10^{-4} and 2×10^{-6} mbar, respectively, when the pressure in the gas cell is 1 mbar.

Notwithstanding the capability for measurement of photoemission spectra at pressures up to 1 mbar, the original construction of the gas cell was not suitable for carrying out experiments under catalytic reaction conditions. To provide the short distance between the sample surface and aperture for the exit of electrons (or a short path of the photoelectrons in zone at the higher pressure), the equipment supplier decreased the dimensions of the gas cell. A section of the cell, together with the standard holder and a typical sample, is shown in Figure 4A; this design size of the sample so that only thin polycrystalline foils can be

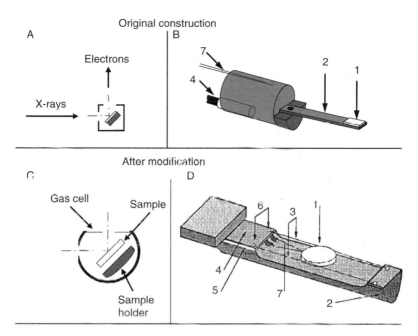

FIGURE 4 Sections of the gas cell, sample holder, and sample and schemes of sample mounting before (A, B) and after (C, D) modernization of the VG ESCALAB HP photo-electron spectrometer for measurements of catalysts in reactive environments (see text): (1) sample; (2) holder; (3) tungsten wires; (4) feedthroughs; (5) insulating ceramic; (6) spot weld; (7) thermocouples.

applied. Another disadvantage of the standard gas cell was the need to use the sample holder with an internal heating device and a thermocouple (Figure 4B). With this design, the sample holder had the same tempera-ture as the sample (or even a slightly higher temperature—as the sample is cooled by radiation). In this case, the material of construction of the sample holder (stainless steel) could catalyze a substantial undesired conversion, because the surface area of the sample holder is much greater than that of the catalyst sample (Figure 4B).

To avoid this limitation, new designs of the sample holder and gas cell were developed (Figures 4C and 4D) (Bukhtiyarov and Prosvirin, 2001; Bukhtiyarov et al., 1994, 2005; Prosvirin et al., 2003). The new holder permitted independent heating of the sample, and the sample tempera-ture was measured with a thermocouple spot welded to it. The sample size was increased to 10 mm in diameter and 1 mm thickness. These changes, as well as changes in the lenses, made possible the use of single-crystal samples and the simultaneous measurement of XP spectra and mass spectra at pressures up to 0.2 mbar. This modified equipment was used for adsorption and catalytic reaction experiments characterizing functioning catalysts.

3.2. XPS Equipment Operated with an Electrostatic Lens System

As mentioned earlier, a basic feature of XPS equipment operated at pressures high enough to allow investigations of catalytic reactions (Figure 5) is the differential pumping between the sample cell and the electron energy analyzer. A criterion for quality of such a spectrometer is the maximum pressure in the sample cell that can be achieved without loss of spectral quality. Several parameters influence the XPS signal intensity (1) the flux of the X-ray source, (2) the transmission through the X-ray window, (3) the absorption of X-rays by the gas in the sample cell, (4) the efficiency of photoionization and scattering of photoelectrons in solids, (5) the scattering of photoelectrons by gas-phase molecules, and (6) the efficiency of photoelectron collection by the spectrometer.

The thickness and area of an X-ray window determine its mechanical stability. For example, a Si_3N_4 X-ray window with an area of 2.5 $\times$ 2.5 mm^2 and a thickness of 100 nm can hold the pressure difference of up to 10 mbar. The transmission through such a window and through thicker and thinner windows is shown in Figure 6; the window attenuates the total XPS signal at most by one order of magnitude in the region of photon energies that are usually used in XPS (200–1000 eV). The X-ray transmission through oxygen and through n-butane is shown in Figure 7. These results show that absorption of X-rays passing through a gas

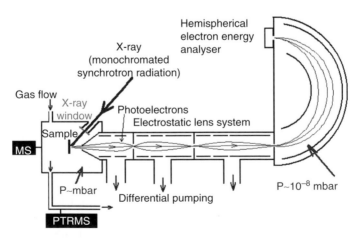

FIGURE 5   XPS equipment with electrostatic lens system.

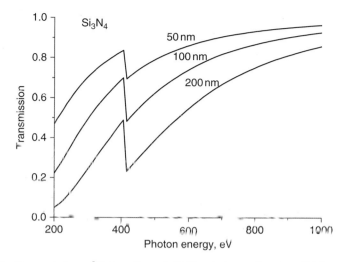

FIGURE 6 Transmission of X-rays through Si_3N_4 windows of various thicknesses (calculated according to Henke et al. (1993)).

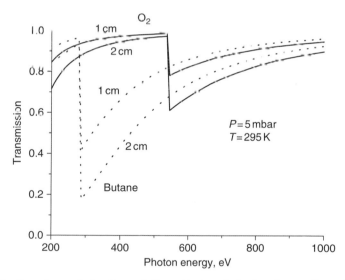

FIGURE 7 X-ray transmission through O_2 and butane (calculated according to Hoffman et al. (1982)).

atmosphere of a path length of a few centimeters will not decrease the signal more than by an order of magnitude.

The influence of photoelectron scattering by a gas on the signal depends on the gas, the pressure, and the path length of photoelectrons in the gas. Scattering of photoelectrons is the principal limitation of the maximum pressure allowable in the sample cell. Because the photoelectron signal decreases exponentially with the path length, the distance between the sample surface and the first differential pumping aperture, defined as z, should be kept to a minimum. However, it is not possible to make this distance less than 1 mm when the aperture radius R is 0.5 mm, because the pressure of a gas near the sample surface depends on the distance/aperture-size ratio. The dependence was estimated by using a molecular flow approach, as shown in Figure 8, where P is the pressure in the sample cell and z is measured in multiples of R. Thus, the pressure near the sample surface would be 99% of P_0 for $z = -4$ or 95% for $z = -2$ or 85% for $z = -1$. Therefore, it is not recommended to go much closer than $z = -2$, which is 1 mm for the aperture diameter of 1 mm.

Cross-sections of photoelectron scattering by hydrogen molecules are shown in Figure 9. The mean free path of low-energy electrons under these conditions is about 1 mm. It can be calculated that the signal S for a distance Z in H_2 at $P = 1$ mbar and $T = 300$ K decreases by three times for low-energy electrons compared to the signal in vacuum, S_{vac}. The signal decreases exponentially with pressure and with distance. Thus, for H_2, the maximum pressure under these conditions should be several mbar. A decrease of the distance could increase the maximum pressure, but, as discussed earlier, this is possible only with a simultaneous decrease of the aperture size, which will cause additional requirements for X-ray

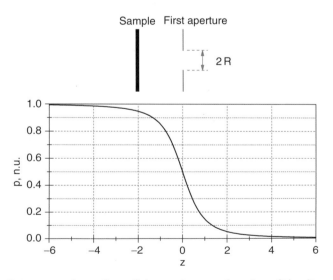

FIGURE 8 Pressure on the surface of the catalyst as a function of the dimensionless distance parameter z, which is the distance from the surface to the aperture divided by the aperture radius R.

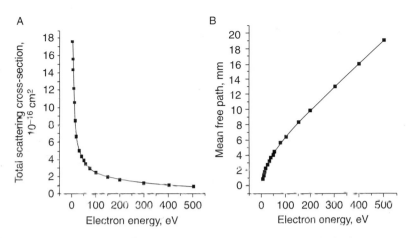

FIGURE 9   Characteristics of photoelectron scattering by molecular hydrogen.
(A) Cross-sections for scattering of photoelectrons by a hydrogen molecule according to
Hoffman et al. (1982); (B) mean free path of photoelectrons in hydrogen at 300 K,
pressure = 1 mbar, calculated by using the cross-sections of panel (A).

focusing. When the gas includes molecules other than H_2, the mean free
path is less and the maximum pressure should be lower.

The efficiency of the photoelectron collection is determined by the
collection solid angle Ω, which is a function of the collection plane
half-angle a: $\Omega = 4\pi \sin^2(a/2)$, which can be well approximated for $a < 10°$
by the function const $\times a^2$. Consequently, it is concluded that at small
angles the photoelectron collection efficiency depends on the collection
plane half-angle squared. Therefore, it is important to achieve as high a
collection angle as possible to achieve the best system sensitivity.

In early experiments with the XPS apparatus used at pressures of the
order of those mentioned earlier (Fellner-Feldegg et al., 1975; Joyner and
Roberts, 1979a; Siegbahn and Siegbahn, 1973; Siegbahn et al., 1981),
photoelectron collection efficiencies were quite low, because no special
electrostatic collection system was applied. The effect of increasing the
collection angle by using the electrostatic lens system is illustrated in
Figure 2. The photoelectron collection angle for the 50-cm-long differential pumping system with an exit-slit radius of 1 mm (the geometrical
sizes were taken from the analyzer Phoibos 150, SPECS GmbH, Berlin)
can be improved from 0.1° to 5° by incorporation of an electrostatic lens
system, which corresponds to the improvement of the collection efficiency by 2.5×10^3 times. This example demonstrates the importance of
the electrostatic collection system.

The scheme of gas flow through the reaction cell is shown in Figure 10.
Gases (or vaporized liquid) are introduced into the sample chamber

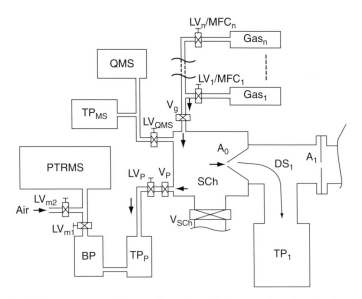

FIGURE 10   Scheme representing gas flow through the reaction chamber (see text for the abbreviations).

(sample cell) SCh through the mass flow controller $MFC_{1, \ldots, n}$ or through the leak valves $LV_{1, \ldots, n}$. The outlet of the system is an aperture A_0. Additional pumping can be provided through the process turbo-molecular pump TP_P. The gas composition in the reaction chamber is monitored by the quadrupole mass spectrometer QMS (Prisma QMS 200 M) and by the proton-transfer reaction mass spectrometer PTRMS (PTRMS instrument produced by IONICON Analytic GmbH, Innsbruck, Austria). The quadrupole mass spectrometer is connected to the sample chamber through the leak valve LV_{QMS}. The leak valve together with the turbo-pump TP_{MS} allows setting of a working pressure in the QMS, which should be in the range of $10^{-7} - 10^{-6}$ mbar. The PTRMS requires an inlet pressure of 1 bar. Therefore, it was connected to the outlet of the backing pump (BP). The outlet gas from the BP can be diluted with air if necessary. The PTRMS can also be connected to the BP of the TP_1 if flow through the process pump TP_P is blocked.

The sample can be heated from the back by using an infrared laser. The radiation of the laser is directed onto the back of the sample through a glass fiber. Disks of SiC or stainless steel are used as heating plates. They were mounted below the catalyst in the case the sample has a high reflectivity. In the initial design the heating was provided by a ceramic electric heating element, which was later replaced by the laser system in order to avoid magnetic fields. The temperature is measured with a thermocouple.

Summarizing this part, we emphasize that the main property limiting the maximum working pressure in the sample cell is the scattering of low-energy photoelectrons by gas-phase molecules. The path of photoelectrons in the gas phase cannot be much shorter than the size of the first aperture, because it otherwise would lead to a decrease in the gas pressure near the sample surface. The aperture diameter cannot be decreased significantly below 1 mm because this value influences the probed area and the collection angle. The possibilities for increasing the maximum pressure are (1) to use a photon source with a higher flux and a tighter focus (which, however, can lead to damage of the sample by the beam) and/or (2) to improve the collection of photoelectrons by using an electrostatic lens system in the differential pumping stages.

4. INTERACTION OF CO WITH Pd(111)

CO on Pd(111) was chosen for an initial XPS investigation with the equipment operated at the Boreskov Institute described earlier, because it has been characterized with numerous vibrational spectroscopy techniques (Bourguignon et al., 1998; Bradshaw and Hoffmann, 1978; Geißel et al., 1998; Hoffmann, 1983; Loffreda et al., 1999; Morkel et al., 2003; Ohtani et al., 1987; Ozensoy et al., 2002; Rupprechter et al., 2002; Tüshaus et al., 1990), allowing comparison of the results with literature data. The main issue addressed by XPS is the possible formation of carbonaceous species as a result of dissociation of the C—O bond in CO or methanol (Section 5). Indeed, in contrast to vibrational spectroscopies (such as SFG, IRAS, and Raman spectroscopy), which are most sensitive to molecular vibrations, XPS is capable of characterizing species regardless of whether they are atomic or molecular.

Figure 11 shows XPS spectra measured during CO adsorption on the palladium single crystal at various temperatures and pressures (the spectra were normalized to the Pd3d integral intensity). CO adsorbed at 400 K and 10^{-6} mbar was chosen as a blank (reference), because only one species, namely, CO bonded on threefold hollow sites, has been shown to be produced under these conditions (Bourguignon et al., 1998; Bradshaw and Hoffmann, 1978; Geißel et al., 1998; Hoffmann, 1983; Loffreda et al., 1999; Morkel et al., 2003; Ohtani et al., 1987; Ozensoy et al., 2002; Rupprechter et al., 2002; Surnev et al., 2000; Tüshaus et al., 1990). In agreement with the literature data, a narrow peak at 285.6 eV with a FWHM of about 1.3 eV in the C1s core-level spectrum gives evidence of the formation of a single species. Assignment of this C1s feature to the aforementioned CO surface species is supported by high-resolution XPS data of Surnev et al. (2000), who investigated the CO adsorption on Pd(111) under UHV

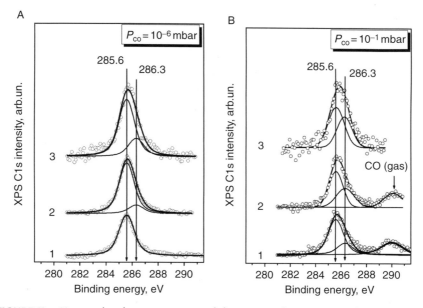

FIGURE 11 C1s core-level spectra measured during CO adsorption on Pd(111) surface at (A) 10^{-6} mbar and (B) 10^{-1} mbar and various temperatures: (1), 400 K; (2), 300 K; (3), 200 K.

conditions by using synchrotron radiation. By analyzing the coverage-dependent sequence of CO phases obtained at 100 K, the authors assigned a C1s feature at 285.6 ± 0.1 eV to CO adsorbed in threefold hollow sites; a feature at 285.85 eV to bridge-bonded CO; and a feature at 286.3 eV to on-top CO.

Decreasing the temperature of adsorption led to an increase in the intensity of the C1s spectra. Furthermore, the spectra became more asymmetric as a result of broadening at the high binding energy (BE) side (Figure 11A, curves 2 and 3). The deconvolution of the C1s spectra into separate components, which were described by the Doniach–Šunjic function after subtraction of Shirley background (Doniach and Šunjic, 1970; Shirley, 1972), indicates that this broadening is the result of the appearance of a new component at 286.3 eV. The same result was obtained by subtracting the spectrum obtained at 10^{-6} mbar and 400 K from the C1s spectra taken at lower temperatures (Kaichev et al., 2003). According to the data of Surnev et al. (Surnev et al., 2000), this feature can be assigned to on-top CO. Bridge-bonded CO, which should also appear increasingly in the spectra with increasing CO coverage, could not be differentiated from the CO bonded at threefold hollow sites, because of the limited resolution of the standard XPS equipment.

The sequences of C1s spectra measured at the same temperatures (400, 300, and 200 K) as the spectra of Figure 11A—but at higher pressure

$(10^{-1}$ mbar)—is shown in Figure 11B. Increasing the CO pressure from 10^{-6} to 0.1 mbar at 400 K increased the amount of on-top bonded CO (Figure 11B). This result indicates that when the higher pressure was applied, XPS allowed the detection of weakly bound species. In UHV, these species could be observed only at temperatures (less than room temperature) far from most realistic catalytic reaction conditions ($T >$ 400 K). Similar results were, as expected, observed when the temperature was decreased. The corresponding XPS measurement at 0.1 mbar and 200 K was used to estimate the ratio of hollow to on-top species at saturation coverage (1.9). This result is in satisfactory agreement with a hollow/ on-top ratio of two arising from the (2×2) CO structure proposed for the saturation coverage (Morkel et al., 2003; Tüshaus et al., 1990).

These data are in agreement with the results of coverage-dependent vibrational spectra (Bourguignon et al., 1998; Bradshaw and Hoffmann, 1978; Hoffmann, 1983; Morkel et al., 2003; Ozensoy et al., 2002; Rupprechter et al., 2002, Tüshaus et al., 1990). At a low CO coverage, a band at 1810–1820 cm^{-1}, which shifted to 1840–1850 cm^{-1} at $\Theta = 0.33$ ML, was observed in the IRAS spectra (Hoffmann, 1983). This band corresponds to the calculated anharmonic frequencies of CO in threefold hollow sites on the palladium surface (1828–1830 cm^{-1}) (Loffreda et al., 1999). When the coverage was increased to values exceeding 0.33 ML, the ν_{CO} band shifted continuously to higher frequencies, and, at 0.5 ML, a $c(4 \times 2)$-CO LEED pattern and a vibrational band were observed at approximately 1920 cm^{-1}. This structure was interpreted as CO adsorbed on bridge sites (Bourguignon et al., 1998; Hoffmann, 1983; Kaichev et al., 2003; Rupprechter et al., 2002). At higher coverages, CO adsorbs on a mixture of bridge and on-top sites (e.g., in a ($c\sqrt{3} \times 5$) structure at 0.6 ML; Tüshaus et al., 1990), and as saturation is approached at a coverage of 0.75 ML, the bridged species transform into threefold hollow and on-top species in a (2×2)-CO structure (fcc and hcp hollow and on-top CO) (Tüshaus et al., 1990).

A similar picture of the transformations of the CO species has also been inferred on the basis of SFG spectra measured in parallel with the XPS spectra (Figure 11) (Kaichev et al., 2003). The SFG spectrum measured at 10^{-6} mbar and 400 K gave evidence of CO adsorbed on threefold hollow sites (1910 cm^{-1}), whereas, at higher CO coverages (10^{-6} mbar at 300 K), mixtures of CO on threefold hollow sites and bridging CO (1935 cm^{-1}), as well as a small amount of on-top CO, were produced on the Pd(111) surface. Increasing the CO pressure to 1 mbar shifted the frequency of the hollow/bridge species to 1948 cm^{-1} and increased the intensity of the on-top species (2083 cm^{-1}). The strong dependence of the CO stretching frequency on the coverage was used for the determination of CO coverages at various temperatures and pressures (Bourguignon et al., 1998; Bradshaw and Hoffmann, 1978; Hoffmann,

1983; Morkel et al., 2003; Ozensoy et al., 2002; Rupprechter et al., 2002, Tüshaus et al., 1990).

Figure 12 is a comparison of the CO coverages as a function of temperature and pressure, as determined both by SFG and XPS (Kaichev et al., 2003). In the analysis of the XPS data, the CO coverage was estimated from the corresponding C1s/Pd3d intensity ratios calculated for each C1s spectrum shown in Figure 11. The CO structure at 10^{-6} mbar and 300 K ($\Theta =$ 0.5 ML) was used as a reference point. As shown in a number of publications (Bradshaw and Hoffmann, 1978; Morkel et al., 2003), these conditions facilitate the formation of the $c(4 \times 2)$ LEED pattern characteristic of $\Theta = 0.5$ CO coverage estimated from the CO stretching frequency in the SFG spectra are in good agreement with these XPS results. This observation is important, because it means that ionization of X-ray irradiated gas-phase molecules does not affect the process of CO adsorption. Indeed, the conditions are characteristic of equilibrium coverage by weakly bonded on-top CO, which is strongly dependent on CO pressure. For example, both methods show that even at 200 K, a pressure of 0.1–1 mbar is necessary to reach the CO saturation coverage of $\Theta = 0.75$ ML. Under typical UHV pressures, this structure is obtained only at approximately 100 K (Morkel et al., 2003; Tüshaus et al., 1990). Then, the coincidence of the quantitative XPS and SFG data indicates that identical adsorbed layers formed in the presence of CO, both in the presence and in the absence of X-rays.

We emphasize that XPS indicated the presence of CO adsorbed only at threefold hollow, bridge, and on-top sites—no signatures of high-pressure species were found.

Palladium carbonyls would produce a C1s XPS signal at 287–288 eV (Barber et al., 1972). This range is not obscured by adsorbed or gas-phase

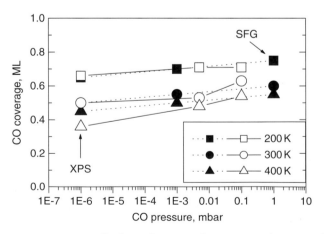

FIGURE 12   CO coverage on Pd(111) as a function of pressure as determined by XPS (full symbols, full lines) and by SFG (open symbols, dashed lines).

CO, and therefore we exclude the presence of these bulk metal carbonyls. The dissociation of CO by the Boudouard reaction (or the decomposition of palladium carbonyls) would lead to carbon deposition (Kung et al., 2000; McCrea et al., 2001) and should produce a feature at 284.0 eV characteristic of graphite or at 284.4 eV characteristic of amorphous carbon. In the case of carbide species, a feature at lower BE (<283.5 eV) would appear. Even if carbon dissolved in the palladium bulk near the surface region, the escape depth of the C1s electrons (about 2 nm) should have been sufficient to allow its detection. The absence of any carbon-related signals indicates that CO does not dissociate at 400 K and approximately 1 mbar, even over the course of several hours. This result is an important argument in the discussion about the possibility of CO dissociation at high pressure.

Although CO does not dissociate on single-crystal palladium surfaces under UHV conditions (Broden et al., 1976; Conrad et al., 1974, 1978; Stara and Matolin, 1994; Weissman et al., 1980), the situation may be different at higher pressures. In a number of investigations of adsorption on platinum and on palladium (Doering et al., 1982; Johánek et al., 2000; Kung et al., 2000; Matolin and Gillet, 1990; Matolin et al., 1991; McCrea et al., 2001; Stara and Matolin, 1994), it was reported that at higher pressures, CO formed metal carbonyls and/or dissociated by the Boudouard reaction (2CO → C + CO_2), leading to carbon deposition. However, our XPS investigation of CO adsorption indicates that this is not the case for the close-packed Pd(111) surface.

We stress that in experiments carried out at pressures such as those mentioned here, attention must be paid to the purity of the CO (Kaichev et al., 2003; Rupprechter et al., 2002). Figure 13 shows C1s and Fe2p spectra measured during CO adsorption at 400 K and various pressures when high-purity CO was used without further purification. In addition to the C1s signal indicative of molecular CO (~285.7 eV), two other features were observed, at 284.1 and 287.7 eV. As mentioned earlier, these signals can be assigned to graphitic/amorphous carbon and to metal carbonyl compounds, respectively. This result indicates that if the gas were not properly cleaned to remove $Fe(CO)_5$ and $Ni(CO)_4$ impurities, the decomposition of these compounds and possible CO dissociation on deposited nickel or iron formed from them can easily produce additional C1s signals. Iron carbonyl was the impurity in the present case, as evidenced by the appearance of Fe2p signals (Figure 13B). At higher CO partial pressures, even stronger iron signals were observed, leading to a decrease in molecular CO and to the appearance of an additional C1s signal originating from iron carbides (283.0 eV).

Cleaning of the CO gas to remove metal carbonyls by cooling the CO container with liquid nitrogen removes all the surface contaminants indicated by C1s spectra, even after treatments at pressures of approximately 1 mbar for 5–6 h. Care has to be taken not to misinterpret such observations as being evidence of CO dissociation on palladium.

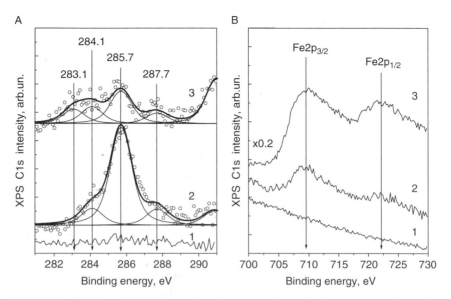

FIGURE 13 (A) C1s and (B) Fe2p core-level spectra obtained when CO was adsorbed at 400 K without purification, at (2) 5×10^{-3} mbar and (3) 0.1 mbar. Spectra of (1) the clean surface are shown for comparison. Results of deconvoluting the C1s spectra are also shown in (A): 283.1, carbide species (FeC_x); 284.1, graphite; 285.7, molecular CO; and 287.7, carbonyl species.

In summary, the results of this section demonstrate that by application of XPS, adsorbate structures and coverages can be obtained at pressures of the order of a millibar. Thus, the pressure range in XPS has been expanded by at least five orders of magnitude. Even at these relatively high pressures, CO structures were found to be similar to those known from UHV investigations, but obtained at low temperatures (<300 K). CO on Pd(111) adsorbs in threefold hollow, bridge, and on-top sites. Bukhtiyarov and coworkers (Bukhtiyarov et al., 1994; Kaichev et al., 2003) did not find any indications of CO dissociation or metal carbonyl formation under the experimental conditions (Bukhtiyarov et al., 2005; Kaichev et al., 2003).

5. DEHYDROGENATION AND OXIDATION OF METHANOL ON Pd(111)

The dehydrogenation (decomposition) of methanol to give CO and H_2 on supported catalysts has attracted much attention because of its practical relevance for methanol-fueled vehicles or heat-recovery techniques

(Cubeiro and Fierro, 1998; Matsumura et al., 1997; Shiozaki et al., 1999; Usami et al., 1998; Wickham et al., 1991, and references cited therein). Although methanol decomposition occurs on palladium-containing catalysts with high selectivities for CO + H_2, the activity is not satisfactory and limited by catalyst deactivation (Shiozaki et al., 1999). The main cause of deactivation is the formation of carbon or carbonaceous species $CH_x(x = 0–3)$ by cleavage of the C–O bond in methanol. In this respect, methanol decomposition on Pd(111) may serve as a simple model system allowing investigation of the various bond scission routes (O—H, C—H, and C—O) governing selectivity and also deactivation by carbonaceous species.

Notwithstanding numerous reports, the exact mechanism of methanol decomposition is still a matter of debate, and the reported results are sometimes rather controversial. Although there is agreement that the dehydrogenation of methanol to give CO and H_2 occurs *via* methoxy (CH_3O) species as the first intermediate, followed by stepwise hydrogen abstraction to give CH_2O, CHO, and CO (Bhattacharya et al., 1988; Chen et al., 1995; Christmann and Demuth, 1982; Davis and Barteau, 1990; Gates and Kesmodel, 1983; Guo et al., 1989; Kok et al., 1983; Kruse et al., 1990; Morkel et al., 2004; Rebholz and Kruse, 1991; Schauermann et al., 2002), the exact mechanism and probability of C–O bond scission is still under discussion.

The scission of this bond on the single-crystal Pd(111) surface was first proposed by Chen et al. (1995) on the basis of XPS, SIMS, and TPD data. In contrast, Guo et al. (1989), who also examined the interaction of methanol with Pd(111) at various temperatures (87–265 K) by using isotopic TPD with a sensitivity limit at 0.1% ML, did not report any dissociation of the methanol C–O bond on palladium. Other authors (Rebholz and Kruse, 1991) found that the dominant route in the methanol decomposition on Pd(111) is the complete dehydrogenation to give CO, and only a small amount of adsorbed CH_3 species (~0.05 ML) was detected. This disagreement was explained by the suggestion that the C–O bond scission on Pd (111), which is considered to be a rather inactive surface for C–O bond cleavage, may proceed on surface defects (Chen et al., 1995, Rebholz and Kruse, 1991; Kok et al., 1983; Kruse et al., 1990). This supposition was confirmed by Schauermann et al. (2002), who reported that both competing pathways in methanol decomposition are observed on supported palladium nanoparticles at high methanol exposures at ambient temperature. They proposed that C–O bond scission occurs preferentially at edges and steps of the palladium nanoparticles, whereas dehydrogenation does not. Another reason for the contrasting reports may be related to the kinetics of the various routes to methanol decomposition. C–O bond scission likely also takes place on a perfect (atomically flat) palladium surface, but at a low rate that is difficult to measure in typical UHV experiments. Consequently, high methanol partial pressures are presumably necessary to produce noticeable amounts of carbon species by C–O bond scission.

To better understand the mechanism of methanol decomposition and to examine the ideas discussed earlier, Morkel et al. (2004) investigated the interaction of methanol with atomically smooth Pd(111) at pressures from 10^{-6} to 10^{-1} mbar and temperatures from 300 to 600 K. XPS, which unambiguously identifies the carbonaceous residues (CH_x) and can be applied from UHV up to mbar pressures, is well-suited for monitoring of methanol decomposition under reaction conditions.

Figure 14 is a comparison of C1s spectra measured during methanol adsorption at 300 K and two different pressures, 10^{-6} and 10^{-1} mbar. All the C1s spectra consist of two features, at approximately 284 and 285.6 eV, with their relative intensities being dependent on pressure. Whereas the feature at approximately 284 eV can be almost unambiguously assigned to carbonaceous species CH_x ($x = 0$–3), assignment of the feature at 285.6 eV requires additional discussion and analysis of literature data (Bhattacharya et al., 1988; Christmann and Demuth, 1982; Davis and Barteau, 1990). Indeed, both CO adsorbed in three-fold and bridge positions and the methoxy group, which is produced *via* scission of the O–H bond in adsorbed methanol molecules, can be responsible for this C1s feature. Unfortunately, overlapping of O1s and Pd3p spectra makes it

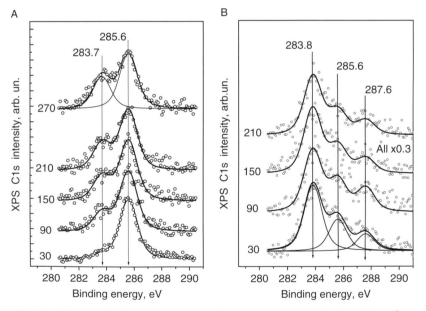

FIGURE 14 C1s core-level spectra measured during exposure of Pd(111) to (A) 10^{-6} mbar and (B) 0.1 mbar of methanol at 300 K.

impossible to use the O1s spectra for discrimination between CO_{ads} and CH_3O_{ads}. Nevertheless, literature data allow us to exclude the methoxy groups from consideration because of their low stability on the palladium surface. As reported in a number of publications (Bhattacharya et al., 1988; Davis and Barteau, 1990; Gates and Kesmodel, 1983), CH_3O_{ads}-originated features, which are clearly observed in HREEL spectra after methanol adsorption on a Pd(111) surface at low temperatures (200–250 K), disappear after heating of the adsorbed layer to room temperature. When methanol was adsorbed at 300 K, only the features originating from CO were observed in the HREEL spectra (Gates and Kesmodel, 1983). On the basis of these data, the authors concluded that the feature observed at 285.6 eV (Figure 14) originated from CO molecules adsorbed on three-fold sites. This assignment was unambiguously confirmed in a recent investigation of methanol adsorption on Pd(111) (Morkel et al., 2004), which showed quantitative agreement in CO coverages determined from the XPS intensity of this C1s feature and from the SFG stretching frequency of adsorbed CO. A C1s feature at 287.6 eV observed in the experiment arose from gas-phase methanol.

Observation of the CH_x and CO_{ads} species as a result of the interaction between methanol and a palladium surface indicates that the methanol decomposition occurs *via* two routes: (1) scission of the C–O bond in methanol with the formation of CH_x species ($x = 0$–3), and (2) methanol dehydrogenation giving CO_{ads}. The contribution of CO bond scission with the formation of carbonaceous species is negligible under UHV conditions, especially at low exposures, but it increases strongly at higher pressures. These XPS results indicate unambiguously that the pathway of methanol C–O bond scission can take place on the atomically smooth surface of palladium, but that high methanol partial pressures are necessary to form a noticeable amount of carbon species produced by the methanol C–O bond scission.

The following model for methanol decomposition on Pd(111) is proposed on the basis of these data. Whereas methoxy CH_3O binds to the surface *via* the oxygen atom, the final product CO is bonded *via* the carbon atom. Consequently, during decomposition, the molecules must turn, producing intermediates in which the C–O bond is no longer perpendicular to the palladium surface. This turning seems to facilitate C–O bond scission as a consequence of a better overlap between the metal valence electrons and the CH_xO orbitals (which weakens the bond). If the C–O bond stays intact, CO and H_2 are produced. As a consequence of the upright adsorption geometry of CO on Pd(111), with the C–O bond perpendicular to the surface, C–O bond scission within the product molecule CO is highly unlikely, as was recently confirmed by combined SFG/XPS investigations (Kaichev et al., 2003).

The contribution of the decomposition route with C–O bond scission also increases with increasing temperature. This conclusion follows from the data of Figure 15, which presents the results of experiments characterizing methanol decomposition on a Pd(111) surface at 10^{-1} mbar and 400 K. Under these conditions CH_x ($x = 0$–3) is the most abundant species in the C1s spectra (Figure 15A). Moreover, the surface coverage by carbonaceous residues exceeds one monolayer. The surface coverages by $CH_{x,ads}$ and CO_{ads} determined from C1s spectra and expressed as fractions of a monolayer are presented in Figure 15B. Notwithstanding high coverage by the carbonaceous species, CO_{ads} is still present at the surface. This fact can be explained by a subsurface location of some carbon that allows some part of the surface to be free for the formation of CO_{ads}.

Similar results was reported in a recent investigation of the interaction of methanol with Pd(111) whereby XPS was combined with SFG (Morkel et al., 2004). Postreaction SFG with CO as probe molecule indicated that CO was presumably no longer able to populate hollow or on-top sites, which probably were physically blocked by CH_x. A single peak at 1920 cm^{-1}, typical of bridge-bonded CO, was observed after methanol decomposition. The preferred binding of CH_x to hollow sites may also reveal its stoichiometry (i.e., the value of x). According to theoretical investigations (Paul and Sautet, 1998; Zhang and Hu, 2002), CH_x ($x = 0$–3)

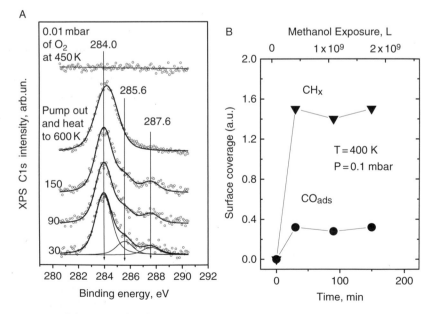

FIGURE 15 (A) C1s core-level spectra measured during exposure of Pd(111) to 0.1 mbar of methanol at 400 K. The quantitative analysis of the XP spectra is shown in (B).

fragments tend to restore their tetravalency on the surface; adsorbed carbon atoms and CH species prefer hollow sites, and CH_2 preferentially binds to bridge sites and CH_3 resides on-top of Pd atoms. Consequently, the partial blocking of hollow sites suggests the presence of carbon atoms and/or CH species. This inference was also supported by the absence of C–H signals in the SFG spectra (Morkel et al., 2004).

Removal of the methanol gas phase (not shown) did not change the C1s spectrum (apart from the CH_3OH gas-phase signal), indicating that there were no (significant) contributions from weakly bonded surface species. However, a reduction of the CO component as a result of partial CO desorption was observed in this case. Further heating the sample to 600 K resulted in complete desorption of CO, whereas the CH_x species was apparently thermally stable (Figure 15A). However, a shift to 284.2 eV points to a structural change of the carbon overlayer.

Thus, XPS and TPRS investigations of Pd(111) clearly show that at elevated pressures methanol decomposition to give CO and H_2 occurs only to a small extent. There is a rapid self-poisoning by adsorbed CO and CH_x at 300 K, whereas at 400 K a rapid formation of CH_x layers prevents a significant conversion. These results indicate that Pd(111) is in fact able to break the C–O bond in methanol, but its observation may require a sufficient methanol impingement rate at temperatures greater than approximately 250 K. To produce CO and H_2 from methanol on palladium catalysts, the CH_x formation must be suppressed or CH_x must be selectively removed from the surface.

To realize the latter goal, we investigated the interaction of CH_x species produced by methanol decomposition at 300 or 400 K (approximately 1–1.5 ML; Figs 4.14 and 4.15) with oxygen. Upon exposure of the carbon- and CO-poisoned Pd(111) surface to 10^{-2} mbar of O_2 at 300 K, no changes were detected by C1s XPS within half an hour (i.e., the two features of CH_x and adsorbed CO maintained their intensities). This result suggests that the CH_x- and CO-covered surface provides no sites for oxygen adsorption. Otherwise, oxygen atoms should react with CO to give CO_2 that would desorb at 300 K. In contrast, all the C1s features quickly disappeared at 400 K in 10^{-1} mbar of O_2. It is evident that CO desorption at approximately 400 K produces vacant sites for dissociative oxygen adsorption, and the resultant species convert the CH_x. This result suggests that deactivation of the palladium surface by accumulation of carbonaceous residues can be prevented by addition of O_2 to the gas phase.

Figure 16 shows the results of TPRS experiments observed when the Pd(111) surface was heated in a mixture of methanol with O_2 from room temperature to 600 K. Mass spectra of all possible products and C1s spectra were measured simultaneously during the experiment. Figure 16A shows the TPRS spectra of O_2, CO, CO_2, H_2, and H_2O, and the variation of the area of C1s spectra is presented in Figure 16A. The data show that no methanol

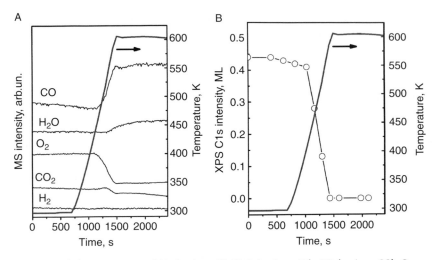

FIGURE 16   (A) TPR spectra of H_2 ($m/q = 2$), H_2O ($m/q = 18$), CO ($m/q = 28$), O_2 ($m/q = 32$), CO_2 ($m/q = 44$), and (B) variation of C1s intensity measured during heating of Pd(111) in a flow of oxygen/methanol (O_2:$CH_3OH = 1$:2.7, molar). The total pressure was about 10^{-2} mbar. The temperature was increased from 300 to 600 K at a heating rate of approximately ~ 0.4 K/s.

oxidation occurred at temperatures up to 450 K, but the reaction occurred at higher temperatures. This conclusion is based on the appearance of CO and H_2O among the reaction products and a decrease of the mass spectrometer signal representing O_2. In the same temperature range, the Pd(111) surface was cleaned of surface species, at least those detectable within the XPS sensitivity limit. Thus, addition of O_2 to methanol allows reactivation of the palladium surface, but the most interesting product, hydrogen, is oxidized to form water under these conditions.

6. ETHENE EPOXIDATION CATALYZED BY SILVER

Silver catalysts for ethene epoxidation have been investigated extensively (Bal'zhinimaev, 1999; Bukhtiyarov et al., 1994, 1999; Campbell, 1985; Campbell and Paffett, 1984; Grant and Lambert, 1985; van Santen and de Groot, 1986; van Santen and Kuipers, 1987), but mostly with high-vacuum surface science techniques and not under reaction conditions. Removal of the reaction gas mixtures by evacuation, which is a usual step in the post-reaction analysis, can lead to the destruction of the active centers on the catalyst surface as a consequence of the removal of, for example, weakly bound species or a change in the surface constituents. Consequently, catalyst surfaces operating under catalytic reaction conditions (pressure > 1 mbar) can be quite different from the surfaces investigated by physical methods under high-vacuum conditions (pressure < 10^{-6} mbar). In part because of the lack of results characterizing the surfaces under reaction conditions, the mechanism of ethene epoxidation and the nature of the epoxidizing oxygen species are still being debated (Campbell, 1985; Campbell and Paffett, 1984; Grant and Lambert, 1985; van Santen and de Groot, 1986). Recent microkinetics modeling investigations (Stegelmann et al., 2004) allowed the elucidation of some aspects of the mechanism. However, experimental investigations of catalysts in the presence of reactants at pressures in the millibar range to provide a direct correlation between the catalytic activity and the surface composition had not been reported.

Consequently, XPS experiments and proton-transfer reaction mass spectrometry (PTRMS) experiments were performed to characterize the ethene epoxidation on silver. Because of its small analysis depth (<1 nm) and its sensitivity to the chemical state of an element, XPS is one of the most powerful methods for investigating the nature of adsorbed species and reaction intermediates. Furthermore, the highest pressure available for XPS (1–2 mbar) is sufficient for observation of ethene oxide among the reaction products (Campbell, 1985; Campbell and Paffett, 1984; Grant and Lambert, 1985). In the experiment, the yield of ethene oxide was measured by using PTRMS (details of the method are given elsewhere; Lindinger et al., 1998). In brief, the ionization of gas-phase molecules for the subsequent mass spectrometric analysis proceeds by the transfer of protons from H_3O^+ ions to the analyzed substance. Consequently, the method deals with a molecular ion that has a greater mass (by one proton) than the ion in the routine mass spectrometric analysis. Only those ions with higher proton affinities than H_3O^+ contribute to the PTRMS spectrum. This ionization method does not fragment the molecules to be analyzed, which simplifies the analysis of the mass spectra. Another advantage of PTRMS is its high sensitivity to organic molecules, such as ethene oxide. Thus, catalytic measurements were combined with characterization of the composition of the catalyst surface.

Figure 17 shows the PTRMS signal of ethene oxide at $P_{C_2H_4} = 0.1$ mbar and $P_{O_2} = 0.25$ mbar measured as a function of temperature and time. Cleaned polycrystalline silver foil (99.99% purity) was used as the catalyst. The concentration of ethene oxide is given in units of ppb, which is the concentration inside the mass spectrometer. Because there was a pressure drop between the reaction cell and the PTRMS sampling volume, the concentrations in the reaction cell were much higher than in the sampling volume, but proportional to the concentrations in the PTRMS. The data show that the PTRMS signal of ethene oxide, which is very small at 370 K, increases with temperature. After cessation of the O_2 flow (in the last part of

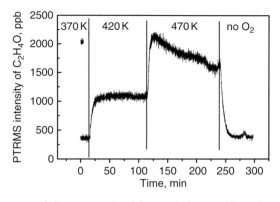

FIGURE 17 Variation of the PTRMS signal from ethylene oxide with temperature, measured in reaction mixtures with $P_{C_2H_4} = 0.1$ mbar $+ P_{O_2} = 0.25$ mbar in the presence of a polycrystalline silver foil. The last parts of the curves were measured after stopping the oxygen flow.

the experiment), the PTRMS signal representing ethene oxide decreased rapidly to the background level. This result indicates that, in the millibar pressure range, silver is active for ethene epoxidation, with the catalytic activity increasing with increasing temperature. From the PTRMS data it also follows that the catalytic activity did not vary with time at 420 K, whereas it decreased with time at 470 K, probably as a consequence of catalyst deactivation. The O1s spectra measured simultaneously with the PTRMS spectrum (Figure 17) are shown in Figure 18. All O1s spectra were normalized to the integrated intensity of the Ag3d$_{5/2}$ peak. The spectra change markedly with temperature.

At 370 K, the silver surface was characterized by a single O1s peak at 530.5 eV (Figure 18, curve 1). One of the most likely candidates for some of the XPS signals measured at this temperature are surface carbonates, $CO_{3,ads}$, which are known for their stability on silver surfaces at ambient temperature (Backx et al., 1983; Barteau and Madix, 1982; Campbell, 1985). Carbonates are easily formed on silver by the reaction of adsorbed oxygen with background CO_2 (Backx et al., 1983), and they are characterized by an O1s feature at approximately 530.5 eV and a C1s feature at approximately 287.5 eV (Barteau and Madix, 1982; Campbell, 1985) (not shown). Therefore, the corresponding features in the spectra are assigned to surface carbonates. This assignment is also supported by the O/C atomic ratio calculated from the XPS peak intensities (2.8–2.9), nearly equal to the stoichiometric ratio (3.0).

At 420 K, carbonates decompose to give adsorbed oxygen and CO_2, which desorbs (Backx et al., 1983; Barteau and Madix, 1982; Campbell, 1985). In agreement with this well-known behavior, the corresponding

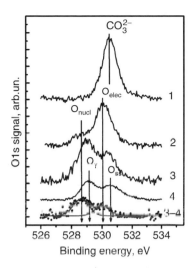

FIGURE 18   The O1s spectra of a silver foil measured in a flowing reaction mixture at $P_{C_2H_4} = 0.1$ mbar and $P_{O_2} = 0.25$ mbar and at various temperatures: (1) 370 K; (2) 420 K; and (3) 470 K. Spectrum (4) was measured at 470 K after the oxygen flow was stopped. The difference spectrum between the measurement in the reaction mixture and in pure ethylene (3, 4) includes peaks that represent species existing on the surface only under reaction conditions.

features disappear from the XPS spectra. The O1s spectrum measured at this temperature exhibits a major peak at 530.0 eV with a shoulder at about 528.5 eV. According to previous investigations (Bukhtiyarov et al., 1994, 1999), binding energies of 528.5 and 530.0 eV are indicative of nucleophilic and electrophilic oxygen, respectively. The terms nucleophilic and electrophilic oxygen were introduced by Grant and Lambert (1985) to designate the nature of the interaction of oxygen with ethene. The nucleophilic oxygen is reactive in the nucleophilic attack of the C–H bond, as the first step of C_2H_4 combustion (van Santen et al., 1980), whereas electrophilic oxygen participates in the electrophilic interaction with the C=C bond of ethene (Grant and Lambert, 1985).

A detailed investigation of the nature of these oxygen species by XPS, XAS, UPS, and XANES has been reported (Bukhtiyarov et al., 2001, 2003). Briefly, nucleophilic oxygen is present in an Ag_2O surface oxide as a result of silver surface reconstruction (Barteau and Madix, 1982; Bukhtiyarov et al., 2001, 2003; Campbell, 1985) under the influence of adsorbed oxygen, whereas electrophilic oxygen is adsorbed on the silver without surface reconstruction. Electrophilic oxygen was identified as adsorbed oxygen by Rocca and coworkers (Rocca et al., 2000; Savio et al., 2002) using

HREELS and by Bukhtiyarov et al. (1999) using angle-dependent XPS in room-temperature experiments characterizing O_2 adsorption on Ag(100) and Ag(111) single crystals. This oxygen species was also found by King and coworkers (Carlisle et al., 2000; Michaelides et al., 2003) in recent STM investigations to be a phase present at low coverages (Θ < 0.25).

When the temperature was increased to 470 K the complex shape of the O1s spectrum suggests that more than two peaks were present in this spectrum (Figure 18, curve 3). The comparison with other O1s spectra supports this suggestion. The O1s spectrum measured with the sample in pure ethene (i.e., after cessation of the O_2 flow) exhibits two peaks, at 529.1 and 530.6 eV (Figure 18, curve 4). The BE of the feature at 529.1 eV is close to that of the so-called O_γ oxygen, which has been extensively investigated by Schlögl and coworkers (Bao et al., 1996; Schedel-Niedrig et al., 1997). This oxygen is reactive in the partial oxidation of methanol to formaldehyde catalyzed by silver. This species represents strongly bound atomic oxygen embedded in the outer layers of silver, with oxygen atoms occupying a fraction of the silver positions in the silver crystal lattice.

The peak at 530.6 eV is assigned to subsurface oxygen embedded in octahedral holes, in accordance with literature data (Bao et al., 1996). The location of these oxygen species in the subsurface region is in line with its lack of reactivity toward ethene. The difference spectrum of the O1s spectra measured at 470 K in the reaction mixture and in pure ethene shows two O1s peaks that are absent from the spectrum recorded with the sample in pure ethene. The peaks at 528.5 and 530.0 eV originate from the nucleophilic and electrophilic oxygen species, respectively.

The spectroscopic characteristics of all the species observed in the experiment are summarized in Figure 18.

A comparison of the catalytic (Figure 17) and spectroscopic (Figure 18) data provides a basis for proposing an explanation for the temperature-induced variation in the catalytic properties of silver in ethene epoxidation catalysis. The low activity of silver at temperatures <420 K is caused mainly by the presence of carbonates and carbonaceous residues on the silver surface that reduce the available silver surface area for the catalytic reaction. When those contaminations are removed from the surface at 420 K, it becomes catalytically active for ethene oxide formation (Figure 17). The silver surface at this temperature incorporates two oxygen species—nucleophilic and electrophilic oxygen. The enhancement of the ethene oxide yield at temperatures >420 K is most likely determined by the Arrhenius dependence of the reaction rate on temperature. An estimate of the Arrhenius contribution to the variation of the reaction rate ($\exp(-E_a/RT_i)$) shows that a temperature increase from 420 to 470 K will result in an increase of the reaction rate by a factor of about 3.8, assuming an activation energy for the epoxidation of 42 kJ/mol; this value of the activation energy has been reported numerous times (Bal'zhinimaev, 1999; Campbell, 1985; Grant and Lambert, 1985).

The decrease of the rate of ethene oxide formation with respect to time observed at temperatures ≥ 470 K can be explained by the accumulation of oxygen species embedded in the silver surface that decrease the surface area available for the formation of the reactive species. Nucleophilic and electrophilic oxygen, which are the major surface species at 420 K, are still present on the silver surface at 470 K; however, they are rapidly removed in the absence of oxygen in the gas phase (see the difference spectrum in Figure 18).

The presence of nucleophilic and electrophilic oxygen on the active silver surface suggests that they participate in the ethene oxidation reaction. This observation is in agreement with the mechanisms of ethene epoxidation proposed by the authors previously on the basis of the experiments with bulk silver (Bukhtiyarov et al., 1994, 1999):

$$O_2 + 4Ag \rightarrow 2(Ag^+)_2-O_{nucl} \tag{1}$$

$$C_2H_4 + (Ag^+)_2 - O_{nucl} \rightarrow \pi\text{-}C_2H_4 - (Ag^+)_2 - O_{nucl} \tag{2}$$

$$O_2 + 2Ag \rightarrow 2O_{elec} - Ag \tag{3}$$

$$\pi\text{-}C_2H_4 - (Ag^+)_2 - O_{nucl} + O_{elec} - Ag \rightarrow C_2H_4O + (Ag^+)_2 - O_{nucl} + Ag \tag{4}$$

$$\pi\text{-}C_2H_4 - (Ag^+)_2 - O_{nucl} + 5(Ag^+)_2 - O_{nucl} \rightarrow 2CO_2 + 2H_2O + 6Ag \tag{5}$$

Electrophilic oxygen (O_{elec}) is responsible for the formation of ethene oxide by oxidation of π-complexes of ethene (step (4)). The role of nucleophilic oxygen (O_{nucl}), which is reactive in total oxidation of ethene only (step (5)), is in the formation of silver ions, Ag^+, the sites for adsorption of ethene molecules as π-complexes (step (2)). The formation of nucleophilic oxygen via formation of the Ag_2O surface oxide (step (1)) has been reported many times (Campbell and Paffett, 1984; Grant and Lambert, 1985; van Santen and de Groot, 1986; van Santen and Kuipers, 1987), but the mechanism of the electrophilic oxygen formation is more debatable. Some authors proposed that the transformation of nucleophilic to electrophilic oxygen is preferred over (Bukhtiyarov et al., 1999) direct adsorption of O_2 on the silver (step (3)). Nevertheless, we were able to show recently that electrophilic oxygen species can be produced as a result of O_2 adsorption on the close-packed Ag(111) surface at room temperature in the absence of nucleophilic oxygen (Bukhtiyarov et al., 1999).

Additional arguments in favor of the hypothesis that electrophilic oxygen epoxidizes ethane follows from the results of experiments carried out in this investigation at two different total pressures of 0.071 and 1.05 mbar but at similar ratios of ethene to oxygen partial pressures (1:10). Figure 19 is a comparison of the PTRMS signals of ethene oxide measured in these experiments; the corresponding O1s spectra taken at 420 K are presented in Figure 20. The conditions of these experiments were chosen on the basis of literature data indicating that ethene oxide appears among the reaction products only in the millibar pressure range (Campbell, 1985; Campbell and Paffett, 1984; Grant and Lambert, 1985). Furthermore, it is known that excess oxygen leads to an enhancement of the ethene oxide yield (Campbell, 1985; Campbell and Paffett, 1984). Figure 19 shows, in agreement with literature data, that ethene oxide did not appear as a product in low-pressure experiments, whereas a total pressure of 1 mbar was high enough for the ethene epoxidation at temperatures >420 K.

Consistent with the data of Figure 17, an increase in temperature enhanced the yield of ethene oxide. Nevertheless, for comparison, the XP spectra measured at 420 K were used, because they allow us to exclude the influence of embedded oxygen species. The main difference in O1s spectra measured at the higher pressure (Figure 20A) is the presence of the feature originating from electrophilic oxygen. Indeed, only nucleophilic oxygen is observed at low pressures. These results indicate that the presence of nucleophilic oxygen, even in rather high concentrations, is not sufficient to produce a surface that is active for ethene epoxidation.

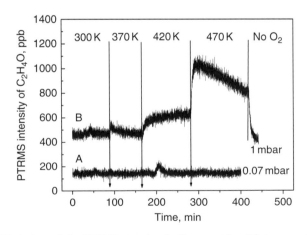

FIGURE 19   Variation of the PTRMS signals of ethene oxide with temperature, measured in reaction mixtures of (A) $P_{C_2H_4} = 0.0065$ mbar and $P_{O_2} = 0.065$ mbar and (B) $P_{C_2H_4} = 0.1$ mbar and $P_{O_2} = 0.95$ mbar in the presence of a polycrystalline silver foil. The last part of the high-pressure curve was measured after stopping the oxygen flow.

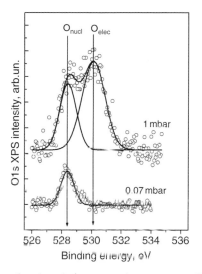

FIGURE 20   O1s spectra of a silver foil measured at 470 K in a flowing reaction mixture at (1) $P_{C_2H_4} = 0.0065$ mbar and $P_{O_2} = 0.065$ mbar and at (2) $P_{C_2H_4} = 0.1$ mbar, $P_{O_2} - 0.95$ mbar.

Moreover, analysis of the XPS and PTRMS data gave a linear correlation between the yield of ethene oxide and the abundance of electrophilic oxygen (Figure 21).

It is concluded that the epoxidation of ethene proceeds *via* a Langmuir–Hinshelwood mechanism involving oxygen species adsorbed in the electrophilic state and ethene, which is chemisorbed on silver ions produced by nucleophilic oxygen. In agreement with this mechanism, the highest yields of ethene oxide were observed for silver surfaces characterized by similar concentrations of nucleophilic and electrophilic oxygen. The data also show that the composition of the active silver surface under catalytic reaction conditions is different from the composition measured when only one reactant is present in the gas-phase, or when the surface is characterized under vacuum. This point underscores the importance of using methods that allow investigation of catalyst surfaces in action.

7. METHANOL OXIDATION CATALYZED BY COPPER

Elemental copper can be used as an unsupported catalyst for the oxidative dehydrogenation of alcohols to their respective aldehydes. There are two main reaction paths: partial oxidation to formaldehyde and total oxidation to carbon dioxide, which is thermodynamically favored. The

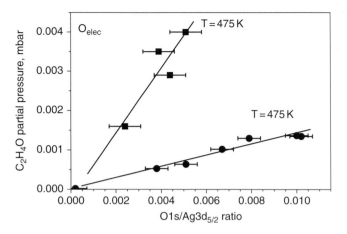

FIGURE 21 Yield of ethene oxide as a function of the abundance of the electrophilic oxygen measured at various temperatures of a silver foil: 425 K (circles) and 475 K (squares).

oxidation of methanol has been investigated by both classical UHV surface-science techniques and synchrotron-based X-ray absorption spectroscopy with catalysts in reactive atmospheres. In room-temperature UHV–X-ray XPS investigations, it was found that methanol reacts with preadsorbed oxygen (O_{ads}) on Cu(110) and polycrystalline Cu to form a methoxy intermediate (Bowker and Madix, 1980; Carley et al., 1996). This species then either decomposes to formaldehyde, which desorbs, or is oxidized to formate (CHOO), which is stable at room temperature, but decomposes at 373 K to give CO_2 and H_2 (Carley et al., 1996). Near-edge X-ray absorption fine structure (NEXAFS) experiments at the oxygen K-edge were performed at temperatures up to 673 K in various methanol-to-oxygen ratios at a total pressure of about 1 mbar (Knop-Gericke et al., 2001). These investigations showed that the formaldehyde yield is correlated with the presence of a suboxide species at the sample surface that could only be detected under reaction conditions and that the catalytically active phase is metallic.

The goal of the experiments that are summarized here was to determine quantitatively the depth-dependent compositions of the surface and near-surface regions, to a depth of a few nanometers, during the catalytic reaction; synchrotron-based XPS was used. The experiments were performed at beamline U49/2 at BESSY in Berlin and at beamline 9.3.2 at the Advanced Light Source in Berkeley (Hussain et al., 1996). The spectral resolution was 0.1 eV at the oxygen K-edge. All spectra were normalized by the incident photon flux, which was measured by using a photodiode with known quantum efficiency. The flows of methanol vapor and oxygen into the sample cell were regulated with MFCs. The combined methanol and oxygen pressure in the cell was 0.6 mbar with a total flow rate of approximately 10 standard $cm^3 min^{-1}$. The sample was a polycrystalline copper foil (99.99% purity) mounted on a temperature-controlled heating stage. O1s, valence band, C1s, Cu3p, and Cu2p photoelectron spectra were measured under various conditions at a fixed methanol-to-oxygen feed ratio of 3:1 (molar) at temperatures from 300 to 720 K, and, in a complementary set of experiments, at a fixed temperature (673 K) with various methanol-to-oxygen feed ratios. In the following, the data from the latter experiments are discussed.

Figure 22A shows the O1s region of photoemission spectra of the copper catalyst observed when the incident photon energy was 720 eV. The four spectra correspond to methanol-to-oxygen ratios in the reactant stream of 1:2, 1:1, 3:1, and 6:1 (molar) and a sample temperature of 673 K. Because the incident X-ray beam irradiated not only the sample surface but also the gas-phase molecules in front of the sample, the spectra show gas-phase peaks alongside the surface peaks. Gas-phase peaks representing all the reactants and products can be distinguished in the spectra at BE values higher than 534 eV. The catalytic activity of the copper foil in the four different gas mixtures can be calculated both from the areas of the O1s and C1s peaks and from mass spectrometry data. The results are summarized in Table 1. The absolute amount of formaldehyde produced in the catalytic reaction was highest when the $CH_3OH:O_2 =$ ratio was 3:1 (molar).

The surface O1s peaks in Figure 22A show a strong dependence on the $CH_3OH:O_2$ ratio. We can distinguish three species at the surface or in the near-surface region under oxidizing conditions ($CH_3OH:O_2 = 1:2, 1:1$) and two species under reducing conditions ($CH_3OH:O_2 = 3:1, 6:1$). When the $CH_3OH:O_2$ ratio was 1:2 (molar), a peak with a BE of 530.3 eV (FWHM 1.0 eV) dominated the spectrum. This peak characterizes Cu_2O, which was confirmed by comparison to the spectrum of a Cu_2O reference sample and with literature values of the BE of Cu_2O (Ghijsen et al., 1988). Furthermore, the valence band spectrum taken right after the O1s spectrum was recorded (Figure 22B, at a photon energy of 262 eV, that is, the same probing depth as for the O1s spectra) also shows a typical Cu_2O spectrum, in agreement with literature spectra (Ghijsen et al., 1988) and our Cu_2O reference spectrum. In addition to the O1s peak of Cu_2O, two smaller peaks at BE values of 529.7 eV (FWHM 1.3 eV) and 531.2 eV (FWHM 1.6 eV) were present under oxidizing conditions. The relative intensities of those peaks increased relative to the Cu_2O peak when the ratio of methanol to oxygen in the reactant stream was changed to 1:1. The valence band spectrum then showed that the surface was characterized by both Cu_2O and metallic areas. This point is evident from the BE gap Δ of 0.9 eV, which is characteristic of Cu_2O, and from the nonzero intensity at the Fermi edge, which is characteristic of a metallic surface. When the sample was characterized under reducing conditions ($CH_3OH:O_2 = 3:1, 6:1$), the valence band spectra

A

B

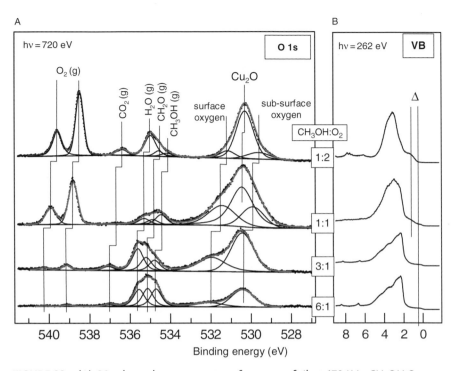

FIGURE 22 (A) O1s photoelectron spectra of a copper foil at 670 K in $CH_3OH:O_2$ reactant streams of 1:2, 1:1, 3:1, and 6:1 molar ratios. Raw data are shown as dots and the result of the fits as solid black lines. The incident photon energy was 720 eV. Gas-phase species are characterized by peaks evident at binding energies > 534 eV; the surface species are characterized by peaks evident at binding energies < 534 eV. In $CH_3OH:O_2 =$ 1:2 and 1:1 (molar) mixtures, the Cu_2O peak dominates the surface XP spectrum. The Cu_2O peak is absent from the spectra taken under reducing conditions ($CH_3OH:O_2 = 3:1$, 6:1 (molar)). In these spectra, the subsurface oxygen peak is the strongest in the surface part of the spectrum. (B) Corresponding valence band spectra. The spectrum recorded under oxidizing conditions ($CH_3OH:O_2 = 1:2$ (molar)) shows a typical shape for a Cu_2O sample with a BE gap Δ of 0.9 eV, thus confirming the observation based on the O1s spectrum in (A). The spectrum recorded with a $CH_3OH:O_2$ ratio of 1:1 (molar) shows that the surface consisted of a mixture of metallic copper and Cu_2O. The spectra measured under reducing conditions ($CH_3OH:O_2 = 3:1$, 6:1 (molar)) exhibit the typical valence band spectrum for metallic copper, which confirms the absence of a Cu_2O peak in the O1s spectra in (A).

showed clearly that the surface was metallic (Hüfner et al., 1973). In the O1s region, two peaks with BE values of 530.4 and 532.0 eV dominated the spectra. Because the valence band spectra show that the surface was metallic, the peak at 530.4 eV cannot be assigned to Cu_2O.

A comparison of the O1s spectra with C1s spectra (not shown) indicates that the peaks at 530.4 and 532.0 eV in the O1s spectra corresponding

TABLE 1 Data characterizing methanol oxidation on a polycrystalline copper surface

CH$_3$OH:O$_2$ molar ratio	CH$_3$OH partial pressure (mbar)	CH$_2$O partial pressure (mbar)	CO$_2$ partial pressure (mbar)	CH$_3$OH conversion	CH$_2$O yield	CO$_2$ yield
1:2	0.053	0.075	0.072	0.73	0.37	0.36
1:1	0.173	0.070	0.027	0.36	0.26	0.10
3:1	0.167	0.179	0.103	0.63	0.40	0.23
6:1	0.307	0.173	0.030	0.40	0.34	0.06

Partial pressures of methanol, formaldehyde, and carbon dioxide as a function of the ratio of oxygen to methanol at 672 K. The values were calculated from the gas-phase peak areas measured by XPS. The methanol conversion (=1 − ($P_{CH_3OH}(T)/P_{CH_3OH}$(297 K)), formaldehyde yield (=$P_{CH_2O}(T)/P_{CH_3OH}$(297 K)), and carbon dioxide yield (=$P_{CO_2}(T)/P_{CH_3OH}$ (297 K)) are also given.

to CH$_3$OH:O$_2$ ratios of 3:1 and 6:1 (molar) are not caused by compounds containing oxygen and carbon. The C1s spectra do not show any surface peaks (the BE of the C1s lines of C–H–O compounds range from 284 to 293 eV (NIST X-ray Photoelectron Spectroscopy Database, Version 3.4 (Web version) http://srdata.nist.gov/xps/)), except under the most methanol-rich condition (CH$_3$OH:O$_2$ = 6:1 (molar)) when there is a surface peak at 284.7 eV with a smaller shoulder at 286.2 eV. From the known O1s/C1s detection sensitivity in the experiment, the authors estimated that if the 284.7-eV peak would be assigned to a C–O compound with a C:O atomic ratio of 1, its O1s peak would have a peak area similar to that of the 530.4 eV peak in the O1s spectrum corresponding to the CH$_3$OH:O$_2$ ratio of 6:1 (molar). Because there was no additional peak in the O1s spectrum at this ratio when compared to the spectrum corresponding to the 3:1 ratio, for example, the C1s peaks in the former spectrum are not likely to have been caused by a C–O compound, but rather by some CH$_x$ compound or pure carbon.

Therefore, none of the likely intermediates of the methanol oxidation reaction, such as methoxy (BE 285.2 eV) or formate (BE 287.7 eV), were observed under reaction conditions (Carley et al., 1996). All features in C1s spectra characterized by values of BE higher than 288 eV are assigned to gas-phase contributions of educts and products of the reaction. The corresponding O1s features at BE values greater than 533 eV are shown in Figure 22A. Because the elemental detection limit in XPS is of the order of 2% (McIntyre and Chan, 1990), we conclude that, if intermediates were present at the surface, their concentrations must have been below that value.

The nature of the O1s peaks at BE 530.4 and 532.0 eV under reducing conditions (Figure 22A) was investigated by using depth profiling by variation of the incident photon energy. Because the mean free path of an electron in a solid depends on its kinetic energy (KE), the escape depth of the photo-electrons varies with the incident photon energy hv, because $KE = hv - BE - \Phi$ (where Φ is the work function) (Rivière, 1990). The depth-profiling measurements (Bluhm et al., 2004) indicated that the 532.0 eV peak intensity decreased with respect to that of the 530.4 eV peak with increasing electron kinetic energy. This result implies that the species characterized by the 532.0 eV peak was located at the surface, whereas the species characterized by the 530.4 eV peak extended into the subsurface region. The exact nature of the surface species marked by the peak at 532.0 eV is not clear. Its BE value is consistent with that of an OH^- species (Au et al., 1979), but it could as well be attributed to oxygen bound by residual impurities at the copper surface. However, this peak did not show any correlation with the catalytic activity of the sample, and it could also be observed under high-vacuum conditions.

To learn more about the nature of the subsurface oxygen species, we estimated its average concentration as a function of depth. The O1s peak area of the subsurface oxygen peak measured at different photoelectron mean free path lengths was compared with the Cu3p peak area measured at the same mean free path lengths. The O:Cu stoichiometry was then calculated from the O1s and Cu3p peak areas at similar values of KE, after correction for the energy-dependent changes of the photoemission cross-sections characterizing Cu3p and O1s (Yeh and Lindau, 1985). The results representing the active catalyst surface shown in Figure 23 indicate that there was an oxygen concentration gradient perpendicular to the surface. In both reducing gas atmospheres, the concentration of subsurface oxygen was highest close to the surface. At our largest probing depth of 15 Å, the Cu:O atomic ratio was about 10, similar to the value found in the NEXAFS experiments characterizing the methanol oxidation on a copper foil under the same reaction conditions (Hävecker et al., 1998). Those NEXAFS experiments were performed in the total electron yield detection mode whereby the probing depth is estimated to be 20 Å, (i.e., similar to the greatest probing depth in our experiments; Abate et al., 1992). It is also noteworthy that when the $CH_3OH:O_2$ molar ratio was 3:1, the O:Cu ratio close to the surface was similar to that of Cu_2O. The valence band spectra in Figure 22B, however, show that even at such an oxygen concentration the sample surface region has still a metallic character. This result is also in good agreement with the earlier mentioned NEXAFS measurements, according to which the O K edge and Cu L edge spectra of the catalytically active surface under reducing conditions showed no resemblance to the spectra of either of the stoichiometric copper oxides Cu_2O or CuO (Hävecker et al., 1998).

The subsurface oxygen peak could be observed only in measurements with the catalyst in reactive atmospheres. O1s spectra taken under reaction conditions and after the oxygen flow was stopped show a direct correlation with the catalytic activity (Figure 24). The upper spectrum

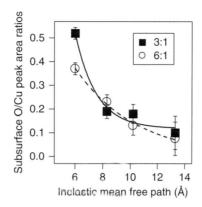

FIGURE 23 XPS data indicating subsurface O-to-Cu ratios in a copper foil under reducing conditions as a function of the inelastic mean free path of the photoelectrons. The ratios shown on the figure are molar $CH_3OH:O_2$ ratios. The concentration of subsurface oxygen close to the sample surface is higher for the oxygen-rich atmosphere ($CH_3OH:O_2 = 3:1$ (molar)). Deeper into the sample the concentration of subsurface oxygen is similar for the two different atmospheres.

(black line) was recorded with the sample in a gas mixture of $CH_3OH:O_2$ in a molar ratio of 6:1 (total pressure = 0.56 mbar) at 673 K. From the presence of the gas-phase peaks of methanol, water, and formaldehyde we conclude that the copper sample was catalytically active. Both surface (BE 531.5 eV) and subsurface oxygen (BE 530.4 eV) are indicated by the spectrum. After the oxygen flow was stopped, the subsurface oxygen peak vanished within 30 s; at the same time, the gas-phase water and formaldehyde peaks disappeared. Figure 24 shows a spectrum (gray line) that was taken 50 s after the oxygen flow was switched off; the difference spectrum is shown at the bottom of the graph. The single peak in the difference spectrum is attributed to subsurface oxygen. The spectra shown in Figure 24 indicate a direct correlation between the catalytic activity of the copper sample and the presence of a subsurface oxygen species in the near-surface region.

For a quantitative correlation between the catalytic activity of the sample and the abundance of subsurface oxygen, Figure 25 shows the CH_2O partial pressure as a function of the peak area of the subsurface oxygen peak for the four $CH_3OH:O_2$ ratios, and also for a separate experiment in which the $CH_3OH:O_2$ ratio was kept constant at 3:1 (molar), but the temperature was varied from 423 to 723 K. There is a linear correlation between the yield of CH_2O and the abundance of subsurface oxygen, in good agreement with the NEXAFS data (Hävecker et al., 1998). Figure 25 also shows that in the absence of subsurface oxygen there was no catalytic activity for the partial oxidation of methanol, in agreement with the spectra shown in Figure 24.

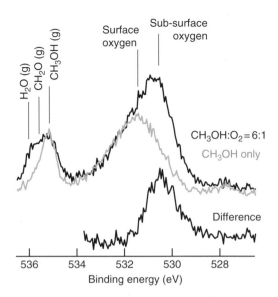

FIGURE 24 Response of the O1s spectra of a polycrystalline copper foil to a fast change in the gas-phase composition. The initial conditions were $CH_3OH:O_2 = 6:1$ (molar) at a total pressure of 0.56 mbar and at a temperature of 670 K (upper black line). In addition to gas-phase methanol, gas-phase water and formaldehyde are evident in the O1s spectrum (i.e., the copper sample is catalytically active). Peaks characteristic of both surface (BE = 531.5 eV) and subsurface oxygen (BE = 530.4 eV) are evident in the spectrum. After cessation of the oxygen flow, the subsurface oxygen peak vanished within 30 s; at the same time, the peaks characteristic of gas-phase water and formaldehyde disappeared. The gray spectrum was taken 50 s after the flow was switched off. The difference spectrum is shown at the bottom of the graph.

We propose that the subsurface oxygen species in our catalytically active copper samples play the role of a cocatalyst. It is not a reaction partner in the reaction (unlike O_{ads} in the model discussed by (Bowker and Madix, 1980; Carley et al., 1996), but, together with the neighboring Cu atoms, forms the active center where the reaction takes place. This model is supported by the results of theoretical investigations of active oxygen species in various metals showing that the Cu— system shows many analogies to the well-investigated silver system (Li et al., 2003; Schimizu and Tsukada, 1993). Detailed investigations of multiple Ag–O configurations (Li et al., 2002, 2003) showed a general trend in metal–oxygen systems: when going from low temperatures and high oxygen partial pressures to high temperatures and low oxygen partial pressures, a series of transitions likely occurs, starting from thick oxides over multiple sandwiches of oxygen–metal–oxygen trilayers, to a thin surface oxide with metal termination, and finally to a pure metal with adsorbates.

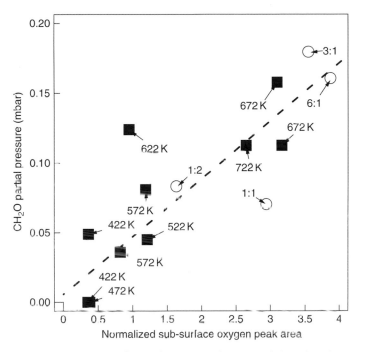

FIGURE 25 Partial pressure of formaldehyde as a function of the subsurface oxygen peak area for the spectra in Figure 22A (open circles), and for a different experiment (black squares) in which the temperature of a copper foil was varied in the range from 420 to 720 K at a constant CH$_3$OH:O$_2$ molar ratio of 3:1 (the total pressure was the same as for the spectra in Fig 22A. The partial pressure of formaldehyde is linearly correlated with the abundance of subsurface oxygen in the near-surface region. The dashed line is a linear fit of the data points.

In the case of silver, the oxygen-poor phase after the oxide-to-metal transition is a trilayer with oxygen in the hollow sites of a metal-terminating layer with a layer of subsurface oxygen (Carlisle et al., 2000). An arrangement of two stacked trilayers would account for the information determined in our XPS experiments. What is assigned in the present work as subsurface oxygen may therefore be similar to an arrangement of trilayers in the terminology used in, for example, Reuter et al. (2002). The approximate stoichiometry of the subsurface species that is found in our work (approximately Cu$_2$O) would also be consistent with this assignment. Recent NEXAFS experiments (Knop-Gericke et al., 2001) showed the same correlation between formaldehyde yield and abundance of a subsurface oxygen species as in these experiments (Figure 25). From the NEXAFS results we conclude that the oxygen in these trilayers would have to be differently bonded to copper than in Cu$_2$O, with a reduced rehybridization between oxygen and copper. It remains a matter of speculation in the absence of a theoretical model of this system whether oxygen surrounded by metal atoms or the local geometric variation of the metal–metal interaction caused by the presence of the oxygen are responsible for the catalytic activity of the Cu–O trilayer ensemble.

In summary, these XPS experiments have shown that the active catalyst surface is metallic copper that contains a subsurface oxygen species. The abundance of subsurface oxygen correlates with the amount of formaldehyde produced in the catalytic reaction. The active surface can be observed only in the experiments involving reactive mixtures in contact with the catalyst. Under conditions such as $T = 673$ K and $P_{total} = 0.6$ mbar, no reaction intermediates could be observed on the catalyst surface.

The measurements have shown that the catalytic activity of the copper sample is determined by its surface properties and also by the composition of the subsurface region.

8. CO OXIDATION CATALYZED BY Ru(0001)

Following the pioneering work demonstrating a high activity of the so-called O-rich Ru(0001) surface for catalysis of CO oxidation (Böttcher and Niehus, 1999a,b; Böttcher et al., 1997), the accepted interpretation is that rutile $RuO_2(110)$, formed on the Ru(0001) surface under realistic oxidation conditions, is the catalytically active phase (Kim et al., 2001; Over et al., 2000, 2001). These findings have been followed by many experimental and theoretical investigations aiming at a detailed characterization and atomic-scale understanding of the RuO_2 formation at various O_2 partial pressures and temperatures and of the mechanism of the CO oxidation. The knowledge of this system gained from numerous spectroscopic and structural investigations represents an essential contribution to the understanding of the so-called "pressure gap" (Blume et al., 2004; Böttcher et al., 2002a,b; Over and Muhler, 2003; Over et al., 2002, 2004).

Theoretical investigations of the oxidation pathway suggested that, after completion of the O-(1×1) adsorbed phase with an oxygen coverage of 1 ML, an important intermediate step is the incorporation of O atoms between the first and second ruthenium layers and the formation of an O_{ads}–Ru–O_{sub} trilayer with 2 ML of oxygen (Reuter et al., 2002; Todorova et al., 2002). This trilayer acts as a metastable precursor state, which virtually decouples from the underlying substrate, thus allowing the successive incorporation of more oxygen and eventually the conversion into a $RuO_2(110)$ lattice structure at a critical thickness of two O–Ru–O trilayers.

Extensive experimental efforts to provide evidence for subsurface oxygen species have met with only modest success (Quinn et al., 2001), but features that can be assigned to subsurface oxygen were observed recently with a combination of photoemission and thermal desorption spectroscopy (Blume et al., 2005).

The most widely accepted mechanisms of CO oxidation on the Ru(0001) surface consider only the structure of the bulk rutile RuO_2 surface, which consists of alternating rows of six-fold and five-fold coordinated Ru atoms. The latter, called coordinatively unsaturated sites (cus), have one perpendicular dangling bond and play a prominent role as active sites in the CO oxidation reaction (Over et al., 2004). According to one mechanism, CO adsorbs on the unsaturated Ru atoms and reacts with surface O atoms in the bridge site, which are restored by dissociative adsorption of O_2 on this site (Böttcher et al., 2002b; Liu et al., 2001; Over et al., 2000, 2004; Reuter and Scheffler, 2003a,b). A second possible mechanism is based on CO and O_2 adsorption on the unsaturated Ru sites, followed by reaction between CO and O atoms (Wang et al., 2002).

The occurrence of both reaction pathways has been demonstrated in recent time-resolved STM investigations (Kim and Wintterlin, 2004), which also showed that the reaction takes place at random positions, because the O and CO diffusion barriers are comparable to the activation energy for the CO + O reaction. The most recent theoretical calculations support these findings fully; they even suggest that the dominating reaction is between the adsorbed CO and O (Reuter et al., 2004).

However, the most disputable issue regarding the oxidation of the Ru (0001) surface is the presence and role of subsurface oxygen in the O/Ru (0001) system and the morphology of the ruthenium surface under realistic oxidation conditions. According to the model suggested by Over and coworkers (Over and Seitsonen, 2002; Over et al., 2002), the surface under oxidation conditions consists of islands of an inactive (1×1) chemisorbed phase and of the active surface RuO_2 (110) phase, the possible coexistence of some buried oxide and subsurface oxygen being of no importance for the reaction. This model suggests that as far as the oxygen exceeds the coverage of the highest loading of the adsorbed phase of 1 ML, the RuO_2 nucleus is formed and the RuO_2(110) film grows progressively in an autocatalytic manner (Over et al., 2002). This scenario, based on the results of investigations of the ruthenium oxidation carried out at temperatures >600 K, also raises questions about the formation of a "surface oxide" acting as a precursor to the RuO_2 rutile phase that was suggested by the DFT calculation (Reuter et al., 2002; Todorova et al., 2002).

Before making catalytic measurements, all oxygen species and oxygen-containing phases that can exist on the Ru(0001) surface were investigated by using postreaction XPS analysis that allowed not only the determination of the conditions of their formation, but also assignments of spectroscopic characteristics to various surface species.

The most recent XPS microscopy and TDS investigations clearly showed that the formation of a rutile RuO_2 phase, starting from an atomically clean Ru(0001) surface, is kinetically hindered at temperatures lower than 500 K and readily occurs at temperatures higher than 600 K (Blume et al., 2004, 2005; Böttcher and Niehus, 1999a,b; Böttcher et al., 2002b). The main reason for such a temperature dependence of the ruthenium oxidation state is the limited incorporation of oxygen, which at temperatures <500 K cannot reach the critical coverage of 4 ML (two rutile layers), because the oxygen penetration below the second layer requires higher temperatures. The evolution of the $Ru3d_{5/2}$ core level and O_2 TD spectra with increasing oxygen loading at temperatures of 450–500 K has provided evidence that the incorporation of O atoms is limited to the top 2–3 ruthenium layers. The most advanced nonoxidic state attained at temperatures below 500 K accommodates 3–4 ML of oxygen (Blume et al., 2005). A similar state, called Ru_xO_y, always developed in the initial stages of ruthenium oxidation at temperatures >600 K, when anisotropic RuO_2 growth was observed. The XPS analysis showed that the local compositions of the coexisting phases continuously changed with the progression of RuO_2 formation in the temperature range 600–770 K and that they are strongly dependent on the temperature (Böttcher et al., 2002b). In the earlier stages of the oxidation, regions with Ru_xO_y layers with thicknesses up to approximately 6 Å (coexisting with the adsorbed phase) are formed and expand with increasing exposure to O_2. The RuO_2 islands nucleate and grow inside this Ru_xO_y precursor, and in a wide exposure range of the oxide islands and the precursor coexist with the area covered by RuO_2 islands and the thickness of the oxide gradually increasing (Böttcher et al., 2002b).

Figures 26A and B show the $Ru3d_{5/2}$ and O1s spectra measured after various exposures of a Ru(0001) catalyst to O_2 at temperatures below and above 500 K. They represent the typical states observed under the low-temperature (<500 K) and high-temperature (>600 K) oxidation conditions. The spectra correspond to the adsorption phases with approximately 0.8 and ≥ 1 ML of oxygen, a "surface oxide" with approximately 2 ML of oxygen formed at temperatures below 500 K, Ru_xO_y, and the RuO_2 phase formed at temperatures above 600 K. The Ru_xO_y and the RuO_2 spectra were taken in the dark and bright regions of the $Ru3d_{5/2}$ map in Figure 26C, which illustrates the complex morphology of the advanced high-temperature oxidation state when the RuO_2 islands nucleate and grow. The contrast in the image corresponds to the intensity of the $Ru3d_{5/2}$ photoelectrons. The RuO_2 islands appear dark, because the RuO_2 phase contains fewer Ru atoms per unit volume than the adsorbed and O-rich intermediate states—and furthermore it is thicker and screens the emission from the metallic ruthenium below.

The components required for deconvolution of the $Ru3d_{5/2}$ spectra are summarized in Table 2. For clarity, the BE of the bulk component, Ru_{bulk}, characterizing the emission from the Ru atoms below the second

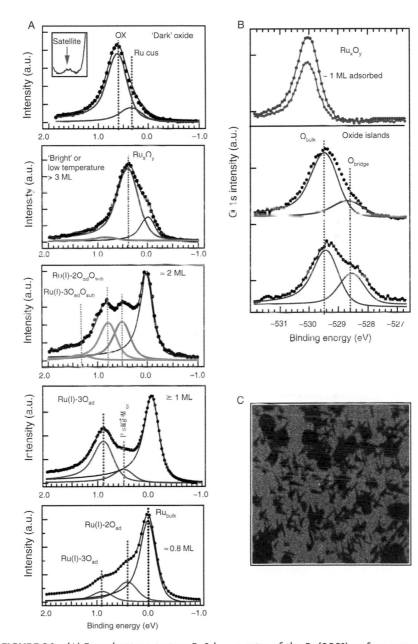

FIGURE 26 (A) From bottom to top: Ru3d$_{5/2}$ spectra of the Ru(0001) surface representing the adsorbed state with approximately 0.8 ML, with the sample in the presence of $\geq$ 1 ML of oxygen and with some oxygen already incorporated below the surface; in the presence of approximately 2 ML of oxygen, with oxygen located on the surface and

ruthenium layer is placed at a zero energy position in the $Ru3d_{5/2}$ panels. This presentation makes it easier to see the absolute energy shifts of the surface, Ru(I), and second layer, Ru(II), components, with respect to the Ru_{bulk}. Because, for the clean surface, the Ru(II) component is positioned very close to the Ru_{bulk} component (Lizzit et al., 2001) and undergoes an energy shift when oxygen starts to become incorporated below the surface (at approximately 1 ML of oxygen), we used a single broader component to account for the Ru(II) and Ru_{bulk} contributions in the spectra from the 0.8-ML adsorbed phase. The $Ru3d_{5/2}$ spectrum of this adsorbed phase requires $Ru(I)-2O_{ads}$ and $Ru(I)-3O_{ads}$ components characterized by features at 0.4 and 0.9 eV, respectively (Lizzit et al., 2001). These components reflect the surface core level shift undergone by the top Ru atoms coordinated with 2 or 3 O adatoms, respectively. The spectrum of the approximately 1-ML adsorbed state has a dominant $Ru(I)-3O_{ads}$ component and another (new) component characterized by a feature at 0.5 eV. As saturation of the adsorbed phase is approached, the component characterized by the 0.5-eV feature always emerges when the Ru(0001) surface has been exposed to O_2 at pressures higher than 10^{-3} mbar. The latter component gains intensity with further exposure to oxygen, as can be seen in the panel for 2 ML, and it is assigned to $Ru(II)-O_{sub}$, accounting for the core level shift of ruthenium from the second layer bonded to the incorporated oxygen.

The other two components of the spectrum recorded at 2.0 ML (at approximately 0.8 and approximately 1.3 eV) are assigned to the additional core level shifts undergone by the Ru(I) bonded to 2 or 3 adatoms, induced by the oxygen incorporated below. The spectrum of the Ru_xO_y precursor from the bright regions of the ruthenium map shown in Figure 26C has a dominant component at approximately 0.4 eV. The attenuation of the Ru_{bulk} indicates that the thickness of this precursor state is approximately 5–6 Å in the presence of 3–4 ML of oxygen. The spectrum of the dark regions includes the two components corresponding to cus-Ru and "bulk" Ru atoms of the RuO_2 phase, shifted by 0.35 and 0.64 eV, respectively. Another distinct feature of the RuO_2 phase is a broad satellite between the $Ru3d_{3/2}$ and $Ru3d_{5/2}$ peaks, which is not present in the spectra of the intermediate phases. The O1s peak of the adsorbed and intermediate oxidation states is at approximately 530.0 eV, but, when the oxygen loading

---

below the surface; in the presence of approximately 3–4 ML of oxygen distributed exclusively within the top three layers (bright areas in the $Ru3d_{5/2}$ map shown in (C)); and RuO_2 phase (dark areas in the $Ru3d_{5/2}$ map shown in (C)). (B) O1s spectra representing the adsorption and intermediate states (top) and RuO_2 (bottom). In the bottom panel, the two O1s spectra illustrate the different relative amounts of the O-bridge component reflecting the degree of the surface order of the oxide. (C) $Ru3d_{5/2}$ image ($64 \times 64 \ \mu m^2$) illustrating the morphology of the surface observed after exposure of Ru(0001) to 10^5 mbar $\times$ s of oxygen at 675 K. The BE of the bulk component, Ru_{bulk}, is placed at zero energy position in the $Ru3d_{5/2}$ panels.

TABLE 2 Characterization of a ruthenium catalyst for CO oxidation: energy positions of Ru3d and O1s components measured for the various states of oxidation of the catalyst

State of catalyst surface	Binding energy (eV) Ru3d$_{5/2}$; O1s	Component	Ref.
Clean	280.1 (0)	Ru$_{bulk}$	Lizzit et al. (2001)
Ru(0001)	279.75 (−0.35)	Ru(I)	Lizzit et al. (2001)
	280.25 (0.15)	Ru(II)	Lizzit et al. (2001)
	280.08	Ru(I)–1O	Lizzit et al. (2001)
Adsorbed	280.48 (0.4)	Ru(I)–2O	Over et al. (2001), Lizzit et al. (2001)
Phase	281.03 (0.93)	Ru(I)–3O	Over et al. (2001), Lizzit et al. (2001)
Oxide	280.71 (0.64)	RuO$_2$-bulk	Over et al. (2001), Böttcher et al. (2002)
RuO$_2$	280.45 (0.35)	cus-Ru	Over et al. (2002)
	283 ((Schütz et al., 1999).92)	satellite	Over et al. (2002)
3–4 ML	280.5 ± 0.05 (∼0.4)	Ru$_x$O$_y$	Blume et al. (2004)
Surface	280.9 ± 0.05 (∼0.5)	Ru(I) 2O$_{ads}$O$_{sub}$	Blume et al. (2004)
Oxide	281.4 ± 0.05 (∼1.3)	Ru(I)– 3O$_{ads}$O$_{sub}$	Blume et al. (2004)
Approximately 1–3 ML	280.6 ± 0.05 (∼0.8)	Ru(II)–O$_{sub}$	Blume et al. (2004)
Adsorbed/ 1–3ML	530.0	O$_{ads}$ and O$_{sub}$	Böttcher et al. (2002)
O1s oxide	529.5	O in RuO$_2$bulk	Over et al. (2001), Böttcher et al. (2002)
	528.7	bridged O	Over et al. (2001)
O1s–CO adsorbed CO	530.8–531.8		Schiffer et al. (1997)

The position of the Ru3d$_{5/2}$ bulk component at 280.1 eV (attributed to emission from the Ru atoms below the second layer) is used as a zero-energy reference in the plot of the Ru3d$_{5/2}$ spectra. The shift of the Ru3d$_{5/2}$ components with respect to the zero-energy reference is given in parentheses.

is higher than 1 ML, its intensity grows. The O1s spectrum of the RuO$_2$ phase has two components reflecting the emission from the bridged O atoms at the RuO$_2$ surface (528.6 eV) and oxygen of bulk RuO$_2$ (529.5 eV).

It is also noted that the weight of the component for the under-coordinated "bridge" oxygen is well defined only when the oxide surface is well ordered. When the probing depth of the XPS measurements is less than three rutile layers (approximately 9 Å), as was the case for the XPS microscopy equipment, the O1s spectra of the Ru_xO_y intermediate state and RuO_2 islands in the advanced high-temperature oxidation state have different lineshapes—but comparable intensities. This result supports the concept that the nucleation of the $RuO_2(110)$ phase occurs when a sufficiently high oxygen content is reached in the subsurface region.

Figure 27 shows an exposure-temperature diagram, constructed on the basis of a large set of O_2-TD and XPS spectra and spectroscopy/microscopy data; these data illustrate the pressure-temperature range in which the adsorbed and intermediate states and the intermediate state and, RuO_2 phase coexist. According to this diagram, depending on the temperature and partial pressures of the reactants, the active phase during catalytic oxidation reactions may not only be the rutile RuO_2 phase, but also the O-rich metallic state that incorporates oxygen.

The simplified ball models of the ruthenium in the various states of oxidation that in the following are called (1) the adsorbed phase (with

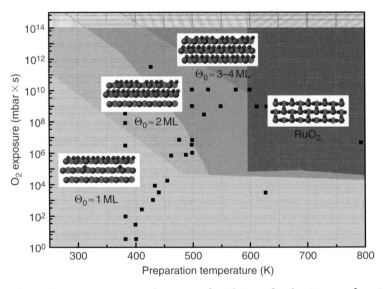

FIGURE 27 Diagram representing the states of oxidation of ruthenium as a function of the O_2 exposure (in mbar × s) and temperature. The patterned areas were determined on the basis of a large set of O_2–TD spectra (Blume et al., 2004; Böttcher and Niehus, 1999a,b). The black squares indicate the (exposure-T) space, when the system was characterized by XP spectroscopy and microscopy (Blume et al., 2005; Böttcher et al., 2002b). The ball models illustrate schematically the distribution of oxygen in the various states of oxidation of the sample.

a maximum coverage of 1 ML), (2) the "surface oxide" (with surface and incorporated oxygen with a total coverage of 1–4 ML), and (3) the RuO_2 phase.

The temperature dependence of the actual oxidation state of the Ru(0001) surface raises the question of whether only the RuO_2 phase is catalytically active. To confirm or correct the suggested mechanism of CO oxidation on a ruthenium catalyst, the evolution of the oxidation state of the catalyst with reaction temperature and the CO_2 production at CO + O_2 pressures comparable to those of realistic catalytic conditions were followed simultaneously. The dynamic response of the O1s and $Ru3d_{5/2}$ core level spectra was used for a precise assignment of the state of oxidation of the catalyst in the course of the reaction—correlated to the corresponding CO_2 yields measured with a mass spectrometer. The already determined $Ru3d_{5/2}$ and O1s core-level spectra summarized in Figure 26 provided the necessary basis for identification of the adsorbed, "incorporated," and oxide states and verification of their roles in the CO oxidation reaction.

The catalytic experiments had the goal of comparing the reactivity of the "surface oxide" with oxygen contents less than 3 ML formed at temperatures below 500 K and the RuO_2 phase formed at higher temperatures. In particular, it was expected that incorporation of oxygen below the surface would play an important role in the reactivity of the oxygen on the surface, which can be unambiguously shown only by data collected under reaction conditions.

The experiments started with a clean Ru(0001) surface and involved measurement of the changes after introduction of 0.1 mbar of CO + O_2 (O_2:CO partial pressure ratio = 1) as the temperature was slowly increased. The excess of oxygen with respect to the reaction stoichiometry provided slightly oxidizing conditions to ensure the formation of the various states of oxidation of the ruthenium. Figure 28A shows the CO_2 yield as a function of the reaction temperature. There is a clear, sharp onset of the reaction at approximately 420 K, with the reaction rate increasing continuously in the temperature range 420–550 K. The selected $Ru3d_{5/2}$ and O1s spectra shown in Figure 29 were measured at different reaction temperatures and represent the milestones in the evolution of the state of oxidation of the catalyst. The first $Ru3d_{5/2}$ and O1s spectra represent the status of the ruthenium catalyst after the onset of CO_2 production. The $Ru3d_{5/2}$ spectrum requires Ru(II)–O_{sub} and Ru(I)–$2O_{ads}$ O_{sub} components, whereas the O1s spectrum is a broad peak centered at 530.0 eV. The $Ru3d_{5/2}$ and O1s spectra resemble those measured for the surface oxide in the presence of approximately 2 ML of oxygen (Figure 29A), but with much less oxygen on the surface, whereas still significant O_{sub} was present (the Ru(II)–O_{sub} component at 0.5 eV is dominant). This result accounts for the relatively low O1s intensity, because, for our probing depth, the signal from the incorporated oxygen between the

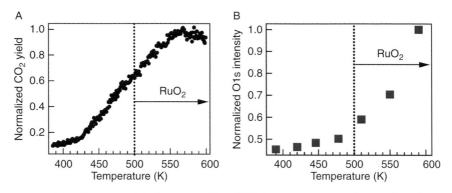

FIGURE 28 Changes in (A) the CO_2 yield and (B) the O1s intensity of the Ru(0001) surface accounting for the total amount of oxygen with increasing reaction temperature. The dashed line indicates the onset of RuO_2 growth. The O1s signal is normalized to the maximum intensity of the final oxide phase. The O1s intensity up to a temperature of approximately 500 K reflects only the surface and subsurface content, whereas at temperatures above 500 K the increase is dominated by the formation of an oxide phase.

first and second layers was attenuated by approximately 60% relative to that of the oxygen on the surface. The $Ru3d_{5/2}$ and O1s spectra underwent negligible lineshape changes in the temperature range 420–480 K, notwithstanding the gradual increase of the CO_2 yield. This observation implies that under operating conditions the catalyst contains a significant amount of subsurface oxygen. Because the ongoing reaction effectively consumes the surface oxygen species, the oxygen concentration on the surface is kept low. The fast dynamics at the surface is confirmed by the absence of a CO-related feature in the O1s spectra at binding energies >531.0 eV (Schiffer et al., 1997). This result indicates that the lifetime of the CO species on the surface before it is converted is shorter than that of the oxygen species.

Because the CO_2 formation reaction cannot disturb significantly the surface composition (equilibrium between CO and O adsorption and reaction is maintained), the observed temperature dependence of the reaction rate is likely related to the subsurface oxygen incorporation, which is facilitated at higher temperatures. As was discussed in earlier investigations (Blume et al., 2005), the increase of the subsurface oxygen content introduces a continuous structural distortion of the lattice, which in turn affects the adsorptive properties of the surface and the reactivities of the O and CO adspecies. A natural consequence of the progressive incorporation of oxygen with further increases in the reaction temperature is the onset of oxide formation, as manifested by the changes in the $Ru3d_{5/2}$ spectra at temperatures exceeding 500 K. These changes reflect the nucleation and growth of RuO_2 islands until a nearly steady-state

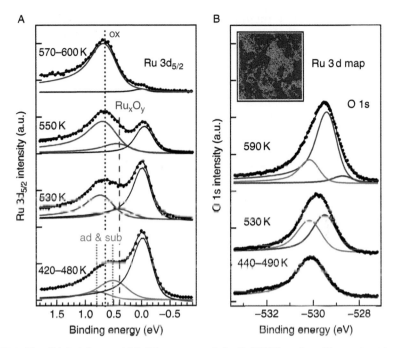

FIGURE 29 (A) Ru3d$_{5/2}$ and (B) O1s spectra of the Ru(0001) surface illustrating the catalyst composition developed during CO oxidation with increasing reaction temperature from 420 to 600 K ($dT/dt = 2$ K/min). Reaction conditions: $P_{CO} = 0.5 \times 10^{-1}$ mbar, $P_{O_2} = 0.5 \times 10^{-1}$ mbar. The Ru3d map illustrates the morphology with RuO$_2$ (blue) and Ru$_x$O$_y$ (green) islands obtained after exposure of the sample to 10^6 l of O$_2$ at 670 K.

composition is reached at temperatures >570 K. The presence of a small feature indicating a Ru$_{bulk}$ component in the Ru3d$_{5/2}$ spectrum indicates that the oxide thickness does not exceed the probing depth attained in the experiment, which was approximately 10 Å.

It also is quite possible that the morphology of the surface is more complex than has been described earlier, consisting of islands of oxide and Ru$_x$O$_y$ precursor, which, would account for the presence of the Ru$_{bulk}$. The Ru3d$_{5/2}$ image inserted in the O1s spectral panel (Figure 29B) illustrates a typical morphology of the ruthenium surface oxidized by exposure to 10^6 l (1 mbar*s) of O$_2$ in the temperature range 620–750 K range (Böttcher et al., 2002b). Because the Ru$_x$O$_y$ component appears close to the cus-Ru component (Table 2), it would be speculative to fit the Ru3d$_{5/2}$ signal by using three components with unknown weights, and instead we allowed broadening of the dominant oxide component at approximately 0.6 eV to account for all three contributions. The coexistence of nonoxidic areas is evidenced by the corresponding

O1s spectrum, which includes evidence of the precursor and oxide components. On the basis of the relative weight of the oxidic O1s components, we estimate that about 80% of the surface should be covered with the RuO_2 islands at 590 K. The significant increase of the O1s intensity with the growth of the oxide phase is not proportional to the absolute oxygen content; the effect of the more open structure of the oxide and the different escape depth of the photoelectrons play an important role as well, but we cannot quantify it. The CO_2 yield in the catalytic reaction reached a flat maximum at about 520–550 K and declined at higher temperatures. The maximum reaction rate could be observed again by reducing the temperature to 550 K, which did not noticeably affect the catalyst surface composition.

The results indicate that the oxygen-rich state of the catalyst formed at temperatures below 500 K (with subsurface oxygen) is also catalytically active; one cannot draw a clear line between the catalytic activity of the oxide and that of the oxygen-rich nonoxide state, because they occur at different reaction temperatures.

9. OUTLOOK

This review gives an overview of the application of XPS in the characterization of surfaces of working catalysts at pressures in the millibar range. The presented examples clearly showed that in situ XPS represents an appropriate tool for the investigation of metastable species formed on the surface of an active catalyst just under reaction conditions. The still limited number of publications in this field shows clearly the disadvantage that only a few groups have access to this method, because the equipment is expensive and only a limited number of groups work at synchrotron radiation facilities. Measurements like those described here are hindered by the lack of infrastructure for handling of flammable gases and liquids at synchrotron radiation facilities. At the synchrotron BESSY in Berlin a beamline and the infrastructure required for the investigation of heterogeneous catalytic processes were constructed recently. The beamline operates in the soft X-ray range of 200-1200 eV and the high pressure XPS set up of the Fritz-Haber-Institute is operated at that beamline permanently. This new experimental tool ISISS (innovative station for in situ spectroscopy) will improve the catalysis community's access to this method. Other synchrotron radiation facilities, like Max-lab (Sweden), Soleil (France), Alba (Spain), Campinas (Brasil) and Brookhaven (USA) are intended to implement XPS at elevated pressures as a new technique as well.

ACKNOWLEDGMENTS

The authors thank the staff of BESSY for help in carrying out experiments. V.I.B. and V.V.K. gratefully acknowledge the Russian Foundation for Basic Research (grants 04-03-32667 and 06-03-33020) for partial support of this work.

REFERENCES

Aballe, L., Barinov, A., Locatelli, A., Heun, S., and Kiskinova, M., *Phys. Rev. Lett.* **93**, 196103 (2004).

Abate, M., Goedkoop, J.B., de Groot, F.M.F., Grioni, M., Fuggle, J.C., Hofmann, S., Petersen, H., and Sacchi, M., *Surf. Interface Anal.* **18**, 65 (1992).

Au, C., Breza, J., and Roberts, M.W., *Chem. Phys. Lett* **66**, 340 (1979).

Backx, C., de Groot, C.P.M., Biloen, P., and Sachtler, W.M.H., *Surf. Sci.* **128**, 81 (1983).

Bal'zhinimaev, B. S., *Kinet. Catal.* **40**, 795 (1999).

Bao, X., Muhler, M., Schedel-Niedrig, T., and Schlögl, R., *Phys. Rev. B* **54**, 2249 (1996).

Barber, M., Connor, J.A., Guest, M.F., Hall, M.B., Hillier, I.H., and Meredith, W.N. E., *Faraday Discuss. Chem. Soc.* **54**, 219 (1972).

Barteau, M.A., and Madix, R.J., *in* "The Chemical Physics of Solid Surfaces and Heterogeneous Catalysis", (D.A. King and D.P. Woodruff, Eds.), Vol. 4, Chapter 4. Elsevier, Amsterdam, 1982.

Bäumer, M., and Freund, H.-J., *Surf. Sci. Rep.* **31**, 231 (1999).

Beitel, G.A., Laskov, A., Oosterbeek, H., and Kuipers, E.W., *J. Chem. Phys.* **100**, 12494 (1996).

Bhattacharya, A.K., Chesters, M.A., Pemble, M.E., and Sheppard, N., *Surf. Sci.* **206**, L845 (1988).

Bluhm, H., Hävecker, M., Knop-Gericke, A., Kleimenov, E., Schlögl, R., Teschner, D., Bukhtiyarov, V.I., Ogletree, D.F., and Salmeron, M., *J. Phys. Chem. B* **108**, 14340 (2004).

Bluhm, H., Ogletree, D.F., Fadley, C.S., Hussain, Z., and Salmeron, N., *J. Phys.Cond. Matter.* **14**, L227, (2002).

Blume, R., Niehus, H., Conrad, H., and Böttcher, A., *J. Phys. Chem. B* **108**, 14332 (2004).

Blume, R., Niehus, H., Conrad, H., Böttcher, A., Aballe, L., Gregoriatti, L., Barinov, A., and Kiskinova, M., *J. Phys.Chem. B* **109**, 14052 (2005).

Boronin, A.I., Bukhtiyarov, V.I., Vishnevskii, A.L., Boreskov, G.K., and Savchenko, V.I., *Surf. Sci.* **201**, 195 (1988).

Böttcher, A., Krenzer, B., Conrad, H., and Niehus, H., *Surf. Sci.* **504**, 42 (2002a).

Böttcher, A., and Niehus, H., *J. Chem. Phys.* **110**, 3186 (1999a).

Böttcher, A., and Niehus, H., *Phys. Rev. B* **60**, 14396 (1999b).

Böttcher, A., Niehus, H., Schwegmann, S., Over, H., and Ertl, G., *J. Phys. Chem. B* **101**, 11185 (1997).

Böttcher, A., Starke, U., Conrad, H., Blume, R., Gregoriatti, L., Kaulich, B., Barinov, A., and Kiskinova, M., *J. Chem. Phys.* **117**, 8104 (2002b).

Bourguignon, B., Carrez, S., Dragnea, B., and Dubost, H., *Surf. Sci.* **418**, 171 (1998).

Bowker, M., and Madix, R.J., *Surf. Sci.* **95**, 190 (1980).

Bradshaw, A.M., and Hoffmann, F.M., *Surf. Sci.* **72**, 513 (1978).

Broden, G., Rhodin, T.N., Bruckner, C.F., Benbow, R., and Hurych, Z., *Surf. Sci.* **59**, 593 (1976).

Bukhtiyarov, V.I., Boronin, A.I., Prosvirin, I.P., and Savchenko, V.I., *J. Catal.* **150**, 262 (1994).

Bukhtiyarov, V.I., Hävecker, M., Kaichev, V.V., Knop-Gericke, A., Mayer, R.W., and Schlögl, R., *Catal. Lett.* **74**, 121 (2001).

Bukhtiyarov, V.I., Hävecker, M., Kaichev, V.V., Knop-Gericke, A., Mayer, R.W., and Schlögl, R., *Phys. Rev. B* **67**, 235422 (2003).

Bukhtiyarov, V.I., Kaichev, V.V., Podgornov, E.A., and Prosvirin, I.P., *Catal. Lett.* **57**, 233 (1999).

Bukhtiyarov, V.I., Kaichev, V.V., and Prosvirin, I.P., *J. Chem. Phys.* **111**, 2169 (1999).

Bukhtiyarov, V.I., Kaichev, V.V., and Prosvirin, I.P., *Top. Catal.* **32**, 3 (2005).

Bukhtiyarov, V.I., and Prosvirin, I.P., *Proc. Eur. Congr. Catal.* 21, (2001).

Bukhtiyarov, V.I., Nizovskii, A.I., Bluhm, H., Hävecker, M., Kleimenov, E., Knop-Gericke, A., and, Schlögl, R., *J. Catal.* **238**, 260 (2006).

Bukhtiyarov, V.I., Prosvirin, I.P., and Kvon, R.I., *Surf. Sci.* **320**, L47 (1994).

Campbell, C.T., *J. Catal.* **94**, 436 (1985).

Campbell, C.T., *Surf. Sci.* **157**, 43 (1985).

Campbell, C.T., and Paffett, M.T., *Surf. Sci.* **139**, 396 (1984).

Carley, A.F., Owens, A.W., Rajumon, M.K., Roberts, M.W., Jackson, S.D., *Catal. Lett.* **37**, 79 (1996).

Carlisle, C.I., Fujimoto, T., Sim, W.S., and King, D.A., *Surf. Sci.* **470**, 15 (2000).

Carlisle, C.I., King, D.A., Bocquet, M.-L., Cerda, J., and Sautet, P., *Phys. Rev. Lett.* **84**, 3899 (2000).

Chen, J.-J., Jaing, Z.-C., Zhou, Y., Chakraborty, B.R., and Winograd, N., *Surf. Sci.* **328**, 248 (1995).

Christmann, K., and Demuth, J.E., *J. Chem. Phys.* **76**, 6308 (1982).

Conrad, H., Ertl, G., Koch, J., and Latta, E.E., *Surf. Sci.* **43**, 462 (1974).

Conrad, H., Ertl, G., and Küppers, J., *Surf. Sci.* **76**, 323 (1978).

Cubeiro, M.L., and Fierro, J.L.G., *J. Catal.* **179**, 150 (1998).

Davis, J.L., and Barteau, M.A., *Surf. Sci.* **235**, 235 (1990).

Dellwig, T., Rupprechter, G., Unterhalt, U., and Freund, H.-J., *Phys. Rev. Lett.* **85**, 776 (2000).

Denecke, R., *Appl. Phys. A* **80**, 977 (2005).

Denecke, R., Kinne, M., Whelan, C.M., and Steinrück, H.-P., *Surf. Rev. Lett.* **9**, 797 (2002).

Doering, D.L., Poppa, H., and Dickinson, J.T., *J. Catal.* **73**, 104 (1982).

Doniach, S., and Šunjic, M., *J. Phys. C* **3**, 285 (1970).

Fellner-Feldegg, H., Siegbahn, H., Asplund, L., Kelfve, P., and Siegbahn, K., *J. Electron Spectrosc. Relat. Phenom.* **7**, 421 (1975).

Fuhrmann, T., Kinne, M., Tränkenschuh, B., Papp, C., Zhu, J.F., Denecke, R., and Steinrück, H.-P., *New J. Phys.* **7**, 107 (2005).

Fuhrmann, T., Kinne, M., Whealan, C.M., Zhu, J.F., Denecke, R., and Steinrück, H.-P., *Chem. Phys. Lett.* **390**, 208 (2004).

Gates, J.A., and Kesmodel, L.L., *J. Catal.* **83**, 437 (1983).

Geißel, T., Schaff, O., Hirschmugl, C.J., Fernandez, V., Schindler, K.M., Theobald, A., Bao, S., Lindsay, R., Berndt, W., Bradshaw, A.M., Baddeley, C., Lee, A.F., et al., *Surf. Sci.* **406**, 90 (1998).

Ghijsen, J., Tjeng, L.H., van Elp, J., Eskes, H., Westerink, J., Sawatzky, G.A., and Czyzyk, M.T., *Phys. Rev. B* **38**, 11322 (1988).

Ghosal, S., Hemminger, J.C., Bluhm, H., Mun, B.S., Hebenstreit, E.L.D., Ketteler, G., Ogletree, D.F., Requejo, F.G., and Salmeron, M., *Science* **307**, 563 (2005).

Goodman, D.W., *Chem. Rev.* **95**, 523 (1995).

Graham, M.D., Kevrekidis, I.G., Asakura, K., Lauterbach, J., Krischer, K., Rotermund, H.H., and Ertl, G., *Science* **264**, 80 (1994).

Grant, R.B., and Lambert, R.M., *J. Catal.* **92**, 364 (1985).

Günther, S., Esch, F., Gregoratti, L., Barinov, A., Kiskinova, M., Taglauer, E., and Knözinger, H., *J. Phys. Chem. B* **108**, 14223 (2004).

Guo, X., Hanley, L., and Yates, J.T., Jr., *J. Am. Chem. Soc.* **111**, 3155 (1989).

Hävecker, M., Knop-Gericke, A., Schedel-Niedrig, T., and Schlögl, R., *Angew. Chem. Int. Ed.* **37**, 1939 (1998).

Hävecker, M., Mayer, R.W., Knop-Gericke, A., Bluhm, H., Kleimenov, E., Liskowski, A., Su, D., Follath, R., Requejo, F.G., Ogletree, D.F., Salmeron, M., Lopez-Sanchez, J. et al., *J. Phys. Chem. B* **107**, 4587 (2003).

Hendriksen, B.L.M., and Frenken, J.W.M., *Phys. Rev. Lett.* **89**, 46101 (2002).

Henke, B.L., Gullikson, E.M., and Davis, J.C., *Atomic Data and Nuclear Data Tables* **54**, 181–342 (1993).

Hoffman, K.R., Dababneh, M.S., Hsieh, Y.-F., Kauppila, W.E., Pol, V., Smart, J.H., and Stein, T.S., *Phys. Rev. A* **25**, 1393 (1982).

Hoffmann, F.M., *Surf. Sci. Rep.* **3**, 107 (1983).

Hüfner, S., Wertheim, G.K., and Wernick, J.H., *Phys. Rev. B* **8**, 4511 (1973).

Hussain, Z., Huff, W.R.A., Kellar, S.A., Moler, E.J., Heimann, P.A., McKinney, W., Padmore, H.A., Fadley, C.S., and Shirley, D.A., *J. Electron Spectrosc Relat. Phenom.* **80**, 401 (1996).

Johánek, V., Stará, I., Tsud, N., Veltruská, K., and Matolín, V., *Appl. Surf. Sci.* **162–163**, 679 (2000).

Joyner, R., and Roberts, M., *Surf. Sci.* **87**, 501 (1979a).

Joyner, R.W., and Roberts, M.W., *Chem. Phys. Lett.* **60**, 459 (1979b).

Kaichev, V.V., Prosvirin, I.P., Bukhtiyarov, V.I., Unterhalt, H., Rupprechter, G., and Freund, H. J., *J. Phys. Chem. B* **107**, 3522 (2003).

Kelly, M.A., Shek, M.L., Pianetta, P., Gür, T.M., and Beasley, M.R., *J. Vac. Sci. Technol. A* **19**, 2127 (2001).

Ketteler, G., Ogletree, D.F., Bluhm, H., Liu, H., Hebenstreit, E.L.D., and Salmeron, M., *J. Am. Chem. Soc.* **127**, 18269 (2005).

Kim, S.H., and Wintterlin, J., *J. Phys. Chem. B* **108**, 14565 (2004).

Kim, Y.D., Over, H., Krabbes, G., and Ertl, G., *Top. Catal* **14**, 95 (2001).

Kinne, M., Fuhrmann, T., Whealan, C.M., Zhu, J.F., Pantförder, J., Probst, M., Held, G., Denecke, R., and Steinrück, H.-P., *J. Chem. Phys.* **117**, 10852 (2002).

Kinne, M., Fuhrmann, T., Zhu, J.F., Tränkenschuh, B., Denecke, R., and Steinrück, H.-P., *Langmuir* **20**, 1819 (2004).

Kinne, M., Fuhrmann, T., Zhu, J.F., Whealan, C.M., Denecke, R., and Steinrück, H.-P., *J. Chem. Phys.* **120**, 7113 (2004).

Kleimenov, E., Bluhm, H., Hävecker, M., Knop-Gericke, A., Pestryakov, A., Teschner, D., Lopez-Sanchez, J.A., Bartley, J.K., Hutchings, G.J., and Schlögl, R., *Surf. Sci.* **575**, 181 (2005).

Knop-Gericke, A., Hävecker, M., and Schedel-Niedrig, T., *Nucl. Instr. Meth. A* **406**, 311 (1998).

Knop-Gericke, A., Hävecker, M., Schedel-Niedrig, T., and Schlögl, R., *Top. Catal.* **15**, 27 (2001).

Kok, G.A., Noordermeer, A., and Nieuwenhuys, B.E., *Surf. Sci.* **135**, 65 (1983).

Kruse, N., Rebholz, M., Matolin, V., Chuah, G.K., and Block, J.H., *Surf. Sci.* **238**, L457 (1990).

Kung, K.Y., Chen, P., Wei, F., Shen, Y.R., and Somorjai, G.A., *Surf. Sci. Lett.* **463**, L627 (2000).

Lindinger, W., Hansel, A., and Jordan, A., *Chem. Soc. Rev.* **27**, 347 (1998).

Liu, Z.-P., Hu, P., and Alavi, A., *J. Chem. Phys.* **114**, 5956 (2001).

Li, W.X., Stampfl, C., and Scheffler, M., *Phys. Rev. B* **65**, 075407 (2002).

Li, W.X., Stampfl, C., and Scheffler, M., *Phys. Rev. B* **68**, 165412 (2003).

Li, W.X., Stampfl, C., and Scheffler, M., *Phys. Rev. Lett.* **90**, 256102 (2003).

Lizzit, S., Baraldi, A., Groso, A., Reuter, K., Ganduglia-Pirovano, M.V., Stampfl, C., Scheffler, M., Stichler, M.C., Keller, C., Würth, W., and Menzel, D., *Phys. Rev. B* **63**, 205419 (2001).

Loffreda, D., Simon, D., and Sautet, P., *Surf. Sci.* **425**, 68 (1999).

Matolin, V., and Gillet, E., *Surf. Sci.* **238**, 75 (1990).

Matolin, V., Rebholz, M., and Kruse, N., *Surf. Sci.* **245**, 233 (1991).

Matsumura, Y., Okumura, M., Usami, Y., Kagawa, K., Yamashita, H., Anpo, M., and Haruta, M., *Catal. Lett.* **44**, 189 (1997).

McCrea, K., Parker, J.S., Chen, P., and Somorjai, G.A., *Surf. Sci.* **494**, 238 (2001).

McIntyre, B.J., Salmeron, M., and Somorjai, G.A., *Science* **265**, 1415 (1994).

McIntyre, N.S., and Chan, T.C., *in "Practical Surface Analysis." Vol. 1: Auger and X-ray Phototelectron Spectroscopy,* 2nd Edition. Briggs, D., Seah, M. P., Eds. Wiley, Chichester, 1990, p. 488.

Michaelides, A., Bocquet, M.-L., Sautet, P., Alavi, A., and King, D.A., *Chem. Phys. Lett.* **367**, 344 (2003).

Morkel, M., Kaichev, V.V., Rupprechter, G., Freund, H.-J., Prosvirin, I.P., and Bukhtiyarov, V.I., *J. Phys. Chem. B* **108**, 12955 (2004).

Morkel, M., Rupprechter, G., and Freund, H.-J., *J. Chem. Phys.* **119**, 10853 (2003).

Narkhede, V., Assmann, J., Muhler, M., *Z. Phys. Chem.*, **219**, 979 (2005).

Neubauer, R., Whealan, C.M., Denecke, R., and Steinrück, H.-P., *J. Chem. Phys.* **119**, 1710 (2003).

Ogletree, D.F., Bluhm, H., Lebedev, G., Fadley, C.S., Hussain, Z.,and Salmeron, M., *Rev. Sci. Instrum.* **73**, 3872 (2002).

Ohtani, H., Van Hove, M.A., and Somorjai, G.A., *Surf. Sci.* **187**, 372 (1987).

Over, H., Kim, Y.D., Seitsonen, A.P., Lundgren, E., Schmid, M., Varga, P., Morgante, A., and Ertl, G., *Science* **287**, 1474 (2000).

Over, H., Knapp, M., Lundgren, E., Seitsonen, A.P., Schmid, M., and Varga, P., *Chem. Phys. Chem.* **5**, 167 (2004) and references cited therein.

Over, H., and Muhler, M., *Prog. Surf. Sci.* **72**, 3 (2003) and references cited therein.

Over, H., and Seitsonen, A.P., *Science* **297**, 2003 (2002).

Over, H., Seitsonen, A.P., Lundgren, E., Smedh, M., and Andersen, J. N., *Surf. Sci.* **504**, L196 (2002).

Over, H., Seitsonen, A.P., Lundgren, E., Wiklund, M., and Andersen, J.N., *Chem. Phys. Lett.* **342**, 467 (2001).

Ozensoy, E., Meier, D.C., and Goodman, D.W., *J. Phys. Chem. B* **106**, 9367 (2002).

Pantförder, J., Pöllmann, S., Zhu, J.F., Borgmann, D., Denecke, R., and Steinrück, H.-P., *Rev. Sci. Instr.* **76**, 014102 (2005)

Paul, J.-F., and Sautet, P., *J. Phys. Chem. B* **102**, 1578 (1998).

Peters, K.F., Walker, C.J., Steadman, P., Robach, O., Isern, H., and Ferrer, S., *Phys. Rev. Lett.* **86**, 5325 (2001).

Pozdnyakova, O., Teschner, D., Wootsch, A., Kröhnert, J., Steinhauer, B., Sauer, H., Toth, L., Jentoft, F.C., Knop-Gericke, A., Paál, Z., and Schlögl, R., *J. Catal.* **237**, 1 and 17 (2006).

Prosvirin, I.P., Tikhomirov, E.P., Sorokin, A.M., Kaichev, V.V., and Bukhtiyarov, V.I., *Kinet. Catal.* **44**, 662 (2003).

Quinn, P., Brown, D., Woodruff, D. P., Noakes, T.C.Q., and Bailey, P., *Surf. Sci.* **491**, 208 (2001).

Rebholz, M., and Kruse, N., *J. Chem. Phys.* **95**, 7745 (1991).

Reuter, K., Frenken, D., and Scheffler, M., *Phys. Rev. Lett.* **93**, 116105 (2004).

Reuter, K., and Scheffler, M., *Phys. Rev. B* **60**, 45407 (2003).

Reuter, K., and Scheffler, M., *Phys. Rev. Lett.* **90**, 46103 (2003).

Reuter, K., Stampfl, C., Ganduglia-Pirovano, M., and Scheffler, M., *Chem. Phys. Lett.* **352**, 311 (2002).

Reuter, K., Stampfl, C., Ganduglia-Pirovano, M.V., and Scheffler, M., *Chem. Phys. Lett.* **352**, 311 (2002).

Rivière, J.C., *in "Practical Surface Analysis." Vol. 1: Auger and X-ray Photoelectron Spectroscopy,* 2nd Edition. (D.Briggs, and M. P.Seah, Eds.),p. 52, Chichester, Wiley, 1990.

Rocca, M., Savio, L., Vattuone, L., Burghaus, U., Palomba, V., Novelli, N., de Mongeot, F.B., and Valbusa, U., *Phys. Rev. B* **61**, 213 (2000).

Rodriguez, J.A., and Goodman, D.W., *Surf. Sci. Rep.* **14**, 1 (1991).

Ruppender, H.J., Grunze, M., Kong, C.W., and Wilmers, M., *Surf. Interf. Anal.* **15**, 245 (1990).

Rupprechter, G., Unterhalt, H., Morkel, M., Galletto, P., Hu, L., and Freund, H.-J., *Surf. Sci.* **502**, 109 (2002).

Rupprechter, G., Unterhalt, H., Morkel, M., Galletto, P., Hu, L., and Freund, H.-J., *Surf. Sci.* **502–503**, 109 (2002).

Savio, L., Vattuone, L., Rocca, M., Buatier de Mongeot, F., Comelli, G., Baraldi, A., Lizzit, S., and Paolucci, G., *Surf. Sci.* **506**, 213 (2002).

Sawhney, K.J.S., Senf, F., and Gudat, W., *Nucl. Instrum. Meth. A* **467**, 466 (2001).

Schauermann, S., Hoffmann, J., Johánek, V., Hartmann, J., Libuda, J., and Freund, H.-J., *Catal. Lett.* **84**, 209 (2002).

Schedel-Niedrig, T., Bao, X., Muhler, M. and Schlögl, R., *Ber. Bunsenges. Phys. Chem.* **101**, 994 (1997).

Schiffer, A., Jacob, P., and Menzel, D., *Surf. Sci.* **389**, 116 (1997).

Schimizu, T., and Tsukada, M., *Surf. Sci. Lett.* **295**, L1017 (1993).

Schütz, E., Esch, F., Günther, S., Schaak, A., Marsi, M., Kiskinova, M., and Imbihl, R., *Catal. Lett.* **63**, 13 (1999).

Shinozaki, K., Hayakawa, T., Liu, Y.Y., Ishii, T., Kumagai, M., Hamakawa, S., Suzuki, K., Itoh, T., Shishido, T., and Takehira, K., *Catal. Lett.* **58**, 131 (1999).

Shirley, D.A., *Phys. Rev. B* **5**, 4709 (1972).

Siegbahn, H., and Siegbahn, K., *J. Electron Spectrosc. Relat. Phenom.* **2**, 319 (1973).

Siegbahn, H., Svensson, S., and Lundholm, M., *J. Electron Spectrosc. Relat. Phenom.* **24**, 205 (1981).

Siegbahn, K., Nordling, C., Johansson, G., Hedman, J., Heden, P.F., Hamrin, K., Gelius, U., Bergmark, T., Werme, L.O., Manne, R., and Baer, Y., "ESCA Applied to Free Molecules." North-Holland, Amsterdam, 1969.

Siegbahn, K., Svensson, S., and Lundholm, M., *J. Electron Spectrosc. Relat. Phenom.* **24**, 205 (1981).

Stampfl, C., Ganduglia-Pirovano, M.V., Reuter, K., and Scheffler, M., *Surf. Sci.* **500**, 368 (2002).

Stara, I., and Matolin, V., *Surf. Sci.* **313**, 99 (1994).

Stegelmann, C., Schiødt, N.C., Campbell, C.T., and Stoltze, P., *J. Catal.* **221**, 630 (2004).

Surnev, S., Sock, M., Ramsey, M.G., Netzer, F.P., Wiklund, M., Borg, M., and Andersen, J.N., *Surf. Sci.* **470**, 171 (2000).

Su, X., Cremer, P.S., Shen, Y.R., and Somorjai, G.A., *Phys. Rev. Lett.* **77**, 3858 (1996).

Szanyi, J., Kuhn, W.K., and Goodman, D.W., *J. Phys. Chem.* **98**, 2978 (1994).

Teschner, D., Pestryakov, A., Kleimenov, E., Hävecker, M., Bluhm, H., Sauer, H., Knop-Gericke, A., and Schlögl, R., *J. Catal.* **230**, 186 and 195 (2005).

Todorova, M., Li, W.X., Ganduglia-Pirovano, M.V., Stampfl, C., Reuter, K., and Scheffler, M., *Phys. Rev. Lett.* **89**, 96103 (2002).

Tränkenschuh, B., Fritsche, N., Fuhrmann, T., Papp, C., Zhu, J.F., Denecke, R., and Steinrück, H.-P., *J. Chem. Phys.* **124**, 74712, (2006).

Tüshaus, M., Berndt, W., Conrad, H., Bradshaw, A. M., and Persson, B., *Appl. Phys. A* **51**, 91 (1990).

Usami, Y., Kagawa, K., Kawazoe, M., Matsumura, Y., Sakurai, H., and Haruta, M., *Appl. Catal. A* **171**, 123 (1998).

Valden, M., Lai, X., and Goodman, D.W., *Science* **281**, 1647 (1998).

van Santen, R.A., and de Groot, C.P.M., *J. Catal.* **98**, 530 (1986).

van Santen, R.A., and Kuipers, H.P.C.E., *Adv. Catal.* **35**, 265 (1987).

van Santen, R.A., Moolhuysen, S., and Sachtler, W.M.H., *J. Catal.* **65**, 478 (1980).

Wang, J., Fan, V.Y., Jacobi, K., and Ertl, G., *J. Phys. Chem. B* **106**, 3422 (2002).

Weissman, D.L., Shek, M.L., and Spicer, W.E., *Surf. Sci.* **92**, L59 (1980).

Wickham, D.T., Logsdon, B.W., Cowley, S.W., and Butler, C.D., *J. Catal.* **128**, 198 (1991).

Würz, R., Rusu, M., Schedel-Niedrig, Th., Lux-Steiner, M.C., Bluhm, H., Hävecker, M., Kleimenov, E., Knop-Gericke, A., and Schlögl, R., *Surf. Sci.* **580**, 80 (2005).

Yeh, J.J., and Lindau, I., *Atomic Data and Nuclear Data Tables* **32**, 1 (1985).

Zhang, C.J. and Hu, P., *J. Chem. Phys.* **116**, 322 (2002).

Zhu, J.F., Kinne, M., Fuhrmann, T., Denecke, R., and Steinrück, H.-P., *Surf. Sci.* **529**, 384 (2003a).

Zhu, J.F., Kinne, M., Fuhrmann, T., Tränkenschuh, B., Denecke, R., and Steinrück, H.-P., *Surf. Sci.* **547**, 410 (2003b).

CHAPTER 5

X-ray Diffraction: A Basic Tool for Characterization of Solid Catalysts in the Working State

Robert Schlögl

Abstract

Powder X-ray diffraction (XRD) provides information about phases and sizes of particles in solid catalysts, including characterization of the constitution and structure of the catalytically active species determined while the catalyst is functioning. This chapter includes a summary of the information provided by XRD, highlights some physical principles of diffraction that are needed to evaluate XRD data, and describes how nanostructural information about catalytically active materials can be determined from the data. Also included are a description of XRD instrumentation and an evaluation of the recent literature regarding experimental strategies and examples of catalysts that have been investigated with this technique. A major conclusion is that XRD data recorded under catalytic reaction conditions are useful for the structural and morphological characterization of solid catalysts, and XRD is augmented enormously when combined with other techniques for elucidating structures of working catalysts at length scales different from that of XRD.

Contents

1. Introduction   274
2. Objectives of XRD in Catalyst Characterization   275
3. Static Versus Dynamic Analysis   284
4. Observations of Catalysts in Reactive Atmospheres   287
5. Cost of Powder Diffraction Under
   Catalytic Conditions   288
6. Aim and Scope of Powder Diffraction Experiments   289

Fritz-Haber-Institut der Max Planck-Gesellschaft, Department of Inorganic Chemistry, 14195 Berlin, Germany

Advances in Catalysis, Volume 52
ISSN 0360-0564, DOI: 10.1016/S0360-0564(08)00005-9

7. Scattering Phenomena   289
8. Nanostructures and Diffraction   296
9. Phases in Working Catalysts   303
10. Applications of XRD to Working Catalysts   307
11. Combination of XRD Data with Auxiliary Information   309
12. Instrumentation   310
13. Case Studies   313
14. Summary and Conclusions   330
     References   332

ABBREVIATIONS

EXAFS	Extended X-ray absorption fine structure spectroscopy
FWHM	Full width at half maximum
GIX	Grazing-incidence X-ray diffraction
HRTEM	High-resolution transmission electron microscopy
LEED	Low-energy electron diffraction
NMR	Nuclear magnetic resonance spectroscopy
NSLS II	PETRA III, Ultrahigh performance synchrotron sources in Brookhaven and Hamburg
PDA	Powder diffraction analysis
SAXS	Small angle X-ray scattering
SLAC	XFEL, Free electron laser sources under construction in 2008 at Stanford and Hamburg
SMSI	Strong metal support interaction
TEM	Transmission electron microscopy
UV–vis	Ultraviolet–visible absorption spectroscopy
XAS	X-ray absorption spectroscopy
XRD	X-ray diffraction (in this text of powders)

1. INTRODUCTION

Powder X-ray diffraction (XRD) is a fundamental technique for the structural characterization of condensed matter. It provides evidence of bulk structures in various dimensions. By coherent scattering, the translational symmetry of a lattice is represented in a diffraction pattern, and the atomic species with their average site occupations are reflected in intensities. In powder diffraction, a full structure analysis has become possible as a result of advances in modeling strategies (Langford and Louer, 1996; McCusker et al., 1999). If the diffracting lattice planes are comparable in their dimensions to the wavelength of the X-rays (i.e., they are nanosized), or if the lattice plane distance is not constant but described by a distribution function (resulting from stress and strain or atomic disorder), then the diffraction profile is modulated and requires profile analysis to extract these parameters. Incoherent scattering of X-rays by the sample yields information about the atomic structure without relying on translational symmetry and gives rise to a modulated background (Vogel et al., 1995) that is rarely investigated in catalysis research.

2. OBJECTIVES OF XRD IN CATALYST CHARACTERIZATION

There is a continuing debate about the usefulness of XRD as a characterization method in catalysis, because heterogeneous catalysis is clearly related to the structures of solid catalyst surfaces. XRD provides surface structural information only for atomically flat surfaces at grazing incidence of radiation from high-brilliance sources.

One argument in favor of the use of XRD in catalysis research is that the dimensions of active particles such as, for example, supported metals, are frequently so small that the number of surface atoms is significant with respect to the number of atoms in the bulk, so that XRD data characterizing such samples should be directly relevant to catalysis. This argument is valid for other bulk-sensitive techniques, but it fails for XRD, because the structural uniformity of the bulk of a material does not pertain to the surfaces of small particles because of restructuring with major consequences on the average arrangement of surface atoms (Conrad, 1992; Gallagher et al., 2003).

Another argument in favor of XRD is that catalytic activity should be related to the defects (or "nanostructure") of a solid catalyst. Active sites are nonequilibrium structures and are therefore not part of the bulk structures that are most commonly determined by diffraction analysis. XRD is, however, in many ways sensitive to the deviations from perfect ordering of the unit cells in a perfect crystal; hence, XRD can be used to detect the nanostructure representing deviations from an ideal crystal.

The importance of surface defects for catalysis was first proposed by Hosemann et al. (1966) in his concept of "paracrystallinity," which, however, could not be substantiated (Borghard and Boudart, 1983) as a general structure–function correlation in catalysis (Ludwiczek et al., 1978). The goal of linking the nanostructure of a catalyst to its function has to be based on a much broader crystallographic understanding of diffraction anomalies (Herein et al., 1996; Herzog et al., 1996) than are manifested in the definition of a paracrystal. Such crystals, which do not seem to exist in catalytic solids, were thought to consist (Hosemann et al., 1966; Ludwiczek et al., 1978; Vogel et al., 2002) of nanoscopic inclusions of a minority structure (between one and a few unit cells in size) embedded in a bulk matrix of different but related lattice symmetry (e.g., iron aluminate in bulk iron).

The most fundamental reason for investigations of solid catalysts by XRD is that surfaces are the terminations of bulk solids, so that their static and dynamic properties are determined by the underlying bulk structure. Knowledge of the ideal and real bulk structure is thus indispensable for the design and understanding of a catalytic material. This point should not lead to the still widely accepted but incorrect view that knowledge of the bulk structure of a solid allows one, by inference, to determine the structure of the surface exposed by cuts through the lattice. Rather, restructuring and relaxation effects (Ammer et al., 1997; Diebold, 2003; Jansen et al., 1995; Saint-Lager et al., 2005; Wang et al., 2003) lead to sometimes massive deviations from idealized cuts through bulk solid structures (Lemire et al., 2004).

Powder diffraction analysis provides fundamental information characterizing catalysts. It determines, to a first approximation, the crystallinity of the solid and, to a second approximation, qualitative and quantitative information about various phases in the catalyst. In the third approximation it tells about the nanostructuring of the phases in the solid. Figure 1 is a summary of the kinds of information about catalytic materials provided by XRD.

Direct surface structural information (Ackermann et al., 2004; Chen, 1996; Gallagher et al., 2003; Stierle et al., 2005) is provided when the

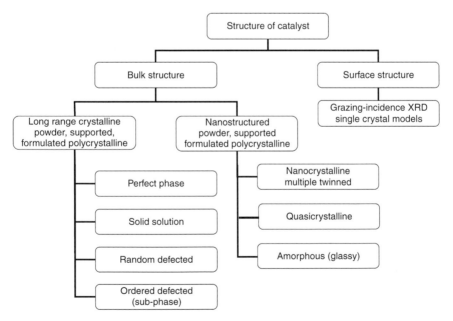

FIGURE 1 Hierarchy of structural information obtained from XRD data characterizing catalytic materials.

samples are atomically flat (model systems) and are characterized by a grazing incidence scattering geometry requiring a synchrotron source. This technique (referred to as GIX, for grazing incidence XRD) is comparable in its information content to LEED (low energy electron diffraction) but conceptually much more convenient in terms of the data analysis, as GIX can be modeled according to a kinematic approximation with no multiple scattering contributions (Ackermann et al., 2004; Chen, 1996; Fujii et al., 2005). The small number of diffraction lines observable from surfaces, however, limits the value of the method for complicated structures, as the number of independent observations is small compared with the number of parameters to be determined.

Bulk structural information can be obtained from XRD data characterizing any form of solid catalytic material; it includes data describing the average crystalline part as well as the noncrystalline part of the solid. In many practical cases, part or all of the catalytic material exhibits a mosaic texture. Such solids consist of a hierarchical ordering of basic structural units that may not be spherical. These agglomerate to form larger particles seen as "crystals," which in turn form aggregates appearing as macroscopic grains of micrometer size.

The relationship between crystallites and particles with respect to XRD is shown in Figure 2. The morphological "crystal" "c" is composed of anisotropic crystallites with dimensions "a,b". The arrows show the difference in dimension detected by XRD (a,b in two dimensions) and by other methods not requiring coherent scattering methods such as electron microscopy or gas adsorption. It is obvious that there may be little relationship between the particle size determined by microscopy or surface area analysis and the

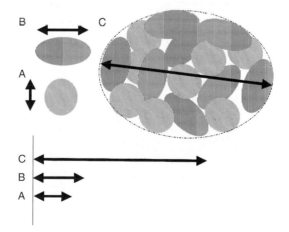

FIGURE 2   Crystallites and crystallinity in powder XRD. The aggregate (dimension "*c*") is composed of anisotropic crystallites with dimensions "*a, b*."

particle size measured by XRD, which characterizes the individual grains of a particle separated by grain boundaries. Platelet-shaped metal oxides and metal sulfides are prominent examples of materials that have quite different sizes determined by the different methods. The differences are less for metals, for which, however, deformation of particles can be caused by the action of the reaction atmosphere (Clausen et al., 1993; Grunwaldt et al., 2000); strong metal–support interactions (SMSI) (Diebold, 2003) can also deform equilibrium shapes of metal particles.

Large, morphologically well-defined crystals of catalytic materials are often agglomerates of primary particles that are nanocrystalline and give rise to only poor or even nonmeasurable XRD patterns because they are so small. Materials with a tendency to have disordered layers (e.g., silicates and metal oxyhydoxides) or materials in which one constituent has a high mobility (e.g., labile oxides) may fall in this category, and the analysis of XRD patterns is challenging as little diffracted intensity in broad and noisy patterns precludes the application of structure-solving analysis tools. For example, Figure 3 shows results characterizing a complex oxide $(MoV)_5O_{14}$ that was prepared as a single crystal (Ovsitser et al., 2002). The XRD data recorded for the $(2kl)$ orientation, which is perpendicular

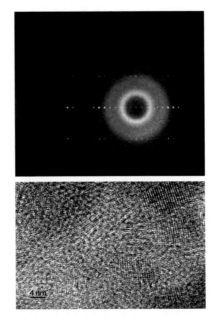

FIGURE 3   Diffraction from a material with strongly anisotropic ordering. The spots indicate the well-developed long-range order coexisting with complete X-ray disorder in the third dimension (rings). The HRTEM image illustrates the real-space information encoded in the diffraction pattern.

to the needle axis of the crystal on an image plate detector, clearly show the coexistence of a well-ordered atomic arrangement in two dimensions and a random orientation in the third dimension. A high-resolution TEM image of the same crystal indicates the random orientation of nanoscopic crystallites (Figure 2), which are large enough to produce the spots in the X-ray diffractogram.

Nanocrystalline materials are characterized by an average size of the coherently scattering crystallites that is small with respect to the coherence length of the radiation used for observation. For most practical cases, the critical dimension in compounds such as oxides is a few unit cells, or a crystallite size of about 1 nm. Such nanomaterials may still have perfect phase ordering, even though they may not be detectable by XRD. Frequent statements about "glassy" materials are inadequate, as glasses— being undercooled fluids—do not exhibit ordering of their basic structural units at any scale. Metals in the form of small particles are often characterized by multiple twinning or other kinetically controlled nonequilibrium shapes (Aslam et al., 2002; Nepijko et al., 2000; Rodriguez et al., 1996; Urban et al., 2000), furthermore, alloys may crystallize in the form of quasicrystals, which provide little structural information in powder diffraction experiments and hence appear to be amorphous. Truly amorphous solids exhibiting no translational symmetry are not common in catalytic materials, with the possible exception (Dutoit et al., 1995; Schaefer and Keefer, 1986) of silicate gels.

The absence of a sharp diffraction pattern characterizing a catalyst is usually an indication of nanocrystalline rather than amorphous structures. Such patterns may change as a result of changes in the environment without a transformation of the material into a crystalline phase. In these cases a change occurs in the molecular motif leading to changes in coordination polyhedra (e.g., tetrahedral to octahedral upon oxidation or hydration) without a growth in crystallite size.

The translationally ordered state characteristic of crystallites that are large with respect to the wavelengths of X-rays represents perfect phases in the sense of thermodynamics, to which phase diagrams apply. Even this idealized state of matter cannot exist without deviations from perfect ordering, however, because of a requirement of thermodynamics (the entropy of the material at equilibrium must be nonzero for the Gibbs free energy to be minimized). Thus, the material will contain a number of deviations from the ideal arrangement, called "defects."

These defects can occur in several classes that are depicted in Figure 4. Point defects may consist of voids (missing atoms) in a lattice. Equally frequently the constituents of the solid or heteroatoms are located at interstitial sites. These defects may be statically distributed giving only indirect evidence of their presence by small deviations of the lattice parameters (a fraction of a percent up to few percent). Frequently,

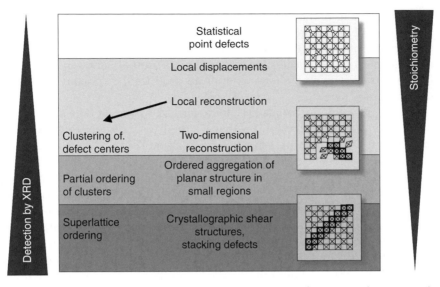

FIGURE 4 Classification of defects. The ordering increases from top to bottom. With increasing ordering the detection sensitivity of XRD increases as does the deviation from the composition of the thermodynamic phase, until it reaches a local minimum for the ordered subphase shown at the bottom. Catalysts often represent the case of the middle scheme and are difficult to characterize by diffraction or by their compositions. This statement also holds also for many catalyst support materials.

order–disorder transitions of the defects occur leading to defect clusters, line defects, or even sub-phases of the parent structure with additional weak lines in the diffraction patterns that cannot be indexed within the parent structure. Defects of a parent structure can become so numerous that the average composition is changing in stoichiometric dimensions. Then the ordered defects define so-called "sub-stoichiometric" phases (or subphases) of the parent phase. An example is the defect subphase $Mo_{18}O_{56}$ arising from weak reduction of MoO_3. Many oxide systems relevant in catalysis exhibit shear structures or Magnelli-type subphases, and metastable alloys tend to restructure into a stable phase and a sub-phase exhibiting together less free energy as the initial alloy.

Such defect-driven structural transformations are effectively investigated by powder diffraction analysis of samples kept in reactive atmospheres. As solid catalysts are dynamic systems, the phase inventory and the defect ordering (real structure) may well change as a result of changes of chemical potential of a constituent in a reactive environment. Some of the changes are irreversible and can be detected by pre- and post-operation analysis of catalysts, but many are reversible and will not be evident in such experiments.

Increasing the temperature of a catalyst will lead to changes in the lattice parameters of crystalline phases as a result of thermal expansion. The volume of the unit cell may change by as much as a few percent as a catalyst is heated from room temperature to its operating temperature (Nagy et al., 1999; Suleiman et al., 2003; Tschaufeser and Parker, 1995; van Smaalen et al., 2005), which is in excess of about 500 K for may reactions involving gas-phase reactants. Such lattice expansions exert a significant effect on the electronic structure of a solid, leading, for example, to energy shifts of the valence electrons relative to the energy levels of adsorbates (Greeley et al., 2002; Topsøe et al., 1997).

Irreversible phase transformations occur as a result of thermal treatments of metastable solids that may enable the self-diffusion of constituents of the solid resulting in segregation, reduction, or restructuring (Nagy et al., 1999). Major modifications of catalytic materials occur upon addition of water of solvation or water in the crystalline structure. The reverse process of dehydration of a catalyst often profoundly changes the positions of, for example, guest species in host lattices such as zeolites or modifies the symmetry of the constituting host lattice by removing spaces in the voids (pores) of the unit cells (Jentoft et al., 2003; Marosi et al., 2000; Palancher et al., 2005; Schnell and Fuess, 1996; Thomas and Sankar, 2001). The importance of such effects in explaining structure–function correlations and deactivation processes calls for the structural analysis of catalysts at operating temperatures.

It is not just effects of temperature that are important in determining the phases present in a catalyst, but also those of the reactive atmosphere. At elevated temperatures and operating pressures, both reactants and products can form compounds with the initial catalyst phases; thus, the initial form of the catalyst would merely represent a precatalyst and not the active phase. The formation of structures including sub-oxides; solid solutions of hydrogen, nitrogen, carbon, or oxygen with metals; and nitrides or carbides may occur. These are drastic examples of such transformations, less drastic are the incorporation of additional atoms as solid solutions. The effects of structural transformations on the electronic structure are still purely understood but the knowledge gained in solid state physics about such phenomena (such as the high-temperature superconductors or the blue light emitting diodes) shed light on an aspect of catalyst characterization that is only in its infancy due to massive experimental and also conceptual difficulties.

An example is the transformation of the technological iron catalyst (Greeley and Mavrikakis, 2005; Herzog et al., 1996; Schlögl, 1991) used for ammonia synthesis (Figure 5). This catalyst has been shown recently to be a sub-nitride under catalytic reaction conditions. At ambient pressure the chemical potential of ammonia is only sufficient to form a phase $Fe_{18}N$ being a sub-nitride of Fe_2N. This structure, which is distinctly different

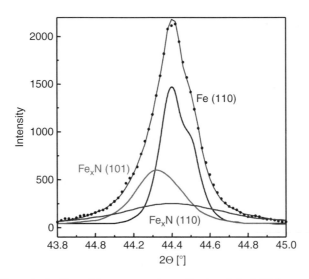

FIGURE 5 Characterization of iron ammonia synthesis catalyst. High-resolution laboratory diffraction indicated a reversible modification of the iron (111) line profile. Under catalytic reaction conditions, a sub-nitride with $x = 15$–18 is present in addition to the bulk iron matrix. The fitting and assignment of the data were substantiated by observations of the line profile during step changes in the composition of the gas atmosphere. Details and references are given in the text.

from that of the alpha iron in the precatalyst, is only barely detectable by XRD. The sub-nitride formation was found in diffraction experiments and later in investigations with single crystals at elevated pressure; it resolved the long-standing issue of the nature (Rayment et al., 1985; Topsøe et al., 1973; Wyckoff and Crittenden, 1925) of the active phase of so-called "ammonia iron" (Alstrup et al., 1997).

The clear possibility of a change of phases resulting from changes in operation conditions of a catalyst is a strong argument for conducting structural analysis of catalysts under working conditions. A practical reason for conducting such investigations by XRD is the instability of the active forms of many catalysts. In the case of highly dispersed metals on supports, one is interested in the phase (Frenkel et al., 2001; Fu et al., 2005; Hills et al., 1999) (e.g., metal or alloy) and in the morphology (Canton et al., 2002; Clausen et al., 1991, 1993; Fagherazzi et al., 2000; Grunwaldt and Clausen, 2002; Riello et al., 2001) of the active material (e.g., the size, shape, and dispersion). These characteristics may change dramatically with changes in the gas atmosphere. For example, in air, many catalysts do not even exist in an activated form. The technological practice of passivation by formation of a thin layer of oxide or covering of the material with a protective coating may modify the active states and make a structural analysis invalid. Thus, in such cases the examination of the catalyst under reaction conditions is the only safe means to a reliable structure determination.

Often it is essential to characterize the formation of a catalytically active state of a highly dispersed phase by XRD under reactive atmospheres. Investigations of reduction and calcination processes, for example (Thomas and Sankar, 2001; Sankar et al., 1991), guided the determination of recipes for catalyst preparation (Günter et al., 2001d; Kirilenko et al., 2005; Ressler et al., 2001, 2002; Wienold et al., 2003) in a complex parameter space.

The target of XRD of catalysts in reactive atmospheres is thus to yield a description as complete as possible of the structure of a working catalyst with all the criteria designated in Figure 1. The experiment is demanding for the determination of phases and even more so if the other structural parameters are to be determined as well. As the structural dynamics of working catalysts can occur on multiple time scales, ranging from seconds to weeks, it is mandatory to ascertain the steady state of a catalyst under working conditions (Epple, 1994).

The literature indicates that only rarely has the ultimate objective of a complete structure determination been achieved for working catalysts. The report presented here includes a discussion of the origin of the apparent difficulties and a summary of detailed practical procedures to approach the final target of the characterization.

Even when this target is reached, it must be kept in mind that XRD can, by the very nature of its basic physics, fall short of describing the structure of a catalyst at all length scales. It misses out on variations of the local structure that are better addressed by EXAFS spectroscopy (Clausen et al., 1993, 1998). Furthermore, XRD is insensitive to the texture and microstructure with dimensions larger than about 10 nm, which are better investigated by electron microscopy or gas adsorption techniques—and the surface structure of a working polycrystalline catalyst is in most cases inaccessible by XRD.

Nonetheless, characterization of catalysts in reactive atmospheres by XRD is a powerful method for obtaining the basic information needed to determine structure-activity correlations. Many phenomena of structural deactivation, either by sintering or recrystallization, are accessible by XRD under reaction conditions. The practice of approximating reacting atmospheres by simple-to-handle proxies (hydrogen for hydrocarbons or dry gases instead of steam-loaded feeds) has to be abandoned.

Even when XRD data are obtained with samples in reactive atmospheres, the data must be complemented by results of other experimental techniques to compensate for the inherent weaknesses of the XRD methods. One approach (Brückner, 2003; Weckhuysen, 2003) is to combine XRD with, for example, EXAFS spectroscopy (Clausen et al., 1993;

Dent et al., 1995; Sankar et al., 2000) or with UV–vis spectroscopy of functioning catalysts. In most cases, separate experiments will have to be conducted, and then it is essential to match as exactly as possible the sample, preparation method, and reaction conditions.

3. STATIC VERSUS DYNAMIC ANALYSIS

The nanostructure of a solid, also referred to in terms such as "defects," "real structure," and "mosaic structure," depends strongly on its environment. As all aspects of the nanostructure may be relevant to catalytic functions, it is common to infer such properties from static determinations of the nanostructure, often carried out at about 300 K and in laboratory or autogeneous atmospheres. This approach neglects the dynamics and assumes incorrectly that the surface catalysis process should not modify the rigid crystalline bulk of a solid.

This point is illustrated in the long history (Herzog et al., 1996; Topsøe et al., 1973; Wyckoff and Crittenden, 1925) of the elucidation of the mode of operation of the iron ammonia synthesis catalyst. At an early stage, it was found (Topsøe et al., 1973; Wyckoff and Crittenden, 1925) that imposition of the working conditions profoundly changed a nanostructured model catalyst. This observation was seemingly forgotten later in the investigation of single-crystal model catalysts (Ertl et al., 1978) and rediscovered much later in high-pressure determinations (5–20 bar of the reactive product and not of the nonreactive nitrogen) of structures of single crystals (Alstrup et al., 1997). Only limited attention was paid to the consequences of the finding that the catalyst structure may change profoundly when the active material is in contact with the reactants at the chemical potentials of their normal operation.

Impressive theoretical progress has been made in the prediction of fundamental modifications of catalyst surface structures and compositions as a function of the chemical potential of the environment in relatively simple cases (Nørskov et al., 2006; Reuter and Scheffler, 2002; Stampfl et al., 2002). This level of dynamic analysis with either single crystals or realistic polycrystalline catalyst materials has not yet been attained experimentally and certainly not in experiments with XRD under reaction conditions. There are no investigations that provide quantitative links between the phases and texture of a catalyst with its performance. All investigations discussed here can at best provide evidence relating the catalytic activity with a phase or a defect structure of a phase.

The first dynamic analysis of a working catalyst to elucidate the universally postulated structure–function correlation as a quantitative relationship still has not been achieved. The fundamental difficulties in relating bulk-sensitive XRD data to surface catalysis are one set of unmet challenges; others are the simple experimental execution and sufficiently accurate structure determinations as a function of operating conditions. Presently, XRD, even of samples in reactive atmospheres, is merely a method of determining phases and defect structures that may be related to catalytic performance. Thus the method must be considered as basic for the elucidation of the bulk dynamics of catalytically active materials, but its full potential for the determination of quantitative structure–function correlations has not yet been achieved.

The literature indicates that another catalysis-related mode of application has become prominent. It has been recognized that the synthesis of active phases can be understood and optimized on the basis of XRD measurements of catalyst synthesis mixtures and determination of phases and transformation kinetics. A prominent set of examples (Bazin et al., 2002; Chen et al., 2005, 2006; Davidson, 2002; Kleitz et al., 2002; Liu et al., 2005; Palancher et al., 2005; Valtchev and Bozhilov, 2004) is grouped around the formation of mesoporous solids from gels with templates. Such data allow an optimum choice of parameters for obtaining a desired phase. The ability of these methods to detect catalytically relevant phases in complex systems is illustrated in Figure 6.

The example (Abd Hamid et al., 2003; Jentoft et al., 2003) is a multiele-ment/multiphase oxide designed for selective oxidation catalysis. The back part of Figure 6 shows a catalyst material suitable for selective oxidation that was investigated at the temperature of operation and in an oxidizing atmosphere (air), as has frequently been done in

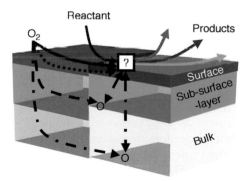

FIGURE 6 Schematic representation of an oxide catalyst with its functional compartments in various structural states for high (back) and low (front) chemical potentials of oxygen. The arrows and the question mark indicate the complex distribution of oxygen in its dual role as a reactant at the surface and as a constituent of the catalyst material in the bulk. Its abundance is controlled by the presence of reducing species in the gas phase leading to a dependence of the results of XRD structural analysis on the availability of reducing gas-phase species. For details and references, see the text.

investigations aiming to approximate catalytic reaction conditions without using the correct reaction atmosphere. Even under these conditions, one encounters a complex scenario of a bulk structure exchanging oxygen with its surface and the gas phase through an intermediate diffusion barrier (Maier, 1987, 1995). If the correct reaction atmosphere is applied, the chemical potential of oxygen will change in all components of the solid, that is, at the surface, in the subsurface diffusion barrier layer, and in the bulk, as indicated in the front part of Figure 6. It was verified (Dieterle et al., 2001; Knobl et al., 2003) that the nanostructuring and even the bulk phase behavior change under these conditions (Ovsitser et al., 2002). As indicated by the different coloring of the material in Figure 6, different components in the solid are affected in different ways by the choice of the chemical potential (composition of the gas phase).

These components of the solid undergo structural rearrangements detectable by XRD, with different kinetics, depending on their location with respect to the active site where the dynamics of restructuring starts, and depending on the preexisting nanostructure that controls the resistance to change driven by the chemical potential of the reactants in the gas phase. Because the chemical potential may vary from grain to grain in a polycrystalline solid, it is to be expected that XRD will indicate distributions of structural properties both laterally and in depth that cannot be resolved by the technique. It is evident that complementary methods applied under reaction conditions (and *ex situ*) are necessary to identify and assign such distributions of structural properties (which give rise to line profile distortions and intensity modulations in the XRD patterns).

Thus, analysis of the nanostructure by integrated XRD measurements is limited with regard to the information it provides about the average catalyst particle. The data may well fail to provide information about the distribution of nanostructures, which can change at various rates, as indicated in Figure 6. Thus, correlations with catalyst performance data may not be meaningful because of the lack of spatial resolution of the structure analysis.

Practical experience shows that distributions of nanostructural properties become narrower as the average size of the crystallite in a phase becomes smaller; metal clusters, for example, tend to be more homogeneous in structure than micron-sized oxide platelets.

Verification of structural adaptations of a catalytic material to changes in the environment can be determined only if the structural analysis and measurements of the catalytic performance are recorded simultaneously with the catalyst under real working conditions. Accidental correlations of catalyst structural changes with kinetics of the catalytic reaction measured in separate experiments are sometimes possible, but they provide no physical insight, because it is not known whether the same structural dynamics operates in the two separate experiments.

Today, most investigations of active structures of working catalysts are carried out in separate measurements of structure and function. This procedure is justified by the necessity to augment the XRD information by other techniques that cannot be applied under the same operational conditions as XRD. The limited sensitivity of XRD to near-surface structures and to highly defective components in a catalyst requires complementary techniques, such as XRD and EXAFS spectroscopy used in combination (Clausen, 1998; Grunwaldt and Clausen, 2002; Sankar et al., 2000) or complementary measurements by TEM (Abd Hamid et al., 2003; Carabineiro et al., 1999; Chen et al., 2006; Hävecker et al., 2003). The combination of XAS and XRD of working catalysts (Andersson et al., 1996; Clausen et al., 1998; Dent et al., 1995; Sankar et al., 2000; Thomas, 1997), with static and transient experiments applied together, today provides the best way to determine catalyst structure–function correlations. In this way it is possible to address the inherent problem of separating dynamical aspects into effects that occur as consequence of thermodynamics and those that are functionally related to the catalytic reaction. (One is reminded of the statement that "everything is interesting, but only little of it is relevant.")

It is not appropriate to state a preference for dynamic versus static XRD methods for investigation of a given catalyst. Both kinds of information are needed, and they are complementary for a complete structural analysis. It is emphasized that a catalyst is dynamic in character and not in equilibrium at ambient observation conditions; thus experiments are needed to characterize working catalysts. It is generally not useful to approximate the reaction environment by convenient mixtures; for example, replacement of hydrocarbons by hydrogen or methanol is not appropriate, nor is it appropriate to omit steam that is either a feed or product component). Steam strongly affects the chemical potential of reductants or oxidants. Furthermore, steam tends to modify the kinetics of solid-state conversions strongly by enhancing structural transformations through hydroxylation of outer and interior surfaces of solids. This caveat has been frequently ignored in reported XRD investigations of functioning catalysts.

4. OBSERVATIONS OF CATALYSTS IN REACTIVE ATMOSPHERES

Because the terms "in situ" and "operando" are used inconsistently to represent investigations of catalysts in reactive atmospheres (Banares, 2005; Weckhuysen, 2003), the latter is avoided here, and use of the former is minimized. Relevant in the present context of structural analysis are investigations of catalysts under dynamic conditions representing those of catalytic application.

Because the physics of XRD requires strong compromises in the choice of the sample environment, it is not realistic to apply actual catalytic reaction conditions in these measurements. Unavoidably, some extrapolations have to be made to link experimental observations to the state of the functioning catalyst in a reactor optimized for performance in the sense of production of desired products. The gas flow and energy transport conditions, for example, must often be chosen to be different in the catalyst testing and in XRD experiments. The catalytic performance is often analyzed in terms of a model of the kinetics, and the results may be used to make a connection with the conditions of the XRD experiment. A discussion of the issues has been published (Schlögl and Baerns, 2004).

5. COST OF POWDER DIFFRACTION UNDER CATALYTIC CONDITIONS

Investigation of working catalysts by XRD is not routine (Langford and Louer, 1996). It requires dedicated instrumentation that is commercially available only in part. Scientific expertise is necessary for the design of equipment and analysis of data. Data from complementary techniques characterizing working catalysts are generally required (Brückner, 2003; Lundgren et al., 2006), as well as data from conventional XRD and usually also from electron microscopy and surface area/pore volume measurements. Such analysis is thus not suitable for large numbers of samples, not least because of the time required to analyze the results.

Nonetheless, the method is invaluable as a basis for determination of routines for thermal treatment and activation of technological catalysts (Dutoit et al., 1995; Enache et al., 2002; Knobl et al., 2003). It is also suitable for defining the ranges of operating conditions of a catalyst with respect to structural deactivation. In favorable cases, the method can even provide direct structure–function correlations for mixtures of phases by exclusion of identified phases as candidates for the active material.

In-depth developments of technological catalysts (Dieterle et al., 2001; Marosi et al., 2000) may benefit greatly from XRD investigations, because the data may allow testing of structural hypotheses and determination of phases that may be necessary or detrimental for the catalyst function. The data may also provide essential information about the mode of operation of structural promoters. The results may be used for designing high-throughput experiments (Corma et al., 2006; Havrilla and Miller, 2005) that test the synthesis and activation conditions using a target phase or its crystallinity as a lead parameter.

As expensive as these experiments may be, among the methods for characterization of functioning catalysts they are probably the most versatile and applicable of any that can support catalyst development. It is thus advantageous to use at least this one method in catalysis development laboratories, because the potential it offers for focusing the development work is substantial and much more applicable to operating catalysts than the conventional *ex situ* XRD analysis that is routinely applied today.

6. AIM AND SCOPE OF POWDER DIFFRACTION EXPERIMENTS

Table 1 is a list of textbooks and their characteristics in which the reader can find the physical background about the XRD technique. The following section includes a brief introduction to the physics (Langford and Louer, 1996; Alexander and Klug, 1974) of the diffraction experiment, intended to familiarize the reader with basic facts of the technique from a practical viewpoint that must be observed when planning experiments and interpreting data obtained under catalytic reaction conditions.

7. SCATTERING PHENOMENA

XRD is a phenomenon of scattering electromagnetic waves with wavelengths of the order of magnitude of atomic sizes at the electrons of the atoms constituting a crystal. The electrons are distributed in periodically fluctuating density maxima (constituting the atoms), and the physical scattering process may be considered as the incoming wave exciting the electron that emits radiation of the same wavelength but in a direction different from that of the incoming wave. The many secondary sources of waves resulting from the many electrons in the crystal give rise to interference phenomena. This interference occurs weakly on the level of atoms and is the reason why there is no "zero intensity" diffraction pattern ("diffuse background") but instead a much stronger diffraction from the periodic array of the atoms in a lattice, giving rise to the sharp maxima known as a diffraction pattern. A diffraction pattern contains, besides the background information, two other independent types of information, namely, the positions of peaks and their intensities (profiles). The pattern of positions is related to the geometry of the crystal lattice, to unit cells, and their arrangements. The intensity distribution relates to atom types and their relative positions within the unit cell, to their charges, their thermal motions, and to the defect structure.

The geometric structural information about the crystal can be extracted from the diffraction pattern by geometric analysis. The concept of a crystal as being a diffraction grating for X-radiation and the relationships between the geometry of the diffraction grating and the pattern of

TABLE 1 Collection of textbooks introducing the physics and data analysis of XRD

Authors	Title	Year	Publisher	Comments
G.H. Stout, L.H. Jensen	X-ray structure determination	1968	Macmillan	Fundamental, precise background
D. McKie, C. McKie	Essentials of crystallography	1986	Blackwell	The ideal crystal
D.D.L Chung, P.W. De Haven, H. Arnold, D. Gosh	XRD at elevated temperatures	1993	VCH Wiley	High-temperature practical work
J.R. Anderson, K.C. Pratt	Introduction to characterization and testing of catalysts	1985	Academic Press	Selected aspects of *ex situ* XRD of catalysts
H.P. Klug, L.E. Alexander	XRD procedures for polycrystalline and amorphous materials	1974	Wiley Interscience	The complete reference for *ex situ* PDA
A. Guinier, G. Fournet	Small angle scattering of X-rays	1955	Wiley	Fundamental account, general
C. Hammond	The basics of crystallography and diffraction	1997	Oxford	Introduction, concepts

diffracted beams is given by the equations of von Laue and Bragg and may be rationalized in the Ewald diffraction sphere construction.

The Bragg equation gives the relationship between the distance of a set of diffracting planes (the planes constructed through the atom positions in the crystal with the relative distance d and the angle between the vector of the incoming radiation of wavelength λ and the diffracting planes Θ).

$$n\lambda = 2d \sin \Theta \qquad (1)$$

As λ and Θ are known from experiment, d can be found and so can be the position of the diffracting lattice plane within the unit cell (as given by the Miller index). The occurrence of a diffraction peak as function of wavelength and diffraction angle can be seen easily from the Ewald construction representing the diffraction condition as the intersection of a sphere with the radius of $1/\lambda$ and reciprocal lattice points representing the diffraction planes from real space in the reciprocal space. An illustration is given in Figure 7 showing the diffraction sphere with the radius $1/\lambda$ in an orthorhombic lattice. The relationship between wavelength and interplanar distances is such that two planes belonging to two different zones simultaneously fulfill the diffraction condition. In an experiment, one would obtain one peak from two sets of atoms contributing to the intensity.

Figure 7 shows that in powder XRD an ambiguity in contributions of lattice planes to diffraction peaks occurs and that a resolution problem

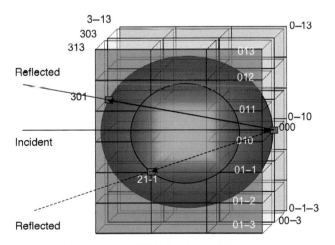

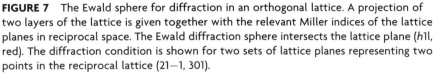

FIGURE 7 The Ewald sphere for diffraction in an orthogonal lattice. A projection of two layers of the lattice is given together with the relevant Miller indices of the lattice planes in reciprocal space. The Ewald diffraction sphere intersects the lattice plane (h1l, red). The diffraction condition is shown for two sets of lattice planes representing two points in the reciprocal lattice (21−1, 301).

exists for wide diffraction angles or small wavelengths. The diagram uses the reciprocal space whereby for dimensions an inverse relationship exists to the real crystal lattice space: small dimensions or changes of dimensions become prominent in reciprocal space, and large dimensions in real space such as large unit cells become small in reciprocal space. This relationship is as simple as stated only for orthogonal lattices; the relationship between real space and reciprocal lattices takes on more complex forms if nonrectangular lattices are considered.

The pattern obtained by rotating the lattice through rotation of the sample or the detector relative to the X-ray beam and thus through the Ewald sphere produces a fingerprint of peak positions that is very useful for phase analysis. The match between a model diffraction pattern as a line diagram of its positions and the experimental observation must, however, be complete. It is obvious from Figure 7 that all reflections must be observed within the range of diffraction angles (or rotation angles between the lattice and the Ewald sphere). If only some reflections fit the positions and others do not, it is quite likely that the phase is not identified.

The observation that some expected peaks do not arise is, however, not necessarily an indication of a mismatch, as the texture of the sample and/ or defects can strongly affect the intensities of diffracted beams. Furthermore, there are symmetry-related systematic absences in the model diffraction pattern that also need to be taken into account. On the basis of a hierarchical search for agreement between reference data sets and observed patterns, automatic assignment programs today allow the identification of phases, even in mixtures, from an analysis of the peak distribution. These assignments fail, however, for strongly textured samples (e.g., samples of pressed platelets or fibers) and for strongly defective samples (substitutional modifications as well as large numbers of point defects).

For the present applications it is relevant to ask how accurate is the intersection of the Ewald sphere with the reciprocal lattice point. Each of these points represents a (series of parallel) lattice planes defined by atom positions in the unit cell. The thermal motion of the atoms expands the ideal plane into a slab. Elements of disorder such as microstrain, chemical impurities in lattice positions (substitution, solid solutions), and interstitial atoms producing "chemical microstrain" also expand the lattice planes effectively into lattice slabs of locally varying thickness.

These elements of disorder will not lead to a shift of a sharp peak. Shifts occur only if defects order in such a way as to modify all unit cells of the crystal in the same way (Figure 4). This effect is frequently encountered in solid solutions (Abd Hamid et al., 2003; Langford and Louer, 1996; Valtchev and Bozhilov, 2004), in which atoms of different sizes occupy the same lattice positions. In many cases there are linear relationships between the chemical composition and the shifts of diffraction peaks translating into modifications of lattice constants.

Because of the inverse relationship of sizes between real space and reciprocal space lattices, the slabs of a real structure reach a substantial thickness in reciprocal space where they occur as spherical or rod-shaped extensions of the lattice point. The diffraction condition becomes softened, and a range of angles (or wavelengths) will fulfill the diffraction condition. Hence, a complex profile will result instead of a sharp peak. This point is illustrated schematically in Figure 8. The Figure indicates that as a consequence of disorder and softened diffraction conditions, additional lattice slabs fulfill the diffraction conditions that are not detected under sharp diffraction conditions. This phenomenon explains the above mentioned ambiguity in phase identification for imperfect crystalline phases. The widened and modified profile is detectable in the normal Θ–2Θ scans as projected line broadening. This broadening β is expressed as the full width at half maximum of a diffracted line corrected for the instrumental broadening

$$\beta = \beta_{\text{instrumental}} + \beta_{\text{physical}}; \quad \beta_{\text{physical}} = 2\delta\Theta \tag{2}$$

and is related to microstrain ε induced by all the factors mentioned above in the following way:

$$\varepsilon = \frac{\delta}{a} = -\cot\Theta\,\delta\Theta : \qquad \beta = -2\varepsilon\tan\Theta \tag{3}$$

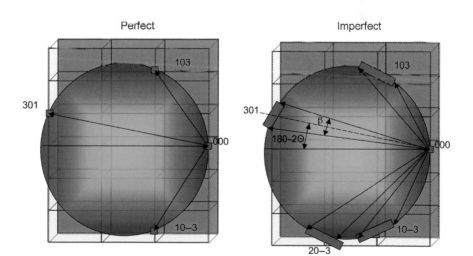

Perfect   Imperfect

FIGURE 8 Comparison of the modification of the Ewald construction when going from ideally translational symmetric lattices to real lattices in which planes degenerate into lattice slabs. The quantity "β" is explained in Equation (2).

Microstrain is defined as the change in length δ over an elementary length *a*, usually the lattice constant, and is given as a percentage of the unit length. Figure 9 illustrates a schematic distinction between uniform and nonuniform strain and their effects on a diffracted line profile. The case of uniform strain (center, Figure 9) must be resolved into tensile (expansion) and compressive (contraction) strain shifting the diffracted line to smaller or larger angles, respectively.

The elements of disorder or defects causing line broadening have to be sharply discriminated from the line broadening arising from the finite size of a diffracting crystallite, a frequent situation in catalysts, and in particular, in diffraction experiment in which gas–solid reactions tend to modify the size distribution of crystallites by breaking larger crystals during ion diffusion or by merging small crystallites through sintering. The line broadening resulting from size effects has nothing to do with disorder (as is often inferred), but instead is a fundamental consequence

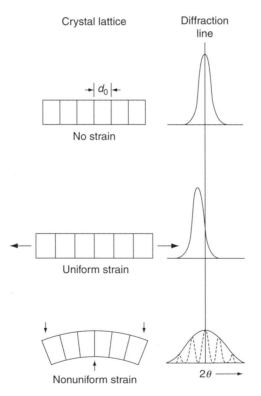

FIGURE 9   Schematic representation of the effect of microstrain defined as the modification of an ideally equidistant lattice (top) by uniform (unidirectional) strain (center) or nonuniform strain (2-D) (bottom). Note the different effects on the diffraction line.

of the interference of diffracted waves from perfect lattice planes. An incoming X-ray wave will create many diffracted wavelets from each atom in the lattice, as the wavelength of X-rays and the size of atoms are comparable. The manifold of the diffracted wavelets leads to strong constructive interferences and hence to readily observable diffraction intensities—and also to a number of destructive interferences in directions deviating strongly from the forward scattering vector. It can be shown by simple geometric arguments that many geometrically identically spaced diffracting objects lead to a strong enhancement of the diffracted beam and a weak background from the residuals of the destructive interferences. As the number of diffracting objects decreases, the contributions of destructive and constructive interferences to the outgoing wave become more nearly equal, with the consequence that the diffraction pattern gets blurred and broadened as the selection between constructive and destructive interferences of the secondary wavelets is lost. Thus, the diffraction from a small object of size s with respect to the wavelength of the radiation λ with few sources of wavelets is a broad asymmetrical peak, giving rise to a range of diffraction angles ($2\delta\Theta$) at which intensity (interference) is observed. This range of angles is equal to the width of the diffracted beam β and hence equal to the full width at half maximum of the observed peak corrected for broadening attributed to the instrumentation and its optical elements.

$$2\delta\Theta = \beta = \frac{\lambda}{s\,\cos\Theta} = \frac{\lambda\,\sec\Theta}{s} \tag{4}$$

Equation (4) is very similar to the often-used Scherrer formula for particle size determinations discussed below. This equation is modified by a constant in the argument to account for the fact that in practice crystallites are not isotropic spheres but may deviate in shape into platelets and needles, leading to anisotropy of the $2\delta\Theta$ values along different directions.

XRD patterns may contain a wealth of nanostructural information in the profile of their diffraction peaks. The diffraction mechanism of a nanostructured sample versus a perfect sample is compared in Figure 8 on the basis of the Ewald construction. It is evident that the intensity distribution of the diffracted beams with respect to the diffraction angle has to be measured with great accuracy (its background included) and must be analyzed with care. The total diffracted intensity over the instrumental and size-dependent background is distributed over a wider range of diffraction angles as the disorder increases, leading to weaker and weaker signals. Weak signals representing small particles are often interpreted as evidence of disordered materials; without a careful intensity calculation, this notion is not justified, as the interference broadening is rarely taken into account. Nevertheless, it is clear from Figure 8 that small and disordered particles may well give such weak lines that great care must be taken to discriminate diffraction from background fluctuations.

The common application of the Scherrer formula in catalyst structure determination is a crude approximation to microstructural analysis. Strain and particle size give rise to the same effects, namely, line broadening, but fortunately causing different variations with diffraction angle, as shown by Equations (3) and (4). The methodologies implied by Williamson–Hall plots and Warren–Averbach profile analyses provide access to the strain and size parameters in the commonly encountered case that both phenomena contribute to an experimental line broadening.

8. NANOSTRUCTURES AND DIFFRACTION

Phenomenologically the term "nanostructure" is often used as a synonym for the term "defect structure." Nanostructure implies an array of microstructural parameters that give rise to various effects that are evident in XRD profiles, as discussed in the previous section. Figure 10 gives a hierarchical representation of the parameters that, taken together, constitute the nanostructural properties of a catalyst. It is obvious that no single analysis technique can address all of these parameters, and hence care must be taken not to over-interpret the results in terms of nanostructure, as only some of the relevant parameters are accessible by XRD.

Figure 10 groups the parameters according to geometry, bulk defects, surface phenomena, and extrinsic modifications. The geometry of a catalyst particle is given by its size, its habitus (meaning the anisotropy or deviation from a spherical shape), and by its pore system. Only for micro- and mesoporous samples is XRD a sensitive tool to determine the pore architecture (Chen et al., 2005; Davidson, 2002; Li and Kim, 2005; Liu et al., 2002; Ohare et al., 1998). In many solids that are more compact than most catalysts, only secondary effects are related to the pores.

The particle habitus is sensitively detected (Nagy et al., 1999) by systematic variations of line intensities with directions in space. Extinctions of lines in platelets and strong deviations of the isotropic intensity distribution in layered structures (Fu et al., 2005; Tarasov et al., 2004; Vistad et al., 2001; Williams and O'Hare, 2006) are frequent observations. Under catalytic reaction conditions it is quite possible that the habitus of a particle may change, in particular if small supported particles are subjected to varying gas atmospheres (Ammer et al., 1997; Andreasen et al., 2005; Bazin et al., 2002; Conte et al., 2006; Grunwaldt et al., 2000). The isotropic size of particles and the size distribution (Aslam et al., 2002; Hills et al., 1999; Konarski et al., 2001; Li and Kim, 2005; Rasmussen et al., 2004; Rodriguez et al., 1996) are among the most frequently analyzed parameters representing catalysts.

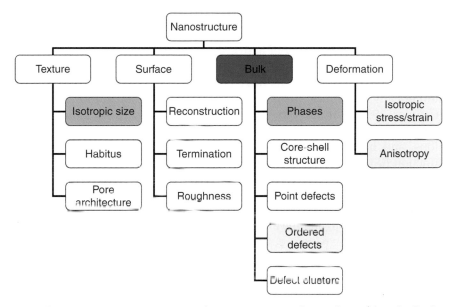

FIGURE 10 Parameters representing the nanostructure of a catalyst. Although all of these parameters can be addressed with specialized diffraction techniques and through sophisticated analysis, there are only a few that are easily accessible by XRD. The normally interpreted bulk information (red) is divided into size and phase information (blue) plus some data characterizing the defect structure (yellow).

The morphological parameters may be compared with size data obtained by electron microscopy (Ackermann et al., 2004; Canton et al., 2002; Chabanier and Guay, 2004; Chang et al., 2004; Conrad, 1992) and gas adsorption (Abd Hamid et al., 2003; Aslam et al., 2002; Clausen, 1998; Kerton et al., 2005; Morcrette et al., 2002; Xu and Mavrikakis, 2003; Zunic et al., 1998) techniques. Systematic deviations in the results obtained with these methods indicate large deviations of particles from the spherical morphology or a distribution of sizes, with a fraction of the particles being within and a fraction being below the detection limit of XRD. In such a case the size parameter s in Equation (4) becomes a probability function of a dimension. If particles are anisotropic, then this function varies with the direction of observation. Because various particle size fractions may exhibit different extrinsic nanostructural parameters (such as different strain induced by metal–support interactions acting strongly on small platelet-shaped particles and not at all on large isotropic particles), it seems highly speculative to try to extract the probability function from a line profile analysis. If such a procedure leads to a distribution function in agreement with that found by electron microscopy or gas adsorption data, then it can be safely concluded that the distribution function of particle morphologies must be small in the given material (i.e., that most particles are of nearly the same shape and size).

As the size distribution of active particles is such a relevant quantity for materials (Alexander and Klug, 1974), including catalysts (Borghard and Boudart, 1983), attempts have been made to derive a procedure for numerically obtaining approximations of the distribution function (Guinier, 1998; Guinier and Griffoul, 1948; Lambert et al., 1962) from a line profile. A line profile corrected for background, support contributions, and instrumental line broadening is represented as an intensity function $I(S)$. The variable S is given by

$$S = \frac{2\sin\Theta}{\lambda} \tag{6}$$

and the desired probability function to find particles with diameters between d and $d + \delta d$ is

$$\frac{P(d)}{d} = \left(\frac{d^2 V(t)}{dt^2}\right)_t \quad \text{with } V(t) = \int I(S)\exp(2\pi/St)dS \tag{7}$$

The function $V(t)$ is proportional to the Fourier transform of the experimentally determined $I(S)$. This method provides realistic distribution functions only if nonstrained particles are present and if the major fraction of particles contributes to the intensity via Bragg diffraction and also if the particles are larger than about 2 nm.

The crudest way of estimating the particle size (D) as an average number from the breadth of a diffraction line is the widely used Scherrer approximation. It may be applied when the instrumental broadening is much smaller than the line profile ($2\Theta > 0.5°$) and when a monomodal size distribution results in a homogeneous line profile. An explicit version of the equation determined by using the breadth of the diffraction line at half height (FWHM, $\beta_{1/2}$) is given as follows:

$$\beta_{1/2} = \frac{K \lambda \; 57.3}{D \cos\Theta} \tag{8}$$

This approximation is sufficient for broad lines and particle diameters <5 nm. The particle size is determined in the direction normal to the lattice plane indicated by the Miller index of the line used for profile determination. The formula featuring a constant accounting for the fact that the breadth is measured in radians (as is 2Θ) is only valid for spherical particles.

If particles become progressively larger, then the influence of their size on the line profile is gradually reduced. In contrast to the sometimes sloppy statements in the literature that "no substantial line broadening" can be detected, it is well possible to determine particle sizes of much larger dimensions than a few nm. For this purpose, however, the definition of a line profile needs to be done more accurately and the corrections in the Scherrer formula for different observation directions and nonspherical shapes need to be considered explicitly. As in many instances TEM data may suggest that catalyst particles are of nonspherical shape, it is useful to make full use of the possibilities of simple XRD-based particle-size determinations during experiments under catalytic reaction conditions.

A suitable description of a peak for advanced Scherrer analysis uses the expression in terms of the centroid and the variance of a peak. For the possible computation of these values, the explicit definitions are given in Equation (9) for the centroid and in Equation (10) for the variance expressed in units of the 2Θ scale. Conventions for a peak profile measured in a step-scan mode are as follows: 2Θ = position of a peak increment, $I(2\Theta)$ = the background corrected intensity in cps for the increment, $\Delta 2\Theta$ = width of the increment:

$$\langle 2\Theta \rangle = \frac{\sum 2\Theta \times I(2\Theta) \times \Delta(2\Theta)}{\sum I(2\Theta) \times \Delta(2\Theta)} \tag{9}$$

$$W = \frac{\sum (2\Theta - \langle 2\Theta \rangle)^2 \times I(2\Theta) \times \Delta(2\Theta)}{\sum I(2\Theta) \times \Delta(2\Theta)} \tag{10}$$

The precise description of the effect of particle size on an infinitely sharp diffraction peak for correctly only cubic materials takes the form of Equation (11) when using Equation (10) for the definition of a peak via its intensity steps (W) and with $\Delta(2\Theta)$ being the range of integration over the peak.

$$W = \frac{K\lambda \, \Delta[2\Theta]}{2\pi D \cos\Theta} \tag{11}$$

Equation (11) uses strongly varying values for the Scherrer constant D, depending on the Miller index of the peak under consideration and on the particle shape. Table 2 is a list of the correction factors from a literature compilation (Alexander and Klug, 1974). The deviations in values are significant in view of the custom in many literature studies of setting the constant equal to unity.

An empirical method that is not related to a rigorous treatment of the convolution of a diffraction profile by size and strain is the Williamson–Hall analysis. This method is suitable for substances characterized by a large number of diffraction peaks and for highly defective samples for which analytical procedures bring upon problems with background definition. The method involves plotting of reciprocal breadth (β^*) (FWHM) in units of the 2Θ scale versus the reciprocal positions (d^*) of all peaks of a phase. The intercept yields the particle size and the slope the "apparent strain" 2η. The required quantities are defined as follows:

TABLE 2   Values of the Scherrer constant for use with Equation (11) and various particle morphologies

Miller indices (hkl)	Particle shape	Scherrer constant
100	Cube	1.000
	Tetrahedron	1.414
	Octahedron	1.651
110	Cube	1.411
	Tetrahedron	1.471
	Octahedron	1.167
111	Cube	1.732
	Tetrahedron	1.802
	Octahedron	1.430
210	Cube	1.342
	Tetrahedron	1.860
	Octahedron	1.477
211	Cube	1.633
	Tetrahedron	1.698
	Octahedron	1.348
221	Cube	1.667
	Tetrahedron	1.732
	Octahedron	1.376
310	Cube	1.265
	Tetrahedron	1.973
	Octahedron	1.566

$$\beta^* = \frac{\beta \, \cos(\Theta)}{\lambda}, \quad d^* = \frac{2 \, \sin(\Theta)}{\lambda} \tag{12}$$

In complicated patterns, groups of peak parameters fitting each a line in the Williamson–Hall diagram may be found. The origin of such groupings is either a strong anisotropy of strain (reported (Compagnini et al., 2001; Nagy et al., 1999; Nepijko et al., 2000; Savaloni et al., 2006) in the silver–oxygen system) or the presence of a phase mixture that was not identified by correct indexing. Similarly, such patterns may be found to represent supported metal samples in which the support-derived peaks are much different in profile from those of the metal particles. In these cases, multiple lines occur in the Williamson–Hall plot.

A principal advantage of all peak deconvolution methods is their applicability, without modification, to both conventional XRD and that characterizing catalysts in reactive atmospheres. Thus, particle morphology data determined in conventional XRD experiments are applicable to

XRD data characterizing the working samples, and the data provide a foundation for understanding the variations of particle sizes and shapes such as occur in processes such as sintering or resulting from changes in reaction environments.

There is a strong motivation for investigation of structural dynamics (Topsøe et al., 1997) of working catalysts by using XRD techniques. The methodology is not as accurate as in-depth EXAFS analysis (Clausen et al., 1998; Grunwaldt and Clausen, 2002; Grunwaldt et al., 2000), but XRD may yield many more observations (including those of laboratory experiments), with characterizations over a broader range of particle sizes. The fundamental limitations stated above regarding single-point morphology analyses also hold for EXAFS analyses.

When XRD is applied to catalysts in reactive atmospheres, however, several pitfalls that are not common in normal XRD can severely affect the results; for example, samples are not spun, so that coarse-grained materials are characterized by broadened and over-structured lines. Furthermore, some materials may move on the sample holder during repeated measurements (particularly in Bragg–Brentano geometries); the resultant changes in the scattering geometry and the movement out of focusing geometries lead to strong modifications of the line profile that have no basis in structure. If the catalyst material consists of larger agglomerates or sinters during operation, then the statistical averaging over grains that is a prerequisite for the existence of a continuous powder diffraction line may no longer be valid. The diffracted intensity leads to a rough pattern arising from various single-crystal spot profiles from the various large crystallites, giving rise to line profiles with several sharp maxima at very close distances. This effect usually strongly modifies the diffraction lines arising from metallic heater strips present in some instruments; these lines cannot serve as internal references because of recrystallization and possible reactions with the gas environment.

A comparison of all the useful methods for determining nanostructure from XRD with the example of silver nanoparticles is given in the literature (Savaloni et al., 2006). The complete and historically correct treatment of the line profiles and the necessary caveats in measuring sufficient data for the treatment is available in Table 1 of the book by Alexander and Klug (1974). This book also includes a critical assessment of the Scherrer analysis and its limitations. A practical example (Günter et al., 2001a–c) of a profile analysis in polycrystalline catalysts deals with the much-studied Cu/ZnO system.

In these and in many other reports, the size analysis is based on a full profile fit determined by using predefined line profiles (Voigt, pseudo-Voigt) and an intensity fitting according to the Rietveldt method (McCusker et al., 1999). This procedure, which requires substantial effort in data analysis, is reliable, provided that no changes in sample geometry occur during the measurements and that no problems arise from a super-position of "amorphous" and crystalline diffraction patterns. In such cases, the less-accurate methods described above are preferred over those determining unrealistically precise values obtained from raw data of insufficient quality.

Alternatively, small angle X-ray scattering (SAXS) may be used for determining particle sizes and their distributions. SAXS is a high-intensity scattering method (Rasmussen et al., 2004; Riello et al., 2001) producing statistically meaningful profiles that can be evaluated in great detail. The inherent disadvantage is the model-dependence of the technique. A model of the solid describing particles and pores in the specimen must be assumed and fitted to the single observation that is the decreasing intensity of the forward-scattered primary beam with increasing scattering angle. Because in many catalysts the matrix or support phase is composed of small crystal-lites with sizes comparable to that of the active phase (Figure 2), it must be expected that both phases contribute to the SAXS signal. An elegant way of avoiding this problem is to scan the wavelength of the X-ray beam so that anomalous small angle scattering makes the signal element-specific. Anom-alous small-angle scattering is a still rarely applied technique, notwithstand-ing the enormous precision of the information it provides compared with other methods (Nagy et al., 1999). By determining the SAXS profile of a catalyst under working conditions at various energies slightly below the absorption edge of the element of interest (this is easy in the case of precious metals supported on oxides of main-group elements), one obtains a scatter-ing function for the metal phase that can be separated from that of the support and from that of the pore system between the particles of interest. Many simplifying assumptions are needed for a viable analysis, such as the assumption that the particles are spherical and homogeneous and that they do not form a dense medium (i.e., that they are isolated).

The basic relationship between the intensity profile $I(s)$ and the parti-cle given by a number of atoms N and number of electrons per atom n and a characteristic radius r is given by the following:

$$I(s) = KNn^2 \exp\left[\frac{-4\pi^2 r^2 s^2}{5}\right] \tag{13}$$

By fitting a large number of such model scattering profiles for particles of different radius r, one obtains a particle size distribution with a high resolution suitable for determining the kinetics of change of the particle size distribution as function of the reaction conditions. This method is highly accurate and transparent, allowing a realistic analysis of the struc-ture and dynamics of supported catalysts under reaction conditions.

The surface structure contributes only limited information to an XRD pattern in standard geometry and to XRD patterns of polycrystalline particles. A measurable contribution comes from surface atoms in the limit of the smallest measurable particles, in which about 200 atoms form the diffracting object. The relative positions of surface atoms with respect to their neighbors and their positions with respect to the bulk equilibrium positions are expected to be different (Suleiman et al., 2003) in truly nano-sized objects (Aslam et al., 2002; Frenkel et al., 2001; Fu et al., 2005; Hills et al., 1999; Roth et al., 2004; Tanori and Pileni, 1995), and hence significant deviations in the diffraction pattern from the expected bulk phase can be detected. The nanostructures may lead not only to modified lattice constants but also to different unit cells and superlattice orderings, as are known to exist from surface crystallography (Ackermann et al., 2004; Narayan, 2005; Saint-Lager et al., 2005). With the present availability of dedicated end stations at synchrotrons, it can be predicted that more surface XRD investigations of catalysts in the functioning state will be reported in the near future—but with the fundamental limitation that the measurements are possible only with homogeneous and very flat model systems (Gunter et al., 1997; Somorjai, 1998) (polycrystalline and single crystal) (Chen, 1996; Ford et al., 2005; Gallagher et al., 2003; Lemire et al., 2004).

The interior volumes of particles are most sensitively probed by XRD. The physical basis for the inherent bulk-sensitivity of the technique is the nature of interference of diffracted waves and the relationship of the X-ray wavelength to the distance of diffracting atoms, as discussed above. The most prominent information about the bulk structure is the average phase composition and the crystallinity. Today, with extensive databases and powerful search-match algorithms, phase identification is fairly reliable, provided that the composition of the sample is known. Phase identifications are complicated by a highly anisotropic habitus of particles and high concentrations of defects or impurities, and these complications are common, leading to significant uncertainties in the identification of complex compounds such as oxides under catalytic reaction conditions. For example, the intercalation into metals of reactant atoms such as carbon, hydrogen, or nitrogen is a prominent cause of phase changes between metal and a carbide, hydride, or nitride, respectively (or mixed phases) (Ackermann et al., 2004; Hills et al., 1999; Jack, 1994; Jack and Jack, 1973; Volkova et al., 2000).

9. PHASES IN WORKING CATALYSTS

In a working catalyst, a variety of phases are often present. Figure 11 discriminates them according to their function. A practical catalyst is frequently in the form of pressed pellets that require some matrix phase along with the powder form of the catalyst. This form frequently consists of a support phase or, in the case of a fused catalyst, of a bulk matrix phase. Dispersed in this matrix are additional phases that are inactive or

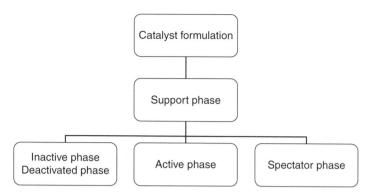

FIGURE 11   Hierarchy of phases in catalysts developed from their function. At the level of the active phase confusion can arise from spectators and unnecessary admixtures or deactivated phases. "Phase cooperation" phenomena are frequently ascribed to beneficial effects these admixtures without much experimental evidence.

deactivated, spectator phases, and the active phase that includes the active sites. It is a key challenge of XRD to assign the various phases to these functions. It is of particular importance to identify the nature of the active phase, which can only be done if a systematic relationship between some measure of the catalyst performance (such as activity) and the abundance of the phase can be found. A suitable method for identification of the active phase is the addition of a catalyst poison to the gas phase flowing over the catalyst that removes the active phase (Busca et al., 1986; Muhler et al., 1990).

A significant problem in phase analysis arises when several phases are present in largely different particle sizes and states of ordering. XRD will then indicate only the phase that is well ordered and occurs as large particles. Besides the well-known case of supported catalysts with a highly dispersed, truly nanosized active phase, such unwanted discrimination can also occur in mixtures of bulk phases. Figure 12 presents an example of hexagonal MoO_3 (Knobl et al., 2003), which is a metastable allotrope of normal orthorhombic MoO_3. The material crystallizes in needles with a well-behaved XRD pattern. When the catalytically most relevant near-surface region of the crystals is characterized by electron diffraction and HRTEM, one finds massive deviations from the XRD information. The figure provides a comparison of the XRD and electron diffraction information, showing no common diffraction at all. The TEM image shows that complex nanostructure with apparently multiple shear structures governs the outermost part of morphologically fully homogeneous needles.

This discrepancy disappears completely when the material is used in oxidation catalysis at temperatures above 723 K, at which a topotactic

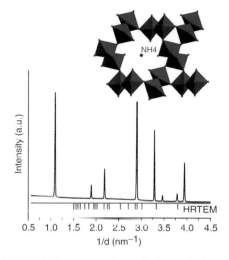

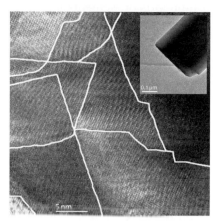

FIGURE 12   Comparison of XRD and electron microscopy for phase analysis of a hexagonal MoO_3 model catalyst. The diffraction information is completely dissimilar despite the well-behaved powder XRD matching perfectly the crystal structure and showing no hints of undetected phases. The needles (SEM inset) are composites of highly crystalline inner parts and nanostructured nonstoichiometric crystallites (main TEM image) possibly precipitating at the end of crystallization but constituting the interface to the reactive gas atmosphere.

phase transformation occurs into orthorhombic MoO_3, with no structural heterogeneity at the crystal edges but with some shear structure detectable throughout the bulk (Ressler et al., 2002, 2003) by EXAFS spectroscopy and XRD. In this homogeneous state the material is a much poorer catalyst than in the heterogeneous state (Abd Hamid et al., 2003).

The active phase is, according to the paradigm of the dynamical catalyst (Ressler et al., 2005; Topsøe et al., 1997; Zunic et al., 1998), not present from the beginning of its life but forms under operating conditions. Thus, kinetics of active phase formation (Enache et al., 2002; Hutchings, 2004; Sankar et al., 1994) may be observed. The progression occurs in gas–solid reactions in fronts throughout solid grains (Maier, 1987, 1995), beginning at the interfaces. In this case, a core-and-shell structure forms (Chui and Chan, 2005; Prakash et al., 2005; Sobal and Giersig, 2005; Wang et al., 2003). Such architectures are the target of synthetic efforts for nanostructured catalysts, as the solid state dynamics allow facile restructuring, and segregation seems to be beneficial to the catalytic function (Chen et al., 2004; Prakash et al., 2005).

Complex morphologies of graded phases that are difficult to obtain by controlled synthesis develop frequently in working catalysts having two compounds or elements in the active phase. Examples are multielement oxides such as Mo–V–W–Ti oxides (Figure 6 and references) in which, according to the chemical potential in the gas phase, more reducing components segregate to the surface (vanadium oxides) and are back-dissolved when the chemical potential becomes more oxidizing (e.g., in air).

The identification of phase grading from XRD alone is not possible; microscopic methods (Chui and Chan, 2005; Dutoit et al., 1995; Hills et al., 1999; Prakash et al., 2005) are needed to determine the locations of the phases that are identified. It is critical to compare the same active state of the catalyst in XRD and in microscopic investigations, as core–shell structures may be the result of phase segregation during sample preparation and transfer from a metastable statistical mix (Conte et al., 2006; Hävecker et al., 2003; Hutchings, 2004; Kiely et al., 1996) into the more stable graded phase mixture (Frenkel et al., 2001; Fu et al., 2005; Hills et al., 1999).

An example of such a graded nanostructure is shown in Figure 13. A model catalyst consisting of nanostructured MoO_3 was prepared by oxidation of MoO_2 nanoparticles with air. This method of physical vapor deposition was used, as highly oxidized molybdenum oxides cannot be evaporated without structural damage. From the TEM image it is apparent that the XRD will only show the well-crystallized MoO_2 nucleus and not even indicate the presence of the disordered halo of oxidized material that incorporates the catalytic function (Wagner et al., 2004).

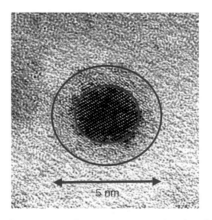

FIGURE 13   Graded oxide nanoparticle. MoO_2 was oxidized with air at 723 K to give a core–shell structure of molybdenum dioxide and possibly molybdenum trioxide that was identified by its different electron energy loss spectrum. No structural description of the highly disordered and catalytically relevant outer oxide shell could be determined, either with XRD or with TEM, (or even with EXAFS spectroscopy) as the signals are dominated by the core structure.

10. APPLICATIONS OF XRD TO WORKING CATALYSTS

XRD experiments can be carried out to characterize gas–solid reactions and, with some limitations, fluid–solid reactions more generally, as long as the fluid contributes little to the pathway of sight for the X-rays. Areas of recent investigation are catalytic gas–solid reactions, electrochemical processes, synthesis procedures involving precipitation and dissolution of solids, temperature-programmed reaction studies of crystallization, and oxidation and reduction of solids. This enumeration covers essentially all phases of the life of a catalyst.

The greatest limitation to the widespread application of XRD of reacting materials is the creation of an appropriate reaction environment that allows the X-rays to penetrate the catalyst and the diffracted X-rays to escape to the detector system. In gas–solid reactions, the greatest challenges are in the provision of the correct gas-phase compositions of the reactants; the addition of condensable components (notably steam and fluid hydrocarbons) presents the challenge of avoiding condensation on the cooling elements of heated cells and on the tubes or pipes in the feed supply and product analysis system.

Substitution of model gases (often diluted hydrogen) to approximate complex gas-phase compositions is an unacceptable simplification for the investigation of phase changes in catalysts, as the relevant details of conditions and of structure depend critically on the exact chemical nature of the reacting environment as outlined above.

A frequent reason for the dependence of catalyst structure on the chemical potential in the gas phase containing all the reactants is the incorporation of molecules or atoms from the reaction mixture into the catalyst phases. Formation of subphases, often only in the near-surface region of the solid, fails to create phases with individual reflections but modifies the reflections of the starting precatalyst phase notably (see previous sections). This complication presents a massive problem in the analysis of working catalysts when significant partial pressures of products are important to the phase formation and when the necessary conversions cannot be reached in the experimental cell. The investigation of ammonia synthesis catalysts when insufficient partial pressures of the product ammonia prevent the formation of the relevant nitride phases is a prominent example of this limitation (Herzog et al., 1996; Walker et al., 1989).

A further challenge is the monitoring of small changes in partial pressures during reaction; these changes may be small because the amount of active material in a transmission experiment or the large dead volume of catalyst (not coming in contact with the gas phase in reflection experiments) may produce only low partial pressures of products (typically in the lower ppm ranges). Besides micro-GC and standard QMS online gas analysis, the technique (Hävecker et al., 2003; Lindinger et al., 1998) of ion–molecule reaction mass spectrometry (PTR–MS) is of significant help in observing reactions leading to medium-sized functionalized hydrocarbon molecules.

In fluid–solid reactors, appropriate contacting of the fluid with the solid is a major challenge in experimental design (Kolb, 2002; Markovic et al., 2000; Roth et al., 2004) for electrocatalytic experiments or processes involving intercalation reactions—the fluid must not weaken the intensity of the X-rays beyond detection limits. Grazing-incidence XRD at single crystals and the application of high-brilliance synchrotron radiation sources are the typical (Ball et al., 2002; Markovic et al., 2000) solutions, but very useful laboratory investigations (Tarasov et al., 2004; Williams and O'Hare, 2006) with lower data quality but sufficient time for observing slow kinetics have also been performed. The choice of the X-ray radiation can be advantageous to some extent: the harder the radiation (the shorter the wavelength), the less interference there is from the reaction environment. The price to be paid is the loss in intrinsic resolution resulting from the contraction of the d-scale with decreasing wavelength. Wavelengths softer than that emitted by iron are not useful except for precision lattice constant determinations or the detection of very small symmetry changes. In these cases, vacuum or helium gas purging are highly recommended for observing useful intensity distributions.

The common copper wavelength is a good compromise between the loss of intensity by absorption from the gas or liquid environment and the precision of the determination of the diffraction pattern that is attainable. The use of hard radiation (e.g., Mo, W, Ag) is advantageous only in limited cases as it helps enormously in the penetration of the X-rays, but it severely limits the resolution of the information essential to phase analysis.

Several investigations have been devoted to the processes of catalyst phase crystallization (Althues and Kaskel, 2002; Brückner, 2003; Kiebach et al., 2005; Vistad et al., 2001). These investigations are characterized by the difficulty that the cell that was used was different from others in affecting the processes of nucleation and growth and may thus have affected not only the kinetics of these processes but also the reaction pathways and even the products. The useful information about the time evolution of a crystalline phase can help greatly in defining reaction parameters such as the reaction time or the pH of a reagent. As phase formation always begins without immediate detection of the nucleation phase (too small particles and/or insufficient long-range order), it is mandatory to complement the diffraction experiments with experiments probing the local ordering, such as NMR or EXAFS spectroscopy.

An XRD experiment can be carried out during functional analysis of a catalyst or during development of a rational synthesis procedure of a precatalyst. Ideally, both aspects of generation and function of a catalyst would be investigated by this method, but such a complete investigation is still lacking.

11. COMBINATION OF XRD DATA WITH AUXILIARY INFORMATION

For functional analysis it is important to know the following information for planning of experiments:

1. The temperature and flow conditions of stable operation of the catalyst;
2. the sensitivity of the kinetics of the catalytic reaction to a change from typical catalytic reactor test conditions to differential reactor test conditions available in XRD ("gaps");
3. the time characteristic of the approach to steady state; and
4. the precatalyst phase composition and a precise determination of its lattice constants for calibration of the experiment.

If this information is available, it is feasible to find explanations of the following typical issues. The precision of the answers will depend greatly on the preexisting knowledge.

The predominant information is the phase inventory during catalyst operation. Besides phase transitions induced by temperature changes, there are typical phase transformations resulting from reactions of the gas phase with the catalyst. The formation of interstitial (Gross et al., 2002; Nagy et al., 1999; van Smaalen et al., 2005; Wienold et al., 2003) compounds (typified by hydrides, carbides, and nitrides), including new phases and solid solutions, or the change in redox state of the catalyst such as metal-oxide transitions are the typical processes encountered in investigations of catalysts under working conditions. This information greatly aids the understanding of the constitution of active sites, as it may allow one to identify rather than to speculate about the active phase of the catalyst. Such information may provide insights into the structural dynamics of active sites that consist of more than a single atom.

Important information emerges from profile analysis (Canton et al., 2002; Günter et al., 2001b; Hartmann et al., 1994; Herzog et al., 1996) of the active phase, including information about particle size, anisotropy, and strain. Evidence of the dispersion of catalytically active phases during operation is essential for evaluation of the catalytic activity. Stability and dynamics of the changes in morphology resulting from sintering can be investigated by XRD. Such data are useful for comparison and calibration with complementary techniques such as gas adsorption (Canton et al., 2002; Fagherazzi et al., 2000; Riello et al., 2001).

12. INSTRUMENTATION

A typical instrument for XRD of a working catalyst consists of a (commercial) powder diffraction instrument, a reaction cell, and the auxiliary equipment for reactant dosing and product analysis. It is desirable to use a computer system that simultaneously controls all these functions or at least records the sample temperature and experiment time together with the diffraction information. A single experiment characterizing a catalyst under working conditions typically produces a large amount of data, the analysis of which consumes much more time than the experiment. Thus, accurate documentation is mandatory and is best supported by automatic routines.

The instrumentation must be constructed to be robust enough to allow continuous operation for many days time on stream without interruption. A typical example is the phase evolution of a partial oxidation catalyst that can easily take 100 h or more under steady-state flow conditions. Computer control of the gas feed and analysis instrumentation is mandatory for stable operation. In cases of nanostructure analysis when profiles of diffraction lines are required, the sheer data acquisition time can well exceed 24 h, during which steady-state operation of the system and the catalyst are mandatory to ensure that meaningful structural parameters are determined.

Another essential consideration is radiation safety. Commercial instruments come with full X-ray radiation safety housings that make it difficult to retrofit piping and cables without loss of the safety certificate. As cells can produce significant stray radiation, care must be taken to properly protect the outside of the diffractometer from such radiation.

The reaction cell must serve the conflicting goals of providing a suitable geometry for diffraction experiments and being a good catalytic reactor. A frequent adaptation is a transmission experiment with a capillary packed with catalyst serving as the reactor and container. This design allows for simultaneous XRD–EXAFS analysis, provides suitable contact between reactant and catalyst, and is conceptually easy to build, in particular, if hot-gas external heating is used as an energy source. The literature (Clausen et al., 1993; Martorana et al., 2003; Palancher et al., 2005; Zunic et al., 1998) provides examples of design and operation of such cells. Because of the strong absorption of many catalysts, their use is most appropriate with synchrotron experiments when sufficient brilliance is available. A key problem with these designs is the tiny volume of catalyst characterized by diffraction relative to the total volume of catalyst contributing to the catalytic activity. Synchrotron X-ray sources usually give much smaller spot sizes than the diameter of the capillary, giving rise to the problem of fluctuating absorption when catalyst grains are moving in the gas stream or when they are undergoing strong restructuring. Not more than one or two grains of material are accounted for in the diffraction analysis, whereas all the grains in the capillary contribute to the catalysis.

An excellent compromise maintaining the small dead volume of the reactor and enhancing the sampling volume of the X-rays for diffraction (Moggridge et al., 1995; Tennakoon et al., 1983; Walker et al., 1989) is found in a cell made from a beryllium body and window with a heater inset capable of being used in Bragg–Brentano reflection geometry. The delicacy of protecting the beryllium body from contact with reactants (almost all of them corrode beryllium) and the difficulties in machining beryllium metal are major drawbacks to this elegant concept.

By far the most popular but least satisfactory compromise is the use of a conventional high-temperature camera fitted with a feed supply and analysis lines. All major instrument vendors sell such equipment, which is characterized by large dead volumes around the sample that can be heated well and kept isothermal. A distant polymer foil separating the cell from air and being sufficiently vacuum- and gas-tight solves the window problem. Figure 14 shows the cell and the internal assembly of the catalyst as a thin powder suspended on a metal strip heated by high-current resistance heating. The large dead volume and the ill-defined gas–solid contact can be estimated. A further drawback to this design is the frequent movement of catalyst particles on the sample holder when the camera is rotated during data acquisition.

A solution to some of these problems is the use of a theta–theta diffractometer (Conte et al., 2006; Kirilenko et al., 2005; Ressler et al., 2005) in which no sample motion occurs but in which the X-ray tube and the detector are rotated around a central cell. Such a cell can be constructed from a ceramic tube holding the catalyst on a frit which is heated externally and which is brought in contact like a fountain from underneath with a gas pipe in the way of a short plug-flow reactor. Here a high-area-X-ray beam analyzes only the top interface between the surrounding gas and the catalyst. This position is also the coldest part of the system, and substantial temperature gradients are possible with a packing of thermally insulating catalysts (often incorporating ceramic supports). Figure 15 shows such an instrument.

None of these cell designs (Palancher et al., 2005; Moggridge et al., 1995) comprises an optimal solution to the XRD technique for characterizing working catalysts. The universality of reflection geometries with commercial cells and laboratory diffractometer facilitates the use of the methods, but the aforementioned drawbacks of all these cells lead to problems with maintaining the geometric integrity of the diffraction

FIGURE 14   A Bühler HTK 1 camera mounted on a STOE diffractometer. The large cooling plate, the substantial surface of heated metal (the central strip serves as a sample holder, length ~5 cm) and the large dead volume preclude an effective gas–solid exchange, lead to poor catalytic performance and thus render difficult the task of detection of products during XRD analysis.

experiments needed for precise data analysis and/or knowledge of the exact sample temperature, which not only affects the catalyst performance but also sensitively modifies the x-axis of the diffractogram, as all crystalline materials exhibit significant thermal lattice expansions. The drastic influence of temperature on peak position is shown in Figure 16, which also exemplifies the effect of a phase transition changing the thermal expansion coefficient. It is obvious that phase identification without proper correction of the diffracted angle scale is erroneous or even false, placing great demands to the precalibration of instrumentation with either reference samples or with a precision low-volume temperature measurement at the X-ray spot.

13. CASE STUDIES

A representative collection of case studies of XRD experiments in catalysis and related fields is summarized in Table 3. Fields related to catalysis are electrochemistry when electrocatalysis is involved, even when energy storage and not chemical conversion is the purpose of the investigation. Other areas concern experiments at elevated pressures aimed at understanding geological issues: these investigations deal with metastable phases, and the transitions are also relevant to nanostructured materials that can exhibit metastable phases which are in many cases bulk high-pressure allotropes of a given compound. Furthermore, few investigations were reported in which the synthesis of the active material was the target of investigation. XRD can be used effectively to characterize the

FIGURE 15   A Paar XRK 900 cell mounted on a STOE theta–theta goniometer. Note the carefully heated piping for the feed. The bottom figure shows the sample holder with a Macor ceramic frit and the gas feed tubing forcing the feed gas through the catalyst bed.

formation of catalytic materials from liquid precursors (in precipitation and polycondensation) or from stable solid precursors (e.g., metals from oxides and nanoparticles from alloys) and thus help provide the basis for design of rational and robust synthesis procedures for catalysts.

The table is organized as follows: Column 2 lists methods used to complement the XRD experiments. It is obvious from the analysis of the literature that most investigations have required (within the same report) the availability of complementary data for the interpretation of the XRD results. By far the most common complementary techniques are EXAFS and XAS to characterize the evolution of short- and long-range ordering simultaneously. The pairing of this very important basic information about the constitution of a catalyst has even led to the construction of a combined synchrotron-based experiment whereby both techniques were used with catalysts in complex reaction atmospheres (Clausen et al., 1993; Dent et al., 1995; Grunwaldt and Clausen, 2002; Sankar et al., 2000).

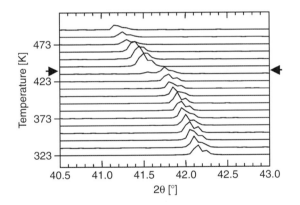

FIGURE 16   Details from a series of diffractograms of a heating series of $RbNO_3$. At 437 K (arrows), a phase transition occurs (a further transition is observed at 497 K). These two fixed points can be used to check the temperature scale of the cell. The slope of peak position versus temperature can also be used to either check the goniometer calibration or control the sample temperature (if the goniometer and the cell are correctly calibrated). For this purpose, hexagonal BN may be used for the thermal expansion, for which an equation such as $c = 6.6516 + 2.74 \times 10^{-4}T$ is commonly used. The experimental schedule for the heating series was as follows: isothermal experiment (heating only between XRD scans to ensure thermal equilibrium): 23 isothermal scans every 10 K between 323 and 543 K ($20-50°$ 2θ, $0.05°$ steps, 5 s/step), 23×53 min $= 20$ h 19 min, wait time (10 min before each scan for equilibration) $23 \times 10 = 3$ h 50 min, heating (298–543 K, 10 K/min) in segments between scans $= 25$ min, total 24 h 34 min. These numbers illustrate the time requirements for equilibrated data collection that can only partly be reduced by using high-brilliance light sources.

TABLE 3 Case studies of powder diffraction experiments under catalytic conditions.

Reference	Methods used to complement XRD	Sample/process	Issues	Postmortem	Characterization in reactive atmosphere	Catalytic reaction atmosphere	Reaction
Ressler et al. (2005)	TG, AS	Heteropolymolybdates	Structural dynamics, defects as active sites	-	+	+	Propane oxidation
Ackermann et al. (2004)	Surface diffraction	Ni (111)	Deactivation by carbide, no reconstruction during catalytic operation	+	+	+	CO hydrogenation to methane
Althues and Kaskel (2002)	TEM, TG–MS	Sulfated zirconia (synthesis of nanoparticles)	Synthesis optimization by crystallization	+	+	n.a.	Butane isomerization
Andreasen et al. (2005)		Mg–Al alloys	Hydrogen release kinetics after storage	+	+	+	Hydrogen storage
Balachandran et al. (1997)	SEM	Perovskites (oxides of Sr, Fe, and Co)	Phase formation in membrane reactors	-	+	+	Oxygen storage for methane-syn gas conversion
Ball et al. (2002)	Surface diffraction	Pd on Pt (111)	Growth from submonolayer to closed films	-	-	n.a.	Electrochemical deposition
Ballivet-Tkatchenko et al. (2006)	NMR, IR	Organostannane complexes	Active phase in a homogeneous reaction	+	-	+	CO_2 fixation to methanol
Belin et al. (2004)	Gas adsorption, IR	Iron oxides	Strain and chemisorption of water	+	+	+	Effect of nanostructuring on water adsorption

(continued)

TABLE 3 (continued)

Reference	Methods used to complement XRD	Sample/process	Issues	Postmortem	Characterization in reactive atmosphere	Catalytic reaction atmosphere	Reaction
Benz et al. (1998)	Mössbauer	Iron oxides	Phase transformation during reduction	+	−	+	Hydrogenation of substituted aromatics
Blagden et al. (2003)		Various species	Crystallization kinetics	+	+	n.a.	Solid phase formation from liquids
Bohne et al. (2004)		Molybdenum oxides, titanium oxides	Phase transformation upon oxidation	+	+	n.a.	Ion implantation, effect of oxygen diffusivity on subsurface phase formation
Breen (1999)	NMR	Bentonite clays	Intercalation of polymers	+	+	n.a.	Intercalation kinetics and ordering
Bueno-Lopez and Garcia-Garcia (2005)	IR	Potassium–carbon	Deactivation by SO_2, phase formation between potassium and sulfur	+	−	+	NO_x reduction by carbon
Carabineiro et al. (2005)	TGA	Vanadium oxides on carbon	Phase transformation during reaction	+	+	+	NO_x reduction by carbon
Carabineiro et al. (2005)	TGA	Manganese oxides on carbon	Phase transformation and melting	+	+	+	NO_x reduction by carbon
Carrell et al. (2002)	XAS, ESR	Manganese complex and cocomplexes in biological photosystem complex	Structure and dynamics of a molecular active site	+	−	+	Water splitting

Reference	Technique	Material	Study				Application
Chabanier and Guay (2004)	Electroanalytics	RuO$_2$, IrO$_2$	Phase formation hydrogen intercalation	+	+	+	Water splitting
Champion et al. (2003)	Dilatometry, TEM	CuO	Reductive sintering of nanoparticles, morphology control	+	+	+	Stoichiometric reduction of oxide
Chupas et al. (2001)	NMR	Fluorination of alumina	Nature of active phase, role of defects	+	+	+	Fluorocarbon dismutation
Ciraolo et al. (1999)	NMR	Zn zeolites	Phase transformation during activation	+	+	+	Dehalogenation of fluorochlorocarbons
Clausen et al. (1991)		Cu/ZnO	Phase formation, nanostructuring	+	+	+	Water gas shift reaction
Clausen et al. (1998)	XAS	Cu/ZnO	Phase formation, nanostructuring short range ordering	+	+	+	Various reactions
Conte et al. (2006)	EPR, Raman	Vanadium phosphorus oxides	Phase transformation, metastable structures	+	+	+	Butane partial oxidation
Cueff et al. (2004)		FeCrAl, Y oxides	Scale formation	+	+	n.a.	High-temperature scale formation, ternary oxides
Davidson (2002)		Mesoporous silica	Structure evolution, role of templates crystallization variables	+	–	n.a.	Synthesis of mesoporous silica with defined properties
Dent et al. (1995)	XAS	Mixed oxide of Cu, Mn, and Co	Mechanism of phase formation of a quaternary oxide	+	+	n.a.	Carbonate-to oxide transformation

(continued)

TABLE 3 (*continued*)

Reference	Methods used to complement XRD	Sample/process	Issues	Postmortem	Characterization in reactive atmosphere	Catalytic reaction atmosphere	Reaction
Diefenbacher et al. (2005)	Optical spectroscopies	Serpentine	Phase stability and transformation	–	+	+	CO_2 sequestration
Dutoit et al. (1995)	IR	Ti–Si xerogels	Phase evolution, synthesis optimization	+	–	n.a.	Mechanism of sol-gel chemistry in mixed oxides
Enache et al. (2002)		Ti, Zr oxide supported Co	Support effects on phase formation, activation protocol optimization	+	+	n.a.	Preparation of Fischer Tropsch catalyst
Epple (1994)	TGA	Various samples	Temperature-programmed phase formation	+	+	n.a.	Method development
Fournier et al. (1992)		Heteropolymolybdates, vanadium substituted	Phase stability, dehydration	+	+	n.a	Partial oxidation of C_3 and C_4 alkanes
Gallagher et al. (2003)	Electroanalysis	Pt_3Sn	Surface reconstruction, structural stability from UHV to fluid, electroadsorption of CO	+	+	+	CO poisoning
Günter et al. (2001)	XAS, TGA	Cu/ZnO	Microstructure of activated copper nanoparticles, detection of strain	+	–	n.a	Methanol synthesis/steam reforming
Günter et al. (2001)	XAS	Cu/ZnO	Phase stability of copper and role of strain, redox behavior	+	+	+	Methanol steam reforming

Reference	Material / Technique	Study				Application
Gross et al. (2002)	NaAlH$_4$	Phase transformation, metal transport	+	⊥	n.a.	Liberation of hydrogen, reversible storage
Gross et al. (2002)	NaAlH$_4$ — SEM	Dynamic and reversible storage	+	+	+	Hydrogen storage cycle
Gross et al. (2002)	TiCl$_3$, NaAlH$_4$	Catalysis of hydrogen storage	+	+	+	Hydrogen storage
Grunwaldt and Clausen (2002)	Cu/ZnO, supported Rh, various supports — XAS	Nonstructural evolution, structure dynamics	+	+	+	Methanol synthesis, water gas shift, various reactions
Grunwaldt et al. (2000)	Cu/ZnO — XAS	Wetting transition during chemical potential change, structural dynamics	+	+	+	Methanol synthesis
Günter et al. (2001)	Cu/ZnO — XAS	Redox behavior during activation, optimization of activation conditions, strain	–	+	+	Methanol synthesis
Hartmann et al. (1994)	Pt/SiO$_2$ — Debye analysis	Structural dynamics during rate oscillations, redox chemistry	+	+	+	CO oxidation
Hata et al. (2003)	Ag/Si — Surface diffraction	Nanostructure of silver particles in a thin film	–	+	n.a.	Cluster synthesis
Schlögl (1991)	Fe	Nature of active phase, microstructure, nitridation	+	⊥	+	Ammonia synthesis
Herzog et al. (1994)	Ag$_x$O$_y$, heteropoly molybdates	Phase transformations, nature of active phase	+	+	+	Partial oxidation of isobutyric acid
Jentoft et al. (2003)	Heteropolymolybdates, vanadium subsituted — IR, UV-vis, XAS	Phase dynamics, structural decomposition	+	+	+	Various partial oxidations and dehydrations

(continued)

TABLE 3 (continued)

Reference	Methods used to complement XRD	Sample/process	Issues	Postmortem	Characterization in reactive atmosphere	Catalytic reaction atmosphere	Reaction
Jung and Thomson (1993)		Iron oxide	Phase transformation active phase	+	+	n.a.	Fischer–Tropsch synthesis
Kaszkur (2001)		Pd/SiO$_2$	Surface restructuring	+	+	n.a.	Adsorption of oxygen, subsurface carbon
Kiebach et al. (2005)	SEM	Organogermanates	Multiple kinetics of nucleation and growth	+	+	n.a.	Solvothermal synthesis
Kim et al. (2004)		CuO, Cu$_2$O	Reduction kinetics, intermediate phases	+	+	+	Reductive activation with hydrogen
Kirilenko et al. (2005)	XAS, TGA	WO$_3$	Phase transformation, redox behavior	+	+	n.a.	Thermolysis of ammonium paratungstate
Klein et al. (1995)	NMR	NaLaY zeolite	Cation migration/ localization of reactant	+	+	+	Xylene isomerzation
Kleitz et al. (2002), Krivoruchko et al. (2001)	TEM, TGA	Zirconium oxophosphate	Mesoporous structuring, template removal	+	+	n.a	Oxidative template removal, phase formation
Knobl et al. (2003)	RAMAN, TGA	Mo$_5$O$_{14}$, vanadium, tungsten substitution	Phase evolution, solidification,	+	–	n.a	Multimethod phase formation
Krivoruchko et al. (2001)		C, iron oxide	Catalytic solid state transformation, melting transitions	+	+	n.a.	Graphitization of amorphous carbon

Reference	Technique	Material		Description			Application
Liu and Ozkan (2005)	XPS	MoO$_x$/Silica titania	+	Phase formation, role of chlorine in phase formation	+	n.a	ODH of ethane, catalyst formation
Liu et al. (2002)	NMR	Mesoporous silica, SBA-1, SBA-3	−	Phase transformation in synthesis	+	n.a.	Phase formation, drying
Ludwiczek et al. (1978)		Fe	−	Real structure, paracrystallinity	−	n.a.	Role of defects in bulk catalysts
Lupo et al. (2004)		ZrO$_2$	−	Phase formation, hydrothermal conditions in autoclave	+	n.a.	Phase formation
Markovic et al. (2000)	Surface diffraction, electrochemistry	Pd on Pt (111)	+	Nanostructure, overpotential, subsurface hydride	+	+	Hydrogen evolution during water splitting
Marosi et al. (2000)		Heteropolymolybdates, vanadium substitution	+	Phase stability, intercalation of substrate	−	+	Partial oxidation of methacrolein
Martorana et al. (2003)		Pt/ceria/zirconia	+	Dynamics of oxygen storage, lattice distortions	+	+	CO oxidation in storage mode of TWC
Martorana et al. (2004)		Pt/ceria/zirconia	+	Kinetics of hysteresis of metal/nonmetal mediated reduction of storage oxide	+	+	CO oxidation storage mode of TWC
Mikulova et al. (2006)	Raman	Zr–Ce–Pr (Nd)–O$_x$	+	Kinetics of reduction, effect of additives, lattice dynamics	+	+	CO oxidation storage mode of TWC
Moggridge et al. (1992)		Mn$_x$O$_y$	+	Phase evolution, effect of structural promoters	+	−	Methane coupling

(continued)

TABLE 3 (continued)

Reference	Methods used to complement XRD	Sample/process	Issues	Postmortem	Characterization in reactive atmosphere	Catalytic reaction atmosphere	Reaction
Moggridge et al. (1995)	XAS Conversion electron detection	Various systems	Bulk vs. surface phase evolution	+	+	–	Various
Morcrette et al. (2002)	XAS Various other	Nb phosphate oxide LiCo oxide	Intercalation, reversibility	+	+	n.a.	Lithium batteries
Morozova et al. (1997)	XAS	NiO, iron oxides	Structure–function correlations, morphology of precursor versus active phase formation	+	+	+	CO hydrogenation
Muhler et al. (1990)	XAS, TEM	Iron oxide Potassium and other promoters	Structural evolution, role of poison, redox behavior	+	+	+	Dehydrogenation of ethylbenzene
Nagy et al. (1999)	SEM, ISS	Ag–O	Restructuring during bulk dissolution	+	+	+	Methanol oxidation
Narayan (2005)	TEM	ZnO–sapphire Ge–Si systems	Epitaxy and other lattice matching concepts	+	+	n.a.	Thin film growth, nanostructuring
Nix et al. (1987)	TEM	Cu-rare earth alloys	Phase formation, hydride formation	+	+	+	Methanol synthesis
O'Mahony et al. (2004)	XPS	Vanadium oxides, phosphoric acid	Phase formation redox behavior	+	+	n.a.	Catalyst synthesis, role of reducing solvent for VPP
O'Mahony et al. (2003)	SEM/EDX	VHPO$_4$	Crystallization kinetics intermediate phases	+	+	n.a.	Catalyst synthesis, VPP
Ohare et al. (1998)		Aluminum phosphates Gallium fluorophosphates	Formation of mesoporous structure, intermediate phases	–	+	n.a.	Synthesis optimization

Reference	Technique	Material/System	Topic				Application
Okada et al. (2004)		C, water, MgO	Phase changes under extreme conditions	+	+	+	Catalytic solid–solid transformation, diamond synthesis
Okada et al. (2002)		C, water, MgO	Phase changes under extreme conditions	+	+	+	Catalytic solid–solid transformation, diamond synthesis
Ovsitser et al. (2002)	TEM, SEM	(MoVW)$_5$O$_{14}$	Activation, defect formation, structure–function correlations	+	−	+	Acrolein oxidation
Palancher et al. (2005)		Sr–X zeolite	Phase deformation	−	+	−	Dehydration of zeolite
Peluso et al. (2003)	DRIFTS	Mn–O systems	Structure function correlation	+	+	+	Total oxidation of ethanol
Pickering et al. (1992)		Li–Ni–oxide	Synthesis conditions Phase evolution	+	+	+	Methane coupling
Pinheiro et al. (2002)		C, Co-alumina	Phase growth Phase distinction between CNT and whiskers	+	+	+	Carbon nanotube growth
Rasmussen et al. (2004)	SAX	Ni–alumina	Sintering mechanism	+	−	+	Distinction between coalescence and atomic migration
Rayment et al. (1985)		Fe	Crystalline versus amorphous state	+	−	−	Ammonia synthesis

(continued)

TABLE 3 (*continued*)

Reference	Methods used to complement XRD	Sample/process	Issues	Postmortem	Characterization in reactive atmosphere	Catalytic reaction atmosphere	Reaction
Ressler et al. (2005)	XAS	Heteropolymolybdate, Nb substituted	Phase stability role of promoter	+	+	+	Propene oxidation
Ressler et al. (2005)	XAS	Heteropolymolybdate, Nb substituted	Phase stability role of promoter	+	+	+	Propene oxidation
Ressler et al. (2001)	XAS	MoO_3	Reduction mechanism	+	+	–	Bronze formation
Ressler et al. (2003)	XAS	MoO_3	Intermediate phases role of defects	+	+	+	Propene oxidation
Ressler et al. (2002)	XAS	MoO_2	Oxidation mechanism comparison to MoO_3	+	+	+	Propene oxidation
Richardson et al. (2004)	SEM	Ni-alumina	Sintering role of structural promoters	+	+	+	Methane steam reforming
Richter and Doppler (1997)		Nickel and copper on supports	Phase evolution, amorphous co-phases	+	–	+	Reductive activation, various reactions
Saint-Lager et al. (2005)		Pd_8Ni_{92} alloy	Dynamics of reconstruction and segregation	+	+	+	Butadiene hydrogenation
Sanchez-Valente et al. (2000)	Mössbauer	$Mg(OH)_2$, Fe hydrotalcite	Phase formation active phase after fluid phase reduction	+	–	+	Catalyst formation
Sankar et al. (1991)	XAS	$Ni-TiO_2$	Metal support interaction	+	+	n.a.	Catalyst formation
Sankar et al. (2000)	XAS	Various metal-microporous material combinations	Phase formation	+	–	–	Various reactions

Reference	Technique	Sample	Description			Application
Schnell and Fuess (1996)		Pt in NaX zeolite	Phase formation, modification of hcs lattice, location contro by pretreatments	+	n.a.	Autoreduction from ammonium complex
Shashkin et al. (1991)		Multiple promoted iron molybdates	Phase formation Active phase identification	+	+	Partial oxidation of isobutylene
Shashkin et al. (1993)	IR	Vanadium molybdates	Identification of hex MoO$_3$ as active phase	+	+	C$_3$, C$_4$ olefin partial oxidation
Shaw et al. (1991)		CeCu$_2$	Phase formatior identification of Cu nanoparticles	+	+	Methanol synthesis
Shmakov et al. (1995)	TEM	Gamma-iron oxide	Vacancy ordering	−	n.a.	Evolution of real structure
Suleiman et al. (2003)		Pd	Lattice dynamics role of morphology of nanoparticles	+	+	Hydrogen storage, loading and unloading
Stierle et al. (2005)	Surface diffraction	Pd	Phase forma ion during oxidation at various pressures	+	+	Metal-to-oxide transformation
Tarasov et al. (2004)		Alumina LDH	Kinetics of Li insertion, role of counterions in aqueous solution	+	n.a.	Li intercalation and its reverse
Tennakoon et al. (1983)	IR	Clays, montmorillonite	Dehydration, intercalatior, phase stability	+	+	Alcohol intercalation at elevated pressures
(Thomas 1997)	XAS	Various samples	phase formation, oxidation states of active sites	+	−	Various reactions
Utsumi et al. (2004)		C, various metal samples	Phase transformation	+	n.a.	Catalytic diamond synthesis
Valtchev and Bozhilov (2004)	TEM, DLS	Faujasite, aluminosilicate gel	Nucleation growth kinetic and mechanism	+	+	Solid formation mechanism

(continued)

TABLE 3 (continued)

Reference	Methods used to complement XRD	Sample/process	Issues	Postmortem	Characterization in reactive atmosphere	Catalytic reaction atmosphere	Reaction
van Smaalen et al. (2005)		Vanadyl phosphates	Thermally induced phase transition, intercalation of oxygen	+	+	−	Thermal load of butane oxidation catalyst
Vistad et al. (2001)		Silicon–aluminum phosphates, SAPO 34	Phase formation role of thermal treatment	+	+	−	Solid formation mechanism, intermediate phases during synthesis of mesoporous solids
Volkova et al. (1998)	TEM, TGA	CuCoO$_2$	Reduction, carbide formation active phases	+	+	+	CO hydrogenation to higher alcohols
Volkova et al. (2000)		Co$_2$C, CuCo alloy	Reduction, carbide formation active phases	+	+	+	CO hydrogenation to higher alcohols
Walker et al. (1989)		Ce–Ru, Ce–Co, Ce–Fe	Nanostructuring, phase segregation	+	+	+	Ammonia synthesis on alloy catalysts
Walters and Scogin (2004)		NaAlH$_4$	Reversible decomposition, role of intermediates and of catalysts	+	+	+	Hydrogen storage
Widjaja et al. (1999)	TEM, XAS	Pd–SnO$_x$ promoters	Reduction of PdO, Pd-support interaction	+	+	+	Low-temperature methane combustion

Reference	Technique	Material	Study			Application
Wienold et al. (2003)	XAS, TGA	Ammonium heptamolybdate	Phase formation ligand removal, sub-oxides	+	+	Propene oxidation, oxidation and reduction of oxide
Wienold et al. (2003)	XAS, TGA	Heteropolymolybdates, vanadium substitution, Cs salts	Phase stability, phase transformation	+	+	Propene oxidation
Williams and O'Hare (2006)	various	Layered double hydroxides (LDH)	Hydration, intercalation	+	−	Various, base catalysis
Yureva et al. (1995)		Cu–ZnO	Phase formation, reduction	+	+	Methanol synthesis
Zunic et al. (1998)	XAS	Various oxides	Phase transformation during steaming, redox behavior	+	+	Catalyst phase formation, relationship to geology

A particular methodology is surface diffraction (Ball et al., 2002; Stierle et al., 2005) of functioning catalysts. This method is limited to adsorbates and ultra-thin layer samples of single-crystal quality. This experimentally demanding technique allows investigation of the structural evolution of catalytically relevant surfaces in cases when the stability of sample complexity precludes the application of quantitative low-energy electron diffraction (which is the standard method of choice for surface-sensitive structure determination). The nature of high-energy X-rays, with penetration depths from many microns to centimeters of solid matter, precludes surface sensitivity. Undesired contributions from beneath the surface are excluded under the conditions of grazing incidence diffraction GIX (Hata et al., 2003; Koga and Takeo, 1996). This technique requires atomically flat samples, a near-perfect alignment of sample and X-ray beam, and, in practice, a synchrotron source when a cell containing a catalyst in a reactive atmosphere is to be used. Besides structural information at the level of cell parameters and atom positions, detailed information about surface termination, roughness, and the strain state of the surface can also be obtained from profile analysis (which, however, requires specialized techniques; Stierle et al., 2005).

An intermediate case (Bazin et al., 2002) between bulk and surface diffraction is reached for nanoparticles when the contribution from surface atoms becomes significant and diffraction analysis in the limit of infinite periodic lattice models inadequately describes the diffraction data. A case study with diamond nanoparticles (Palosz et al., 2002) describes elegantly the possibilities and limitations of diffraction analysis of such samples; there is a focus on the nonperiodic structure such as strain and disorder induced by the dominant presence of a nonideal surface termination.

In column 3 of Table 3, the material under investigation is listed. A wide variety of samples have been investigated, with little preference for any family of materials. This observation underlines the universal character of XRD, which is characterized by few material and process limitations. The same is valid for the principal issues motivating the experiments.

Column 4 of Table 3 is a list of a wide span of issues addressed in the case studies.

One critical point is emphasized here. In many investigations, one finds interpretations of the relevance of the results for the determination of structure determination of the catalytically active site. This approach is guided by the single-crystal approach in which it is implicitly assumed that all sites on a given surface contribute to the catalytic function. In recent years it has become clear, however, that often only a small fraction of sites on a given surface are catalytically active. Hence, it is highly speculative to use the averaging information of diffraction data to draw uncritical conclusions about the structure of the active site.

A prominent example of this issue (Carrell et al., 2002) is the elucidation of the structure of the active site in biological water-splitting catalysts whereby X-ray analysis has provided an opportunity for controversial and conflicting pictures of the active site that were only resolved by the combination of XRD with spectroscopic experiments. Diffraction analysis of "active sites" is only applicable when a large number of active sites contribute to the termination of the active catalyst and when all these sites are characterized by sufficient long-range ordering. Bulk diffraction is by and large a poorly sensitive method for identifying high-energy nonequilibrium sites in a material. The best use of the method for this purpose is still to determine the average metastability of a sample by its real structure and try to correlate this parameter with catalytic function. The dream of some authors (Bañares, 2005; Brückner, 2003; Weckhuysen, 2002) to arrive by XRD at quantitative structure–function correlations can hardly be fulfilled by investigation of the average bulk structure—but may emerge indirectly by correlation of defect properties with performance. Important as the detailed structural information of a functional material clearly is, it is essential that this information be used with care for elucidating the active sites. XRD of functioning catalysts should not be overestimated in this respect, as it frequently has been in the reports collected in Table 3.

In column 5 of this table, results are collected regarding the availability of structural information in the deactivated state of the sample. Such information is invaluable to contrast the active state with the less active or deactivated equilibrium state of the system. Notwithstanding the available information, such a comparison is only rarely given in the literature.

It is usually considered sufficient to report the structural modifications before and after reaction experiments. The ample evidence of the differences provides an overwhelming justification for applying XRD to functioning catalysts and not relying simply on XRD of the fresh catalyst, as has been done in the vast majority of catalytic investigations with XRD.

Column 6 of Table 3 is a list of diffraction experiments carried out with catalysts under working conditions and the simultaneous determination of the conversion of the feed. Not all of the investigations claiming catalytic relevance have included such experiments; in some cases only comparisons of the structure before and after reaction were reported. The isolation and removal of the catalyst from the reactor may well affect the structure of a reactive material.

Column 7 of Table 3 is a summary of the application of a reactive atmosphere. Such atmospheres were not always applied, particularly not when catalyst formation and phase transformations were the subjects of interest or when formation or activation of a catalyst were investigated. In a few cases, "proxy" atmospheres were applied, usually to circumvent complex analytical efforts. This practice is not recommended, because each feed plus product mixture will have a different effect on the structure of a catalyst. For example, either H_2 or CO is a poor substitute for hydrocarbons, and inert gases cannot replace water vapor as a diluent in reactive atmospheres. These substitutions are among the more common ones that may give rise to substantial deviations from structural details that occur under reaction conditions.

Column 8 of Table 3 is a list of the reactions. In addition to the wide range of catalytic transformations with gas-phase reactants, the list includes a representative selection of noncatalytic gas–solid reactions characterized by phase transformations combined with redox reactions of the solid. Iron catalysts were investigated in the presence of a liquid environment, and the results show that there is no fundamental problem with the use of X-rays penetrating a thin film of a liquid electrolyte or a solvent.

In summary, the compilation of relevant case studies shows that XRD of working catalysts is a widely applicable technique. It gives rich and useful information about synthesis and activation of catalysts as well as deactivation by structural transformations. The pertinent question about the structure of the active sites is not accessible directly by this method despite such claims in the literature. It must be pointed out that this shortcoming of a technique involving characterization of samples in reactive atmospheres is common to all methods when one is concerned with high-performance catalysts in which the active sites are a small fraction of the active surface. Model systems do a better job in this respect, provided that they are active for the reaction of interest and not only in proxy reactions.

14. SUMMARY AND CONCLUSIONS

XRD is a useful yet seldom applied technique for characterization of solid catalysts in the functioning state. Its merits emerge from the unambiguous determination of phases, their dynamics, and their relevant nanostructures under operating conditions. This information should be the basis of every attempt to determine the structure and function of a catalytic material. Speculations about structure and function are much more frequent than XRD investigations of the catalysts under working conditions, leaving many open questions about the nature of active phases in solid catalysts.

A common hesitancy regarding the application of this method stems from the inherent bulk sensitivity of XRD. The physics of diffraction does not improve the sensitivity of the method in cases of nanostructured and disordered samples. These inherent disadvantages in many cases are not compensated by EXAFS spectroscopy, the method that is usually considered to be complementary to XRD; EXAFS spectroscopy, besides being characterized by the same inherent bulk sensitivity as XRD, has the disadvantage of providing much more indirect structural information beyond the unambiguous elucidation of local coordination.

In view of the substantial efforts required for structural investigations of functioning catalysts, XRD and EXAFS spectroscopy should routinely be conducted as tandem experiments. Only then is the maximum of information limited only by the physics of the experiments. The arguments about bulk sensitivity are invalid for micro- and mesoporous samples and for truly nanostructured supported catalysts (with dimensions <2 nm), for which the numbers of bulk and surface atoms become comparable. In the many other cases of catalysts, the knowledge of the active phase, its lattice constants, strain state, and the possible formation of a solid solution with reactants are invaluable for any reasoning about the active state. Furthermore, the bulk nanostructure controls the evolution and dynamics of the surface structure through the gradient of surface free energy to the bulk lattice energy and through the defects that may become the activation centers for structural dynamics and surface reordering.

Inherent advantages of XRD for catalyst characterization include its applicability at high pressures and the potential for minimization of radiation damage and interference from beam chemistry by choosing wavelengths with minimum interaction cross section when a synchrotron is used as light source. To ensure that these advantages are met, a range of instrumentation and radiation sources must be available. The concept of maximum brilliance for maximum diffracted intensity demanding synchrotron sources is not helpful in the characterization of many samples; optimized detectors and extended experiment times under steady-state conditions are generally recommended. At many existing synchrotron sources in the world, excellent opportunities exist for catalysis-related powder diffraction experiments with samples in reactive atmospheres, but these are usually strongly inhibited by the current practice of beam time allocation policies and partly by the lack of suitable end stations allowing realistic chemical reactions to be carried out and monitored. In the near future, the advent of ultra-high-brilliance radiation sources such as free electron lasers (at SLAC, XFEL) and much-improved synchrotron sources (NSLS II, PETRA III) will open new possibilities for time-resolved XRD allowing not only the time-resolved evolution of structural dynamics but elucidation of details of nanostructure of a working catalyst without beam damage as consequence of the short exposure times. Extremely high-quality diffraction patterns and EXAFS data acquired in sub-microsecond time regimes should provide new insights into unperturbed structural dynamics of catalysts.

XRD in dedicated laboratory environments provides and will provide a wealth of useful information, because they allow operation with few time constraints and unsurpassed experimental flexibility. Synchrotron sources, on the other hand, provide unique opportunities to combine XRD with EXAFS spectroscopy, which together provide an enormous advantage as long as the conduct of the experiments is not constrained by beam time limitations. Bulk transformations under reaction conditions are typically slow, being characterized by time scales of hundreds of hours, and are therefore prohibitive for standard user operations at synchrotrons. It is also difficult to handle many catalytically relevant reactants safely at synchrotrons, so that XRD investigations are limited to a few reactants, in contrast to the situation in most catalyst characterization laboratories.

XRD could play a much greater role in providing fundamental understanding of catalysts than it does today, if only the community could agree to operate a few competence centers at synchrotron facilities with dedicated end stations and appropriate beam-time allocation policies. Furthermore, catalyst characterization centers should all operate laboratory XRD instruments for investigation of catalysts in reactive atmospheres. These could become much more popular if equipment manufacturers would provide more suitable infrastructure for experimentation. More user demand could trigger this evolution, because all the necessary technology is available, although not in a system-integrated form.

The scientific basis for widespread use of XRD in catalysis science is well developed, as is the understanding of the possibilities and limitations. The understanding of the relevance of bulk/nanostructural details for the control of surface structuring is, albeit clearly established, not yet prominently communicated in the community. The debate about methodology for characterization of catalysts in reactive atmospheres versus during catalytic reaction with analysis of the products is by and large not helpful to this communication issue. The growing need to understand catalyst synthesis and operation will drive the evolution of many techniques for characterization of solid catalysts as they function, and, among these techniques, XRD is a powerful method providing fundamental information about the nature and constitution of the catalytically active materials.

REFERENCES

Abd Hamid, S.B., Othman, D., Abdullah, N., Timpe, O., Knobl, S., Niemeyer, D., Wagner, J., Su, D., and Schlögl, R., *Top. Catal.* **24**, 87 (2003).

Ackermann, M., Robach, O., Walker, C., Quiros, C., Isern, H., and Ferrer, S., *Surf. Sci.* **557**, 21 (2004).

Alexander, L.E., and Klug, H.P., "X-ray diffraction procedures." 2nd edn. (reprint) ed., Wiley, New York, 1974.

Alstrup, I., Chorkendorff, I., and Ullmann, S., *J. Catal.* **168**, 217 (1997).

Althues, H., and Kaskel, S., *Langmuir* **518**, 201 (2002).

Ammer, C., Meinel, K., Wolter, H., and Neddermeyer, H., *Surf. Sci.* **377–379**, 81 (1997).

Andersson, M., Jansson, K., and Nygren, M., *Catal. Lett.* **39**, 253 (1996).

Andreasen, A., Sorensen, M.B., Burkarl, R., Möller, B., Molenbroek, A.M., Pedersen, A.S., Andreasen, J.W., Nielsen, M.M., and Jensen, T.R., *J. Alloy Comp.* **404**, 323 (2005).

Aslam, M., Gopakumar, G., Shoba, T.L., Mulla, I.S., Vijayamohanan, K., Kulkarni, S.K., Urban, J., and Vogel, W., *J. Colloid Interface Sci.* **255**, 79 (2002).

Balachandran, U., Dusek, J.T., Maiya, P.S., Ma, B., Mieville, R.L., Kleefisch, M.S., and Udovich, C.A., *Catal. Today* **36**, 265 (1997).

Ballivet-Tkatchenko, D., Chambrey, S., Keiski, R., Ligabue, R., Plasseraud, F., Richard, P., and Turunen, H., *Catal. Today* **115**, 80 (2006).

Ball, M.J., Lucas, C.A., Markovic, N.M., Stamenkovic, V., and Ross, P.N., *Surf. Sci.* **518**, 201 (2002).

Banares, M.A., *Catal. Today* **100**, 71 (2005).

Bazin, D., Guczi, L., and Lynch, J., *Appl. Catal. a* **226**, 87 (2002).

Belin, T., Millot, N., Villieras, F., Bertrand, O., and Bellat, J.P., *J. Phys. Chem. B* **108**, 5333 (2004).

Benz, M., van der Kraan, A.M., and Prins, R., *Appl. Catal. A* **172**, 149 (1998).

Blagden, N., Davey, R., Song, M., Quayle, M., Clark, S., Taylor, D., and Nield, A., *Cryst. Growth Des.* **3**, 197 (2003).

Bohne, Y., Shevchenko, N., Prokert, F., von Borany, J., Rauschenbach, B., and Möller, W., *Vacuum*, **76**, 281–285 (2004).

Borghard, W.S., and Boudart, M., *J. Catal.* **80**, 194–206 (1983).

Breen, C., *Appl. Clay Sci.* **15**, 187 (1999).

Brückner, A., *Catal. Rev.-Sci. Eng.* **45**, 97 (2003).

Bueno-Lopez, A. and Garcia-Garcia, A., *Energy & Fuels* **19**, 94 (2005).

Busca, G., Cavani, F., Centi, G., and Trifiro, F., *J. Catal.* **99**, 400 (1986).

Canton, P., Fagherazzi, G., Battagliarin, M., Menegazzo, F., Pinna, F., and Pernicone, N., *Langmuir* **18**, 6530 (2002).

Carabineiro, S.A., Fernandes, F.B., Vital, J.S., Ramos, A.M., and Fonseca, I.M., *Appl. Catal. B* **59**, 181 (2005).

Carabineiro, S.A., Silva, I.F., Klimkiewicz, M., and Eser, S., *Mater. Corr.* **50**, 689 (1999).

Carrell, T.G., Tyryshkin, A.M., and Dismukes, G.C., *J. Biol. Inorg. Chem.* **7**, 2 (2002).

Chabanier, C., and Guay, D., *J. Electroanal. Chem.* **570**, 13 (2004).

Champion, Y., Bernard, F., Guigue-Millot, N., and Perriat, P., *Mater. Sci. Eng. A* **360**, 258 (2003).

Chang, H.Y., Cheng, S.Y., and Sheu, C.I., *Acta Mater.* **52**, 5389 (2004).

Chen, H.D., *Mater. Chem. Phys.* **43**, 116 (1996).

Chen, H.R., Gao, J.H., Ruan, M.L., Shi, J.L., and Yan, D.S., *Micropor. Mesopor. Mater.* **76**, 209 (2004).

Chen, H.R., Wang, X.M., Shia, J.L., Xiao, P., and Yan, D.S., *J. Mater. Res.* **20**, 42 (2005).

Chen, L.F., Norena, L.E., Navarrete, J., and Wang, J.A., *Mater. Chem. Phys.* **97**, 236 (2006).

Chui, Y.H., and Chan, K.Y., *Chem. Phys. Lett.* **408**, 49 (2005).

Chupas, P.J., Ciraolo, M.F., Hanson, J.C., and Grey, C.P., *J. Am. Chem. Soc.* **123**, 1694 (2001).

Ciraolo, M.F., Norby, P., Hanson, J.C., Corbin, D.R., and Grey, C.P., *J. Phys. Chem. B* **103**, 346 (1999). *Fresenius J. Anal. Chem.* **349**, 247 (1994).

Clausen, B.S., *Catal. Today* **39**, 293 (1998).

Clausen, B.S., Grabaek, L., Steffensen, G., Hansen, P.L., and Topsøe, H., *Catal. Lett.* **20**, 23 (1993).

Clausen, B.S., Grabaek, L., Steffensen, G., Hansen, P.L., and Topsøe, H., *Catal. Lett.* **20**, 23 (1993).

Clausen, B.S., Steffensen, G., Fabius, B., Villadsen, J., Feidenhansl, R., and Topsøe, H., *J. Catal. V* **132**, 524 (1991).

Clausen, B.S., Topsøe, H., and Frahm, R., *Advan. Catal.*, **42**, (1998), 315–344.

Compagnini, G., Fragala, M.E., D'Urso, L., Spinella, C., and Puglisi, O., *J Mater. Res.* **16**, 2934 (2001).

Conrad, E.H., *Prog. Surf. Sci.* **39**, 65 (1992).

Conte, M., Budroni, G., Bartley, J.K., Taylor, S.H., Carley, A.F., Schmidt, A., Murphy, D.M., Girgsdies, F., Ressler, T., Schlögl, R., and Hutchings, G.J., *Science* **313**, 1270 (2006).

Corma, A., Moliner, M., Serra, J.M., Serna, P., Diaz-Cabanas, M.J., and Baumes, L.A., *Chem. Mater.* **18**, 3287 (2006).

Cueff, R., Buscail, H., Caudron, E., Riffard, F., Issartel, C., Perrier, S., and El Messki, *J. Phys. IV* **118**, 307 (2004).

Davidson, A., *Curr. Opin. Colloid Interface Sci.* **7**, 92 (2002).

Dent, A.J., Oversluizen, M., Greaves, G.N., Roberts, M.A., Sankar, G., Catlow, C.R.A., and Thomas, J.M., *Physica B* **209**, 253 (1995).

Diebold, U., *Surf. Sci. Rep.* **48**, 53 (2003).

Diefenbacher, J., McKelvy, M., Chizmeshya, A.V.G., and Wolf, G.H., *Rev. Sci. Instrum.* **76**, 15103 (2005).

Dieterle, M., Mestl, G., Jäger, J., Uchida, Y., Hibst, H., and Schlögl, R., *J. Mol. Catal. A* **174**, 169 (2001).

Dutoit, D.C.M., Schneider, M., and Baiker, A., *J. Catal.* **153**, 165 (1995).

Enache, D.I., Rebours, B., Roy-Auberger, M., and Revel, R., *J. Catal.* **205**, 346 (2002).

Epple, M., *J. Thermal Anal.* **42**, 559 (1994).

Ertl, G., Huber, M., and Thiele, N., *Z. Naturforsch.* **34a**, 30 (1978).

Fagherazzi, G., Canton, P., Riello, P., Pernicone, N., Pinna, F., and Battagliarin, A., *Langmuir* **16**, 4539 (2000).

Ford, D.C., Xu, Y., and Mavrikakis, M., *Surf. Sci.* **587**, 159 (2005).

Fournier, M., Feumijantou, C., Rabia, C., Herve, G., and Launay, S., *J. Mater. Chem.* **2**, 971 (1992).

Frenkel, A.I., Hills, C.W., and Nuzzo, R.G., *J. Phys. Chem. B* **105**, 12689 (2001).

Fujii, Y., Komai, T., and Ikeda, K., *Surf. Interface Anal.* **37**, 190 (2005).

Fu, W.Y., Yang, H.B., Chang, L.X., Li, M.H., Bala, H., Yu, Q.J., and Zou, G.T., *Colloids Surf. A* **262**, 71 (2005).

Gallagher, M.E., Lucas, C.A., Stamenkovic, V., Markovic, N.M., and Ross, P.N., *Surf. Sci.* **544**, L729 (2003).

Greeley, J., and Mavrikakis, M., *J. Phys. Chem. B* **109**, 3460 (2005).

Greeley, J., Nørskov, J.K., and Mavrikakis, M., *Ann. Rev. Phys. Chem.* **53**, 319 (2002).

Gross, K.J., Sandrock, G., and Thomas, G.J., *J. Alloy. Comp.* **330**, 691 (2002).

Gross, K.J., Thomas, G.J., and Jensen, C.M., *J. Alloy. Comp.* **330**, 386 (2002).

Grunwaldt, J.D., and Clausen, B.S., *Top. Catal.* **18**, 37 (2002).

Grunwaldt, J.D., Molenbroek, A.M., Topsøe, N.Y., Topsøe, H., and Clausen, B.S., *J. Catal.* **194**, 452 (2000).

Guinier, A., *Cryst. Res. Technol.* **33**, 543 (1998).

Guinier, A., and Griffoul, R., *Acta Crystallogr.* **1**, 188 (1948).

Günter, M.M., Bems, B., Schlögl, R., and Ressler, T., *J. Synchrotron Radiat.* **8**, 619 (2001a).

Günter, M.M., Ressler, T., Bems, B., Büscher, C., Genger, T., Hinrichsen, O., Muhler, M.H., and Schlögl, R., *Catal. Lett.* **71**, 37 (2001b).

Günter, M.M., Ressler, T., Jentoft, R.E., and Bems, B., *J. Catal.* **203**, 133 (2001c).

Günter, M.M., Ressler, T., Jentoft, R.E., and Bems, B., *J. Catal.* **203**, 133 (2001d).

Gunter, P.L.J., Niemandsverdriet, J.W.H., Ribeiro, F.H., and Somorjai, G., *Catal. Rev.-Sci. Eng.* **39**, 77 (1997).

Hartmann, N., Imbihl, R., and Vogel, W., *Catal. Lett.* **28**, 373 (1994).

Hata, A., Akimoto, K., Horii, S., Emoto, T., Ichimiya, A., Tajiri, H., Takahashi, T., Sugiyama, H., Zhang, X., and Kawata, H., *Surf. Rev. Lett.* **10**, 431 (2003).

Hävecker, M., Mayer, R.W., Knop-Gericke, A., Bluhm, H., Kleimenov, E., Liskowski, A., Su, S., Follath, R., Requejo, F.G., Ogletree, D.F., Salmeron, M., Lopez-Sanchez, J.A., et al., *J. Phys. Chem. B* **107**, 4587 (2003).

Havrilla, G.J., and Miller, T.C., *Rev. Sci. Instrum.* **76**, 0602201–1 (2005).

Herein, D.J., Find, B., Herzog, H., Kollmann, R., Schmidt, R., Schlögl, R.T., and Timpe, *"On the Relation between Structure and Reactivity in the Carbon Oxygen Reaction."* ACS Symposium Series, (1996).

Herzog, B., Herein, D., and Schlögl, R., *Appl. Catal. A-Gen.* **141**, 71 (1996).

Herzog, B., Ilkenhans, T., and Schlögl, R., *Fresenius J. Anal. Chem.* **349**, 247 (1994).

Hills, C.W., Nashner, M.S., Frenkel, A.I., Shapley, J.R., and Nuzzo, R.G., *Langmuir* **15**, 690 (1999).

Hosemann, R., Freisinger, A., and Vogel, W., *Ber. Bunsenges.* **70**, 797 (1966).

Hutchings, G.J., *J. Mater. Chem.* **14**, 3385 (2004).

Jack, D.H., and Jack, K.H., *Mater. Sci. Eng.* **11**, 1 (1973).

Jack, K.H., *J. Appl. Phys.* **76**, 6620 (1994).

Jansen, R., Brabers, V.A.M., and van Kempen, H., *Surf. Sci.* **328**, 237 (1995).

Jentoft, F.C., Klokishner, S., Kröhnert, J., Melsheimer, J., Ressler, T., Timpe, O., Wienold, J., and Schlögl, R., *Appl. Catal. A-Gen.* **256**, 291 (2003).

Jung, H., and Thomson, W.J., *J. Catal.* **139**, 375 (1993).

Kaszkur, Z. in *Epdic 7: European Powder Diffraction*, Pts 1 and 2, Vol. 378–383, pp. 314-319 (2001).

Kerton, O.J., McMorn, P., Bethell, D., King, F., Hancock, F., Burrows, A., Klely Ellwood, S., and Hutchings, G., *Phys. Chem. Phys.* **7**, 2671 (2005).

Kiebach, R., Schäfer, M., Porsch, F., and Bensch, W., *Z. Anorg. Allg. Chem.* **631**, 369 (2005).

Kiely, C.J., Burrows, A., Hutchings, G.J., Bere, K.E., Volta, J.C., Tuel, A., and Abon, M., *Faraday Discuss.* **105**, 103 (1996).

Kirilenko, O., Girgsdies, F., Jentoft, R.E., and Ressler, T., *Eur. J. Inorg. Chem.* 2124 (2005).

Kleitz, F., Thomson, S.J., Liu, Z., Terasaki, O., and Schüth, F., *Chem. Mater.* **14**, 4134 (2002).

Klein, H., Fuess, H., and Hunger, M., *J. Chem. Soc. Faraday Trans.* **91**, 1813 (1995).

Kim, J. Y., Hanson, J. C., Frenkel, A I., Lee, P. L., Rodriguez, J. A., *J. Phy. Cond. Matt.* **16**, S3479–S3484 (2004).

Knobl, S., Zenkovets, G.A., Kryukova, G.N., Ovsitser, O., Niemeyer, D., Schlögl, R., and Mestl, G., *J. Catal.* **215**, 177 (2003).

Koga, K., and Takeo, H., *Rev. Sci. Instrum.* **67**, 4092 (1996).

Kolb, D.M., *Surf. Sci.* **500**, 722 (2002).

Konarski, P., Iwanejko, I., Mierzejewska, A., and Diduszko, R., *Vacuum* **63**, 679 (2001).

Krivoruchko, O.P., Shmakov, A.N., and Zaikovskii, V.I., *Nucl. Instrum. Methods Phys. Res. Sect. A* **470**, 198 (2001).

Lambert, M., Lefebvre, S., and Guinier, A., *C. R. Hebd. Seances Acad. Sci.* **255**, 97 (1962).

Langford, J.I., and Louer, D., *Rep. Progr. Phys.* **59**, 131 (1996).

Lemire, C., Meyer, R., Henrich, V.E., Shaikhutdinov, S., and Freund, H.J., *Surf. Sci.* **572**, 103 (2004).

Lindinger, W., Hansel, A., and Jordan, A., *Chem. Soc. Rev.* **27**, 347 (1998).

Liu, C., and Ozkan, U.S., *J. Phys. Chem. A* **109**, 1260 (2005).

Liu, J., Yang, Q.H., Kapoor, M.P., Setoyama, N., Inagaki, S., Yang, J., and Zhang, L., *J. Phys. Chem. B* **109**, 12250 (2005).

Liu, M.C., Sheu, H.S., and Cheng, S.F., *Chem. Comm.* 2854 (2002).

Li, Y.Z., and Kim, S.J., *J. Phys. Chem. B* **109**, 12309 (2005).

Ludwiczek, H., Preisinger, A., Fischer, A., Hosemann, R., Schönfeld, A., and Vogel, W., *J. Catal.* **51**, 326 (1978).

Lundgren, E., Mikkelsen, A., Andersen, J.N., Kresse, G., Schmid, M., and Varga, P., *J. Phys.-Condens. Matter* **18**, R481 (2006).

Lupo, F., Cockcroft, J.K., Barnes, P., Stukas, P., Vickers, M., Norman, C., and Bradshaw, H., *Phys. Chem. Phys.* **6**, 1837 (2004).

Maier, J., *J. Electrochem. Soc.* **134**, 1524 (1987).

Maier, J., *Prog. Solid State Chem.* **23**, 171 (1995).

Markovic, N.M., Lucas, C.A., Climent, V., Stamenkovic, V., and Ross, P.N., *Surf. Sci.* **465**, 103 (2000).

Marosi, L., Cox, G., Tenten, A., and Hibst, H., *Catal. Lett.* **67**, 193 (2000).

Martorana, A., Deganello, G., Longo, A., Deganello, F., Liotta, L., Macaluso, A., Pantaleo, G., Balerna, A., Meneghini, C., and Mobilio, S., *J. Synchrotron Radiat.* **10**, 177 (2003).

Martorana, A., Deganello, G., Longo, A., Prestianni, A., Liotta, Macaluso, Pantaleo, Balerna, and Mobillo, *J. Solid State Chem.* **177**, 1268 (2004).

McCusker, L.B., Von Dreele, R.B., Cox, D.E., Louer, D., and Scardi, P., *J. Appl. Crystallogr.* **32**, 36 (1999).

Mikulova, J., Rossignol, S., Gerard, F., Mesnard, D., Kappenstein, C., and Duprez, D., *J. Solid State Chem.* **179**, 2511 (2006).

Moggridge, G.D., Rayment, T., and Lambert, R.M., *J. Catal.* **134**, 242 (1992).

Moggridge, G.D., Schroeder, S.L.M., Lambert, R.M., and Rayment, T., *Nucl. Instrum. Methods Phys. Res. Sect. B* **97**, 28 (1995).

Morcrette, M., Chabre, Y., Vaughan, G., Amatucci, G., Leriche, J.B., Patoux, S., Masquelier, C., and Tarascon, J.M., *Electrochim. Acta* **47**, 3137 (2002).

Morozova, O.S., Krylov, O.V., Kryukova, G.N., and Plyasova, L.M., *Catal. Today* **33**, 323 (1997).

Muhler, M.H., Schütze, J., Wesemann, M., Rayment, T., Dent, A., Schlögl, R., and Ertl, G., *J. Catal.* **126**, 339 (1990).

Nagy, A.J., Mestl, G., Herein, D., Weinberg, G., Kitzelmann, E., and Schlögl, F., *J. Catal.* **182**, 417 (1999).

Narayan, J., *Metall. Mater. Trans. A* **36A**, 277 (2005).

Nepijko, S.A., Ievlev, D.N., Schulze, W., Urban, J., and Ertl, G., *Chemphyschem* **1**, 140 (2000).

Nix, R.M., Rayment, T., Lambert, R.M., Jennings, J.R., and Owen, G., *J. Catal.* **106**, 216 (1987).

Nørskov, J.K., Scheffler, M., and Toulhoat, H., *MRS Bull.* **31**, 669 (2006).

O'Mahony, L., Curtin, T., Zemlyanov, D., Mihov, M., and Hodnett, B.K., *J. Catal.* **227**, 270 (2004).

O'Mahony, L., Henry, J., Sutton, D., Curtin, T., and Hodnett, B.K., *Catal. Lett.* **90**, 171 (2003).

Ohare, D., Evans, J.S.O., Francis, R.J., Halasyamani, P.S., Norby, P., and Hanson, J., *Micropor. Mesopor. Mat.* **21**, 253 (1998).

Okada, T., Utsumi, W., Kaneko, H., Turkevich, V., Hamaya, N., and Shimomura, O., *Phys. Chem. Miner.* **31**, 261 (2004).

Okada, T., Utsumi, W., Kaneko, H., Yamakata, M., and Shimomura, O., *Phys. Chem. Miner.* **29**, 439 (2002).

Ovsitser, O., Uchida, Y., Mestl, G., Weinberg, G., Blume, A., and Schlögl, R., *J. Mol. Catal. A* **185**, 291 (2002).

Palancher, H., Pichon, C., Rebours, B., Hodeau, J.L., Lynch, J., Berar, F., Prevot, S., Conan, G., and Bouchard, C., *J. Appl. Crystallogr.* **38**, 370 (2005).

Palosz, B., Grzanka, E., Gierlotka, S., Stel'makh, S., Pielaszek, R., Bismayer, U., Neuefeind, J., Weber, H.P., Proffen, T., Von Dreele, R., and Palosz, W., *Z. Kristallogr.* **217**, 497 (2002).

Peluso, M.A., Sambeth, J.E., and Thomas, H.J., *React. Kinet. Catal. Lett.* **80**, 241 (2003).

Pickering, I.J., Maddox, P.J., and Thomas, J.M., *Chem. Mater.* **4**, 994 (1992).

Pinheiro, J.P., Schouler, M.C., Dooryhee, E., *Solid State Commun.* **123**, 161 (2002).

Prakash, A., McCormick, A.V., and Zachariah, M.R., *Nano Lett.* **5**, 1357 (2005).

Rasmussen, F.B., Sehested, J., Teunissen, H.T., Molenbroek, A.M., and Clausen, B.S., *Appl. Catal. A* **267**, 165 (2004).

Rayment, T., Schlögl, R., Thomas, J.M., and Ertl, G., *Nature* **315**, 311 (1985).

Ressler, T., Timpe, O., and Girgsdies, F., *Z. Kristallogr.* **220**, 295 (2005).

Ressler, T., Timpe, O., Girgsdies, F., Wienold, J. and Neisius, T., *J. Catal.* **231**, 279 (2005).

Ressler, T., Wienold, J., and Jentoft, R.E., *Solid State Ionics* **141**, 243 (2001).

Ressler, T., Wienold, J., Jentoft, R.E., and Girgsdies, F., *Eur. J. Inorg. Chem.* 301 (2003).

Ressler, T., Wienold, J., Jentoft, R.E., and Neisius, T., *J. Catal.* **210**, 67 (2002).

Reuter, K., and Scheffler, M., *Phys. Rev. B* **65** (2002).

Richardson, J.T., Scates, R.M., and Twigg, M.V., *Appl. Catal. A* **267**, 35 (2004).

Richter, K. and Doppler, P., *Solid State Ionics* **101**, 687 (1997).

Riello, P., Polizzi, S., Fagherazzi, G., Finotto, T., and Ceresara, S., *Phys. Chem. Phys.* **3**, 3213 (2001).

Rodriguez, A., Amiens, C., Chaudret, B., Casanove, M.J., Lecante, P., and Bradley, J.S., *Chem. Mater.* **8**, 1978 (1996).

Roth, C., Martz, N., and Fuess, H., *J. New Mater. Electrochem. Syst.* **7**, 117 (2004).

Saint-Lager, M.C., Jugnet, Y., Dolle, P., Piccolo, L., Baudoing-Savois, R., Bertolini, J.C., Bailly, A., Robach, O., Walker, C., and Ferrer, S., *Surf. Sci.* **587**, 229 (2005).

Sanchez-Valente, J., Millet, J.M.M., Figueras, F. and Fournes, L., *Hyperfine Interact.* **131**, 43 (2000).

Sankar, G., Rao, C.N.R., and Rayment, T., *J. Mater. Chem.* **1**, 299 (1991).

Sankar, G., Rey, F., Thomas, J.M., Greaves, G.N., Corma, A., Dobson, B.R., and Dent, A.J., *JCS Chem. Comm.* 2279 (1994).

Sankar, G., Thomas, J.M., and Catlow, C.R.A., *Top. Catal.* **10**, 255 (2000).

Savaloni, H., Gholipour-Shahraki, M., and Player, M.A., *J. Phys. D* **39**, 2231 (2006).

Schaefer, D.W., and Keefer, K.D., *Phys. Rev. Lett.* **56**, 2199 (1986).

Schlögl, R., in *"Ammonia Synthesis."* (J. R. Jennings, Ed.), p. 19. Plenum Press, New York, (1991).

Schlögl, R., Baerns, M., in "Basic Principles in Applied Catalysis", (A.W. Castleman, J.P. Toennies, F.P. Schüfer, and W. Zinth, Eds.), pp. 321–360. Springer Verlag, Berlin Heidelberg New York, 2004.

Schnell, R., and Fuess, H., *Ber. Bunsen Ges. Phys. Chem. Phys.* **100**, 578 (1996).

Shashkin, D.A., Shiryaev, P.A., Kutyrev, M.Y., and Krylov, O.V., *Kinet. Catal.* **34**, 302 (1993).

Shashkin, D.P., Maksimov, Y.V., Shiryaev, P.A., Matveev, V.V., *Kinet. Catal.* **32**, 1079 (1991).

Shaw, E.A., Rayment, T., Walker, A.P., Lambert, R.M., Gauntlett, T., Oldman, R.J., and Dent, A., *Catal. Today* **9**, 197 (1991).

Shmakov, A.N., Kryukova, G.N., Tsybulya, S.V., Chuvilin, A.L., and Solovyeva, L.P., *J. Appl. Cryst.* **28**, 141 (1995).

Sobal, N.S., Giersig, M., *Aust. J. Chem.* **58**, 307 (2005).

Somorjai, G.A., *MRS Bulletin* **23**, 11 (1998).

Stampfl, C., Ganduglia-Pirovano, M.V., Reuter, K., and Scheffler, M., *Surf. Sci.* **500**, 368 (2002).

Stierle, A., Kasper, N., Dosch, H., Lundgren, E., Gustafson, J., Mikkelsen, A., and Andersen, J.N., *J. Chem. Phys.* **122**, 44706 (2005).

Suleiman, M., Jisrawi, N.M., Dankert, O., Reetz, M.T., Bahtz, C., Kirchheim, R., and Pundt, A., *J. Alloys Compounds* **356**, 644 (2003).

Tanori, J., and Pileni, M.P., *Adv. Mater.* **7**, 862 (1995).

Tarasov, K.A., Isupov, V.P., Chupakhina, L.E., and O'Hare, D., *J. Mater. Chem.* **14**, 1443 (2004).

Tennakoon, D.T.B., Schlögl, R., Rayment, T., Klinowski, J., Jones, W., and Thomas, J.M., *Clay Miner.* **18**, 357 (1983).

Thomas, J.M., *Chem. Eur. J.* **3**, 1557 (1997).

Thomas, J.M., and Sankar, G., *J. Synchrotron Radiat.* **8**, 55 (2001).

Topsøe, H., Dumesic, J.A., and Boudart, M., *J. Catal.* **28**, 477 (1973).

Topsøe, K., Ovesen, C.V., Clausen, B.S., Topsøe, N.Y., Nielsen, P.E.H., Tornqvist, E., and Nørskov, J.K., *Stud. Surf. Sci. Catal.* **109**, 121 (1997).

Tschaufeser, P., and Parker, S.C., *J. Phys. Chem.* **99**, 10609 (1995).

Urban, J., Sack-Kongehl, H., Weiss, K., Lisiecki, I., and Pileni, M.P., *Cryst. Res. Technol.* **35**, 731 (2000).

Utsumi, W., Okada, T., Taniguchi, T., Funakoshi, K., Kikegawa, T., Hamaya, N., and Shimomura, O., *J. Phys. Condens. Matter* **16**, S1017 (2004).

Valtchev, V.P., and Bozhilov, K.N., *J. Phys. Chem. B* **108**, 15587 (2004).

van Smaalen, S., Dinnebier, R., Hanson, J., Gollwitzer, J., Bullesfeld, F., Prokofiev, A., and Assmus, W., *J. Solid State Chem.* **178**, 2225 (2005).

Vistad, O.B., Akporiaye, D.E., and Lillerud, K.P., *J. Phys. Chem. B* **105**, 12437–12447 (2001).

Vogel, W., Dhayagude, D., Chitra, R., Sen, D., Mazumadar, S., Urban, J., and Kulkarni, S.K., *J. Mater. Sci.* **37**, 4545–4554 (2002).

Vogel, W., Duff, D.G., and Baiker, A., *Langmuir* **11**, 401 (1995).

Volkova, G.G., Plyasova, L.M., Kriger, T.A., Zaikovskii, V.I., and Yureva, T.M., *Kinet. Catal.* **39**, 706 (1998).

Volkova, G.G., Yurieva, T.M., Plyasova, L.M., Naumova, M.I., and Zaikovskii, V.I., *J. Mol. Catal. A* **158**, 389 (2000).

Wagner, J.B., Abd Hamid, Othman, D., Timpe, O., Knop-Gericke, A., Niemeyer, D., Su, D.S., and Schlögl, R., *J. Catal.* **225**, 78 (2004).

Walker, A.P., Rayment, T., and Lambert, R.M., *J. Catal.* **117**, 102 (1989).

Walters, R.T., and Scogin, J.H., *J. Alloy. Comp.* **379**, 135 (2004).

Wang, D., Penner, S., Su, D.S., Rupprechter, G., Hayek, K., and Schlögl, R., *J. Catal.* **219**, 434 (2003).

Wang, Z.H., Choi, C.J., Kim, J.C., Kim, B.K., and Zhang, Z.D., *Mater. Lett.* **57**, 3560 (2003).

Weckhuysen, B.M., *Chem. Comm.* 97 (2002).

Weckhuysen, B.M., *Phys. Chem. Phys.* **5**, 4351 (2003).

Widjaja, H., Sekizawa, K., and Eguchi, K., *Bull. Chem. Soc. Japan* **72**, 313 (1999).

Wienold, J., Jentoft, R.E., and Ressler, T., *Eur. J. Inorg. Chem.* 1058 (2003).

Wienold, J., Timpe, O., Ressler, T., *Chem.-a Eur. J.* **9**, 6007–6017 (2003).

Williams, G.R., and O'Hare, D., *J. Mater. Chem.* **16**, 3065 (2006).

Wyckoff, R.W.G., and Crittenden, E.D., *J. Am. Chem. Soc.* **47**, 2866 (1925).

Xu, Y., and Mavrikakis, M., *J. Phys. Chem. B* **107**, 9298 (2003).

Yureva, T.M., Plyasova, L.M., Kriger, T.A., and Makarova, O.V., *Kinet. Catal.* **36**, 707 (1995).

Zunic, T.B., Steffensen, G., and Villadsen, J., *in* "EPDIC 5, Pts 1 and 2, Materialsl Science Forum." R. Delhez, and E.J. Mittemeier, Eds.), Vol. 278–2, Transtec Publ., Zürich, p. 270. 1998.

CHAPTER 6

Characterization of Catalysts in Reactive Atmospheres by X-ray Absorption Spectroscopy

Simon R. Bare* and Thorsten Ressler[†]

Abstract

X-ray absorption spectroscopy (XAS) is a powerful method for probing the average local electronic and geometric structures of catalysts in the working state. Element-specific data can be obtained over a wide range of temperatures (room temperature to >1000 K), and pressures (from subambient to well more than 100 bar). Because the specimen can be investigated under such a wide range of conditions, XAS has become one of the most frequently applied techniques for structural characterization of working catalysts. Often there is no other way to obtain the structural information provided by this technique that allows one to develop specific structure-activity relationships in catalysis. XAS is applicable to a broad range of elemental concentrations (from tens of ppm to wt% levels), and therefore is used to characterize both high-surface-area supported catalysts and bulk catalysts (e.g., oxides). This review is focused on the application of XAS to the investigation of catalysts in the working state. The designs of the experimental reaction cells, which are essential for the XAS analysis of catalysts, are comprehensively reviewed. Specific examples illustrate the application of the technique to the investigation of catalysts at both steady-state and under dynamic conditions. The examples are chosen to demonstrate the wide variety of catalysts that can be investigated. Although XAS has been used for almost 40 years to characterize catalysts in the working state, methodology, equipment, and applications are still being advanced. Some recent major developments (as a result of

* UOP LLC, a Honeywell Company, Des Plaines, IL 60016, USA
† Technische Universität Berlin, Fachgruppe Anorganische und Analytische Chemie, Institut für Chemie, D-10623 Berlin, Germany

Advances in Catalysis, Volume 52
ISSN 0360-0564, DOI: 10.1016/S0360-0564(08)00006-0

improvements in optics of the X-ray beam lines, new synchrotron radiation sources, advanced detectors, and software) are therefore highlighted.

Contents

1. Introduction   342
   1.1. Previous Reviews   345
   1.2. Scope and Outline   346
   1.3. X-Ray Absorption Spectroscopy   347
2. Importance of XAFS Spectroscopy to Catalyst
   Characterization in Reactive Atmospheres   349
   2.1. Introduction   349
   2.2. Structure Determination   349
   2.3. Transformation of One Species to Another   355
   2.4. Oxidation State   359
   2.5. Supported Metal Cluster Size and Shape   363
   2.6. Electronic Structure   365
3. Catalytic Reactors that Serve as Cells   369
   3.1. Introduction   369
   3.2. Cells as Reactors   373
   3.3. Summary   400
4. XAFS Spectroscopy of Samples in Reactive
   Atmospheres Under Static Conditions   404
   4.1. Introduction   404
   4.2. Hydrogenation   406
   4.3. Gold Cluster Catalysts for CO Oxidation   411
   4.4. Other Catalytic Reactions   418
5. XAFS Spectra Measured Under Dynamic Conditions   428
   5.1. Introduction   428
   5.2. Instrumentation for TR-XAFS Spectroscopy   429
   5.3. Cells for TR-XAS Investigations   431
   5.4. Analysis of TR-XAFS Data   431
   5.5. Applications of TR-XAFS Spectroscopy   432
   5.6. Correlation between Average Metal Valence and
        Catalytic Selectivity   434
   5.7. Solid-State Kinetics of Reduction and
        Reoxidation of MoO_3 With Propene and O_2 438
   5.8. Concluding Remarks regarding Time-Resolved
        Modes   443
6. Look to the Future   446
7. Concluding Remarks   456
Acknowledgment   456
References   456

ABBREVIATIONS

BM	bending magnet
CCD	charge coupled device
CPO	catalytic partial oxidation
CS	crystallographic shear
DTA	differential thermal analysis
DXAFS	dispersive X-ray absorption fine structure
EXAFS	extended X-ray absorption fine structure
FT	Fourier transform
GC	gas chromatography
HERFD	high energy resolution fluorescence detection
ID	insertion device
IR	infrared
PCA	principal component analysis
PEEK	poly-ether ether ketone
QEXAFS	quick extended X-ray absorption fine structure
RIXS	resonant inelastic X-ray scattering
SAXS	small angle X-ray scattering
SCR	selective catalytic reduction
SEXAFS	surface extended X-ray absorption fine structure
S/N	signal/noise
STEM	scanning transmission electron microscopy
SWNT	single wall nanotubes
TEM	transmission electron microscopy
TG	thermogravimetry
TOF	turn over frequency
TPR	temperature programmed reduction
TR	time resolved
UHV	ultra-high vacuum
UV	ultra-violet
UV-vis	ultra-violet visible
WAXS	wide angle X-ray scattering
XAS	X-ray absorption spectroscopy
XAFS	X-ray absorption fine structure
XANES	X-ray absorption near edge structure
XRD	X-ray diffraction
N	degeneracy of a scattering path
C-N	coordination number
R	the half path length. For a single scattering path, this is the bond length between the absorber and backscatterer atoms

$\Delta\sigma^2$ sigma squared factor, is the mean-square displacement of
          the half path length and represents the stiffness of the
          bond for a single scattering path
ΔE_0 inner potential correction
S_0^2 passive electron reduction factor

1. INTRODUCTION

The catalytic properties of a material are determined by its composition
and structure. These properties then define its physical and electronic
structure. It is clear that both the physical and electronic properties may
change when the catalyst is in the working state. Thus, the ideal characterization of a catalyst involves measurement of these properties during the
catalytic reaction. One goal of catalyst characterization science is then to
measure these properties as the catalyst works and to relate these properties to the catalytic activity and selectivity. If these properties are
measured, and their relationship with catalytic activity is understood,
then knowledge will be obtained that should allow the researcher to
design these properties into the catalyst. In other words, if the structure
of the working catalyst is probed by using some physical or chemical
technique and some physical property of the catalyst is found to change
or scale directly with catalyst performance, then insight is gained about
the active site. The next step then is to understand that specific physical
property, and the factors that control it. Once these pieces of information
are known, it should then be possible to design the desired properties into
the material, and thereby develop an improved version of the catalyst, a
new catalyst for a new process, or whatever the research objective calls for.
    Clearly there are many methods that can be used to probe a catalyst's
structure under reaction conditions, as documented in other chapters in
this series (*Advances in Catalysis*, Volumes 50 and 51, and this volume). In
this chapter we focus on X-ray absorption spectroscopy (XAS), including
X-ray absorption near edge structure (XANES) and extended X-ray
absorption fine structure (EXAFS) spectroscopy, sometimes simply
referred to as X-ray absorption fine structure (XAFS) spectroscopy. In
this review the term XAFS will be used generically, but EXAFS and
XANES will be used when the information is specifically related to the
extended or near edge structure, respectively.
    XAFS spectroscopy provides element-specific information about the
local chemistry and physical structure of the element under investigation.
XANES provides information about the chemical state of the element,
including the oxidation state, and sometimes the local geometry (via
selection rules), and EXAFS provides quantitative information about the

distance from the absorbing atom to neighboring atoms and the coordination number and type of the neighboring atoms.

A major reason why XAFS spectroscopy has become a critically useful probe of catalyst structure is the fact that it is easily adapted to characterization of samples in reactive atmospheres. The X-ray photons are sufficiently penetrating that absorption by the reaction medium is minimal. Moreover, the use of X-ray- transparent windows on the catalytic reaction cell allows the structure of the catalyst to be probed at reaction temperature and pressure. For example, the catalyst may be in a reaction cell, with feed flowing over it, and normal online analytical tools (gas chromatography, residual gas analysis, Fourier transform (FT) infrared spectroscopy, or others) can be used to monitor the products while at the same time the interaction of the X-rays with the catalyst can be used to determine critical information about the electronic and geometric structure of the catalyst.

The modern era of EXAFS spectroscopy began in 1971 with the pioneering paper by Sayers et al (1971) in which they elucidated the fundamental physics of the technique. It was only three years after publication of this paper that the first EXAFS paper mentioning the application to catalysis was published (Lytle et al., 1974). From that date until today, the growth in the use of EXAFS for the understanding of catalyst structure has been impressive. An indication of the historical trend of the number of publications concerning EXAFS and catalysis (identified by using a SciFinder search of EXAFS and Catalysis) is shown in Figure 1. There was steady growth in the number of publications during the 1980s and 1990s, and currently the average number of publications is approximately 150 per year.

Before some of the reasons for this rapid growth are discussed, it is worthwhile revisiting history, as it is often easy to forget some of the important observations and predictions from decades ago.

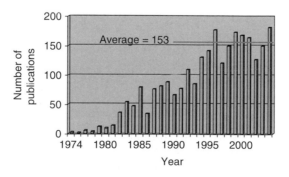

FIGURE 1 Number of publications per year identified by using the search terms of "EXAFS" and "Catalysis" in SciFinder.

Before the advent of modern EXAFS spectroscopy, and well before the use of synchrotron radiation as the X-ray source for the technique, Robert van Nordstrand published a paper in 1960 entitled "The Use of X-ray K-edges in the Study of Catalytically Active Solids" (Van Nordstrand, 1960). He made the following statements in this forward- thinking paper: "... for study of the chemistry of catalysts and other noncrystalline systems this technique XAS may have a role comparable to that of X-ray and electron diffraction in crystalline systems." How true those words have become, as will be highlighted in this review. He further added, "... some features making X-ray *k*-edge spectroscopy especially attractive for catalyst investigations are: remarkable sensitivity to the chemical condition of the catalyst ..." and "... *the ability of X-rays to penetrate matter, such as reactor windows, catalyst supports and catalyst deposits.*" This remarkable "chemical sensitivity" combined with the ease of using the technique with samples in reactive atmospheres has led to the marked growth rate in the application to catalysis, and now such investigations have become relatively common.

In one of the first papers on the application of XAFS spectroscopy to catalysis (Lytle et al., 1974) is the statement: "... these results demonstrate that the EXAFS technique can be a powerful tool for studying catalysis in order to determine the precise structural relationship between catalytically active sites and the surrounding atoms." It is exactly this precise structural relationship that is the critical kind of information needed if true structure-property relationships are to be developed.

The continuous growth in the popularity of XAFS spectroscopy in catalyst characterization science is a result of several factors. First among these, of course, has to be the belief that the information determined from the technique is truly useful and provides understanding, and therefore worthwhile. The other factors relate to the ability to collect and analyze the XAFS data. The number of synchrotron radiation sources, and the corresponding availability of spectroscopy beam lines, have increased dramatically over the last decade or so, and more synchrotron radiation sources are being built. In 2006 alone, new sources were under construction in the United Kingdom (Diamond), France (Soleil), Spain (ALBA), and elsewhere (http://www.lightsources.org/cms/). Moreover, there are plans for the construction of yet more sources (e.g., NSLS-II in Brookhaven, USA, and PETRA-III in Hamburg, Germany). It appears that the saturation point in the number of facilities and beam lines has not yet been reached—as more sources and beam lines are built, more catalysis researchers come to the facilities and conduct XAFS measurements.

In tandem with the increased number of synchrotron radiation sources, and therefore increased beamtime, is the increased reliability and stability of the sources themselves, and the ease of operation of the beam lines. These advances have enabled researchers to be able to plan and conduct the necessary experiments. Coupled with the increased ability to collect the XAFS data have been significant advances in user-friendly data analysis software and the associated theoretical codes. There are many XAFS data analysis packages now available for both processing and modeling the experimental data. (e.g., Athena and Artemis (Ravel and Newville, 2005), which are an interface to IFEFFIT (Newville, 2001),WinXAS (http://www.winxas.de/), EXAFSPAK (http://www-ssrl.slac.stanford.edu/exafspak.html), SixPack (http://www-ssrl.slac.stanford.edu/~swebb/sixpack.htm), EXCURV (http://srs.dl.ac.uk/XRS/Computing/Programs/excurv97/intro.html), and others (a listing of XAFS analysis packages is available at http://xafs.org/Software). The associated *ab initio* theoretical codes (e.g., FEFF (Rehr and Albers, 2000) (http://leonardo.phys.washington.edu/feff/), GNXAS (http://www.aquila.infn.it/gnxas/), and others) have not only provided a deeper quantitative understanding of XAFS, but also allowed researchers to quickly and thoroughly model experimental results, which previously had not been possible.

The result of all of these advances has allowed *in situ* XAFS spectroscopy to become a workhorse technique in catalyst characterization science, and also provided opportunities for the application to new areas. Indeed, the evolution of X-ray spectroscopy has not ended—recently there have been several significant new advances that are discussed briefly at the end of this chapter.

1.1. Previous Reviews

XAS, and particularly its application to catalysis, has been the subject of several previous reviews and books. In 1988, Koningsberger and Prins published the book "X-ray absorption: principles, applications, techniques of EXAFS, SEXAFS and XANES" (Koningsberger and Prins, 1988). In this monograph there is a thorough description of the technique together with a chapter on its application to catalysis. Iwasawa in 1996 published "XAFS for catalysts and surfaces" (Iwasawa, 1996), which focused solely on XAFS spectroscopy as applied to catalyst characterization. This volume includes a chapter by Bazin, Dexpert, and Lynch about measurements of catalysts in reactive atmospheres, and several other chapters allude to examples of such characterization. Recently a book entitled "*In situ* Spectroscopy of Catalysts" (Weckhuysen, 2004) was published that contains three chapters focused on XAFS of catalysts in reactive atmospheres: one on XANES, one on EXAFS, and one on time-resolved XAFS.

An excellent reference to the latest innovations and results of such XAFS characterization of catalysts is available in the proceedings of the International XAFS Society (http://www.i-x-s.org/) conferences, which are held every three years (see XAFS XIII proceedings (2007) for the 2006 meeting, XAFS XII proceedings (2005) for the 2003 meeting, and XAFS XI proceedings (2001) for the 2000 meeting).

1.2. Scope and Outline

As the purpose of this review is to focus on characterization of catalysts in the working state, we have selected investigations in which it was demonstrated that the catalyst was indeed in an active state during the measurement. This restriction implies that the catalytic activity was actually measured simultaneously with the XAFS data, substantially limiting the types of catalysts and reactions that can be investigated by the technique, as is to be discussed later. There is a plethora of papers in the literature that include reports of catalysts subjected to some type of treatment, followed by measurement of XAFS data; these papers are not reviewed here.

A typical example of this type of treatment is hydrogen reduction of a supported metal catalyst. The catalyst would be placed in an XAFS cell, subjected to reducing conditions, and then cooled (often to subambient temperatures) for collection of XAFS data. However, in the vast majority of these studies the implicit assumption was made that the actual active phase was characterized, but as the specific catalyst under investigation was not simultaneously tested, this assumption was not verified. Nonetheless, highly relevant structural information can be learned about a catalyst during its activation, or by comparing the XAFS of a fresh and a spent catalyst, or a good and a poor catalyst, or the effects of catalyst reduction temperature, loading of active metal, etc. These investigations fall outside the scope of this review.

We begin with a summary of the importance of XAFS spectroscopy in catalyst characterization science, using examples from the literature to illustrate each point. This introduction to the field includes all types of XAFS spectroscopy of catalysts in reactive atmospheres and is not restricted to investigations in which activity data were measured simultaneously with catalyst performance. This section is meant to familiarize the reader with the types of relevant information that can be provided by XAFS data.

This section is followed by a thorough review of the cells used for XAFS data collection, as this instrumentation is critical to obtaining meaningful data. The cell designs include those that are suited both to relatively simple catalyst pretreatment and those suited to XAFS characterization accompanied by online product analysis.

In the second part of the review, specific examples of applications are presented, with the first section on XAFS measurements made under static conditions and the second on those made in a time-resolved manner under dynamic conditions. We end the review with an outlook to the future, where advances in time, spatial, and energy resolution are highlighted.

1.3. X-Ray Absorption Spectroscopy

Virtually all prior reviews and textbooks on XAFS spectroscopy begin with a derivation of the XAFS equation, coupled with a theoretical background. Given the depth and excellence of some of these reviews, we do not repeat this work here. The reader is referred to these articles (Iwasawa, 1996; Koningsberger and Prins, 1988; Teo, 1986; Teo and Joy, 1981) and to online tutorials (http://xafs.org/); only a brief synopsis of the X-ray absorption process is given here, together with the pertinent parameters that are measured in an XAFS experiment.

In X-ray absorption a photon is absorbed by the atom giving rise to a transition of an electron from a core state to an empty state above the Fermi level. To excite an electron in a given core level the photon energy has to be equal or greater than the energy of the core level. This results in a new channel in the absorption when the photon energy is scanned from below to above this core-level energy. XAFS is the modulation of the X-ray absorption coefficient (μ) at energies near and above an X-ray absorption edge.

The X-ray absorption spectrum is conventionally divided into two regions: the XANES and theEXAFS regions. The XANES region includes the preedge, the edge, and features within approximately 50 eV above the absorption edge, and the EXAFS region includes all features in the spectrum 50 eV above the edge.

After normalization to the edge jump and background subtraction, the EXAFS region is described by $\chi(k)$, which is the sum of all the contributions from all scattering paths of the photoelectron (Kelly et al., 2008; Stern, 1978; Stern and Heald, 1983):

$$\chi(k) = \sum_i \chi_i(k). \tag{1}$$

Each path can be written in the form shown here,

$$\chi_i(k) \equiv \frac{(N_i S_0^2) F_{\text{eff}_i}(k)}{k R_i^2} \sin\left(2k R_i + \varphi_i(k)\right) e^{-2\sigma_i^2 k^2} e^{\frac{-2R_i}{\lambda(k)}}, \tag{2}$$

with

$$R_i = R_{0i} + \Delta R_i \tag{3}$$

and

$$k^2 = \frac{2m_{\text{e}}(E - E_0 + \Delta E_0)}{\hbar}. \tag{4}$$

The terms $F_{\text{eff}_i}(k)$, $\varphi_i(k)$, and $\lambda(k)$ are the effective scattering amplitude of the photoelectron, the phase shift of the photoelectron, and the mean free path of the photoelectron, all of which can be theoretically calculated by a computer program such as FEFF (Rehr and Albers, 2000). The term R_i is the half path length of the photoelectron, that is, the distance between the absorber and a coordinating atom (for a single-scattering event). The value of R_{oi} is the half path length used in the theoretical calculation, which can be modified by ΔR_i. The remaining variables, described below, are usually determined by modeling the EXAFS spectrum. Equation (4) is used to express the excess kinetic energy of the photoelectron in wavenumber, k, by using the mass of the electron m_e and Planck's constant $\hbar$.

$N_i S_0^2$ is a term that modifies the amplitude of the EXAFS signal. The subscript i denotes that this value can be different for each path of the photoelectron. For a single scattering path, N_i represents the number of coordinating atoms within a particular shell (at the same radial distance from the absorber). For multiple scattering, N_i represents the number of identical paths. The passive electron reduction factor, S_0^2, usually has a value between 0.7 and 1.0. It accounts for the slight relaxation of the remaining electrons in the presence of the core hole vacated by the photoelectron.

The term $F_{eff_i}(k)$ is the effective scattering amplitude. For a single scattering path it is the atomic scattering factor used in X-ray diffraction. For a multiple scattering path it is the effective scattering amplitude described in terms of the single scattering formalism (Rehr and Albers, 2000). The term accounts for the element sensitivity of EXAFS spectroscopy.

$1/R_i^2$ represents the contribution from a shell of atoms at distance R_i which diminishes with increasing distance from the absorber atom.

The term $\sin(2kR_i + \varphi_i(k))$ accounts for the oscillations of the EXAFS signal with a phase given by $2kR + \varphi_i(k)$. The path of the photoelectron is given by $2R_i$ (the distance from the absorber to the neighboring atom and back again), which is multiplied by its wavenumber (k) to determine the phase; $\varphi_i(k)$ is the phase shift of the photoelectron caused by the interaction of the photoelectron with the nucleus of the absorber atom and by the interaction of the photoelectron with the nuclei of the coordinating atoms in the photoelectron path. It is the sine term of the EXAFS equation that makes the FT of the XAFS signal so useful in data analysis: the FT of the EXAFS peaks at distances related to R_i.

Because all of the coordinating atoms in a shell are not fixed in position at exactly R_i, the term σ^2 in $e^{-2\sigma_i^2 k^2}$ accounts for the disorder in the interatomic distances. The term has contributions from dynamic (thermal) disorder and static disorder (structural).

$e^{\lambda(k)}$ This exponential term depends on $\lambda(k)$, the mean free path of the photoelectron—the distance the photoelectron travels after excitation.

ΔR_i represents the change in the interatomic distance relative to the initial path length R_i.

ΔE_i is related to the change in the photoelectron energy. It is used to align the energy scale of the theoretical spectrum to match the measured experimental spectrum.

2. IMPORTANCE OF XAFS SPECTROSCOPY TO CATALYST CHARACTERIZATION IN REACTIVE ATMOSPHERES

2.1. Introduction

XAS can be used in several different ways to determine local structural information about catalysts in reactive atmospheres. This structural information may be static or dynamic; it may be geometric or electronic. The depth of information that can be ascertained is often dependent upon the type of catalyst, for example, supported metal nanoclusters versus bulk or surface oxides. It may also be controlled by some property of the catalyst, for example, the concentration of the element in the catalyst that is being investigated. In this section a few examples are provided to highlight the importance and relevance of XAFS in catalyst characterization. The examples are focused on (1) structural information characterizing samples in reactive atmospheres, (2) transformation of one species to another, (3) oxidation state determination, (4) determination of supported metal cluster size and shape, and (5) electronic structure. These examples illustrate the type of information that can be learned about the catalyst from XAFS spectroscopy.

2.2. Structure Determination

Elucidation of the structure of a solid catalyst is paramount to any understanding of its activity. Without such information, inferences about its activity would be speculation. Often it is instructive to determine the structure of a catalyst after a treatment such as oxidation, reduction, or exposure to a reactant, or with the catalyst in a particular state; it may be helpful to compare a fresh catalyst with a spent or a regenerated catalyst. XAFS spectroscopy used in this manner is "static"; the structure of the catalyst is determined in a specific well-defined state determined by the treatment and gas environment during the measurement. Two such examples are discussed here: the determination of the location of the isomorphous substitution of a heteroatom (tin) into a zeolite framework (zeolite beta), and the structure of dispersed rhenium oxide supported on γ-Al_2O_3.

A key objective in heterogeneous catalysis is the development of catalysts that are 100% selective for the desired reaction. One path toward this goal is to synthesize catalysts with uniform isolated active sites. The engineering of these catalysts, especially for selective oxidation reactions, can be achieved by using open nanoporous oxidic solids (e.g., zeolites) in which the active site is isolated and uniform in the solid. XAFS spectroscopy is a key characterization tool for determining the local atomic environments of these sites, and the example described here beautifully illustrates the point.

Silico–alumina zeolites are an important class of catalyst, serving both as solid acids and as supports in bifunctional catalysts. The acidity of the zeolite can be modified by substituting a heteroatom for the Si or Al atoms in the zeolitic framework. Whenever a framework substitution is attempted, the first question is always whether the heteroatom is indeed in the framework, or instead exists as an extra-framework species. Then, if it can be demonstrated that the heteroatom is in the framework, the question arises as to the exact crystallographic site in the lattice where the substitution has occurred. Detailed knowledge of the site (the so-called T-site in a zeolite) is needed for a complete characterization of the catalyst.

For tin substituted in zeolite beta, it was possible to determine all of this information by using EXAFS spectroscopy (Bare et al., 2005). Sn-beta zeolite is an efficient and stable catalyst for the Baeyer–Villiger oxidation of saturated and unsaturated ketones by hydrogen peroxide, producing the desired lactones with >98% yield (Corma et al., 2001; Renz et al., 2002). In the earlier papers describing these catalysts, no detailed explanation was given for the near 100% selectivity in this demanding reaction, and therefore a detailed structural characterization was conducted using XAFS spectroscopy. To avoid any effects of adsorbed moisture on the catalyst, the Sn-beta zeolite was treated in dry air at 798 K prior to cooling to room temperature to collect the EXAFS data. The Sn K-edge EXAFS data together with the best fit are shown in Figure 2. The best-fit values of the parameters characterizing the zeolite are provided in Table 1. The EXAFS data were modeled by substituting a Sn atom into the beta-zeolite structure at sites T3, T5, or T9. These sites are taken as representatives of each of the three distinct groups, because these sites span all the possible configurations of T-atoms in zeolite beta (Bare et al., 2005). Analysis of the data for all three models gave an Sn–O bond length of 1.91 Å, consistent with tetrahedral Sn(IV), and a coordination number of four. Thus, for these parameters to be true, the tin must be in the zeolite beta framework and not present in extra-framework tin oxide clusters. When the models and data for tin in each of the three sites were compared in the range of 2–5 Å, they were found to be statistically different, with the model for tin in the T5 site clearly providing the best fit of the data (Bare et al., 2005). Moreover, it was shown that there is a Sn–Sn shell at 5.0 Å, indicating that the substitution of the Sn atoms in the T5 sites is always paired: if a T5 site is occupied by Sn, then the T5 site on the opposite side of the six-ring in

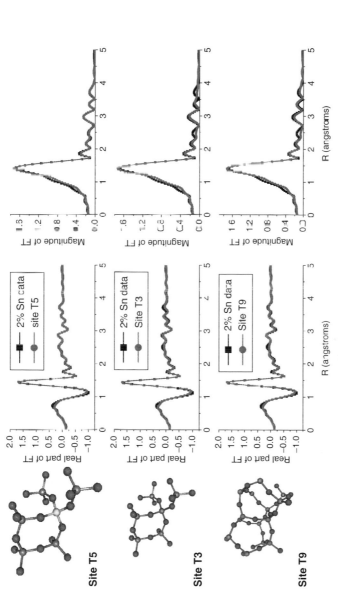

FIGURE 2 EXAFS fit results for Sn K-edge data of Sn-β zeolite. Each row represents one of the three possible sites for tin in the β-structure. The schematic representations of the atoms included in the model surrounding each of these T-sites are shown in the first column. The T5-site is made up of a 4- and a 5-membered ring, the T3-site is made up of a single 4-membered ring, and the T9 contains several 5- and 6-membered rings. The real part and the magnitude of the FT of the EXAFS data characterizing the Sn-β zeolite sample are shown in the middle and final columns (black curve). The best-fit data are compared with the model for tin substitution into each of the various T-sites (red curves).

TABLE 1 Best-Fit values of EXAFS parameters characterizing Sn-β zeolite.

Path[a]	N[b]	R (Å)[c]	σ^2 ($\times 10^{-3}$ Å^2)[d]
Sn–O$_1$–Sn	4	1.906 ± 0.0001	3.1 ± 0.2
Sn–O$_1$a–O$_1$b–Sn triangle	12	3.252 ± 0.009	0.0 ± 0.1
Sn–Si$_1$–Sn	3	3.499 ± 0.007	4.0 ± 0.8
Sn–O$_1$–Si$_1$–Sn	6	3.570 ± 0.007	4.0 ± 0.8
Sn–O$_1$–Si$_1$–O$_1$–Sn	3	3.640 ± 0.007	4.0 ± 0.8
Sn–Si$_2$–Sn	1	3.855 ± 0.096	14.9 ± 13.1
Sn–O$_1$-Si$_2$–Sn	2	3.876 ± 0.096	14.9 ± 13.1
Sn–O$_1$–Si$_2$–O$_1$–Sn	1	3.896 ± 0.096	14.9 ± 13.1
Sn–O$_1$–Sn–O$_1$–Sn	4	3.815 ± 0.003	12.3 ± 0.7
Sn–O$_2$–Sn	2	4.475 ± 0.015	5.3 ± 2.0
Sn–O$_3$–Sn	2	4.528 ± 0.015	5.3 ± 2.0
Sn–O$_4$–Sn	2	4.626 ± 0.015	5.3 ± 2.0
Sn–Si$_3$–Sn	3	4.033 ± 0.049	12.0 ± 5.1
Sn–Sn–Sn	1	4.992 ± 0.044	9.3 ± 4.8

[a] Path i describes the scattering path of the photoelectron included in the model. For example Sn–O$_1$–Sn is a scattering path from the absorbing Sn atom to the nearest oxygen atom (O$_1$) and then back to the absorbing Sn atom.

[b] N describes the degeneracy of the scattering path. For paths that include only one scattering atom, Sn–O$_1$–Sn, N is simply the number of O$_1$ atoms. These values were not determined in a fit of the data but taken from the crystal structure of the β-zeolite.

[c] R is the half path length. For a single scattering path, this is the bond length.

[d] σ^2 is the mean-square displacement of the half path length and represents the stiffness of the bond for a single scattering path. Additional parameters determined in the model include an energy shift ΔE of 0.52 ± 0.20 eV and the percentage of the tin with the nanoparticulate SnO$_2$ structure of 4 ± 2%, with the assumption that the remaining tin is in the β-zeolite structure. A k-range of 2.5–13.0 Å^{-1} and an r-range from 1 to 5 Å were used, resulting in 27 independent points per data set. The model includes a total of 16 parameters. This table lists more values than parameters determined in the fit to the data because many of the values were mathematically related.

the zeolite is also always occupied by Sn. The final structural model accounting for the substitution is shown in Figure 3.

What emerges from this detailed EXAFS analysis is, first, that the tin is indeed substituted into the zeolitic framework. In many cases this kind of information is all that can be obtained from such an analysis—a first-shell fit in which the bond lengths and coordination numbers are consistent with a framework species versus a nonframework one. However, in this example it was possible to analyze higher-shell data, up to a distance of 5 Å, and thereby to determine the site in the zeolite framework where the tin is substituted. It is believed that the unique selectivity of this catalyst in Baeyer–Villiger oxidation reactions is a consequence of the occupation of specific crystallographically well-defined sites by tin in the framework of the zeolite in a spatially uniform manner.

The chemistry of rhenium supported on alumina is of technological importance in heterogeneous catalysis. Rhenium oxide dispersed on

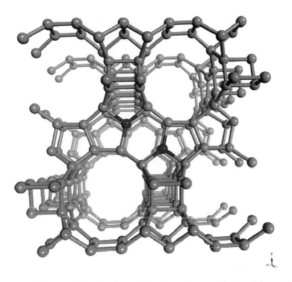

FIGURE 3   Representation of the Sn-β zeolite structure as derived from the EXAFS data, viewed along the b-axis (for clarity, the oxygen atoms are not shown). The only distribution of Sn atoms consistent with the EXAFS data is one in which pairs of Sn atoms occupy opposite vertices of the six-membered rings. A pair of T5 sites (red), with the required 5.1 Å separation is shown, representing one possible Sn-pair within the β-zeolite structure. This Sn-pair distorts two of the 12-membered ring channels as viewed from the [100] direction and all four 12-membered ring channels as viewed from the [010] direction. This distortion is either direct by the replacement of silicon by tin, or by the expansion of the neighboring SiO_4 tetrahedra.

high-surface-area Al_2O_3 has been shown to have high activity and high selectivity in olefin metathesis (Mol and Moulijn, 1975; Moulijn and Mol, 1988; Sibeijn et al., 1994), and bimetallic rhenium platinum clusters are the active metal species in petroleum reforming catalysts.

There are many reports of both the application and characterization of the rhenium in such catalysts. However, the chemistry of rhenium alone on high-surface-area γ-Al_2O_3 has been investigated much less. Rhenium is an oxophilic metal, and its chemistry is therefore expected to be dominated by its propensity for forming strong metal–oxygen bonds. The catalyst is typically prepared by using incipient wetness impregnation of the Al_2O_3 with a solution of perrhenic acid. The catalyst is then dried and calcined. In many cases the active form of the catalyst is formed after reduction in H_2. However, a key question to address in the catalyst preparation step is what is the form of the rhenium on the support after the impregnation, drying, and calcining?

To address this question, Re L_3-edge EXAFS data were collected after loading the sample into a reactor and drying by incremental heating in a flow of 20% O_2 and 80% helium to 798 K followed by a dwell at 798 K, and

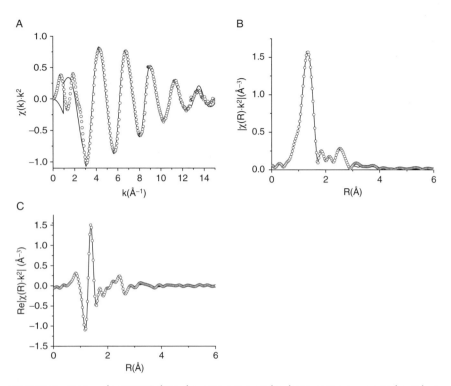

FIGURE 4   Re L₃-edge EXAFS data characterizing oxidic rhenium in a supported catalyst. (A) EXAFS data (symbols) and model fit (line). (B and C) Magnitude and real part of FT of EXAFS data (symbols) and model fit (line). The FT range of data is from 3.1 to 12.7 $\mathring{A}^{-1}$. The fit range is from 1.0 to 3.5 $\mathring{A}$. The model was fit to the data using k-weightings in the FT of 1, 2, and 3. The FT using only k-weight of 2 is shown.

TABLE 2   Best-fit parameters characterizing supported rhenium oxide[a].

Path	C–N	R ($\mathring{A}$)	σ^2 ($\times 10^{-3}$, $\mathring{A}^2$)
Re–O$_1$	3.3 ± 0.1	1.72 ± 0.01	2 ± 2
Re–O$_2$	0.7 ± 0.1	2.09 ± 0.01	2 ± 2
Re–Al	1.0 ± 0.1	3.04 ± 0.01	4 ± 2

[a] The fit of the data was done in the range of $\Delta R = 1.0$–3.4 $\mathring{A}$, with k-weightings of 1, 2, and 3. C–N is the coordination number, R the bond length, and σ^2 the sigma squared parameter.

cooling to room temperature (Yang et al., in preparation(a)). These conditions were chosen to mimic the calcination step. EXAFS data were then collected to characterize the oxidized, dried sample at room temperature.

The averaged EXAFS data of the dried, oxidized catalyst and the best fit to the data are plotted in Figure 4. The best-fit values are given in Table 2.

Thus, it can be seen that the data are well modeled by including four oxygen atoms in a distorted tetrahedral arrangement (3 short, 1 long). These distances and local geometry are consistent with the formation of a perrhenate $[ReO_4]^-$ species on the Al_2O_3 surface, with the rhenium in the +7 oxidation state. Moreover, there is a well-defined Re–Al scattering path at 3 Å. This result implies that the $[ReO_4]^-$ species is anchored to the Al_2O_3 surface through a Re–O–Al linkage (Bare et al., unpublished). Thus, the EXAFS data show that the rhenium is present in a highly dispersed oxide phase with the rhenium in a well-defined geometric environment. Such data are critical to monitoring the synthesis of this catalyst.

These examples are two of many in the literature illustrating how XAFS spectroscopy has been used to obtain detailed structural information about the active site —the species present on the catalyst surface after some pretreatment but prior to catalytic reaction. This type of XAFS analysis is ideally suited to samples in which there is a well-defined bonding arrangement between the species of interest and the support *and* all of the species are the same. Often there is no other way to obtain this information.

2.3. Transformation of One Species to Another

In the preparation and activation of a catalyst, it is often the case that the chemical form of the active element used in the synthesis differs from the final active form. For example, in the preparation of supported metal nanoclusters, a solution of a metal salt is often used to impregnate the oxide support. The catalyst is then typically dried, calcined, and finally reduced in H_2 to generate the active phase: highly dispersed metal clusters on the oxide support. If the catalyst contains two or more metals, then bimetallic clusters may form. The activity of the catalyst may depend on the metal loading, the calcination temperature, and the reduction temperature, among others.

It is then instructive to use spectroscopic tools to follow the structural evolution of the active phase. In this manner it may be possible to determine the optimum conditions for the synthesis of the precise active phase that is desired, and therefore it may be possible to control these conditions to achieve the desired phase. XAFS spectroscopy is a powerful tool that can be used to track the structural transformation of one species to another during the evolution of the active phase.

The example chosen here to illustrate this point is the sulfiding of a Ni-W hydrocracking catalyst. Hydrocracking technology is important for producing high-value naphtha or distillate products from a wide range of petroleum refinery feedstocks. Supported Ni-W catalysts are attractive in hydrotreating of heavy oil because of their high hydrogenation activities when the catalysts are sulfided (Ahuja et al., 1970; Kabe et al., 1999; Stanislaus and Cooper, 1994). This family of catalysts has been the focus of intensive research over many years. The hydrocracking catalysts are typically prepared from metal salts that are deposited onto a high-surface-area support. Following calcination, these metal salts form a dispersed oxide phase or phases, which are then converted to the corresponding metal sulfide in a sulfidation process by passing a mixture of H_2S and H_2 over the catalyst. The mechanism by which the metal oxide clusters are sulfided, and the ease and extent to which they are sulfided, are affected by many parameters during the preparation phase. The ability to monitor the sulfidation of the two metals can provide important information about what affects the formation of the active phase and the ensuing activity.

For example, the data shown in Figures 5 and 6 provide a comparison of the W L_3-edge and Ni K-edge EXAFS, respectively, recorded during the sulfiding of two different hydrocracking catalysts that were prepared by different methods, denoted as Prep A and Prep B. Other than the method of preparation, the catalysts were identical. In this experiment the temperature was ramped to 683 K with the catalyst in a flow of 10% H_2S/90% H_2 (Yang et al., 2007). Figure 5 shows the $\chi(k) \cdot k^3$ FT of all the W L_3-edge EXAFS spectra collected during the sulfidation of the samples. The numbers next to the spectra denote the sample temperature at the absorption edge during the scan. The first spectrum is that of the initial calcined (oxidized) catalyst. The only observable peak in this initial spectrum is at 1.40 Å, and it is assigned to the contribution from the W–O bonds. The intensity of the first W–O peak in the FT gradually decreased with increasing temperature of the sample in the flow of H_2S/H_2. This change is a result of the oxidic tungsten being progressively sulfided. Note that the intensity of this peak at any given temperature was always greater in Prep B than in Prep A. The second major peak in the FT of the data appears at a distance of approximately 2.0 Å. This peak is at the same distance as the first-shell (W–S single scattering) of WS_2. Thus, the presence of this peak in the FT of the data indicates the presence of W–S bonds as WS_2 platelets begin to form in the catalyst. This W–S signal first appeared at about 558 K in Prep A, and at 598 K in Prep B. A third, intermediate, peak is also evident in the W L_3-edge EXAFS spectra. This peak, at 1.9 Å, grew in intensity in both data sets, and then diminished. In earlier investigations, this signal was assigned to an oxysulfide phase in which the first oxygen shell is partially replaced by sulfur (Sun et al., 2001). The qualitative trends in these FT's of the W L_3-edge EXAFS spectra show changes that can be interpreted within the chemistry of the material; as the oxidized catalyst becomes sulfided, the first oxygen shell peak at 1.40 Å diminishes and the second W–S peak at 2.0 Å becomes more pronounced.

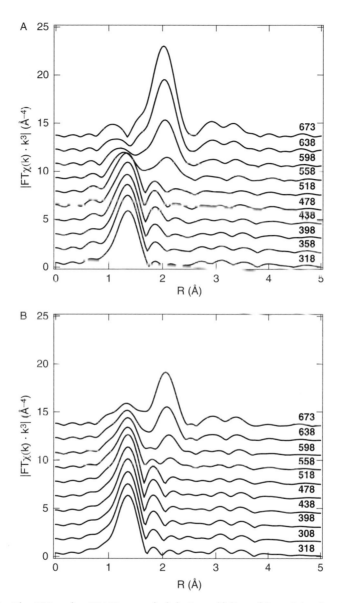

FIGURE 5   The W L$_3$-edge EXAFS recorded during sulfiding of Prep A (top) and Prep B (bottom). The spectra have been offset by 1.5 for clarity. The temperature during each scan is indicated on the figure (in K). See text for details.

Similar to the W-edge data, the Ni K-edge data are characterized by the same types of plots. Figure 6 shows the magnitude of the FT of the $\chi(k) \cdot k^3$ FT of the Ni EXAFS spectra collected during the sulfidation. The

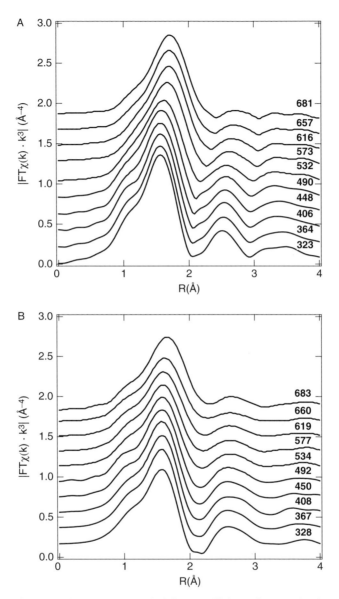

FIGURE 6 The Ni K-edge EXAFS recorded during sulfiding of Prep A (top) and Prep B (bottom). The spectra have been offset by 0.2 for clarity. The temperature during each scan is indicated on the figure (in K). See text for details.

first spectrum is that of the oxidized catalyst, and shows a prominent peak at 1.56 Å that is assigned to oxygen atoms neighboring the nickel. The change in the spectra with increasing temperature is more subtle for the Ni K-edge EXAFS data than the W L_3-edge data. There is a shift in the peak to greater distance, with the first-shell peak position gradually shifting from 1.56 to 1.67 Å at about 573 K during the sulfidation of sample Prep A. This same trend was observed for sample Prep B, but at substantially higher temperatures, that is, exceeding 623 K. The peak shift corresponds to the difference in the bond distance typical of Ni–O of approximately 2.1 Å in the oxidized catalyst and the bond distance typical of Ni–S of approximately 2.3 Å in the sulfided catalyst. The shift is not as pronounced as that indicated by the W L_3-edge data, as the change in average bond length between Ni–O and Ni–S is less than 0.2 Å, whereas the comparable change for tungsten is 0.58 Å.

The spectral changes in the Ni K-edge FT EXAFS data are subtle; however, the changes in the Ni K-edge XANES data are striking. Figure 7 shows Ni K-edge XANES data characterizing the temperature-programmed sulfiding of samples Prep A and Prep B. The nickel oxide and nickel sulfide phases are easily distinguishable. Previous investigations have shown that the white line of the Ni K-edge XANES indicates the ionic bonding between nickel and oxygen (Cattaneo et al., 1999). As oxygen atoms are replaced by sulfur atoms, the white line intensity decreases as a consequence of the more covalent bond between nickel and sulfur (Louwers et al., 1993; Lytle et al., 1979). From these data it can be determined that in the Prep A sample, both nickel and tungsten were almost fully sulfided as the temperature reached 683 K. However, in sample Prep B only about 60% percent of the nickel and tungsten were sulfided as the temperature reached 683 K (Yang et al., 2007).

This example illustrates the type of information that can be learned about the transformation of one species to another by XAS measurements of catalysts. There are many other examples that could be cited. Often, it is only the XANES that is measured, but sometimes the quantitative EXAFS data are presented. In the large majority of cases the EXAFS information has been valuable in allowing understanding of the structure of the catalyst, and often this information could not be obtained by any other method.

2.4. Oxidation State

XAFS spectroscopy has been used frequently to determine the oxidation state of an element as a catalyst is subjected to treatments such as reduction, oxidation, or exposure to some reactant. It is the near edge of the X-ray absorption spectrum, the XANES region that is usually used for these measurements. As mentioned in the introduction, the first to recognize the value of XANES for this type of investigation was van Nordstrand (1960), and his report appeared well before the advent of synchrotron radiation sources.

The features of a XANES spectrum that are typically used to determine the average oxidation state of an element are the absorption edge position

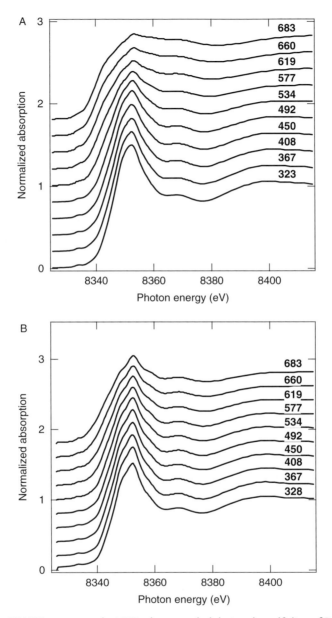

FIGURE 7 XANES spectra at the Ni K-edge recorded during the sulfiding of Prep A (top) and Prep B (bottom). The temperature recorded during each scan is indicated on the figure (in K). See text for details.

(energy) or the intensity of the white line above the absorption edge. The method that is used depends on the element and absorption edge. An example of the use of edge energies to determine metal oxidation states is

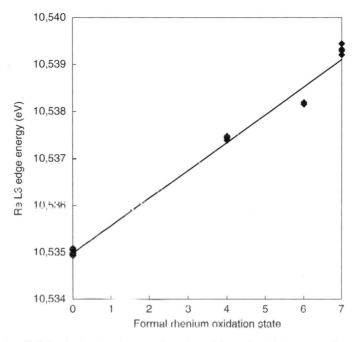

FIGURE 8   Shift in the Re L$_3$-edge as a function of formal oxidation state for a series of bulk rhenium oxides.

the work of Wong et al. (1984) with a series of vanadium oxides. They found a linear relationship between the edge position and the formal vanadium oxidation state.

There are many other such examples, illustrated by the results of Figures 8 (Bare, S.R., unpublished data) and 9 (Ressler et al., 2002). The latter example is interesting in that instead of the edge position as a monitor of the average molybdenum oxidation state, the absolute energy position of a feature in the XANES region was used (indicated by an arrow in Figure 9).

To illustrate how such information might be used in the characterization of catalysts, we return to the rhenium oxide on Al$_2$O$_3$ samples discussed in Section 2.2. When this dispersed perrhenate species is reduced and the Re L$_3$-edge XANES recorded as the sample temperature is ramped from room temperature to 773 K, then, by use of the absorption edge position, the data readily determine the temperature at which the reduction occurs and the extent of reduction. Typical data obtained in such an experiment are shown in Figure 10. Both an edge shift and a decrease in the white line intensity were observed as the catalyst was heated in H$_2$. Both of these changes are related to the reduction of the rhenium from the +7 species to the metallic state. By the use of suitable standards (Figure 8), the amount of oxidized

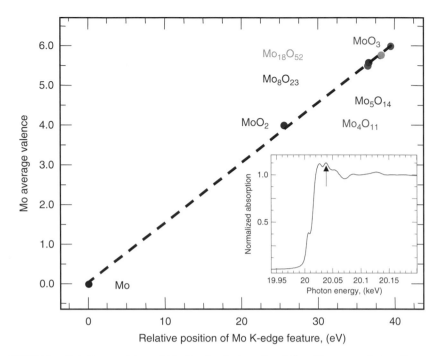

FIGURE 9   Relative position of Mo K-edge feature (as defined by arrow in inset) as a function of the Mo valence for a series of Molybdenum oxides (Ressler et al., 2002). Reprinted from (Ressler et al., 2002), Copyright 2002, with permission from Elsevier.

rhenium remaining as a function of temperature was determined (Figure 11). The onset of reduction at approximately 473 K is clearly indicated, with a maximum rate at approximately 543 K. After the initial rapid reduction, there was a continuous slow reduction as the temperature increased to 773 K. Such reduction information is useful for comparing catalysts prepared by various techniques, or as a function of metal loading, and it serves as a guide for determining when the reduction to the metal is complete—or whether the resulting oxidation state after a given pretreatment is an intermediate one.

Often it is not possible to determine such information by any other characterization method. This fact, combined with the elemental specificity of the method, the fact that the edge position can be determined accurately, and the high intensity of the edge features, has made this use of XANES popular for characterization of changes in catalysts in reactive atmospheres. There are now many reported examples of changes in oxidation state of an element as a function of reduction temperature determined by XANES (for some recent examples see Becker et al., 2007; Gamarra et al., 2007; Haider et al., 2007; Jentoft et al., 2005; Martinez-Arias et al., 2007; Reed et al., 2006; Safonova et al., 2006; Saib et al., 2006; Silversmit et al., 2006).

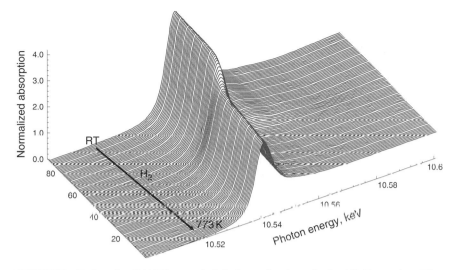

FIGURE 10 Re L_3-edge XANES recorded during reduction of a Re/Al_2O_3 catalyst. The spectra were recorded as a function of temperature from room temperature to 773 K.

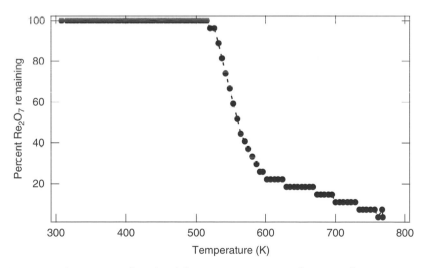

FIGURE 11 The amount of oxidized rhenium remaining as a function of temperature calculated from the data in Figure 10.

2.5. Supported Metal Cluster Size and Shape

Metal clusters supported on refractory oxides are used extensively in catalysis for the production of chemicals and petroleum-derived transportation fuels. Catalysts in this class typically have metal loadings of less than a few weight percent, with the metal clusters (nanoparticles) being less than 2–3 nm in diameter (and more usually 0.5–2 nm diameter). Common supports are Al_2O_3, SiO_2, zeolites, or aluminosilicates with surface areas of several hundred square meters per gram. Thus, a property of such catalysts is that the metals are highly dispersed on the support, with a larger fraction (>0.8) of the total metal atoms on the surface of the cluster. More than one metal can be present, with the resulting formation of bimetallic clusters.

From the earliest days of dedicated synchrotron facilities and the advent of modern XAFS spectroscopy, the power of the technique has been realized for the quantitative determination of local structural parameters representing the clusters. There continue to be many publications per year representing the application of XAFS spectroscopy to probe some structural or electronic property (Section 2.6) of catalysts in this class. The types of element-specific local atomic-level structural details of the clusters that can be learned are the following: average metal–metal coordination numbers, which lead to average cluster size (especially if data from higher shells are analyzed), interatomic distances, thermal and static disorder, cluster shape, and desegregation or segregation in bimetallic clusters. In the large majority of such catalysts, the highly dispersed metal clusters are highly reactive and readily reoxidized upon exposure to the atmosphere. This reactivity necessitates that the catalysts be activated in the XAFS spectroscopy cell to allow determination of these structural parameters.

One of the key aspects of using XAFS spectroscopy for characterizing supported metal clusters is that the average coordination number is a strong function of cluster size for clusters <2 nm in diameter. This point is illustrated in Figures 12 and 13, which show the average first-shell coordination number and the number of Pt atoms, respectively, as a function of cluster size for cuboctahedral platinum clusters. The ability of XAFS spectroscopy to determine the average metal cluster size, and sometimes shape, by use of the average first-shell coordination number was realized early (Greegor and Lytle, 1980) and the method was expanded by Kip et al. (1987), and further refined as anharmonic effects were accounted for (Clausen et al., 1993, 1994).

An example of how the average cluster size affects the XAFS data is illustrated in Figure 14 (Frenkel et al., 2001). The figure is a summary of Pt L_3-edge EXAFS data characterizing reduced platinum clusters on a carbon support. The average cluster diameters are 24, 45, and 81 Å for the samples S1, S2, and S3, respectively. As the average platinum particle size decreased, the signal intensity of the $\chi(k)$ data decreased (Figure 14A), and the amplitude of the magnitude of the FT also decreased (Figure 14B). In this work, however, the authors took the analysis one step further than a simple single-shell analysis. By analyzing the EXAFS up to the fifth coordination shell, the authors were able to determine not only the average cluster size, but also the cluster shape and morphology. Thus, they

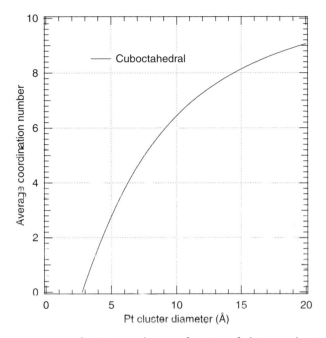

FIGURE 12 Average coordination number as a function of platinum cluster diameter for cuboctahedral clusters.

were able to determine that the platinum clusters were present on the carbon support as hemispherical cuboctahedra. This type of analysis is dependent on the fact that the effect of cluster size on the coordination number is much stronger for more distant shells than for the first shell (Benfield, 1992; Frenkel, 1999; Jentys, 1999).

The use of XAFS spectroscopy to probe the structures of dispersed metal clusters is expected to continue to be a major application of the technique in catalysis, primarily because this information cannot be determined in any other manner. It is clear that, when XAFS spectroscopy, which provides structural information about the bulk average of the clusters present, is combined with electron microscopy (STEM and TEM), then a more complete picture of the cluster structure emerges.

2.6. Electronic Structure

XAFS spectroscopy, by default, probes the electronic structure of the specific element in the catalyst being studied, because the XANES spectral shape reflects the density of empty states (Bart, 1986; Fuggle and Ingelsfield, 1992). An important question about the local electronic state

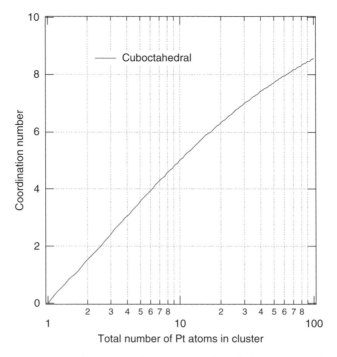

FIGURE 13   Average coordination number versus number of platinum atoms in the cluster.

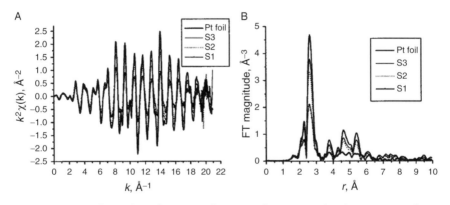

FIGURE 14   Size-dependent changes in the Pt L_3-edge EXAFS of carbon-supported platinum nanoparticles. Representative k^2-weighted EXAFS (A), and magnitude of the FT (B), representing the samples described in the text (Frenkel et al., 2001). Reprinted with permission from (Frenkel et al., 2001). Copyright 2001 American Chemical Society.

of a catalyst component is the number of d-electrons. As the electronic structure of the element drives the chemistry, it would be important if one could derive the number of occupied d-electron states from XANES, or at least have some relative measure of the density of d-states to allow comparisons of catalysts or of a catalyst in two different states, for example, fresh and spent. Because of the dipole selection rule ($\Delta l = \pm 1$) that operates in XAFS spectroscopy, one would need to use 2p (or 3p) core states to probe the d-band (Brown et al., 1977). In the case of 3d-systems, this limitation poses a problem, because the 2p core states occur in the soft X-ray range between 400 and 900 eV (the field of soft XAFS spectroscopy is growing rapidly, both experimentally and theoretically, but it is outside the scope of this review, see, e.g., Knop-Gericke et al. (2004), and references cited therein). The 1s core states are excited into 4p and higher p-orbitals. As such, K-edges do not probe the occupation of the d-states. However, the preedge intensity at the K-edge probes the 3d–4p mixing but not the 3d occupation. In the case of the 4d and 5d states, the L_3- and L_2-edges are separated by relatively large energies that allow for the separate determination of the L_3 and L_2 white line intensities.

The use of XANES spectroscopy to determine the d-band occupancy (or more correctly the unfilled d-band) in pure metals and supported catalysts was discovered early on by Lytle et al. (1976, 1979).

Thus, for a metal with a filled d-band, one would expect a simple step function at the L_3-edge, whereas those elements with a high density of unoccupied d-orbitals would show a pronounced peak (white line) at the absorption edge. Such a correlation is indeed found, as shown in the now famous data of Figure 15, which shows the L_3-edge XANES for the 5d transition elements rhenium through gold (Meitzner et al., 1992). This

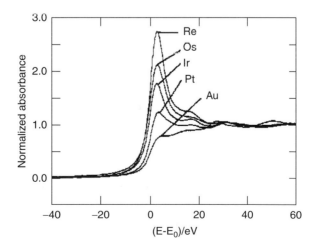

FIGURE 15   XANES spectra of the L_3-edges of the 5d transition metals. The spectra have been shifted so that the absorption edges align to a common energy (Meitzner et al., 1992). Reprinted with permission from (Meitzner et al., 1992). Copyright 1992 American Chemical Society.

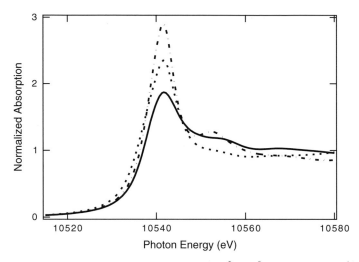

FIGURE 16   Re L$_3$-edge XANES of rhenium metal (Re^0)—5d^5 (solid); ReO$_2$ (Re^{4+})—5d^1 (dashed); and NH$_4$ReO$_4$ (Re^{7+})—5d^0 (dash-dot). The XANES spectra have been aligned in energy to allow a more direct comparison of the white line intensities.

general observation is also found for a given metal in different oxidation states, as demonstrated for rhenium oxides in Figure 16.

Together with these general empirical observations, there have been attempts to quantify the d-occupancy on the basis of XANES data (Mansour et al., 1984). A word of caution is needed for these types of investigations, as the absolute intensity of the absorption edge (white line) can be influenced by self-absorption effects in XAS (Stern and Kim, 1981), and it is essential to demonstrate that the intensities being measured are not influenced by the concentration of element in the sample. In addition, there could be effects associated with adsorbates on the dispersed metal clusters.

A well-documented example is the effect of adsorbed hydrogen atoms on the Pt L$_3$/L$_2$ XANES of nano-clusters of platinum (Ankudinov et al., 2001a). In this case, the intensity of the white line is suppressed and the spectrum broadens to higher energies as hydrogen chemisorbs on the platinum clusters. It was shown that the spectral changes are a result of changes in the atomic background contribution, and to reduced Pt–Pt scattering at the edge. This example illustrates that one has to be very careful to either remove any adsorbed hydrogen after the reduction of the catalyst, or at least to be aware of this phenomenon.

The use of XANES spectroscopy to determine electronic structure has been demonstrated for supported bimetallic alloys (Haller, 1996; Meitzner et al., 1988; Reifsnyder and Lamb, 1999; Sinfelt, 1983; Viswanathan et al., 2002). A specific example is shown in Figure 17. In this figure, the Pt L$_2$-

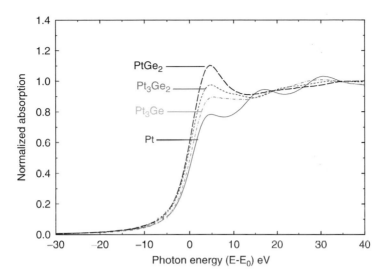

FIGURE 17 Pt L_2-edge XANES of a series of bulk Pt-Ge alloys. The value of E_0 was set to be 13,272.0 eV. The spectra are shown as a function of increasing germanium content from the bottom to the top.

edge XANES data characterizing a series of bulk platinum–germanium alloys are shown. As the germanium content increases, there is increased charge transfer from the d-band of platinum to germanium, increasing the number of d-holes and resulting in an increase in the white-line intensity. Once such a series of spectra of reference compounds has been acquired, the results can be used for comparison with the XANES spectrum of a supported catalyst. For example, Meitzner et al. (1988) used this method to determine the extent of alloying in a PtSn/γ-Al$_2$O$_3$ catalyst. By comparing the intensity of the Pt L_3-edge XANES characterizing the bimetallic sample with that characterizing a reference compound, they concluded that the composition of the bimetallic clusters was most likely PtSn$_5$. Similarly, in the PtGe/γ-Al$_2$O$_3$ catalyst, the Pt L_2-edge XANES can be compared with that of the actual catalyst (Figure 18). In this example, the Pt L_2-edge XANES of the reduced catalyst most closely resembles that of Pt$_3$Ge$_2$, in agreement with the detailed fitting of the corresponding EXAFS data (Bare, S.R., unpublished data).

3. CATALYTIC REACTORS THAT SERVE AS CELLS

3.1. Introduction

The successful design of a catalytic reactor that also serves as a cell for XAS is critical to obtaining XAFS data of catalysts in the functioning state. There are no commercially available cells for such applications. Therefore,

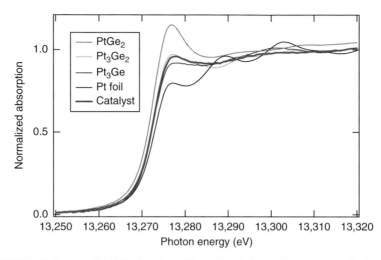

FIGURE 18   Pt L_2-edge XANES of reduced Pt-Ge/γ-Al$_2$O$_3$ catalyst compared with those of the bulk reference alloys.

each research group has fabricated their own cells, often specifically designed for a particular catalyst or catalytic reaction. As there has not been a comprehensive review of these reaction cells, we included a detailed chronological listing of them here, with the focus on cells for investigating heterogeneous catalysts.

Each cell design has its own set of advantages and disadvantages, and the information content desired from the experiment drives the designs of the cell. There are two broad classifications of the cells: (1) those that approximate plug-flow reactors—these are attempts to mimic the flow conditions in a true catalytic reactor and (2) cells that are not plug-flow reactors, and often include a pressed wafer of catalyst with substantial gas by-passing. We designate this type of cell as a nonplug-flow cell. Each of these categories of cell designs is discussed below.

The overall purpose of the reactor cell is to protect the environment surrounding the catalyst such that the catalyst can be characterized in a well-controlled state, whether it is that of some precursor or that of the working catalyst. To accomplish the desired goals, the XAFS spectroscopy cells have been designed to allow a flow of reactants over, or through, the catalyst sample while the reactor is brought to a given temperature and pressure.

Several key design issues must be considered for applications of reactors as XAS cells. First, the experiment has to be conducted in a safe and environmentally responsible manner. Specific safety concerns include the maximum operating temperature and pressure, the materials of construction, and compatibility with the chemicals and gases being used. Moreover, the whole experimental arrangement, of which the cell is only one component, must be operated safely, with appropriate standard operating procedures and emergency (worst-case scenario) procedures.

Some key questions are the following: What is the X-ray energy range that the cell will be used for? What is the maximum temperature range desired? What is the maximum operating pressure? What detection scheme will be used: transmission or fluorescence, or a combination of both? What chemicals will the cell be exposed to (the choices will affect the choice of materials of construction). What is the X-ray beam size? What sample size is needed? What resources are available for the design and construction? Is cost an issue? Can gaskets and O-rings be tolerated? Does the catalyst bed temperature need to be measured? Does the temperature profile of the catalyst bed need to be uniform (radially and axially)? These and other issues must be considered in any XAFS spectroscopy cell design.

A critical component of the cell is the X-ray-transparent window that allows the X-ray beam to impinge on the sample and the transmitted or fluorescent X-rays to be detected. Typical window materials that have been used are polyimide (Kapton®), beryllium, quartz, diamond, polyester (Mylar®), and titanium. Table 3 shows estimates of the thicknesses of window materials for various X-ray energies from 5 to 25 keV, determined on the basis of the assumption that 25% of the X-rays are absorbed by the window material.

The chemical and physical properties of each of these window materials vary widely. For example, polyimide is flexible, semitransparent, and chemically inert, but it has an upper working temperature of 673 K (for information about the properties of Kapton® see http://www2.dupont.com/Kapton/en_US/assets/downloads/pdf/summaryofprop.pdf).
Beryllium is stiff, has a low density, high thermal conductivity, and a moderate coefficient of thermal expansion; it can be machined and is very stable mechanically and thermally. It also retains useful properties at both elevated and cryogenic temperatures. However, it does require a few safety-related handling requirements that are well documented (for detailed environmental safety and health information about beryllium see http://www.brushwellman.com). Nonetheless, as is stated in the Brush Wellman literature (for detailed environmental safety and health information about beryllium see http://www.brushwellman.com), "handling beryllium in solid form poses no special health risk."

The list in Table 3 is not exhaustive; other typical window materials include Mylar®, boron nitride, Teflon®, and aluminum. Each of these materials can be modified for additional chemical resistance. For example, Kapton® film can be purchased coated with graphite or Teflon®, and beryllium windows can be coated with DuraCoat™. There is no one window material that is suitable for all applications. The optimum choice

TABLE 3   Calculated total window thickness for several common window materials as a function of X-ray energy.

Window material	Window density (g/cm³)	Energy of X-rays (keV)	Window thickness needed to meet criterion of 25% absorption (μm)
Beryllium	1.848	5.0	360
Beryllium	1.848	6.0	620
Beryllium	1.848	7.0	960
Beryllium	1.848	8.0	1390
Beryllium	1.848	9.0	1880
Beryllium	1.848	0.0	2410
Beryllium	1.848	15.0	5080
Beryllium	1.848	20.0	6930
Beryllium	1.848	25.0	8030
Polyimide	1.43	5.0	80
Polyimide	1.43	6.0	140
Polyimide	1.43	7.0	220
Polyimide	1.43	8.0	330
Polyimide	1.43	9.0	460
Polyimide	1.43	10.0	630
Polyimide	1.43	15.0	1930
Polyimide	1.43	20.0	3720
Polyimide	1.43	25.0	5530
Quartz	2.649	5.0	8
Quartz	2.649	6.0	13
Quartz	2.649	7.0	20
Quartz	2.649	8.0	30
Quartz	2.649	9.0	42
Quartz	2.649	10.0	57
Quartz	2.649	15.0	187
Quartz	2.649	20.0	427
Quartz	2.649	25.0	785
Diamond	3.51	5.0	43
Diamond	3.51	6.0	75
Diamond	3.51	7.0	120
Diamond	3.51	8.0	179
Diamond	3.51	9.0	254
Diamond	3.51	10.0	346
Diamond	3.51	15.0	1017
Diamond	3.51	20.0	1856
Diamond	3.51	25.0	2613

(continued)

TABLE 3   (*continued*)

Window material	Window density (g/cm³)	Energy of X-rays (keV)	Window thickness needed to meet criterion of 25% absorption (μm)
Titanium	4.51	5.0	n/a
Titanium	4.51	6.0	n/a
Titanium	4.51	7.0	2
Titanium	4.51	8.0	3
Titanium	4.51	9.0	4
Titanium	4.51	10.0	6
Titanium	4.51	15.0	18
Titanium	4.51	20.0	40
Titanium	4.51	25.0	76

The values were calculated by using the Elam database and total cross section (Elam et al., 2002). Note that 25% absorption by the window material corresponds to $\mu d = 0.288$, and that the thickness of each window in a pair should be half the value shown for the total thickness.

of window material depends on the required chemical and physical properties and the intended X-ray energy.

There are typically three forms of solid catalyst used for XAFS spectroscopy investigations of catalysts in reactive atmospheres: pressed wafers, loose powders, and meshed particles. Each of these forms has advantages and disadvantages (Grunwaldt et al., 2004). For example, a pressed wafer is the ideal form of a sample for spectroscopy, but it may impose mass transfer limitations. It is also possible that compacting the sample may result in a significant concentration gradient in the wafer that could affect the performance of the catalyst and also its structure, as the structure of the catalyst can be changed by the high pressures necessary to produce self-supporting wafers. Researchers should ensure that the design is appropriate for the information content that is desired and be aware of the disadvantages of the chosen methodology.

3.2. Cells as Reactors

One of the first designs of a cell for XAFS spectroscopy of catalysts in reactive atmospheres was published by Lytle et al. (1979) and it is still available commercially (http://www.exafsco.com). This elegant design (Figure 19) incorporates many features that have subsequently been shown to be critical in many newer designs. The design allows for water cooling of the furnace block and for both an inert-gas flush and evacuation of the cell. The design incorporates Soller slits and a fluorescent X-ray ion chamber detector and amplifier for those experiments that require fluorescence detection. A particularly appealing feature of the design is

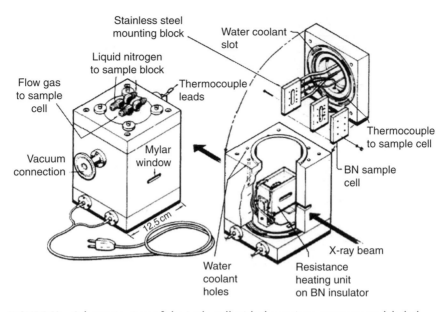

FIGURE 19   Schematic view of the Lytle cell with the various components labeled (Lytle et al., 1979). Reprinted with permission from (Lytle et al., 1979). Copyright 1979, American Institute of Physics.

that it accommodates two different types of sample, pressed wafers and powders. A cylindrical insert uses a self-supporting pressed wafer of the catalyst, and it can be used for soft or hard X-rays with the sample at atmospheric pressure. A boat-type sample holder insert is used with the catalyst in powder form, with X-ray energies >7 keV, and at pressures up to 100 bar. The maximum operating temperature is quoted to be approximately 1000 K. Significant drawbacks of this design are that the catalyst temperature is not measured, and there is significant gas bypassing. Nevertheless, this design has been used by many groups around the world, with publication of many data representing XAFS spectra of catalysts, particularly supported metal and bimetallic catalysts. In the majority of the reported investigations, the catalyst was activated by reduction in hydrogen, and subsequent XAFS data were collected to determine the structure of the activated (reduced) form of the catalyst. The Lytle cell has also been used in experiments to determine structure–catalytic property relationships by monitoring the products exiting the reactor while the XAFS data were collected in a time-resolved manner (Coulston et al., 1997).

Following the pioneering investigations of Lytle et al., the 1980s saw several other groups using XAFS spectroscopy for characterization of catalysts in reactive atmospheres. In 1981, the Haldor Topsøe group reported an investigation of cobalt–molybdenum hydrodesulfurization catalysts (Clausen et al., 1981). The oxidic catalyst precursor samples were sulfided in the cell so that XAFS measurements could be made to characterize the sulfided form of the material. Their publication gives no details on the cell design other than the statement that the catalyst was pressed into a self-supporting wafer 16 mm in diameter that was held in a "specially designed cell equipped with X-ray transparent windows." Clearly, the cell was constructed of materials compatible with the hydrogen sulfide used in the sulfiding, and the sample could be heated to 673 K. Hydrodesulfurization catalysts were also the subject of the first investigation in which the reactor effluent was monitored with a gas chromatograph at the same time that the XAFS data were being recorded (Betta et al., 1984). This cell, shown in Figure 20, allowed the recording of XAFS data characterizing a three-phase catalytic reaction at temperatures up to 700 K and pressures up to 140 bar. The X-ray beam entered and exited the cell through thin (0.51 mm) beryllium windows. The catalyst was wetted by a liquid reactant (benzothiophene) film trickling down the catalyst while reactant hydrogen dissolved in the liquid reacted with the

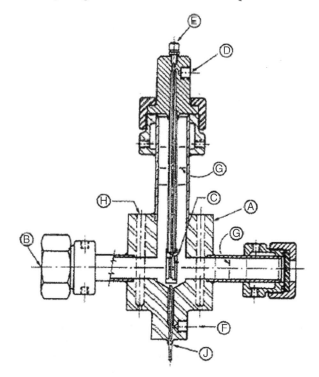

FIGURE 20   Cross-sectional view of the Dalla Betta cell (Betta et al., 1984). (A) Main body; (B) beryllium windows; (C) catalyst sample holder; (D) gas inlet; (E) liquid inlet; (F) gas and liquid outlet; (G) convection baffles; (H) hole for cartridge heater; (J) thermocouple. Reprinted with permission from (Betta et al., 1984). Copyright 1984, American Institute of Physics.

benzothiophene at the catalyst surface (the catalyst was pressed into a self-supporting wafer). The authors reported the safety aspects of their design, having examined the beryllium windows for signs of chemical attack, and they conducted hydrostatic pressure testing of the assembled cell prior to use.

The cell design of Tohji et al. (1985) was similar to that of Lytle et al. (1979). The former authors provided scant detail, but the sample cell was fabricated from boron nitride, which was clamped together by stainless steel mounting blocks. Gas entered and exited the cell through small holes. The maximum operating temperature of the cell was 1300 K, and (although not stated) the catalyst is presumed to have been in powder form. The distribution of the gas flowing through the catalyst bed is not clear from the design. The authors presented XAFS data characterizing the structure of a Cu/ZnO methanol synthesis catalyst following calcination and hydrogen reduction in the cell. The following year the group of Ichikawa et al. (1986) reported a cell incorporating a self-supporting wafer. Little detail was provided, but the authors stated that the wafers were held in a Pyrex® sample cell that had 500-µm thick Kapton® windows. The cell allowed them to conduct catalyst treatments in the cell with H_2 and CO at elevated temperatures.

Although a large majority of the reported XAFS spectroscopy cells operate at a maximum pressure of 1 bar, there are several reports of cells that are capable of operation at significantly higher pressures. For example, Neils and Burlitch (1989) reported a cell designed to investigate the Cu/ZnO methanol synthesis catalyst; it could be operated at 68 bar and 433 K. The authors provided significant details of the cell design (Figure 21). The reaction chamber was fabricated from two 4.75 in.

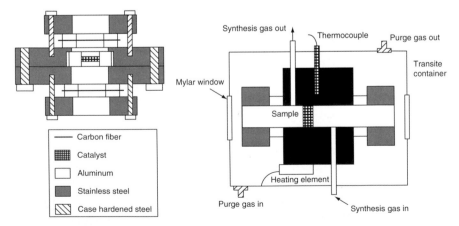

FIGURE 21 Cell design of Neils and Burlitch (1989) showing the placement of windows and catalyst sample holder (left), and the cell in the heating chamber (right). Reprinted from (Neils and Burlitch, 1989), Copyright 1989, with permission from Elsevier.

diameter stainless steel ultra-high vacuum (UHV) Conflat® flanges that were bolted together. One of the flanges was machined to provide the gas inlet and outlet. The critical component of the design is the window material, which needs to be both X-ray transparent and hold the 68 bar pressure. To solve the design problem, the researchers fabricated their own windows from adhesive-coated carbon fiber cloth, which was supported by aluminum disks with slots milled in them. The sample holder held a pressed wafer of the catalyst. The whole assembly was placed in a box heated by cartridge heaters. The investigators pointed out that the design has some limitations, the foremost of them being the maximum operating temperature of only 453 K. The windows were also thermally shock sensitive as a consequence of the differing coefficients of thermal expansion. However, the reactor operated well over several heating and cooling cycles at 453 K and 68 bar.

A popular design that has subsequently been used by several groups worldwide was reported by Kampers et al. (1989). This design (Figure 2?) includes a self-supporting wafer sample in a rectangular opening of dimension 4 × 18 mm housed in a stainless steel holder (a) (the cell parts are designated by letters in Figure 22). The holder fits precisely into a stainless steel heater and cooler cylinder (b). Within this cylinder a coiled heating wire enables heating of the sample to 773 K. This cylinder is hollow so that liquid nitrogen, or another coolant, can be added to cool the sample. The cylinder is welded onto a flange (f) of a stainless steel chamber. An O-ring seal ensures a vacuum-tight connection between the flange and chamber. The chamber is equipped with beryllium windows for transmission of the X-ray beam. A gas inlet and outlet are placed diagonally over the sample. To minimize heating and cooling of the beryllium windows, water flows through the space between the double walls of the chamber. The sample temperature of this cell can be varied from 77 to 773 K (the actual catalyst bed temperature is not measured and must be calibrated). The cell operates at ambient pressure, but can also be evacuated to allow XAFS data to be recorded under moderate vacuum (1×10^{-4} mbar).

A design based on a similar concept was published by Castner et al. (1990) the following year. This cell (Figure 23) also incorporates a pressed catalyst wafer, but heating was by 500-W halogen bulbs. The cell could operate from 77 to 923 K and from high vacuum (10^{-8} mbar) to 1 bar; the window material was Mylar®.

A cell using similar design concepts was reported by Asakura and Iwasawa (1989). Their cell was equipped with beryllium windows sealed with O-rings and was stated to be capable of holding pressures up to 30 bar. Little detail was provided, but gases could enter the cell through a valve arrangement, and the sample could be cooled to 100 K by the use of liquid nitrogen. Resistance heating was used to provide a maximum operating temperature of 733 K.

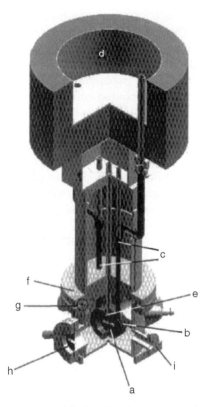

FIGURE 22   Cell (Kampers et al., 1989) that functions as a continuously stirred tank reactor: (a) sample holder; (b) heater/cooler cylinder; (c) hollow tubes for coolant and electrical wires; (d) reservoir for coolant liquid; (e) thermocouple; (f) flange enclosing *in situ* chamber; (g) O-ring; (h) X-ray transparent window; (i) gas inlet/outlet. Reprinted with permission from J.A. van Bokhoven, T. Ressler, F.M.F. de Groot, and G. Knopp-Gericke, in "*In situ* Spectroscopy of Catalysts", B.M. Weckhuysen, Ed., published by American Scientific Publishers (2004). Copyright © American Scientific Publishers.

The cell designs described above are in general not ideal for providing simultaneous measurements of spectra and kinetics of the catalytic reaction, because each cell has a large dead volume, and the catalyst is a pressed wafer, which may not be fully accessible to the reactants because of mass transfer limitations.

These issues were partially addressed by a design reported by Clausen and Topsøe (1991). These researchers developed a small reactor cell, which allows simultaneous measurements of catalytic activity and XAFS spectra. They stated that real reaction conditions were established, with catalyst temperatures up to 773 K and pressures up to approximately 100 bar. Details of the design were not published, but the authors stated that the cell consisted of stainless steel walls with two beryllium windows for transmission of the X-ray beam. A key

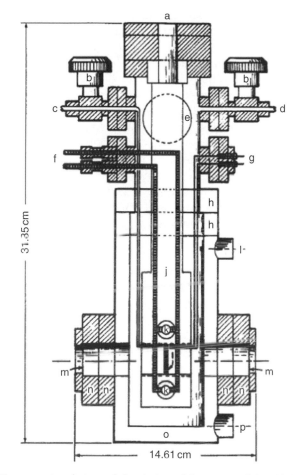

FIGURE 23   Cross-sectional view of the design of Castner et al. (1990). (a) opening for liquid nitrogen, (b) valves for gas flow in and out, (c) gas inlet tube, (d) gas outlet tube, (e) evacuation port, (f) heating leads to halogen lamps, (g) thermocouple, (h) conflat® flange, (i) outlet for water jacket, (j) quartz Dewar, (k) 500-W halogen lamps, (l) self-supporting sample wafer, (m) Mylar® windows, (n) conflat® flange, (o) water jacket, (p) inlet for water jacket. Reprinted with permission from (Castner et al., 1990). Copyright 1990 American Chemical Society.

characteristic of the cell is that it had a low dead volume, as the reactor was completely filled with catalyst powder. The optimum absorber thickness required for XAFS measurements was obtained by using an appropriate dilution material. Precautions were taken to protect the beryllium windows by avoiding direct contact with the reaction gases. The entire reactor cell was enclosed in a heater to ensure nearly uniform catalyst temperature. A versatile gas handling unit controlled the gas flow, gas composition, and total pressure in the cell. The investigators stated that during several catalytic test experiments reliable kinetic data were obtained.

In the same timeframe, Evans et al. (1990) developed an XAFS spectroscopy cell for the investigation of air-sensitive supported organometallic catalysts for propene metathesis. In this cell, the powder sample was supported on a silica frit through which the gases pass. The frit was mounted on a stainless steel block that also incorporated the heating elements and thermocouple and was held between two beryllium windows 4 mm apart. This sample holder was in turn supported in an outer housing with small beryllium windows for transmission XAFS spectroscopy and a larger window for fluorescence measurements. The cell could be operated under high vacuum or at atmospheric pressure.

In an early application of energy dispersive XAFS spectroscopy, Keegan et al. (1991) performed time-resolved measurements by using a "stage of the type commonly used for microscopy" to hold the catalyst sample in the beam. The design consisted of a small furnace with Kapton® windows that held a pressed wafer of the catalyst. Unfortunately no other detail is provided in the report of this cell design.

Zhang et al. (1991) published a relatively simple transmission XAFS cell in 1991. Their design incorporated a cylindrical body with sealed Kapton® windows at each end through which the reaction gas flowed. A heating coil was wrapped around the cylinder, and a liquid nitrogen vessel provided cooling of the sample. The sample was mounted on an insert and sealed on the holder with Kapton® tape. This insert was placed inside the cylindrical cavity. It is not clear from the description provided how the gas in the cylindrical cavity made contact with the catalyst in the sample holder.

The same year saw the publication of the first true plug-flow cell design by Clausen et al. (1991). This publication focused on a cell designed specifically for XRD measurements of catalysts, combined with online measurement of conversions of the catalytic reaction occurring in the cell; the same design was later applied by the same group for XAFS investigations (Clausen et al., 1993). The Clausen et al. cell design was an attempt to address the issues of (1) difficulties of operating at high temperatures and pressures, (2) inhomogeneity in sample and cell temperature (e.g., because the windows were cooled), (3) the presence of a large dead volume, (4) poorly defined gas flow patterns, e.g., gas bypass, and (5) difficulties in ensuring that the catalyst that is active is the same catalyst that is being probed by the X-ray beam. The reactor design is simple and elegant in concept, as shown in Figure 24. The cell has typically been a quartz (or other glass) capillary tube connected to stainless steel tubing via compression fittings. The authors pointed out that the tube could also be fabricated from other low absorbing materials, such as

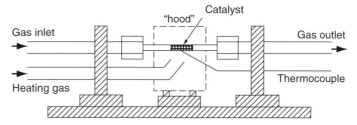

FIGURE 24 Schematic drawing of combined XRD/XAFS cell incorporating a quartz capillary (Clausen et al., 1991). Reprinted from (Clausen et al., 1991), Copyright 1991, with permission from Elsevier.

carbon or silicon, and they stated that the dimensions of the tube depend on the specific application but are typically 0.4 mm outside diameter with 0.01 mm wall thickness. The catalyst powder (106–150 μm sieved particles were mentioned in the publication) is loaded between two pieces of quartz wool. Heating is accomplished by a stream of hot air or nitrogen passing over the capillary tube. A small Kapton® hood covers the tube to maximize temperature uniformity of the catalyst bed. A thermocouple, placed about 1 mm from the sample is used to measure the catalyst temperature. Temperature control, via a Eurotherm controller, is achieved by either regulating the flow rate of the heating gas or the power to the electric resistance heater. The authors measured the axial temperature profile and stated that the catalyst bed temperature in the range 373–673 degrees varied less than 3 K over a 2 cm zone. Rapid temperature ramps were possible because of the small mass of the reactor; for example, the catalyst could be heated from room temperature to 673 K in a few seconds. Any potential effects of pressure drop were also discussed in the evaluation of the design. The reactor was used at temperatures up to 725 K and pressures up to 50 bar.

Although this design addresses all of the issues mentioned by the authors, the downside has to be the lack of structural consistency of the quartz or other glass tubes that are the heart of the design. For example, the authors stated that "from a safety point of view... small imperfections in the glass may substantially reduce the load at which the glass breaks. It is therefore not possible *a priori* to give an exact upper limit for the conditions at which specific tubes are safe to use."

The impact of the paper reporting this cell design should not be underrated, as it began a pattern of use of a whole new type of cell for XAFS spectroscopy of catalysts under reaction conditions. The cell enabled simultaneous measurements of X-ray absorption spectra and catalytic activity of the same material under conditions that are ideal for catalysis: plug flow. However, although data obtained with a capillary-type cell will provide a more accurate measurement of catalyst performance, the spectroscopic data could be compromised.

First, if a capillary cell is used, the X-ray beam size by default has to be small—at least in the vertical direction. So, if the experiments are conducted on a nonfocused beam line then the X-ray flux irradiating the sample is lower and this will degrade the signal/noise ratio (S/N). Typically in the pressed-wafer cell designs the size of the wafer is large to allow a large X-ray beam size to be used.

Second, the use of meshed particles versus a pressed wafer will typically lead to nonuniformity of X-ray absorption thickness. This can be directly observed by placing an X-ray sensitive camera behind the sample: a sample of a powder pressed into a wafer is spectroscopically more uniform than a catalyst bed of meshed particles. Naturally the contrast becomes more extreme as the meshed particles become larger. Moreover, if the sample is spatially nonuniform then severe constraints are placed on the positional stability of the X-ray beam. Any motion of the position of the X-ray beam will then probe different thicknesses of the sample, with direct consequences on the measured S/N. From the perspective of XAFS spectroscopy, any nonuniformity of the sample thickness could directly affect the accuracy of the measurement of the amplitude of the X-ray absorption coefficient. It is the amplitude that contains information about the coordination number and site disorder. As has been discussed elsewhere (Koningsberger and Prins, 1988), these amplitude distorting effects are given the general heading of "thickness effects." In brief, a thickness effect occurs when part of the incident X-ray beam is not attenuated by the sample. In the case of meshed particles this would be in the form of pinholes in the sample.

Third, there may be a concentration gradient of reactants and products along the length of the catalyst bed. If the structure of the catalyst depends upon the composition of the gas phase, then an average of the various structures will be measured. There is little discussion of this topic in the literature of XAFS spectroscopy of working catalysts. An extreme example of structural variations within a sample is discussed in Section 6, where there is a discussion of XAFS spatially resolved spectra recorded to allow direct observation of the axial distribution of phases present. If the XAFS data are not measured with spatial resolution, then it is recommended that XAFS data be measured under differential conversion conditions. However, if the aim of the experiment is to relate the catalyst structure directly to that in some industrial catalytic processes, then differential conversion conditions will only reflect the structure of the catalyst at the inlet of the bed. To learn about the structure of the catalyst near the outlet of the bed, the reaction has to be conducted at high conversions. If it is anticipated that this operation will lead to variations in the catalyst structure along the bed, then the feed to the micro-reactor should be one that mimics the concentration of reactants toward the downstream end of the bed (i.e., products should be added to the reactants).

The design of a cell that is compatible with more than one technique was illustrated by Couves et al. (1991). They briefly presented a cell design that allowed the collection of both XAFS spectra (in a dispersive geometry) and X-ray diffractograms. The cell operated at atmospheric pressure and included a pressed wafer of the catalyst. Little detail was provided, but the authors stated that the cell consisted of a custom-built Kanthal heating element embedded in a pyrophyllite block with a recess for the pressed sample. The sample could be heated to 1073 K in a flow of gas. More detail of the design was published by Dent et al. (1995). In essence, there is an outer container fabricated from aluminum that has the necessary Kapton® windows and water-cooled end caps. A pyrophyllite heat shield fits inside the outer container, and a wire-wound heater, into which the sample holder fits, is placed inside this heat shield. The furnace was capable of operation from 373 to 1473 K.

In the same year, the group reported another design (Sankar et al., 1995) allowing combined XAFS/XRD, but there was no provision for gas flow. According to this design, the sample is sealed in a 1.0 mm-diameter quartz capillary with a 10-μm wall thickness. The sealed capillary was placed in a specially designed furnace to allow transmission of the X-rays.

A transmission XAFS spectroscopy cell with a relatively long path length (10–12 mm) was briefly described by Mccaulley (1993). The powder catalyst was placed in a Kapton®-sealed sample holder. This holder had entrance and exit holes for gas flow that were press-fitted against matching holes in the body of an aluminum block incorporating cartridge heaters. There was no requirement for the gas to flow through the catalyst bed in this design, and the thermocouple measured the temperature of the aluminum block. The cell body was placed in containment housing. The maximum operating temperature and pressure were limited by the Kapton® tape (473 K and 1 bar).

A series of cell designs were published by Moggridge et al. (1995). These designs allowed only fluorescence yield XAFS data to be collected. The initial design was basically a beryllium cup heated by a cartridge heater and had a maximum operating temperature of 623 K, limited by the sealing materials. The authors determined that the primary heat loss mechanism is via thermal conduction, and so they increased the thermal resistance between the sample and the seals. These changes allowed use of the cell at a temperature of 873 K in air and at 773 K in 50 bar of H_2. The authors were concerned about the inertness of the beryllium, but stated that a passivating layer of oxide is typically formed and, provided that wet environments are avoided, it is remarkably inert. Although the cell was used to investigate many catalytic reactions, it was unsuitable for some, especially those carried out at temperatures exceeding 773 K and in the presence of corrosive gases.

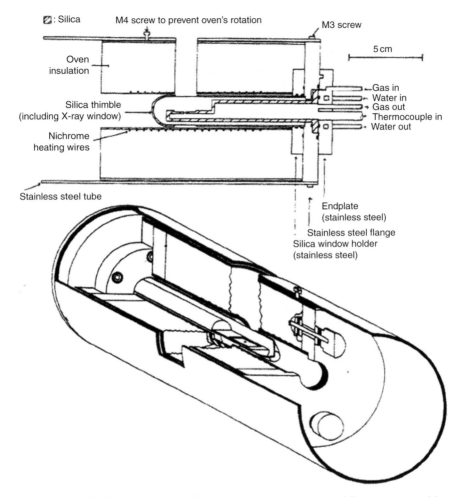

FIGURE 25 All-silica environmental cell for XAS measurements of fluorescence yield at temperatures up to 1273 K with sample in an atmosphere of corrosive gas (Moggridge et al., 1995). Reprinted from (Moggridge et al., 1995), Copyright 1995, with permission from Elsevier.

To address these concerns, the authors designed an all-silica reaction cell (Figure 25). The maximum operating temperature of this cell is approximately 1273 K, and it is limited only by the furnace windings and the silica window. This is the first published example of the use of silica as a window material. Although the thickness of the window was not stated, it likely limited the usable photon energy range.

Some of these designs are rather complex, requiring considerable machining and other specialized parts, and they are therefore relatively expensive, in addition to being physically large. These issues were

X-ray beam path

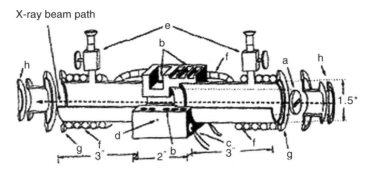

FIGURE 26   Cell design of Jentoft et al. (1996): (a) Sample holder; (b) liquid-nitrogen well; (c) cartridge heater; (d) thermocouple well; (e) gas inlet and outlet valves; (f) cooling-water line; (g) flange; (h) beryllium windows. Reprinted with permission from (Jentoft et al., 1996). Copyright 1996, American Institute of Physics.

specifically addressed by Jentoft et al. (1996). Their cell was made from a 316 stainless steel block that contained a sample mounted in a circular metal disc; the block was heated by 120-W cartridge heaters (Figure 26). Welded to the block were 3-in.-long stainless steel tubes at the ends of which X-ray-transparent windows (beryllium, Kapton®, or others) were mounted with quick-connect vacuum couplings. The O rings in the quick connect fittings were thermally protected by cooling water. The cell is capable of operation at temperatures up to 773 K and can be cooled to near liquid nitrogen temperature. The catalyst is a powder pressed into a slot in the stainless steel disc. There is a significant amount of dead space at the ends of the cell. A disadvantage of this design is the large mass of the stainless steel block, which affects temperature control. Furthermore, the thermocouple is not placed inside the catalyst bed but instead is placed a few millimeters away. Nevertheless, these reaction cells are relatively inexpensive, and the low cost allows the use of multiple cells, which leads to productivity improvements as simultaneous treatments of several samples could take place while XAFS data are being collected on another sample.

Bazin (1996) published a treatise on XAFS measurements of catalysts in the working state. He described a series of cells that his group developed over the years. The cells range from relatively simple to more complex. Unfortunately, in this work there are only schematic descriptions given for the cell designs; details are lacking. For example, the authors described a high-temperature and high-pressure cell but gave little detail other than stating that the reactor was made of stainless steel with entrance and exit windows made of beryllium, rated up to 50 bar. The sample holders were initially fabricated of boron nitride, but the authors stated that these repeatedly broke and were replaced with aluminum sample holders, but with the obvious limitation of lower maximum operating temperature. The holder was sealed with graphite sheets. Machined channels in the sample holder allowed the gas to flow from the entrance line attached to the mounting block, through the catalyst, and back to the exit line attached to the mounting block. Although it was not directly stated by the authors, we infer that the form of the catalyst used in their measurements was a powder that filled the aluminum sample holder.

The initial design of Sankar et al. (1995) that allowed collection of combined XAFS and XRD data was further modified in a design published in 1997 (Shannon et al., 1997). This was a simple design comprising a stainless steel body through which a slot was cut to allow the recording of both transmission and energy dispersive XRD data. This metallic body supported the catalyst wafer, which was sandwiched between two 50-µm-thick Kapton® windows. Heating was achieved by either a band-heater clamped around the external diameter of the cell or a jet of hot air directed onto the sample area. Limitations of this design include inhomogeneous heating and indirect temperature measurements, in addition to the limitation of the maximum temperature imposed by the Kapton®.

Another design incorporating a pressed wafer form of catalyst was reported by Meitzner et al. (1998). A key feature of this design was that the gas flow was forced through the pellet with essentially no bypassing, while at the same time ensuring that the temperature was uniform, with an adequate preheat zone to ensure that the reactants were at the same temperature as the catalyst. The cell comprised three major components: the sample insert, the beam path tube, and a jacket for heating and cooling (Figure 27). The sample wafer is held on a shelf on the end of the insert and retained by a ring that is threaded on the outside. The heating and cooling jacket is an aluminum block that completely surrounds the sample region. Heat is supplied by cartridge heaters, and a maximum temperature of 873 K was routinely obtained. If the XAFS data need to be recorded at subambient temperatures, cooling is by liquid nitrogen flowing through channels bored in the aluminum block. The design, as published, was only for transmission XAFS experiments, but the authors described a modification that would allow fluorescence detection.

The advantage of simplicity is illustrated in the design reported by Pettiti et al. (1999); this was assembled mostly with commercial parts. The cell allowed thermal treatments at temperatures up to 823 K, in either a reducing or oxidizing atmosphere, and it could be used in transmission and fluorescence modes of detection. The body of the cell was a stainless steel six-way cross. One flange of the cross carried the heater and the opposite flange the sample holder. The heater was a simple design of Ni–Cr wire wrapped around a silica cylinder. The sample holder was

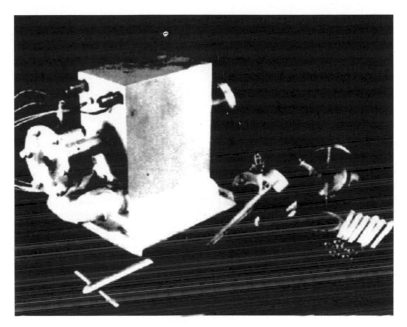

FIGURE 27 Photograph of the cell design of Meitzner et al. (1998) with insulation and sample insert removed. Reprinted with permission from (Meitzner et al., 1998). Copyright 1998, American Institute of Physics.

held inside this heater cylinder. Liquid nitrogen could be used for cooling. The clear disadvantage of this design is the very large volume of the cell compared with the catalyst volume. Nevertheless, it was effective for treatments of catalysts prior to XAFS measurements.

Barton et al. (1999) briefly described their version of a plug-flow quartz capillary reactor cell. They used quartz capillaries with 0.9 mm diameter and 0.1 mm wall thickness. The heating was by a copper heat sink that was in turn heated by cartridge heaters. The quartz tube was packed with meshed particles of catalyst.

Reaction conditions encountered in heterogeneous catalysis are quite diverse, and thought must be given to the design of reaction cells to suit a particular application. An excellent example is the cell designed for the investigation of catalysts under DeNOx catalysis conditions by Revel et al. (1999). This design focused on the gas flow through the powdered catalyst bed. The cell was designed to ensure that there was no preferential circulation of the reactant gases in order to avoid the possibility that part of the catalyst would be under gas flow and part would not. This goal was achieved by using sintered stainless steel tubes above and below the catalyst bed. The design, showing the gas inlet and outlet, is illustrated in Figure 28. Unfortunately, in this design the thermocouple could not be

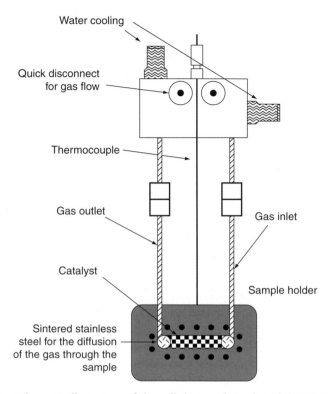

FIGURE 28   Schematic illustration of the cell design of Revel et al. (1999). Reprinted from (Revel et al., 1999), Copyright 1999, with permission from Elsevier.

placed in the catalyst bed; instead it was placed 2 mm from the sample. This placement meant that temperature calibrations were necessary. A furnace, wound with resistance wire, encircles the sample holder, and the whole arrangement is placed in a water-cooled thermal shield. The maximum working temperature is 823 K, limited by the thermal and mechanical weakness of the carbon foil used to seal the sample compartment.

Liu and Robota (1999) described a plug-flow reactor cell used to determine the mechanism of nitric oxide selective catalytic reduction (SCR) by hydrocarbons catalyzed by CuZSM-5 (Figure 29). The authors made the important point that any reactor design that will also be useful as a spectroscopic cell has to ensure a substantial conversion of the catalytic reaction while incorporating only a small amount of catalyst and producing sufficient product for analysis. The reactor must also sustain any stress caused by repeated heating and cooling. Their compact design incorporated a removable sample holder comprising two parts: a fixed tube and a matching threaded head. A 0.05-mm gap is left between them when the cell is assembled. An alumina filter membrane disk, with greater than 60% opening of straight channels, is placed in this gap.

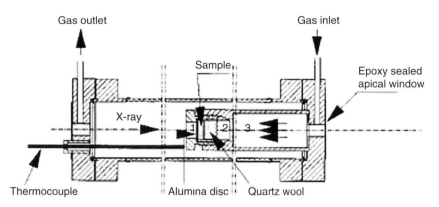

Gas outlet

Gas inlet

Sample

Epoxy sealed
apical window

X-ray

Thermocouple   Alumina disc   Quartz wool

FIGURE 29   Design of the plug-flow reactor cell of Liu and Robota (1999). Reprinted with permission from (Liu and Robota, 1999). Copyright 1999 American Chemical Society.

During the experiment, a layer of catalyst powder (or sieved meshed particles) is packed into the center tube holder. The alumina membrane disc holds the powder sample at the downstream side while allowing the reactant to flow through with minimal resistance. At the upstream side, a small amount of quartz wool is inserted to support the powder. This whole sample holder assembly is fastened to the outer connector and this whole tube assembly is enclosed in the quartz tube that is surrounded by a resistive heating unit. At each end of the tube, X-ray transparent windows are mounted. The thermocouple is placed close to the sample holder. The long gas inlet tube is heated by the tubular furnace, ensuring that the gas is preheated before it reaches the catalyst sample. The upper working temperature is 973 K, which is only limited by the materials of construction.

Schneider et al. (2000) described a cell for characterization of automotive postcombustion catalytic converters. Given the nature of the catalysts (incorporating low loadings of the active metals), the data had to be collected in fluorescence mode by use of a solid-state detector. Thus, the detector had to be close to the sample in order give a large solid angle. The heart of the design is a sample holder made of pyrolytic boron nitride that holds the powdered catalyst. This holder is a pipe that allows a high flux of the reactive gases to go through the sample as in the case of industrial converters. The researchers stated, however, that when operating with the necessary gas flow, the boron nitride became more and more permeable to hydrogen or helium with increasing temperature, and thus they used carbon foils to seal the cell. They described the process for fabricating the pyrolytic boron nitride holders by chemical vapor deposition. The furnace surrounding the boron nitride has Nichrome® heater windings for high-temperature operation. The whole assembly is placed inside the water-cooled stainless steel jacket, as shown in Figure 30.

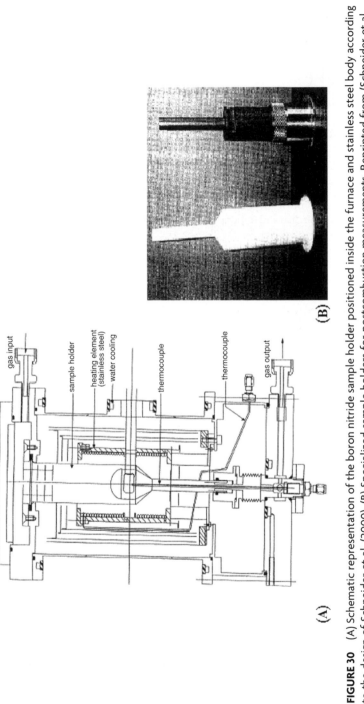

FIGURE 30 (A) Schematic representation of the boron nitride sample holder positioned inside the furnace and stainless steel body according to the design of Schneider et al. (2000). (B) Specialized sample holders for postcombustion measurements. Reprinted from (Schneider et al., 2000). Copyright 2000, with permission from IOS Press.

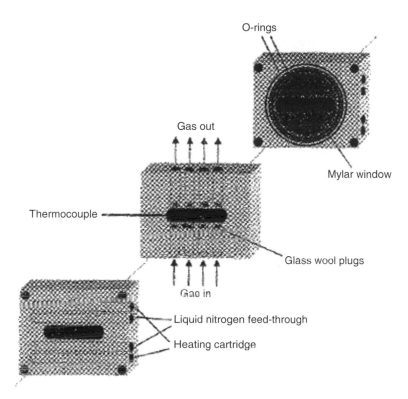

FIGURE 31 Schematic view showing the two end pieces and the cell body according to the design of Odzak et al. (2001). Reprinted with permission from (Odzak et al., 2001). Copyright 2001, American Institute of Physics.

A once-through flow reactor that holds a powder catalyst was described by Odzak et al. (2001). Their design approximates an ideal plug-flow reactor which for many reactions operates almost isothermally with differential conversions while also providing high-quality XAFS spectra. The design is illustrated in Figure 31. The reactor body is a stainless steel block heated by cartridge heaters; the block also incorporates channels for cooling by liquid nitrogen. The gas contacts the catalyst sample by flowing from top to bottom through four 1/8-in. diameter ports. The cell is sealed with Mylar® windows held in place by stainless steel end pieces and sealed with two O-rings. A thermocouple is located within the catalyst bed to monitor the temperature in the reaction zone. The authors stated that the cell allows collection of XAFS spectra with gas flow rates from 10 to 500 ml per minute and temperatures from 230 to 470 K. A limitation of this reactor design is the upper working temperature imposed by the O-rings and window material. Furthermore, the gas flow through the sample bed is not uniform.

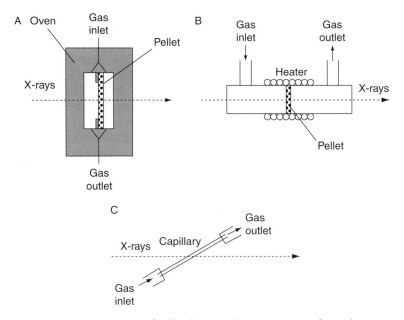

FIGURE 32 Various categories of cells allowing characterization of samples in reactive atmospheres: (A) Pressed wafer with gas flowing on both sides of the wafer. (B) Pressed wafer with gas forced to flow through the wafer. (C) Plug-flow reactor consisting of quartz capillary (Grunwaldt et al., 2004). Reproduced by permission of the PCCP Owner Societies.

A key paper was published by Grunwaldt et al. (2004). Instead of describing a new design of an XAFS spectroscopy cell the paper discusses some of the important advantages and limitations of various cell designs, particularly with respect to measurements of time-resolved kinetics and spectra. The authors' major point is that if the reaction is fast, then the cell design should be such that the kinetics are not affected, and if they are, the effect is understood and acknowledged. Grunwaldt et al. divided the designs of cells into three general categories (Figure 32): (1) a pressed wafer cell in which the gas bypasses the catalyst, (2) a pressed wafer cell in which the gas passes through the wafer, and (3) a plug-flow reactor (specifically a quartz capillary tube cell). The focus of the work was the evaluation of film diffusion, pore diffusion, and the effectiveness factor (determined by the Thiele modulus) representing the cells of type (1) and (3), and how these characteristics could affect the spectroscopic observations. The authors showed that for these two cell designs the rate of reduction of an oxide is higher in the capillary cell than in the pressed wafer, with the difference attributed either to mass transfer limitations or different flow conditions.

However, the authors noted that both cells can give similar results during reaction if the ramp rate is low, and the pore diffusion of the molecules sufficiently fast. Another interesting point discussed by Grunwaldt et al. is the effect of boron nitride as a diluent in the pressed wafer (in XAFS experiments, boron nitride, which consists of low-Z elements, is often added to make the sample more X-ray transparent). If boron nitride is used in pressed wafer pellets, then pellets with significantly lower porosity are obtained and the low porosity in turn affects the rate of mass transfer. This result suggests that high-porosity materials should be used as diluents rather than the low-porosity boron nitride. Clearly, as the authors pointed out, in any investigation when the activity of the catalyst is being measured at the same time that spectroscopic informa tion is being collected, it is critical to ensure that the spectroscopic data are obtained from the catalyst fraction that is actually active. They also showed by the use of an X-ray sensitive camera, that the transmission of the X-ray beam is uniform through a pressed wafer, whereas there is some inhomogeneity in the transmission through a sieved fraction of catalyst in a capillary cell. They stated that this inhomogeneity may be minimized by use of a sieved fraction with smaller particle size.

The field of XAFS spectroscopy reactor cell design continues to be as innovative as ever. In 2005, there were at least five papers published describing novel cell designs. For example, another cell-within-a cell-design was described in that year by Jentoft et al. (Jentoft et al., 2005). This design (Figure 33) comprised an internal powder catalyst bed reactor in a helium-filled temperature-controlled outer enclosure. This cell was specifically designed to record XAFS data characterizing powered catalyst samples in fluorescence mode, while simultaneously recording data for a reference foil in transmission mode. The motivation was to collect XAFS data from powders rather than pressed wafers, as some materials undergo structural changes when subjected to the high pressures necessary to form wafers. The inner reactor body is fabricated from stainless steel. The powdered catalyst (0.3–0.9 cm^3) is supported on a glass frit, the gases flow through the catalyst bed, and a Kapton® window serves as the X-ray-transparent medium. The temperature is controlled by a thermocouple placed in the catalyst bed. The maximum operating temperature was stated to be 723 K.

The growing interest in the use of supercritical fluids in heterogeneous catalysis has led to the design of an XAFS spectroscopy cell specifically for such systems. Grunwaldt et al. (2005) reported the design of a batch reactor cell for investigation of catalytic reactions in supercritical carbon dioxide, illustrated in Figure 34. This novel 10 ml batch reactor has dual X-ray paths so that when the beam passes through the middle window of the reactor the liquid phase in the reactor is probed, and when the beam goes through the bottom window the solid phase is probed. The design parameters are a maximum temperature of 493 K and a maximum pressure of 150 bar. The cell is lined with a polyether–ether–ketone (PEEK) insert for chemical

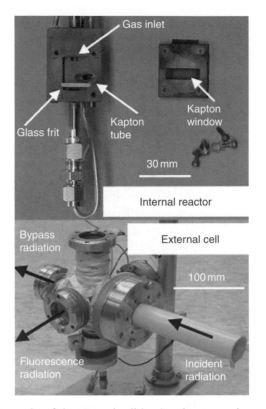

FIGURE 33 Photographs of the internal cell (top) and outer enclosure (bottom) of the powder cell of Jentoft et al. (2005). Reprinted with permission from (Jentoft et al., 2005). IOP Publishing Ltd.

resistance, and it is heated with cartridge heaters. A thermocouple is placed inside the reactor, and there is a provision for stirring with a magnetic stirrer. Some limitations of this design are the relatively long path length, which limits X-ray energies to >7 keV, and the maximum operating temperature imposed by the use of a thermoplastic insert.

The correlation of spectroscopic data between model and real catalysts has always been a concern in catalyst characterization. Weiher et al. (2005) tried to address this issue with a cell design that was compatible with both model catalysts (e.g., submonolayer amounts of metals deposited on a silicon wafer) and real catalysts such as high-surface-area supported metals. Moreover, they also wished to have a design in which plug-flow conditions existed for the powder catalyst experiments.

Their solution of these design problems is illustrated in Figure 35. A base mount, onto which different sample holders fit, incorporates the heating and cooling, and the thermocouple. The separation of the sample

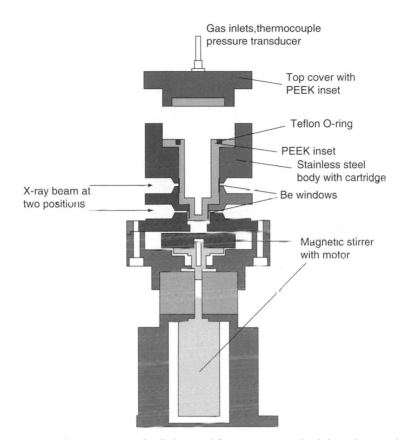

FIGURE 34 Schematic view of cell designed for monitoring of solid catalysts in the presence of supercritical fluids (Grunwaldt et al., 2005). The two X-ray beam paths are shown. Reprinted with permission from (Grunwaldt et al., 2005). Copyright 2005, American Institute of Physics.

holder and heating and cooling block provides flexibility for adapting the design to various experimental needs. The maximum sample temperature achievable with this design is 900 K, with a maximum heating rate of 10 K/min. For investigations of model catalysts, the wafer is held on an aluminum support plate and sealed with Kapton® film. The gas is distributed over the surface of the catalyst via a nozzle. A capillary is placed above the sample surface to withdraw gas to a mass spectrometer or gas chromatograph. The cell has a low dead volume. In experiments with a powder catalyst, the sample is held in a slot machined in an aluminum frame, and it is sealed with aluminum or Teflon® foils. A thermocouple placed just below the catalyst measures the temperature. Plates machined with various sizes of slots can be used to hold various volumes of catalyst. The cell has worked reliably in the pressure range

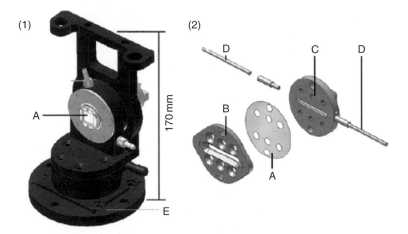

FIGURE 35 (1) Heating block of the Weiher XAFS cell (Weiher et al., 2005): (A) heating cartridge with thermocouple; (E) rotation stage for sample alignment. (2) Cell insert: (A) X-ray transparent window; (B) window seal; (C) sample holder; (D) gas connections. Reprinted with permission from (Weiher et al., 2005). Copyright 2005 Wiley-Blackwell.

10^{-5} to 3 bar and the temperatures range 100–900 K. Given that the materials of construction of the cell are all metal, the design has an advantage over the glass capillary cells in that it is more robust. However, the design does have both temperature limitations (associated with the Teflon® or aluminum window) and pressure limitations (associated with the window). It also has some chemical compatibility concerns associated with the aluminum from which it is machined.

Girardon et al. (2005) discussed a transmission XAFS spectroscopy cell that is compatible with characterization of catalysts in the working state and with online analysis of reaction products. The cell consists of several plates of stainless steel and boron nitride linked together with graphite seals. The catalyst powder is held in a recessed channel in a central boron nitride plate. The cell is heated with cartridge heaters, and the thermocouple is placed in a channel in the central boron nitride block close to the catalyst bed. Gas flow is through the catalyst bed from top to bottom. The volume of this bed (0.45 cm^3) is fixed by the dimensions of the recessed channel in the boron nitride plate—but it can be adjusted by having several different plates of different dimensions. The authors claimed that the cell is leak free and operational at temperatures up to 623 K in O_2 and 673 K in H_2, all at atmospheric pressure.

Kawai et al. (2006) extended the design of a high-pressure cell to accommodate operating conditions of 50 bar and 723 K. The cell had low volume, with flat windows that were chemically, structurally, and thermally stable, fabricated from a novel material, cubic boron nitride, which was synthesized by the authors by sintering hexagonal boron nitride. The windows were 0.8 mm thick. The catalyst was in the form of a pressed wafer, and the path length of the cell was 3 mm. The cell body was heated by cartridge heaters. It was not stated in the publication whether the thermocouple was placed in the catalyst bed.

A simple cell design for XAFS spectroscopy of liquids was reported by Maris et al. (2006). This cell was specifically designed for investigation of solid catalysts in the presence of liquid saturated with H_2 at 40 bar. The catalyst powder (particle size 53–75 µm), was hand-pressed into a stainless steel sample holder. The windows were fabricated from 1-mm-thick PEEK disks. The overall assembly was sandwiched together by stainless-steel flanges. There was no mention in the report of a thermocouple, and the heating was simply via electrical heating tape wrapped around the cell. The temperature uniformity of the catalysts bed in this design was not discussed, but this has to be a concern.

Bare et al. (2006) described two cell designs, one for transmission XAFS and one for fluorescence measurements, whereby all of the critical components could be purchased commercially, with little machining required. The designs are simple, robust, and relatively low in cost. The basis of the designs is a quartz tube with windows fabricated from an appropriate material. The tube is heated by a clamshell furnace. The catalyst powder (with or without diluent) is hand-pressed into a quartz sample holder that is inserted into the quartz tube. Thus, a spectroscopically uniform sample is obtained without the high pressure compaction that is necessary to obtain a self-supporting wafer. The cell has been operated at temperatures from 80 to 1373 K, but the maximum working pressure is only approximately 1 bar.

The same group also described a plug-flow reactor cell (Bare et al., 2007) that is capable of operation at temperatures up to 873 K and a pressure of 14 bar. The concept is similar to that of the quartz capillary cells, but the reactor tube is fabricated from beryllium, as shown in Figure 36. The thermocouple is embedded within the catalyst bed, thereby ensuring that the actual catalyst temperature is measured. The tube is heated by a controlled flow of hot air or nitrogen. An advantage of this cell over the quartz tube cells is that the material of construction, beryllium, is robust, and its strength is not sensitive to small scratches. Moreover, the inside diameter of the tube can be fabricated to any given size so that the reactor tube can be sized appropriately for the absorption coefficient of the element under investigation. The beryllium walls allow use of the reactor at substantially lower energies than are applicable with the quartz tube reactors. A disadvantage is that the chemistry being conducted in the reactor has to be compatible with beryllium.

Hannemann et al. (2007) designed a versatile cell that is compatible with both transmission and fluorescence XAFS spectroscopy, XRD, and both gas and liquid environments; it also allows for online monitoring of

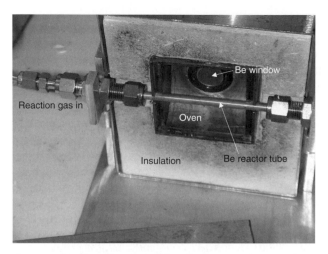

FIGURE 36   Photograph of beryllium tube plug-flow reactor (Bare et al., 2007) and oven with one-half of the oven removed to show placement of the tube. Reprinted from (Bare et al., 2007), Copyright 2007, with permission from Elsevier.

products by mass spectrometry or IR spectroscopy. The design is a "cell within a cell" format, with a small compartment to hold the catalyst, preferably in the form of sieved powder (typically 100–200 μm), although pressed wafers can also be used; the catalyst is held inside an outer body that contains Kapton® windows. A schematic diagram of the cell is shown in Figure 37. The sample is held at the appropriate angle for optimum geometry for fluorescence or transmission, and there are appropriate windows in the outer body for measuring either transmitted (or diffracted) and fluorescence photons. The sample can be heated to 973 K or cooled to liquid-nitrogen temperature. The window material of the insert can be fabricated from graphite, Kapton®, or aluminum. Given that the reaction cell is only a small component of the overall design, several different inserts could be used, depending on the desired thickness of the sample required to optimize the XAFS measurements. The gas flow dynamics of the sample compartment are designed to mimic a laboratory plug-flow reactor. No mention was made of the pressure rating of the cell, but it is assumed to be limited by the thickness and size of the windows on the cell insert.

A recent example of a cell designed for a particular catalytic chemistry is that of Hass et al. (2007). The focus of the investigation was the structure of a $AgIn/SiO_2$ catalyst during the hydrogenation of acrolein. The corrosive nature of acrolein necessitated that the reactor be fabricated from stainless steel and that no O-rings be used. A screw-cap design, similar to that of Meitzner et al. (1998), was used both for attaching the windows and holding the catalyst in the reactor. However, Hass et al. modified the design so that the catalyst could be used in the form of a powder, which is

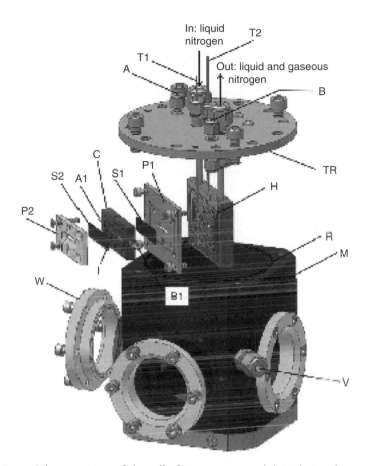

FIGURE 37 Schematic view of the cell of Hannemann et al. (2007): A and A1, gas inlet for reactants; B and B1, gas outlet for products; C, reaction cell; H, heating plate; I, inset; M, main body with Kapton® windows (W) and seals (R) and outlet for vacuum pump (V); P1 and P2, plates for assembly of cell; S1 and S2, seals for cell; T1 and T2, thermocouples. A Teflon® O-ring (TR) may be optionally used. Reprinted with permission from (Hannemann et al., 2007). Copyright 2007 Wiley-Blackwell.

held between two quartz frits. The cell, for transmission geometry only, is capable of operation at temperatures up to 773 K and pressures of 20 bar.

While XAFS spectroscopy is a powerful catalyst characterization method it is clear that XAFS spectroscopy combined with other complementary techniques offers the possibility of providing a more complete understanding of catalyst structure. The designs of cells that allow both XAFS and XRD data to be collected have already been described (above). More recently, the combination of two, or even three, other complementary techniques to XAFS spectroscopy has now been successfully demonstrated. Beale et al. (2005) briefly described a design combining UV-vis spectroscopy, laser Raman spectroscopy, and XAFS spectroscopy. Not only were three techniques coupled, but the design ensured that the same area on the catalyst was probed by all three methods. More detail of the design was given by Tinnemans et al. (2006) (Figure 38). The heart of the design is a custom-made glass capillary that holds the catalyst, in powder form, in such a way that optimum absorption lengths are obtained for the three techniques. The capillary is held in a furnace, and the two optical spectroscopy techniques are coupled to the reactor via glass fibers.

This group has further developed multitechnique cells, and designed a cell for SAXS/WAXS and XAFS, and SAXS/WAXS/XAFS and UV–vis spectroscopy (Beale et al., 2006; Grandjean et al., 2005). They reported use of this cell to investigate hydrothermal crystallization processes of inorganic catalysts such as CoAPO-5. The synthesis cell design was simple in concept; essentially a mini-autoclave, with a usable volume of 2 mm^3, and mica windows, heated by an insulated aluminum block.

In the preceding few years there has been a growth in the use of parallel screening and in the use of microstructured reactors in catalysis research; thus, it is not surprising that research groups have designed spectroscopic cells to characterize catalysts in these types of reactors. Sankar et al. (2007) recently published the design of a microstructured reactor cell for XAFS investigations. The microstructured reactor was fabricated by photolithography, deep reactive ion etching, and anodic bonding. The design incorporated is a single microchannel 8 mm wide and 120 μm deep. The catalyst was a silver film coated on the reaction channel, and the reactor was sealed with a glass cover with inlet and outlet holes. The reactor was heated with a custom-made heating assembly. The authors showed EXAFS data at the Ag K-edge recorded at 773 K during the oxidative dehydrogenation of methanol to give formaldehyde.

The coupling of a microreactor array in combination with an X-ray camera was used by Grunwaldt et al. (2007) to record the XAFS spectra of ten catalysts simultaneously. In this feasibility exercise, the arrangement comprised a spectroscopic cell with ten sample compartments (2 or 5 mm in thickness) that were filled with meshed and sieved catalyst particles. There was little detail provided about the heating and gas flow arrangement, but the cell could be heated (although the preliminary data were all recorded after flowing either a H_2–helium or O_2–helium mixture at room temperature).

3.3. Summary

Since the earliest days of the investigation of catalysts by XAFS spectroscopy, it has been realized that specialized equipment is needed to make the measurements. The cell designs span a huge range, and even to this

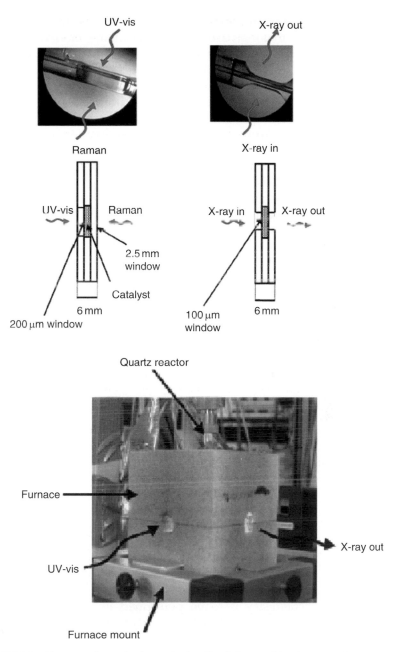

FIGURE 38 Photographs and schematic details of the combined UV–vis/Raman/XAFS cell of Tinnemans et al. (2006). Reprinted from (Tinnemans et al., 2006), Copyright 2006, with permission from Elsevier.

day papers are still being published with new modifications and improvements. New materials are becoming available; modern synchrotron beam lines often have focused X-ray beams allowing the cell designs to be compact; and XAFS spectroscopy is being applied to new catalytic materials and reactions necessitating specialized designs. Moreover, the ability to apply more than one spectroscopic technique to the sample in the cell is providing insight into the structure and chemistry of the catalyst. We believe that there will be continual innovation in cell design over the next decade.

It is evident that there are many designs of cells compatible with the measurement of XAFS data characterizing a working catalyst. There is however no single design that is suitable for the investigation of all catalysis research problems; and there is no single "best" design. Instead each of the designs discussed earlier have certain advantages and disadvantages—and how these rank against each other depends on the research problem. In an attempt to provide guidance we list below the most important criteria that should be considered when choosing or developing a design.

- Are the XAFS data going to be collected in transmission or fluorescence yield mode? By the nature of the experiment transmission cells are simpler in design than those for fluorescence. The complexity increases further if the cell has to be capable of allowing both transmission and fluorescence XAFS spectra to be collected.
- In which X-ray energy range are the XAFS data to be collected? The X-ray energy places constraints on the thickness and material of the window. For example, if the cell is to be operated for the collection of XAFS data at energies >20 keV, the window material can be thicker and more absorbing than for a cell used for the collection of XAFS data at energies <7 keV (Table 3).
- What are the maximum and minimum operating temperatures and pressures? A cell that is to be operated at atmospheric pressure at 473 K can be far simpler in design than one operated at 50 bar and 1000 K. These parameters place constraints on the window materials (e.g., Kapton® vs. beryllium), the sealing mechanism of the cell (e.g., can O-rings be tolerated, will water cooling of the seals be necessary?), and the thicknesses of the reactor walls and windows.
- What catalytic chemistry is to be performed in the cell? This will determine the choice of materials of construction, particularly the windows.
- Does the research require that accurate catalytic activity data be collected simultaneously with the XAFS data? If the answer is yes, then this is the major reason for using a plug flow-type cell (e.g., a capillary),

as these mimic laboratory microreactors. However, as discussed earlier, there are compromises with regards to XAFS spectral quality. Pressed wafers of the catalyst are the most uniform spectroscopically but may result in poor gas diffusion.

- Are the XAFS data to be collected on an insertion device (ID) or bending magnet (BM) beam line, and what X-ray focusing optics are available? The beam line optics should also be taken into account in the design of the cell. If the XAFS data are to be collected on an ID beam line, then, given the naturally small X-ray beam size (less than 1 mm^2), the whole reaction cell (or at least the sample size) can be significantly smaller. If the sample is nonuniform, data quality will be more affected on an ID beam line than a BM beam line.
- Does the catalyst need to be heated (or cooled) rapidly? If it is necessary to heat the catalyst more than a few degrees per minute, then the overall thermal mass of the cell has to be taken into account. For example, the quartz capillary-type cells can be heated (and cooled) rapidly (tens of degrees per minute), whereas cells with a large mass have heating rate limitations. Fast heating and cooling can be desirable to take optimum advantage of available beamtime.
- Is cost an issue? The cost could influence the design by affecting whether the researcher could have more than one cell available, as a spare in case of technical problems, or to have one in use collecting XAFS data while another is being prepared offline.
- Is XAFS the only spectroscopy that will be used with the cell? If the cell needs to be compatible with more than one technique then this has to be one of the major design criteria.
- What type of catalysis is going to be performed? Clearly, the entire design principle of the cell is determined by whether the reactants are in the gas or liquid phase, or whether a slurry is present, etc. Questions regarding the corrosive nature of the reactants and products need to be addressed from a safety standpoint.
- Does the researcher have access to a skilled mechanical designer and machine (or glass) shop? Many of the designs discussed earlier require such support. Otherwise, the simpler designs are preferred.
- How important is it to measure the temperature of the catalyst bed? If the reaction is highly exothermic or endothermic, it is essential to have an integral thermocouple.

Considering the number of design parameters, it is not necessarily a simple process of choosing a particular design. It is likely that compromises will have to be made, with deviations from the optimal design.

4. XAFS SPECTROSCOPY OF SAMPLES IN REACTIVE ATMOSPHERES UNDER STATIC CONDITIONS

4.1. Introduction

The number of papers reporting XAFS data used to determine the structure of a catalyst under reaction conditions is only a small subset of the total number of XAFS papers in catalysis science. In this subset of papers, the ones including catalytic activity data measured simultaneously with the XAFS data are a small fraction. The two major reasons for the low number of such investigations are believed to be the limiting requirements of the reactor design coupled with the reaction chemistry, and the temperature dependence of the σ^2 factor in the EXAFS equation. Regarding the former, if the conversion is high, then there will be significant concentration gradients across the catalyst bed, a situation encountered commonly in oxidation reactions. In the large majority of cases, the X-ray beam probes most or all of the catalyst bed, giving an average signal that may not be representative of the catalyst structures, because the gas phase composition often affects the catalyst structure. Thus, ideally the reactor is operated with differential conversions, often in the range of a few percent. In this manner it can be assured that the gas-phase composition is uniform and any structural changes can be related directly to the catalytic conditions. If the catalyst structure depends on the concentration of product, then the feed could be adjusted to include a known concentration of the product. Another potential problem could be the presence of hot or cool spots in the catalyst bed, expected when the reaction is strongly exothermic or endothermic.

An example of how a large concentration gradient in the reactor affects the catalyst structure was recently demonstrated for the partial oxidation of methane catalyzed by Rh/Al_2O_3 (Grunwaldt et al., 2006). This reaction was conducted at 556 K, the temperature at which there is light-off, with subsequent total conversion of the methane. In experiments with a relatively small (1 mm) X-ray beam, it was shown that the rhodium changed from an oxidized species to a reduced species, depending on the position in the reactor. The authors were able to collect a 2-D axial image of the rhodium species (Section 6).

When the goals include measurement of kinetics (Section 5), the challenge of the reactor design and conduct of the catalytic reaction is even greater. Some of the key issues have been discussed recently (Grunwaldt et al., 2004).

In the EXAFS equation, there is an exponential term, $\exp(-2\sigma_j^2 k^2)$, that damps the EXAFS signal. This term describes the disorder and includes the mean square disorder in the particular bond length. It represents the fluctuation in bond length by thermal motion and the structural disorder

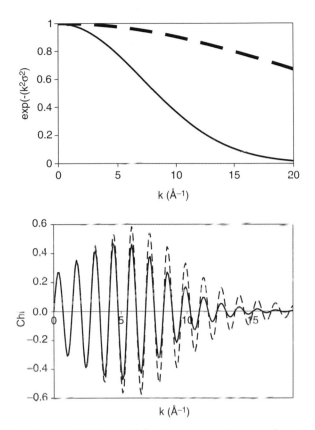

FIGURE 39   Top: Exponential decay of the sigma squared term as function of wave vector, k, in the backscattering function for two experimentally encountered values of σ_j^2: 0.01 (solid), and 0.003 (dashed). Bottom: Effect of the larger σ^2-term (solid) on the EXAFS oscillations. Dashed line is with $\sigma^2 = 0.003$. Assumed bond length = 2.3 Å.

in the material defining thermal and static disorder. This factor decays with increasing values of the wave vector, k, according to the $\exp(-2\sigma_j^2 k^2)$, as shown by the top graph in Figure 39. Two values of σ_j^2 are indicated for an absorber–backscatterer pair having a typically measured mean square disorder in bond length (dashed line, $\sigma^2 = 0.003$) caused by structural disorder or by high temperature (solid line, thermal disorder, $\sigma^2 = 0.01$). The effect of the higher disorder term on the EXAFS oscillations is shown in the bottom figure in Figure 39. The results show that the EXAFS signal decays rapidly with increasing k. The loss of signal with increasing values of k therefore severely limits the data range over which the signal can be Fourier transformed—and which therefore can be used in the fitting.

Although the appropriate reaction conditions combined with an appropriate reactor design can be chosen to overcome issues involved with the catalytic chemistry, there is nothing that can be done to overcome the temperature dependence of the σ^2 factor. This will always be a limitation of EXAFS data collected at elevated reaction temperature.

However, although the number of parameters that can be used in the fitting of EXAFS data is strongly dependent on the usable k-range, XANES data do not suffer from this limitation. If we assume a reasonable value of 0.005 for σ^2, and use a value of k of 2.56 (25 eV above the edge, a typical range for XANES; $k = (2m(E - E_0)/\hbar^2)^{1/2}$, or $k = (0.2625 \times [E - E_0])^{1/2}$)), it is evident that there is only a weak temperature dependence: $\exp(-2\sigma_j^2 k^2) = \exp(-2(2.56)^2 \times 0.005) \sim 1$, so that XANES spectra can be measured at high reaction temperatures. Sometimes knowledge of some basic chemical information about the catalyst (e.g., metal oxidation state or local geometry) that can be determined from the XANES may be sufficient to better understand the workings of the catalyst.

In this section the literature representing several different catalytic reaction chemistries is summarized. The reactions include hydrogenation, CO oxidation catalyzed by supported gold, and others.

4.2. Hydrogenation

A good example of XAFS data recorded under reaction conditions with simultaneous analysis of the reaction products is the work of the Gates group in alkene and arene hydrogenation catalyzed by supported metal clusters (Alexeev et al., 2005; Argo and Gates, 2002, 2003; Argo et al., 2002; Argo et al., 2003, 2006; Bhirud et al., 2004). The goal of the work was to elucidate any patterns of reactivity as a function of the size of the reactant hydrocarbon (ethene, propene, or toluene) together with the size and nuclearity of the metal clusters to better understand how metal cluster properties affect the hydrogenation catalysis. The experiments were conducted with the cell described previously (Odzak et al., 2001) (Figure 31). The supported metal clusters were prepared from well-defined metal carbonyl precursors and decarbonylated in the cell to form nearly monodisperse clusters (all the clusters had essentially the same size). For example, to generate Ir_4 clusters on γ-Al_2O_3, the precursor $Ir_4(CO)_{12}$ was used. Focusing only on the clusters for which data were recorded during catalysis, the Gates group demonstrated by using EXAFS and other spectroscopic techniques (particularly IR spectroscopy) that the clusters Ir_4, Ir_6, and Rh_6 can be prepared on γ-Al_2O_3 and MgO supports, with a typical metal loading of 1 wt%. In each case, site-isolated tetrahedral Ir_4 and site-isolated octahedral Ir_6 and Rh_6 can be prepared on the supports after careful decarbonylation. Typically, the EXAFS data characterizing the supported metal clusters were recorded at low temperatures (by use of liquid nitrogen as a coolant, resulting in a sample temperature of ~77 K), and sometimes under vacuum (i.e., far from conditions of catalysis). However, the Gates group has also shown that meaningful EXAFS data can be recorded under catalytic hydrogenation conditions. For example, the effect of cluster size on the rate of ethene hydrogenation by supported iridium clusters was investigated by Argo et al. (2003). The catalytic activity was evaluated over a temperature range of 273–335 K and at a total pressure of 1 bar of reactants (e.g., 0.243 bar of H_2, 0.239 bar of C_2H_4, and a balance of He). To investigate the reaction mechanism, various partial pressures of H_2 and of hydrocarbon reactant were used. Under their reaction conditions, the only product of ethene conversion was identified as ethane. The EXAFS parameters characterizing the Ir_4/γ-Al_2O_3 catalyst during ethene hydrogenation catalysis at 298 K are shown in Table 4.

The conversion of ethene ranged from 65% to almost complete conversion (third column of Table 4). The EXAFS data show that under reaction conditions the Ir_4 tetrahedra were maintained intact, as evidenced by the essentially unchanged Ir–Ir first-shell coordination number of nearly three and the lack of higher-shell Ir–Ir scattering contributions. The inference is that these monodisperse Ir_4 clusters are indeed responsible for the catalysis. In each data set, recorded at a particular reaction condition, a small Ir–C scattering signal was identified, indicating the presence of a steady-state concentration of adsorbed hydrocarbon on the clusters. Unfortunately, the data were not of sufficient quality to distinguish between various potential hydrocarbon reaction intermediates. Similar data were obtained for the Ir_6/γ-Al_2O_3 catalyst during ethene hydrogenation catalysis: the Ir_6 octahedra were maintained. In this case, however, no Ir–C scattering could be fit to the EXAFS data. Although the gross features of the metal framework of the Ir_4 and Ir_6 clusters remained intact during the catalysis, there were changes in the structural parameters. At a constant partial pressure of ethene, and constant temperature, the Ir–Ir distance characterizing both Ir_4/γ-Al_2O_3 and Ir_6/γ-Al_2O_3 increased approximately in proportion to the partial pressure of H_2, as shown in Figure 40. Concomitant with the increase in bond length with H_2 partial pressure there was an increase in the catalytic activity. The researchers also showed that the catalytic activity of the supported Ir_4 clusters was several times that of the Ir_6 clusters under the same reaction conditions.

This work was extended to other unsaturated hydrocarbons (propene and toluene), other supports (MgO), and other clusters (Rh_6) (Argo et al., 2006). As an example of the type of information that was learned in this work, Table 5 shows the EXAFS fit parameters characterizing the MgO-supported Ir_4 clusters during toluene hydrogenation catalysis as a function of reaction temperature.

TABLE 4 EXAFS best-fit parameters characterizing the Ir_4/γ-Al_2O_3 catalyst during ethene hydrogenation at 1 bar and 298 K (Argo et al., 2003).

| | | | Absorber–backscatterer pair | | | | | | | | | | | | | | | |
| | | | Ir-Ir | | | | Ir-O | | | | Ir-C | | | | Ir-Al | | | |
P_{H_2} (Torr)	$P_{C_2H_4}$ (Torr)	X	N	R (Å)	$\Delta\sigma^2$ ($\times10^3$ Å²)	ΔE_0 (eV)	N	R (Å)	$\Delta\sigma^2$ ($\times10^3$ Å²)	ΔE_0 (eV)	N	R (Å)	$\Delta\sigma^2$ ($\times10^3$ Å²)	ΔE_0 (eV)	N	R (Å)	$\Delta\sigma^2$ ($\times10^3$ Å²)	ΔE_0 (eV)
131	209	0.65	3.3	2.65	5.6	1.9					0.5	2.01	-1.6	7.2	0.2	1.86	0.0	-18.8
							1.4	2.63	0.2	-1.3					0.2	3.23	-2.0	4.5
191	205	0.77	3.3	2.66	5.5	-0.1	0.2	2.03	-0.6	7.3	0.4	1.86	-5.0	18.6	0.2	1.59	-0.7	15.6
							1.1	2.65	-1.7	-2.5								
228	203	0.95	3.3	2.67	5.4	-2.0	0.6	2.04	9.5	1.1	0.2	1.92	-1.9	-0.5	0.2	3.24	0.3	4.5
							0.9	2.67	-4.4	-4.8								
292	203	0.98	3.3	2.68	5.4	-4.2	0.4	2.23	2.9	-13.8	0.3	1.93	-1.0	-1.0	0.2	3.22	-3.0	8.8
							0.8	2.68	-4.5	-5.4								

N, coordination number; R, absorber–backscatterer distance; $\Delta\sigma^2$, sigma squared factor; ΔE_0, inner potential correction; X, conversion of ethene.

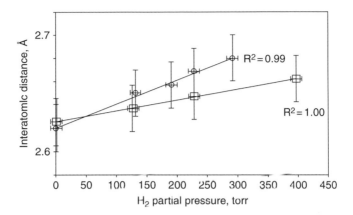

FIGURE 40 Dependence of the Ir–Ir distance of Ir₄ (upper) and Ir₆ (lower) on the partial pressure of H₂ during ethene hydrogenation catalysis at 298 K (Argo et al., 2003). Reprinted with permission from (Argo et al., 2003). Copyright 2003 American Chemical Society.

The online measured conversion of toluene ranged from 28% at 293 K to 86% at 323 K. All of the EXAFS data are consistent with the Ir₄ tetrahedral clusters remaining intact during the catalysis, and thus the inference that the catalytically active species are well approximated as Ir₄ tetrahedra on the alumina support (there was no activity measured with the alumina alone). On the basis of these data, numerous correlations of structural parameters with catalytic activity (represented by turnover frequency, TOF) were determined for hydrogenation of toluene, ethene, and propene. Plots of these correlations are shown in Figure 41. In each case, the average Ir–Ir bond length increased with increasing TOF, and the σ^2 factor characterizing the Ir–Ir scattering decreased with increasing rate of reaction—but the magnitudes of the increase and decrease were dependent upon the specific hydrocarbon reactant. For each reactant hydrocarbon, the coordination number characterizing the metal–support–oxygen (Ir–O$_{short}$) decreased with TOF, and there was an increase in the Ir–O$_{long}$ nonbonding metal–oxygen coordination number.

The data from such EXAFS studies show that in all cases the metal clusters remained intact and are catalytically active. However, as shown, for example, in Figure 41, the clusters did remain bonded to the oxide support, but underwent slight rearrangements to accommodate the reaction intermediates. For example, it was shown that as the concentration of adsorbed reaction intermediates (e.g., π-bonded alkenes, alkyl species) on the clusters increased, the cluster frames expanded, and the clusters flexed away from the support.

These studies are excellent examples illustrating how the structure of the reactive transition metal cluster is probed directly under catalytic reaction conditions. From these data, correlations can be drawn regarding

TABLE 5 EXAFS best-fit parameters characterizing Ir_4/MgO catalyst during toluene hydrogenation catalysis (Argo et al., 2006).

			Metal-backscatterer pair															
			Ir–Ir				Ir–O				Ir–Mg				Ir–C			
T (K)	Atmosphere	X	N	R (Å)	$\Delta\sigma^2$ ($\times10^3$ Å^2)	ΔE_0 (eV)	N	R (Å)	$\Delta\sigma^2$ ($\times10^3$ Å^2)	ΔE_0 (eV)	N	R (Å)	$\Delta\sigma^2$ ($\times10^3$ Å^2)	ΔE_0 (eV)	N	R (Å)	$\Delta\sigma^2$ ($\times10^3$ Å^2)	ΔE_0 (eV)
293	He	–	2.9	2.62	6.8	1.1	1.1	2.05	–4.4	1.0	0.6	2.45	1.5	0.7	0.5	1.91	–7.9	–11.2
							0.5	3.33	–5.2	–9.0								
293	Toluene + H₂; catalysis	0.28	3.1	2.68	5.7	–1.6	0.4	2.02	–4.8	–1.9					0.6	3.28	–7.3	–5.4
306	Toluene + H₂; catalysis	0.53	3.0	2.68	4.8	0.0	1.0	2.63	–3.7	1.4	0.6	2.57	–1.7	–3.8	0.8	3.27	–7.3	0.6
							0.4	1.99	–4.4	0.2								
323	Toluene + H₂; catalysis	0.86	3.0	2.68	4.8	0.0	1.0	2.59	–4.3	–4.7	0.7	2.57	–1.7	–6.8	0.8	3.26	–7.6	6.4
							0.3	1.99	–4.7	0.0								
323	He	–	3.0	2.67	4.5	2.3	1.1	2.60	–4.3	–7.7	0.5	2.55	–0.5	14.3	1.0	3.28	–6.9	–0.6
							0.3	1.98	–4.5	3.6								
							1.7	2.62	–4.3	–0.5								

N, coordination number; R, absorber-backscatterer distance; $\Delta\sigma^2$, sigma squared factor; ΔE_0, inner potential correction; X, conversion of toluene.

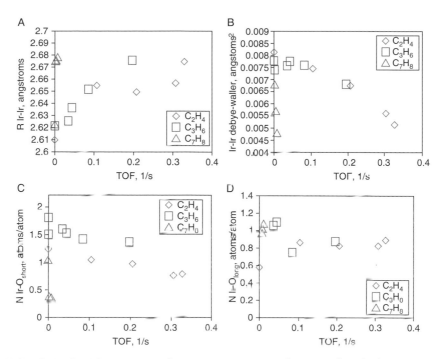

FIGURE 41 EXAFS parameters characterizing Ir$_4$/MgO during catalytic hydrogenation of ethene, propene, and toluene (see Argo et al., 2006 for experimental details). (A) Ir–Ir bond distance correlated with average catalytic activity (TOF), (B) Ir–Ir σ^2 factor compared with TOF, (C) Ir–O$_{short}$ coordination number correlated with TOF, (D) Ir–O$_{long}$ coordination number correlated with TOF. Reprinted with permission from (Argo et al., 2006). Copyright 2006 American Chemical Society.

the relative rates of reaction of the different metals, of clusters of different sizes of the same metal, and the relative rates of reaction of different molecules on the same metal. One critical difference in the investigations discussed here compared to many other EXAFS investigations of supported metal clusters is that the structure of the clusters is precisely controlled (by the gentle decarbonylation of the metal carbonyl precursor clusters), and therefore it is relatively straightforward to derive structure-activity correlations that are at the heart of any spectroscopy measurement of a working catalyst.

4.3. Gold Cluster Catalysts for CO Oxidation

Since Haruta's initial report (Sanchez et al., 1997) of the unexpected activity of supported gold catalysts for low-temperature CO oxidation, there has been a resurgence of research and interest in gold-mediated catalysis. Supported gold clusters have since been found to be active in a number of other catalytic reactions, including propene epoxidation, partial oxidation of hydrocarbons, hydrogenation of unsaturated hydrocarbons, and reduction of nitrogen oxides (Deng et al., 2007; Haruta, 1997). These discoveries were somewhat surprising, as gold is generally assumed to be an inert (noble) metal (Hammer and Nørskov, 1995). Several explanations have been proposed for this unique activity. It has been attributed to the size of the gold clusters (Bamwenda et al., 1997; Haruta et al., 1993), quantum size effects (Valden et al., 1998), a high density of defect sites (Mavrikakis et al., 2000; Xu and Mavrikakis, 2003), the existence of special sites at the metal–support interface (Bond and Thompson, 2000; Costello et al., 2003; Davis, 2003; Fu et al., 2003), the presence of a cationic gold species, the presence of metallic gold particles (Haruta and Date, 2001), and electron-rich gold (Yoon et al., 2005), among others. Given that the structure of the gold clusters (cluster size and shape—and oxidation state of gold) is thought to be critical to the activity of these supported catalysts, and that the cluster size is typically of the order of a few nanometers, it is not surprising that XAFS spectroscopy has been used to probe these catalysts in reactive atmospheres, and in particular for the oxidation of carbon monoxide. This reaction is almost ideal for an experiment characterizing the catalyst in the working state: The chemistry is straightforward, $CO + {}^1/_2O_2 \rightarrow CO_2$, with CO_2 being the only product, the reactants and products are all present in the gas phase, and the reaction temperature is quite mild.

Gold supported on MgO was one of the first catalysts reported to exhibit high activity for CO oxidation, and it was also the focus of the first XAFS paper representing a working supported gold catalyst (Guzman and Gates, 2002). In this work the catalyst, 1 wt% Au supported on MgO, was subject to various activation treatments, and measurements were also made during carbon monoxide oxidation catalysis. The conditions chosen for the measurements were $P_{CO} = 0.014$ bar, $P_{O_2} = 0.014$ bar, and a reaction temperature of 373 K. Under these conditions, a steady-state conversion of 3% was reported (after a high initial activity). The results of the Au L_3-edge EXAFS analysis under these conditions are given in Table 6.

These data show that under conditions of steady-state catalysis, the gold is present in small metallic clusters with an average diameter of $\sim$30 Å, containing $\sim$400 Au atoms each. The same average gold particle size was obtained starting from a catalyst in which the gold had deliberately been prepared as isolated Au$_6$ clusters, and also from a catalyst prepared by more traditional means with an initial average cluster size of $\sim$30 Å. The corresponding XANES data are consistent with a mixture of both cationic and zerovalent gold. From an XAFS standpoint, it is noted that the structure (electronic and geometric) of the gold clusters at steady-state

TABLE 6 Au L_3-edge EXAFS best-fit results for the Au/MgO catalyst in H_2, O_2, and CO at 373 K and a total pressure of 1 bar (Guzman and Gates, 2002).

	Treatment gas											
	H_2				O_2				CO			
Backscatterer	N	R (Å)	$\Delta\sigma^2$ (×10³ Å²)	ΔE_0 (eV)	N	R (Å)	$\Delta\sigma^2$ (×10³ Å²)	ΔE_0 (eV)	N	R (Å)	$\Delta\sigma^2$ (×10³ Å²)	ΔE_0 (eV)
Au first shell	7.2	2.79	8.54	1.60	10.4	2.84	8.50	1.87	9.2	2.85	4.52	1.54
Au second shell	3.5	3.85	8.21	9.78	3.1	4.05	3.98	10.44	4.0	4.05	6.23	3.31
Support O_2	1.1	2.14	6.69	2.55	2.4	2.17	6.63	0.33	1.0	2.16	5.72	2.63
Support O_1	0.8	2.83	5.33	1.42	2.5	2.70	8.74	5.35	0.8	2.75	4.75	2.32
Support Mg	0.8	2.72	6.42	1.35	0.8	2.73	5.47	1.63	0.9	2.71	6.51	1.27

was different from that of the initial catalyst, demonstrating the need for characterization of the catalyst under reaction conditions.

This initial work was further expanded to probe the gold cluster size and relative amount of cationic and zerovalent gold as a function of the partial pressures of CO and O_2 in contact with the MgO-supported clusters (Guzman and Gates, 2004). Although the clusters on these supports are sufficiently simple to allow correlation with catalytic activity, they are about two orders of magnitude less active than the most active CO oxidation catalysts. In all cases the average gold cluster size remained constant ($\sim$30 Å), but what did change was the relative amount of the cationic and metallic gold species, as shown in Figure 42. Cationic and zerovalent gold were always present together, indicating that both are likely present in the catalytically active species, which remain to be defined (but some authors have suggested the presence of cationic gold at the gold–support interface). As the partial pressure of CO was increased, the reaction rate decreased together with the relative amount of cationic gold. These data are plotted in Figure 43 as a function of the rate of reaction. The data show that at higher concentrations of the cationic gold species (corresponding to lower partial pressures of CO in the reactants) there is higher activity, thereby demonstrating the dual role of the CO—it is both a reactant and a reducing agent that converts the Au(I) into Au(0), thereby lowering the catalytic activity.

To further investigate the role of cationic gold, catalysts were specifically prepared to contain only mononuclear gold species and no detectable gold clusters (Fierro-Gonzalez and Gates, 2004). Specifically, dimethyl acetylacetonate Au(III) was adsorbed on NaY zeolite to give a catalyst with 1 wt% gold loading. By EXAFS spectroscopy it was shown that in the fresh catalyst there was no measurable agglomeration of the

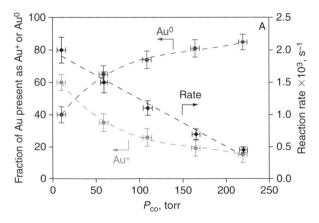

FIGURE 42   Effect of partial pressure of CO on the fraction of gold present as Au(I) and as Au(0) at $P_{O_2} = 11$ Torr (equals 14.7 mbar) together with the rate of the CO oxidation reaction per total gold atom (Guzman and Gates, 2004). Reprinted with permission from (Guzman and Gates, 2004). Copyright 2004 American Chemical Society.

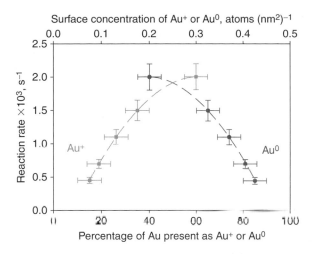

FIGURE 43 Correlation of the catalytic activity for CO oxidation with the percentage of cationic and zerovalent gold in MgO-supported catalyst (Guzman and Gates, 2004). Reprinted with permission from (Guzman and Gates, 2004). Copyright 2004 American Chemical Society.

gold species, and the EXAFS data were well represented assuming mononuclear Au(III) species present on the catalyst. When this catalyst was exposed to the reactant mixture ($P_{CO} = P_{O_2} = 0.016$ bar, balance He) at 298 K, an initial CO conversion of 40% was measured, and the conversion decreased to a steady-state value of 5% after 15 min time on stream. There was no evidence from EXAFS spectroscopy of Au–Au scattering contributions during reaction, indicating that the gold remained mononuclear and site-isolated (i.e., there were no detectable nanoclusters of gold formed). The corresponding XANES data show that the gold was reduced from the initial Au(III) species to a Au(I) species at steady state, but there was no evidence for any Au(0) species. These data demonstrate that active gold catalysts for low-temperature CO oxidation can be prepared so that there is only cationic gold present. However, as noted by the investigators (Fierro-Gonzalez and Gates, 2004), this catalyst is significantly less active than those containing a mixture of cationic and zerovalent gold.

These investigations were expanded to include time-resolved EXAFS experiments (Fierro-Gonzalez and Gates, 2005). The experimental conditions for the catalysis in this paper are similar to those in the earlier work. However, the substantially higher initial CO conversion of 60% was now measured, which declined to 40% within 20 min, whereas in the earlier work the initial conversion was 40%, which declined to 5% after 15 min. No explanation was given for this discrepancy, but it may be evidence for some potential problems with sample packing or gas flow issues in the cell.

This example brings up an interesting point regarding XAFS analysis of catalysts in the working state. Rarely, if ever, is there an example in the literature in which reproducibility of the catalytic performance is given, never mind the reproducibility of the structural analysis. Unfortunately, presumably because of the manner in which these data are collected (beam time assigned for a few days at a time every few months), time is of the essence and, given the complexity of the experiments, it is unlikely that many researchers prioritize time for repeat measurements.

Nevertheless, the authors (Fierro-Gonzalez and Gates, 2005) showed that the Au–O distance and the Au–O coordination number changed with time on stream when the feed was switched between inert helium and the reacting $CO + O_2$ mixture (Figure 44). With the aid of IR spectra recorded under the same conditions, the authors interpreted the EXAFS data to imply that there was a change in the gold–support interface during CO oxidation catalysis. Chemical bonds formed between the supported gold and CO, resulting in gold carbonyl species. The formation of these new bonds occurred at the expense of the gold–support–oxygen bonds. Thus, a gold complex bonded by only one oxygen atom of the support was now formed.

Work on supported gold catalysts for CO oxidation was also reported by Overbury et al. (2006); the catalyst was Au/TiO_2. Although this EXAFS work does not fit into the strict definition being used in this section, it

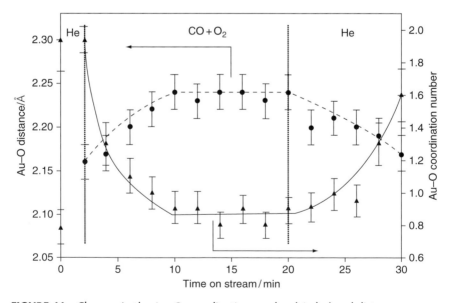

FIGURE 44 Changes in the Au–O coordination number (circles) and distances (triangles) determined from time-resolved EXAFS spectra characterizing Au/NaY zeolite catalyst (Fierro-Gonzalez and Gates, 2005). Reprinted with permission from (Fierro-Gonzalez and Gates, 2005). Copyright 2005 American Chemical Society.

does make some valuable points that are relevant for EXAFS investigations of catalysts in working atmospheres. In their work they used gold loadings of 4.5 and 7.2 wt% on a Degussa P25 TiO_2 support; activity measurements were made with samples in the cell at reaction temperature, but the catalyst was subsequently cooled to 135 K for the collection of the EXAFS data to minimize thermal disorder (σ^2). The cell was operated, according to the authors, as "a continuously recirculating stirred tank reactor that did not interfere with the XAFS measurements." The authors made the point that the online measurements recorded with the catalyst in the cell were valuable for demonstrating that the samples were active for CO oxidation near room temperature, the measurements could not be used for determination of accurate activity data. Overbury et al. noted that the measured conversion did not change strongly with reaction temperature or with heat treatment. This result was in sharp contrast to the offline activity measurements. They surmised that the online measurements did not accurately represent the true activity of the catalyst, possibly because of mass transfer limitations associated either with the flow patterns in the cell or the pressed wafer catalysts that were required for the EXAFS measurements. One other difference between the two reactor configurations is that in the EXAFS reactor cell, the catalyst was diluted with boron nitride powder, whereas in the conventional laboratory reactor, quartz chips were used for better transfer of the heat of reaction. Nevertheless, the authors combined the EXAFS data with the offline activity measurements to draw correlations between average gold particle diameter and the reaction rate, as shown in Figure 45. The reaction rate (turnover frequency) decreased with particle size d, approximately as $d^{-1.8}$ for the catalyst containing 7.2 wt% Au, and as $d^{-1.0}$ for the catalyst containing 4.5 wt% Au, for mean particle sizes in

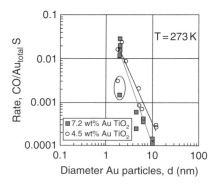

FIGURE 45 Activity data at 273 K characterizing Au/TiO_2 catalysts. Best power law fits to the data are shown for each catalyst. The rate of CO conversion is calculated per total amount of Au (atomic rate) (Overbury et al., 2006). Reprinted from (Overbury et al., 2006), Copyright 2006, with permission from Elsevier.

the range 2–10 nm. This work is a good example illustrating how valuable structural data characterizing catalysts under reaction conditions were obtained by EXAFS spectroscopy, and the online activity measurements demonstrated that the catalyst was active, but because of compromises with the reactor design and limitations associated with the form of the sample, it was not possible to extract XAFS measurements truly representing the working catalyst. This work also highlights the need for a comparison of activity data from offline activity measurements (conventional laboratory reactors) with those from the online activity measurements.

More recent XAFS work on supported gold catalysts for CO oxidation represents an attempt to elucidate the structure of the active gold species by exposing the catalyst to various CO oxidation conditions. For example, Weiher et al. (2006) conducted a series of EXAFS measurements of gold supported on Al_2O_3, TiO_2, and SiO_2. They collected catalytic activity data measured offline, and then correlated their XAFS measurements with these activity data. In all of the catalysts the EXAFS data were found to be consistent with the presence of only metallic gold particles, but the most active catalyst (Au/Al_2O_3) had the smallest gold clusters. Although there was no measurable change in the EXAFS spectra, the authors did observe changes in the XANES depending on the reaction conditions, which they ascribed to adsorption of CO on the metallic gold particles.

An interesting twist to the spectroscopic data was reported by van Bokhoven et al. (2006). They used high energy-resolution fluorescence detection (HERFD) (Section 6) to sharpen the features in the Au L_3-edge XANES characterizing an Au/Al_2O_3 catalyst. HERFD XANES spectra were recorded under various gas atmospheres (e.g., 1% CO, 20% $O_2/$ He, or He). The authors reported that carbon dioxide was detected by residual gas analysis, but activity data were not reported. They interpreted their data by asserting that there is charge transfer from a reduced gold particle to oxygen and thus the O_2 molecule is activated. A similar conclusion was reached by Weiher et al. (2007) on the basis of standard fluorescence yield EXAFS spectroscopy characterizing an Au/TiO_2 catalyst.

4.4. Other Catalytic Reactions

There are many reports of XAFS spectroscopy used to characterize catalysts under working conditions. This section is not meant to be a comprehensive review of all of these publications, but instead is meant to provide a window into the literature of some of the types of catalysts and catalytic chemistry that have been documented. The emphasis is on investigations published in the last few years. What is clear to the reader of these papers, however, is the difficulty in deciding whether these investigations truly merit being described as representing spectra recorded under reaction conditions. What we mean by this statement is that in many cases, although the authors stated that the catalyst was active and that the XAFS data were recorded in a flow of the reactant mixture, there was no activity data presented, so that the reader is left to surmise whether the catalyst was indeed active and forming product. Catalytic data are often reported in these publications, but they were recorded offline with a completely different reactor configuration. Clearly, this is a subject that demands more care and attention by the practitioners in the field, and some correlation between the offline and online measurements needs to be shown.

Lamberti et al. (2002, 2003) investigated a 5 wt% $CuCl_2/Al_2O_3$ catalyst by using time-resolved XANES spectroscopy at the Cu K-edge during the catalytic oxychlorination of ethene with hydrochloric acid and oxygen to give vinyl chloride. They collected Cu XANES spectra by using dispersive XAS as the catalyst was subjected to a heating cycle from 373 to 623 K in a flow of ethene, O_2, and HCl. The results of their investigation are summarized in Figure 46, which shows the Cu(II) fraction in the catalyst as a function of reaction temperature together with the O_2 conversion, which was measured simultaneously by use of an online mass spectrometer. It

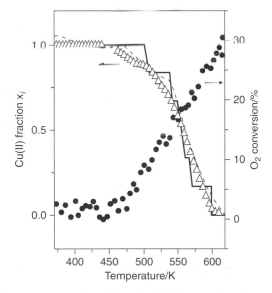

FIGURE 46   Left axis, amount of Cu(II) in $CuCl_2/Al_2O_3$ catalyst determined by three different methods (for details see Lamberti et al., 2003) during ethene oxychlorination as a function of reaction temperature. Right axis, the corresponding catalytic activity measured by simultaneously determining the oxygen conversion. Reproduced by permission of the PCCP Owner Societies.

was determined that at the lowest reaction temperature, 373 K, copper was present only as Cu(II), and the catalyst was inactive. As the temperature increased to between 470 and 490 K, the O_2 conversion increased and the reduction of Cu(II) was observed. The Cu(II) reduction was complete at 600 K. The authors claimed that their result demonstrates that at the typical commercial oxychlorination temperature (490–530 K), Cu(I) dominates on the catalyst surface, and thus that the rate-determining step is the reoxidation of CuCl to form a copper oxychloride.

Haller et al. in a series of publications (Chen et al., 2004a,b; Ciuparu et al., 2004a,b, 2005; Lim et al., 2005) investigated the growth of single-walled carbon nanotubes (SWNT) synthesized by CO disproportionation using small metallic cobalt clusters. The cobalt clusters were prepared by the controlled reduction of cobalt ions isomorphously substituted into the silica framework of a MCM-41 mesoporous molecular sieve. The authors have extensively investigated the Co K-edge XAFS of such catalysts during the initial activation (reduction in H_2 at temperatures up to 993 K), and during SWNT synthesis conditions (1023 K, 2–6 bar CO partial pressure). For example, Figure 47 shows the Co K-edge XANES spectra recorded during catalyst prereduction in H_2 and subsequent SWNT synthesis conditions. The authors were able to show that the prereduction treatment in H_2 at temperatures below 973 K increased the density of electrons at the Fermi level of the oxidized cobalt species, preserving the local tetrahedral environment, and only subsequent exposure of the prereduced catalyst to CO at 1023 K produced completely reduced cobalt clusters. The authors also provided evidence for the formation of an intermediate Co(I) species during reduction of the

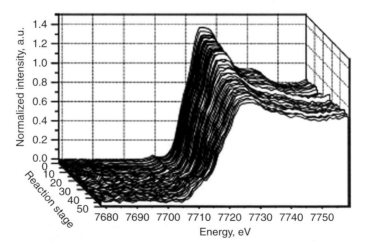

FIGURE 47 Cobalt K-edge XANES spectra recorded during prereduction in H_2 and SWNT synthesis at 1023 K and 6 bar of CO pressure (Ciuparu et al., 2005). Reprinted with permission from (Ciuparu et al., 2005). Copyright 2005 American Chemical Society.

Co–MCM-41 catalyst in H_2 at temperatures up to 993 K. It is this interme-diate species that preserves the tetrahedral environment in the silica framework and provides the resistance to complete reduction to the metal in the presence of H_2. The Co(II) species is resistant to reduction in pure CO; the intermediate Co(I) species is more reactive in CO, likely forming cobalt carbonyl-like compounds with high mobility in the MCM-41. These mobile species are the precursors of the metal clusters that grow the carbon nanotubes. Controlling the rates of each step of this two-step reduction process is a key to controlling the sizes of the cobalt metal clusters formed in the cobalt MCM-41 catalysts.

Ethene polymerization is a major industrial process, and so it is not surprising that catalysts used in this reaction have been the focus of XAFS investigations. For example, Groppo et al. (2005) studied the chromium K edge XAFS of a Phillips catalyst during activation and ethene polymer-ization. The Cr/SiO_2 catalyst was first reduced in CO at 623 K, and subsequently ethene polymerization was investigated at 298 K and at 373 K. Unfortunately, in this investigation the loading of chromium in the catalyst was quite high (4–8 times the normal loading used in indus-trial applications), and so there was a mixture of chromium species in the catalyst. This complication hampered the quantitative determination of the active sites involved in the polymerization reaction. Although the ethene polymerization reaction was conducted in the XAFS spectroscopy cell, the XAFS data were collected after and not during the reaction. Nevertheless, the authors concluded that Cr(VI) species were reduced by ethene during the polymerization reaction and that the isolated Cr(II) species are the active sites. Unfortunately, there was no report of a correlation of the number of these sites with catalytic activity, nor a detailed EXAFS analysis of the structure of these sites.

The SCR of NO in the presence of O_2 and hydrocarbons has attracted much recent attention because of its significant environmental impact (Amiridis et al., 1996; Parvulescu et al., 1998). Many different catalysts for this reaction have been investigated. Recent studies have highlighted the use of XAFS spectroscopy to probe the structure of the active sites. In the first of these, Caballero et al. (2005) characterized a Cu/ZrO_2 catalyst during SCR conditions using methane, propane, and propene. Their publication is some-what misleading, as the XAFS data that they reported were not collected under reaction conditions, even though their interpretation leads the reader to believe that the species they measured are those present under reaction conditions. Moreover, they did not show any catalytic data. They recorded Cu K-edge XANES spectra during reaction, for example, heating the catalyst to 773 K in a hydrocarbon/NO/O_2 mixture, but they then cooled the catalyst to room temperature in the reaction mixture to record their XAFS data. Figure 48 shows the relative concentrations of the various copper species in the Cu/ZrO_2 catalyst (which had been pressed as a self-supporting

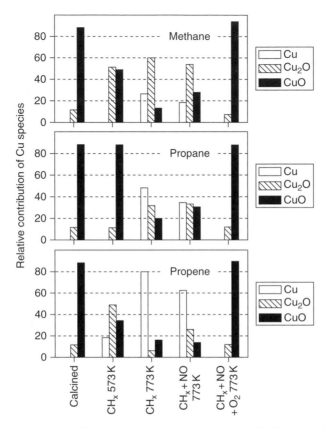

FIGURE 48 Percentage of the various copper species in the Cu/ZrO$_2$ catalyst under various SCR of NO reaction conditions using methane (top), propane (middle), and propene (bottom), as reductant (Caballero et al., 2005). Reprinted from (Caballero et al., 2005), Copyright 2005, with permission from Elsevier.

wafer—XAFS data were recorded in transmission mode) after various treatments (e.g., after SCR with 1000 ppm of NO, 2500 ppm of hydrocarbon and 3% of O$_2$, diluted in helium at 773 K). These data show that the concentration of reduced species was related to the reactivity of the hydrocarbon (CH$_4$ < C$_3$H$_8$ < C$_3$H$_6$). The analysis showed that most of the copper remained as Cu (II) when the sample was heated in the SCR mixture, although some Cu(I) persisted (10%) under these conditions. The authors stated that the presence of the Cu(I) is interesting because processes based on a Cu(II)/Cu(I) redox cycle have been proposed for the intermediate steps in the reduction of NO with hydrocarbons in exhaust combustion.

XAFS was used to probe the structure of an Ag/Al$_2$O$_3$ catalyst during SCR of NO, with the emphasis on an understanding of the promotional effect of hydrogen (Breen et al., 2005), which is unique to silver-containing catalysts. This publication is another example in which insufficient experimental detail is provided for the reader to determine whether the catalytic data were actually measured while the XAFS data were being recorded. Mention was made of the same gas feed system used for the XAFS cell, but it is ambiguous if activity data similar to those determined in the laboratory reactor were obtained. Nevertheless, the authors fed reactant (720 ppm NO, 4.3% O_2, 0.53% H_2O, 7.2% CO_2, 543 ppm C_8H_{18}, balance He) to the catalyst in the cell while the XAFS data were being recorded. The XAFS spectra were collected at a temperature of 498 K, whereby the NO conversion was 60% when H_2 was added to the feed, although the catalyst was inactive when there was no coreductant or when CO was used as a reductant. The fit of the EXAFS data indicates the formation of $Ag_n^{\delta+}$ clusters under SCR conditions in the presence of H_2, where n was estimated to be ~ 3. However, the catalytic activity cannot be explained by the formation of these clusters alone, as the XAFS data are consistent with the presence of the same silver clusters when either CO or no coreductant was used, and when the catalyst was inactive. The authors concluded that although silver clusters may have formed during the SCR, and could be the active sites, other influences, such as the direct chemical effect of the coreductant with the active site, are more important in determining the activity of the catalyst. Other researchers (e.g., Shibata et al., 2004), also used XAFS spectroscopy to probe silver-containing catalysts used for SCR, but the XAFS data were not recorded under reaction conditions. However, these authors were able to infer a structure–activity correlation for Ag–MFI catalysts, as depicted in Figure 49., which shows an increase in reaction rate as a function of Ag–Ag coordination number determined in the EXAFS analysis. The most active cluster was estimated to be Ag_4^{2+}. In this case it would be of great interest to repeat these measurements with the catalyst under working conditions to determine whether these structures change under reaction conditions, and indeed whether there is any correlation between the presence of these structures and changes in activity.

Continuing with the NO abatement theme, we cite another catalyst that has been used, Cu–ZSM-5, which is highly active for the decomposition of NO to N_2 and O_2 (Iwamoto et al., 1986). This catalyst has been much investigated, and a recent XAFS study at the Cu K-edge during NO decomposition is an example of characterization of the active copper site (Groothaert et al., 2003). The controversy in the literature regarding this catalyst surrounds the detailed structural characteristics of the active copper site, hence the belief that XAFS measurements under reaction conditions can provide insight. Groothaert et al. (2003) collected Cu K-edge EXAFS data under several conditions related to this catalytic chemistry. The fits to the EXAFS data recorded under various conditions are summarized in Table 7.

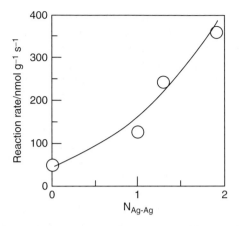

FIGURE 49   NO reduction rate at 573 K in the presence of 0.5% H_2 for propane-SCR on various silver-form zeolites as a function of the Ag–Ag coordination number characterizing $Ag_n^{\delta+}$ clusters (Shibata et al., 2004). Reprinted from (Shibata et al., 2004), Copyright 2004, with permission from Elsevier.

TABLE 7   EXAFS best-fit values of the Cu–ZSM-5 catalyst (Groothaert et al., 2003).

Ab-Sc pair[a]	N	R (Å)	$\Delta\sigma^2$	ΔE_0 (eV)
Cu–ZSM-5 treated in He at 393 K, measured at LN[b]				
Cu–O	4.0	1.96	−0.001	−4.2
Cu⋯Al	0.7	3.18	0.008	−12.6
Cu–ZSM-5 treated in He at 773 K, measured at 773 K				
Cu–O	2.3	1.92	0.006	2.7
Cu⋯Cu	0.5	2.84	0.015	1.4
Cu–ZSM-5 in NO/He at 773 K, measured at 773 K				
Cu–O	3.4	1.92	0.008	−0.8
Cu⋯Cu	1.4	2.79	0.017	6.8
Cu–ZSM-5 in NO/He at 773 K, measured at LN				
Cu–O	3.5	1.95	0.001	−4.6
Cu⋯Cu	1.5	2.87	0.013	−1.0

[a] Ab, absorber; Sc, scatterer.
[b] LN, liquid nitrogen temperature (specific temperature not provided by the authors).

The authors showed that during treatment in helium at 773 K, all of the Cu(II) was reduced to Cu(I). They concluded that mononuclear Cu(II) entered the zeolite during the ion-exchange procedure and that a mixture of Cu(I) monomers and Cu(I) pairs were formed during the dehydration in helium. During NO decomposition at 773 K, Cu(I) was partly oxidized to Cu(II). Concomitantly, an increase in the Cu–O coordination number from 2.3 to 3.4 was observed (Table 7). With additional information from

UV–Vis spectroscopy, the authors were able to interpret the XAFS data (both XANES and EXAFS) as resulting from a bis(μ-oxo)dicopper species. This $[Cu_2-\mu-(O)_2]^{2+}$ moiety was inferred to be present in oxygen-activated overexchanged Cu–ZSM-5, with a Cu–Cu distance of 2.87 Å and a coordination number of one during NO decomposition. Using this structural information, the authors postulated that the catalytic cycle involves a Cu (I)···Cu(I) pair that reacts with two NO molecules to form N_2 and bis(μ-oxo)dicopper, which reconstitutes the Cu(I)···Cu(I) pair after desorption of O_2. Although this detailed XAFS investigation indeed does shed light on the structure of the active copper species present during NO decomposition, it does suffer from the limitation that there was no mention of the activity measured in the XAFS cell. Instead, catalytic activity data were measured offline in a laboratory reactor, whereas the XAFS data were collected in a flow of NO at reaction temperature; the reader is left to assume whether the catalyst was indeed active.

XAFS investigations of catalysis in the presence of liquid-phase reactants, particularly, oxidation reactions in the presence of supported palladium-containing catalysts, are the focus of several papers by Grunwaldt and coworkers (Grunwaldt and Baiker, 2005; Grunwaldt et al., 2003a,b, 2006). In each of these investigations, the catalytic activity was measured online and reported together with the XAFS data. In this manner it was demonstrated that the structural information derived from the XAFS measurements was of an active material. In the first investigation, an XAFS study of a 5 wt% Pd/Al$_2$O$_3$ during aerobic oxidation of cinnamyl alcohol in an organic solvent. The conversion and product distribution of the oxidation reaction in the XAFS reactor cell were compared with those observed with a laboratory slurry reactor. Predictably, the conversion in the XAFS cell was only ∼7%, compared with close to 100% conversion in the slurry reactor. Although this paper focused on the feasibility of conducting such XAFS experiments with liquid-phase reactants, and therefore provided little detail about the analysis of the XAFS data, the authors did report that the palladium was metallic under reaction conditions (328 K, toluene solution of cinnamyl alcohol saturated with 5% O_2) and that there was essentially no change in the XAFS spectra—and therefore the size and chemistry of the supported palladium clusters, whether O_2 was cofed or not.

Grunwaldt et al. (2003b) reported XAFS measurements recorded during palladium-catalyzed alcohol oxidation in supercritical CO_2. A commercial shell-impregnated catalyst consisting of 0.5 wt% Pd on alumina was used for benzyl alcohol oxidation (to benzaldehyde) in supercritical CO_2 with pure O_2 as oxidant. The conditions were 353 K and 150 bar. The results are summarized in Table 8. The authors reported only Pd XANES data, not EXAFS data, and thus the analysis is limited to information about the average oxidation state of the palladium.

TABLE 8   Results of the aerobic oxidation of benzyl alcohol in supercritical CO_2 at 150 bar and 353 K (Grunwaldt and Baiker, 2005).

Experimental conditions	Reaction rate (TOF) ($mol_{product}$ $mol_{Pd}^{-1} h^{-1}$)	Selectivity	Relative Pd^0/Pd^{2+} ratio from XANES
$c_{alcohol} = 0\%$, $c_{O_2} = 0\%$	–	–	0.92/0.08
$c_{alcohol} = 1.89\%$, $c_{O_2} = 0.95\%$	$1518 h^{-1}$	96.3%	0.88/0.12
$c_{alcohol} = 1.89\%$, $c_{O_2} = 1.89\%$	$1894 h^{-1}$	95.8%	0.86/0.14
$c_{alcohol} = 0\%$, $c_{O_2} = 1.89\%$	–	–	0.67/0.33
$c_{alcohol} = 1.89\%$,[a] $c_{O_2} = 1.89\%$	$1285 h^{-1}$	94.4%	0.77/0.23
$c_{alcohol} = 1.89\%$,[b] $c_{O_2} = 1.89\%$	$1555 h^{-1}$	95.3%	0.81/0.19

[a] After treatment of the catalyst in 1.9% O_2/CO_2 and again with alcohol for 0.5 h.
[b] After treatment of the catalyst in 1.9% O_2/CO_2 and again with alcohol for 2 h.

   The relative amounts of Pd(0) and Pd(II) were estimated by using a linear combination fit of the spectrum of the catalyst reduced in H_2 (taken as Pd(0)) and the as-prepared catalyst (taken as Pd(II)). The changes in the XANES spectrum as a function of the various reaction conditions are subtle, but there is a measurably greater amount of reduced palladium present when the alcohol is present compared to when there is only O_2 present in the supercritical CO_2 solvent.

   This work was expanded in 2006, when the benzyl alcohol oxidation was conducted with cyclohexane as the solvent (Grunwaldt et al., 2006). The authors set out to address whether the active phase of the catalyst was oxidic or metallic palladium. The reaction conditions were as follows: temperature 323 K; reactant mixture, 200 μl of benzyl alcohol in 100 ml of cyclohexane saturated with O_2. The catalytic activity was determined online by monitoring the intensity of the C=O band in the benzaldehyde with IR spectroscopy. If the catalyst was not reduced in the XAFS reactor (by the H_2-saturated cyclohexane), then no catalytic activity was measured, and the palladium remained oxidic. If the catalyst was pre-reduced, a highly active catalyst was obtained, and the XAFS data were found to be consistent with the presence of only metallic palladium. Thus, the conclusion was reached that palladium oxide exhibits hardly any activity at 323 K, whereas metallic palladium particles are much more active.

   The oxidation states of antimony, vanadium, and molybdenum in a "MoVSbNbO" propane-to-acrylic acid catalyst were probed by XANES spectroscopy (Safonova et al., 2006). The conditions during catalysis in the XAFS spectroscopy cell were as follows: temperature, 653 K; feed ratio typically $O_2/C_3H_8/He = 10/5/8$. The measured propane conversion of ~16% is noted as being slightly lower than that obtained in a laboratory fixed-bed reactor (19%). This difference was attributed to the fact that the gases did not pass thorough the catalyst bed, coming in contact with only part of it. It was shown that, under reaction conditions, in the so-called M1 phase of this complex oxide catalyst, the antimony and vanadium undergo changes in their oxidation states depending on the propane/O_2 ratio, but there was no change in the molybdenum oxidation state. These changes occurred simultaneously, with the same kinetics. These results were interpreted as evidence that different types of oxygen (from the hexagonal channels and from the MO_6 octahedral network) must be involved in the catalytic process, and their relative contributions are different.

As a final example, we consider XAFS data recorded during the catalytic partial oxidation (CPO) of methane (to CO and H_2) with 0.5 and 2.5 wt% Ir/Al_2O_3 catalysts (Grunwaldt et al., 2002). The catalysts were first reduced in H_2, and small clusters of metallic iridium were obtained (smaller clusters on average were obtained with the lower loadings of iridium). Ir L_3-edge XANES data recorded (in transmission mode) for the 2.5 wt% Ir catalyst are shown in Figure 50 for temperatures both below and above the ignition temperature (593 K). The corresponding catalytic activity data, measured online with a mass spectrometer, are shown in Figure 51. The CPO reaction was found to ignite at 593 K, and

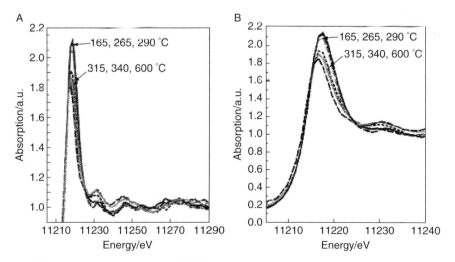

FIGURE 50   Selected Ir L_3-edge XANES spectra recorded at temperatures below and above the ignition temperature during the CPO catalyzed by 2.5 wt % Ir/Al_2O_3 (Grunwaldt et al., 2002). With kind permission from Springer Science+Business Media: (Grunwaldt et al., 2002) Fig. 5, copyright 2002 Plenum Publishing Corporation.

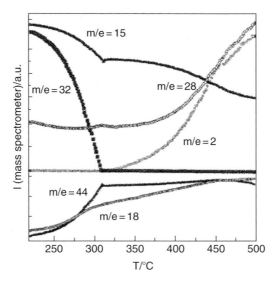

FIGURE 51 Mass spectrometer signals recorded simultaneously with the Ir L₃-edge XANES spectra shown in Figure 50 (Grunwaldt et al., 2002). With kind permission from Springer Science+Business Media: (Grunwaldt et al., 2002) Fig. 4, copyright 2002 Plenum Publishing Corporation.

the ignition temperature appears to be independent of the average iridium cluster size. However, the structure of the clusters changes significantly at the ignition point. At temperatures below the ignition point, the iridium was partly (2.5 wt% Ir) or almost fully (0.5 wt% Ir) oxidized, and at temperatures above the ignition point, the iridium was abruptly reduced. The catalytic and structural changes were observed to be fully reversible.

5. XAFS SPECTRA MEASURED UNDER DYNAMIC CONDITIONS

5.1. Introduction

Distinguishing time-resolved XAFS (TR-XAFS) investigations from XAFS investigations of samples in reactive atmospheres as summarized earlier, in particular under changing reaction conditions, is difficult and has to be somewhat arbitrary. In the following, we focus on XAFS investigations aiming at (1) understanding the structural evolution of a solid catalyst under dynamic conditions, and (2) indicating properties of the catalyst not available from investigations under stationary conditions (e.g., kinetics of reduction and reoxidation). In the literature cited here, not only was the local structure of a material investigated under relevant reaction conditions, but properties of the reaction were also elucidated, such as reaction intermediates or the reaction kinetics. These points distinguish the investigations described in this section from those described earlier, including investigations of materials under stationary conditions or of thermal treatments of materials or temperature-programmed reduction or oxidation under noncatalytic reaction conditions.

Several recent publications include reviews of parts of the literature related to TR-XAFS (Dent, 2002; Grunwaldt and Clausen, 2002; Newton et al., 2002; Ressler et al., 2002; Shido and Prins, 1998). The reactivity of solids and the kinetics of solid-state reactions constitute important subjects in heterogeneous catalysis research. Although most heterogeneously catalyzed reactions proceed on the surface of catalytically active phases, the active phases and their surfaces may be influenced or even determined by the underlying catalyst bulk. Investigations of the preparation procedure and the structure of a catalyst under dynamic conditions have the following characteristics in common: not just the geometric and electronic structure of the individual phases involved are of interest, but also the transition of one phase into another—hence, the mechanism of the kinetics of this reaction. Naturally, structural evolution and kinetics of phase transitions can only be obtained from time-resolved investigations. By use of XAFS in a time-resolved mode, investigations of the bulk structural response of a catalyst to rapidly changing dynamic reaction conditions can be performed with a sub-second time resolution and, thus, they may result in reliable structure-activity relationships.

5.2. Instrumentation for TR-XAFS Spectroscopy

The majority of conventional XAFS beam lines operating in the hard X-ray regime (>2.5 keV) are dedicated to transmission and fluorescence XAFS spectroscopy. A double crystal monochromator is used to monochromatize the radiation coming from a BM or an ID. To obtain reliable data in the step-scanning mode, a total measuring time of at least several minutes is required. In the quick EXAFS (QEXAFS) (Frahm, 1988, 1989) technique, the stepper motors of the monochromator crystals are driven continuously with a simultaneous read-out of the detectors. The measuring time is decreased to several tenths of a second. Channel-cut monochromators adjusted by piezo crystals can further reduce the measuring time by a factor of ~100 (Richwin et al., 2002). A disadvantage of this latter approach is the restricted spectral range associated with the finite translation of the piezo crystals. However, further developments may increase the time resolution available with the QEXAFS technique, with improved stability and data quality: new detectors (Kappen et al., 2002), improved monochromator crystal benders (Zaeper et al., 2001), and QEXAFS at a third-generation X-ray source (Als-Nielsen et al., 1995). QEXAFS offers versatility as experiments can be performed using various detection modes (transmission, fluorescence, electron yield, etc.) with a wide variety of materials (solid, liquid, or diluted samples, enhanced surface sensitivity, etc.). Furthermore, an increased number of reports can be found in which TR-QEXAFS has been combined with other experimental techniques (see later text).

Another spectrometer design that can be used for TR-XAS investigations is based on the energy-dispersive XAFS (DXAFS (Hagelstein et al., 1997)) technique. The first experiment combining dispersive optics and XAFS was performed by Kaminaga et al. (1981); they used a plane Si(111) monochromator in transmission geometry on an X-ray tube and recorded the intensity distribution with a position-sensitive proportional counter. The measuring time for a single absorption spectrum amounted to several hours. With application of the advantages of synchrotron radiation, the necessary measuring time per spectrum has been decreased by several orders of magnitude. In the most common energy-dispersive equipment, the incoming synchrotron radiation is focused by a bent monochromator crystal in either reflection or transmission geometry. The curvature of the bent crystal results in a variation of the Bragg angle over the crystal surface, and, hence, a defined position–energy correlation on the detector. The sample is located in the polychromatic focal point.

With the energy-dispersive design, an entire absorption spectrum is collected at once, without requiring any movement of the spectrometer components. Therefore, the superior stability of the DXAFS equipment permits reliable detection of small changes in the experimental spectra.

The energy range available for XAFS analysis using the energy-dispersive arrangement is restricted by the Bragg angle used and, hence, by the edge energy of the absorber to be measured. Generally, at edge energies below $\sim$10 keV, the energy range available with a Si(111) crystal is not sufficient for a detailed EXAFS analysis. Furthermore, DXAFS investigations are restricted to the transmission mode, which poses certain concentration limitations on the materials that can be investigated. This point is of particular importance in combination with a heavily absorbing matrix. Although the energy-dispersive design poses higher demands on the homogeneity of the samples investigated, reliable EXAFS data of sufficient quality can be obtained in the DXAFS mode with a time resolution in the sub-second range. The photo diode arrays currently in use at some energy-dispersive XAFS spectrometers offer a high dynamic range, but only a moderate minimum integration time because of the necessary read-out time after the exposure. Furthermore, photodiode arrays suffer from radiation damage.

Therefore, a new detector design, based on a scintillating screen lens coupled to a CCD camera, was developed and has been successfully employed in numerous applications (Koch et al., 1995). Beyond the established TR-XAFS designs, the feasibility of performing "ultrafast" TR-XAFS experiments in the femtosecond range by use of the brilliant synchrotron radiation provided by potential fourth-generation X-ray sources has been evaluated in several recent publications (Bressler et al., 2002; Chen, 2001; Tomov et al., 1999).

5.3. Cells for TR-XAS Investigations

A variety of flow-through cells or batch reactors that can be used for absorption measurements with X-ray photons in the energy range from 2.5 to 30 keV were reviewed in Section 3, above. In principle, the various cells designed for XAFS investigations of samples under reaction conditions can also be employed for TR-XAFS measurements (Clausen et al., 1991). However, the application of DXAFS equipment requires improved sample homogeneity compared to the conventional scanning monochromator in order to obtain undistorted data. DXAFS investigations carried out with capillary-style reactors, for instance, which have been shown to be well-suited for TR investigations in the QEXAFS mode, suffer from this experimental constraint. On the other hand, cells that permit characterization of chemical reactions in solution can be successfully employed by use of stopped-flow techniques to perform time-resolved investigations (Inada et al., 1997; Scheuring et al., 1996; Thiel et al., 1993). Cells designed for fluorescence yield measurements are restricted to the QEXAFS mode.

5.4. Analysis of TR-XAFS Data

To determine the unique information available from time-resolved experiments under dynamic conditions (which is not readily available from experiments carried out under stationary conditions), several data analysis features are particularly desirable for TR-XAFS investigations in heterogeneous catalysis (Ressler, 1998). First, capabilities to handle a large number of spectra measured during a single time-resolved experiment (often consisting of several hundred scans) are most important for the analysis of TR-XAFS data. The analysis of every single spectrum by hand is hardly feasible, in particular, because many steps of the analysis have to be performed identically for each spectrum to obtain comparable results. Instead, automated procedures are required that can reliably apply certain data reduction steps to each spectrum in the large TR-XAFS data set (Ressler, 1998).

Moreover, TR-XAFS data reduction needs to be performed even at an early stage of a TR-XAFS experiment. To use the available beamtime as efficiently as possible, it is crucial to monitor the progress of a TR experiment by a continuous analysis of the measured spectra. (Compilations of XAFS software can be found at http://www.esrf.fr/computing/

expg/subgroups/theory/xafs/xafs_software.html   and   http://ixs.iit.
edu/catalog/XAFS_Programs].

Furthermore, quantitative structural phase analysis, for instance, is important for investigations of solid catalysts, because one frequently has to deal with more than one phase in the active or precursor state of the catalyst. Principal component analysis (PCA) permits a quantitative determination of the number of primary components in a set of experimental XANES or EXAFS spectra. Primary components are those that are sufficient to reconstruct each experimental spectrum by suitable linear combination. Secondary components are those that contain only the noise. The objective of a PCA of a set of experimental spectra is to determine how many "components" (i.e., reference spectra) are required to reconstruct the spectra within the experimental error. Provided that, first, the number of "references" and, second, potential references have been identified, a linear combination fit can be attempted to quantify the amount of each reference in each experimental spectrum. If a PCA is performed prior to XANES data fitting, no assumptions have to be made as to the number of references and the type of reference compounds used, and the fits can be performed with considerably less ambiguity than otherwise. Details of PCA are available in the literature (Malinowski and Howery, 1980; Ressler et al., 2000). Recently, this approach has been successfully extended to the analysis of EXAFS data measured for mixtures containing various phases (Frenkel et al., 2002).

Moreover, using the quantitative phase information obtained from the analysis of the XANES region, a combination of more than one theoretical EXAFS function can be refined to one experimental spectrum. Thus, the short-range structure of individual phases in mixtures of phases can be determined, which is also particularly important for investigations of solid catalysts and solid-state chemistry (Ressler et al., 2001). Hence, in addition to using the XANES region to elucidate phase composition and average metal valence states, a detailed analysis of the EXAFS region can indicate the evolution of the short-range structure under reaction conditions.

A structural analysis is not readily available from most conventional techniques used to investigate solid-state kinetics (TG/DTA, TPR, etc.), which makes TR-XAFS a powerful complementary tool to investigate the reactivity of solids.

5.5. Applications of TR-XAFS Spectroscopy

The majority of investigations carried out with time-resolved XAFS spectroscopy concern (heterogeneous) catalysis research, focusing on both structure—reactivity relationships on the basis of investigations of catalysts under reaction conditions and evolution of catalyst structures under preparation conditions. The use of the two available operating modes (DXAFS and QEXAFS) is equally distributed among these investigations. However, in only a few cases have the two modes been used in a complementary manner or has the choice of the mode been explained.

TR-DXAFS spectroscopy has been used to investigate bulk metal or metal oxide catalysts and supported metal catalyst. Recently, TR-DXAFS data were used for a analysis of the kinetics of the reactions proceeding in and on bulk catalysts (Ressler et al., 2002, 2003; Wienold et al., 2003) and on supported catalysts (Evans and Newton, 2002; Evans et al., 2002; Newton et al., 2001, 2002), whereas, previously, TR-DXAFS investigations had mainly focused on elucidating reaction intermediates (Bottger et al., 2000; Hagelstein et al., 1994; Lamberti et al., 2002; Ressler et al., 1997, 2000; Sankar et al., 1992). Evidently, owing to the limited quality of the DXAFS data collected in the various investigations, only few publications include reports of detailed EXAFS analyses of the catalytically active material under reaction conditions that exceed the first coordination shell. Moreover, although the equipment employed in the investigations cited would have provided a sub second time resolution, these investigations did not fully exploit this potential and mainly used a time resolution in the range of seconds.

Most of the TR-XAFS investigations employing the conventional double-crystal monochromator design have used the QEXAFS mode to obtain an improved time resolution (on average, in the range of minutes per scan) relative to what is attainable with the step-scanning mode. TR-QEXAFS investigations were performed with bulk metal and metal oxide catalysts (Gunter et al., 2001; Reitz et al., 2001) to investigate catalyst sulfidation (Cattaneo et al., 2000, 2001; Geantet et al., 2001; Sun et al., 2001), and to investigate supported catalysts (Clausen et al., 1993; Grunwaldt et al., 2000, 2002; Oudenhuijzen et al., 2002; Ovesen et al., 1997). Owing to the superior data quality in the investigations cited compared to that of the TR-DXAFS investigations mentioned earlier, a detailed structural analysis of the TR-QEXAFS data was performed. However, possibly because of the inferior time resolution compared with that in the TR-DXAFS investigations, only a few publications include reports of a detailed analysis of the kinetics, and most investigations focused on the identification of reaction intermediates.

In addition to characterizing a catalytically active material under dynamic reaction conditions, the method has been used to characterize the structural evolution taking place during the preparation of active catalysts from precursor materials. Either a DXAFS spectrometer (Fiddy et al., 2002; Hatje et al., 1994; Sankar et al., 1992; Sankar et al., 1993; Shido et al., 2002; Thomas et al., 1994; Yamaguchi et al., 2000, 2002) or the QEXAFS mode at a conventional XAFS beam line (Cimini and Prins, 1997; Wienold et al., 2003) was employed. Because a controlled thermal treatment of a precursor material permits a better tailoring of the time resolution required, a time resolution in the range of minutes was employed in many of the investigations, resulting in sufficient data quality for a detailed EXAFS analysis. Again, only a few of the reports include any analysis of the kinetics of the TR data and, instead, most focused on elucidating reaction intermediates.

The structural evolution of solids during various solid-state reactions with no direct relationship to heterogeneous catalysis constitutes the subject in the second largest fraction of investigations employing TR-XAFS spectroscopy. A comparable number of investigations were carried out with DXAFS or QEXAFS modes to monitor the evolution of phase compositions and the short-range order structure of the materials during solid–gas reactions (Ressler et al., 2000, 2001; Rodriguez et al., 1999), solid–solid reactions (Chauvistre et al., 1994; Grunwaldt et al., 2001; Ressler et al., 2000; Rodriguez et al., 1999; Rumpf et al., 2002), phase transformations (Dacapito et al., 1993; Douillard et al., 1996; Rodriguez et al., 1998; Rodriguez et al., 2000), thermal treatments (Choy et al., 2001; Hilbrandt and Martin, 1997; Pickup et al., 2000; Troger et al., 1997; Walton and Hibble, 1999; Walton et al., 1998), and solid combustion syntheses (Frahm et al., 1992).

The third largest fraction of investigations in which TR-XAFS spectroscopy was used falls in the area of electrochemistry, with a comparable number of investigations employing DXAFS (Allen et al., 1995; Guay et al., 1991; Millet et al., 1993), or conventional QEXAFS modes (Lutzenkirchen-Hecht and Frahm, 2001) or QEXAFS in the reflection mode (Hecht et al., 1996, 1997). To date, there are only few reports indicating that TR-XAFS spectroscopy has been used to characterize reactions in the liquid phase (Epple et al., 1997) or under environmentally relevant conditions (Villinski et al., 2001). Several additional short reports of the use of TR-XAFS spectroscopy can be found in the proceedings of the biannual International Conference on XAS (XAFS VII Proceedings, 1993; XAFS VIII Proceedings, 1995; XAFS IX Proceedings, 1997; XAFS X Proceedings, 1999; XAFS XI Proceedings, 2001).

5.6. Correlation between Average Metal Valence and Catalytic Selectivity

Time-resolved X-ray absorption spectra of an activated $H_5[PV_2Mo_{10}O_{40}]$ oxidation catalyst were recorded to determine correlations between the dynamic structure and the catalytic selectivity of the material (Ressler and Timpe, 2007). In addition to experiments carried out under steady-state conditions, time-resolved XAFS measurements at the Mo K-edge were performed under changing reaction conditions (with a time resolution of $\sim$30 s per spectrum) (Figure 52). Therefore, the gas-phase composition was isothermally switched from a reducing atmosphere (propene) to an

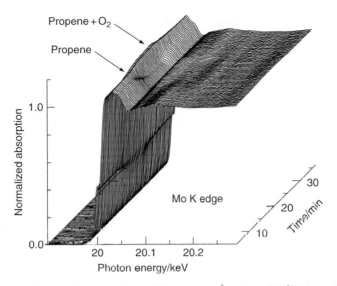

FIGURE 52   Evolution of Mo K-edge XANES spectra of activated $H_5[PV_2Mo_{10}O_{40}]\cdot3H_2O$ during isothermal switching of the gas phase from oxidizing conditions (propene/O_2) to reducing conditions (propene) and back (Ressler and Timpe, 2007). Reprinted from (Ressler and Timpe, 2007), Copyright 2007, with permission from Elsevier.

oxidizing atmosphere (propene and O_2) (complete exchange took place in about 20 s in the cell used).

The time-resolved XAFS investigations permitted monitoring of the structural response of the catalyst to the changing reaction conditions. Analysis of the time-resolved XAFS data revealed the solid-state kinetics of the bulk structural changes upon reduction and reoxidation of the catalyst. The evolution of the average molybdenum valence under reducing and oxidizing conditions at various temperatures is depicted in Figure 53. Upon switching from propene and O_2 to propene at 673 K, the average molybdenum valence declined from 6 to 5.94. Apparently, at 673 K, the reduction of the molybdenum centers in the lacunary Keggin ions is mostly limited to the surface of the accessible crystallites. Upon switching to oxidizing conditions (propene and O_2) at 673 K, the mostly surface-reduced catalyst was rapidly reoxidized.

Under reducing conditions (propene), increases in the reaction temperature were accompanied by a more pronounced decrease of the average valence of the molybdenum centers in the activated catalyst (Figure 53B and C). The more reduced catalyst at 723 K exhibited a prolonged reoxidation behavior (Figure 53B). The corresponding extent of the reoxidation curve could be simulated with a solid-state kinetics model assuming three-dimensional diffusion to be rate limiting (Figure 54).

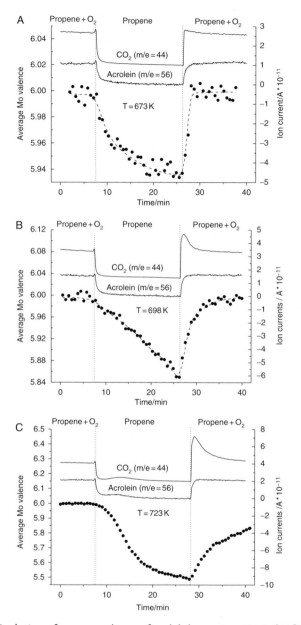

FIGURE 53 Evolution of average valence of molybdenum in activated $H_5[PV_2Mo_{10}O_{40}]\cdot$ $13H_2O$ during changes in gas-phase composition (propene to propene $+$ O_2) together with the corresponding evolution of acrolein and CO_2 in gas phase at 673 K (A), 698 K (B), and 723 K (C) (Ressler and Timpe, 2007). Reprinted from (Ressler and Timpe, 2007), Copyright 2007, with permission from Elsevier.

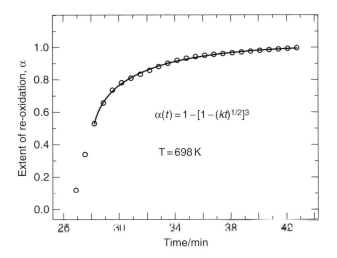

FIGURE 54 Refinement of a solid-state kinetics model for three-dimensional solid-state diffusion $(\alpha(t) = 1 - [1 - (kt)^{1/2}]^3)$ to the experimental extent of reoxidation of activated $H_5[PV_2Mo_{10}O_{40}]\cdot13H_2O$ at 698 K (Ressler and Timpe, 2007). Reprinted from (Ressler and Timpe, 2007), Copyright 2007, with permission from Elsevier.

An apparent activation energy of about 90 kJ/mol corroborated the assumption of the oxygen diffusion limitation in the catalyst bulk.

Intriguingly, the selectivity of the catalyst at the various reaction temperatures exhibited a pronounced correlation with the degree of reduction and the solid-state kinetics of the reoxidation process. After the reductive treatment in propene and switching back to propene and O_2 at 673 K, the rapid reoxidation of the catalyst was accompanied by a rapid increase in the concentration of both acrolein and CO_2 in the gas phase (Figure 53A). This result indicates that the entirely reoxidized catalyst quickly regained its activity and selectivity in propene oxidation. Conversely, with an increasing reaction temperature (698 and 723 K, Figure 53B and C) the partially reduced catalyst exhibited a bulk diffusion limited reoxidation rate that coincides with an increased production of CO_2 and, hence, a reduced selectivity of the catalyst. The production of CO_2 decreased with increasing reoxidation of the catalyst and, after complete reoxidation, the catalyst reached the same activity and selectivity as prior to the switching experiments.

The correlation between electronic structure and catalytic performance corroborates the assumption that the selectivity of the catalyst is governed by the electronic structure of the surface. The latter in turn appears to be determined by the electronic defect structure of the underlying bulk. In the investigation of the activated $H_5[PV_2Mo_{10}O_{40}]$ catalyst, reaction conditions that favored a conventional redox mechanism with fast reduction and diffusion-limited reoxidation led to low selectivity.

Conversely, reaction conditions that maintained a rapid reoxidation and a small number of Mo^{5+} centers in the catalyst resulted in an increased selectivity. Hence, it may be concluded that in a process that involves diffusion of oxygen ions in the catalyst bulk and a prolonged lifetime of partially reduced $V^{4+}-Mo^{5+}$ metal sites, total oxidation of propene dominates. On the other hand, catalytic oxidation of propene proceeding on an oxidized $V^{4+}-Mo^{6+}$ active site at the surface of the catalyst yields an improved selectivity for partial oxidation products.

5.7. Solid-State Kinetics of Reduction and Reoxidation of MoO_3 With Propene and O_2

Molybdenum trioxide constitutes an active model catalyst for the oxidation of propene in the presence of gas-phase O_2 at temperatures above approximately 600 K (Grzybowska-Swierkosz, 2000). Reduction of MoO_3 in propene and oxidation of MoO_2 in O_2 were investigated by time-resolved XAFS spectroscopy combined with mass spectrometry (Ressler et al., 2002). Reduction and reoxidation of MoO_{3-x} are of particular interest because they constitute the two fundamental transformations of the so-called redox mechanism for partial oxidation of alkenes on molybdenum oxide catalysts.

Temperature-programmed and isothermal TR-XAFS and XRD experiments were performed to elucidate the phases present during the reactions and the structural evolution of the catalyst and also to indicate the solid-state kinetics of the processes. The evolution of Mo K-edge spectra of MoO_3 and the various kinetics of the reduction and reoxidation are displayed in Figure 55.

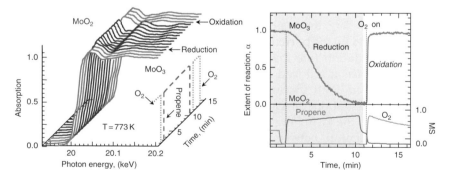

FIGURE 55 Left: Evolution of Mo K-edge XANES spectra of MoO_3 during isothermal switching of the gas phase from oxidizing (O_2) to reducing conditions (propene). Right: The extent of reaction α reflects the different kinetics of the reduction and reoxidation of MoO_3 at 773 K.

In addition to elucidating ordered or disordered intermediate phases present during a chemical reaction, TR-XAFS data are most suitable to determine the kinetics of the reaction in the solid or the liquid phase. Figure 56 shows the extent of reduction (α) obtained from XAFS experiments during isothermal reduction of MoO_3 in 5 vol% propene at 723 K, 10 vol% propene at 673 K, and 10 vol% propene at 698 K (Ressler et al., 2002). From the α trace at 673 K in 10 vol% propene, a deviation from a symmetric sigmoidal trace is evident. The acceleratory regime of the reduction at 673 K (up to $\alpha \approx 0.3$) can be described by a power rate law ($\alpha \sim t^2$), whereas

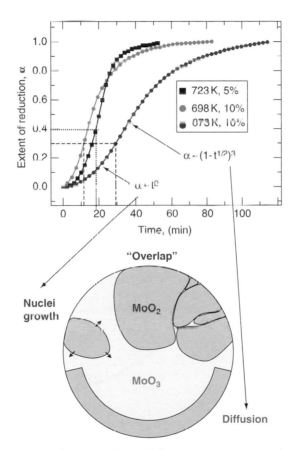

FIGURE 56   Extent of reduction α obtained from XAFS experiments during isothermal reduction of MoO_3 in 5 vol % propene at 723 K, in 10 vol % propene at 673 K, and in 10 vol% propene at 698 K. The changes in the rate-limiting step from an $\alpha \sim t^2$ rate law (power law) to an $\alpha \sim 1 - (1 - t^{1/2})^3$ rate law (three-dimensional diffusion) in the experiments at 673 K in 5 vol % ($\alpha \approx 0.30$) and at 723 K in 10 vol % propene ($\alpha \approx 0.4$) are indicated by a dashed and a dotted line, respectively (Ressler et al., 2003). Reprinted from (Ressler et al., 2003), Copyright 2003, with permission from Elsevier.

the deceleratory regime of the reduction can be described by a "three-dimensional diffusion" rate law ($\alpha \sim 1 - (1 - t^{1/2})^3$). The point of change between the two rate laws (i.e., a change in the rate-limiting step) is indicated in Figure 57. On the basis of the isothermal reduction experiments it was inferred that the point of change between the two rate laws is approximately independent of the reaction temperature, but it varies with the reactant concentration.

The solid-state kinetics of the reduction of MoO_3 in propene exhibits a change in the rate-limiting step, both as a function of temperature and as a

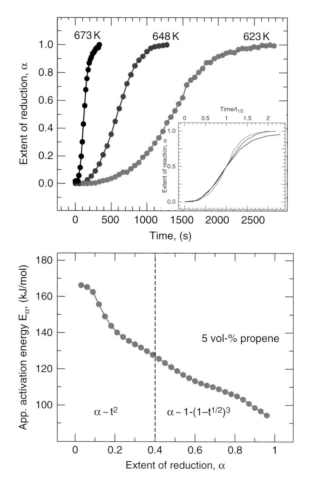

FIGURE 57 Top: Evolution of extent of reduction α during isothermal reduction of MoO_3 in 5 vol % propene at 623, 648, and 673 K obtained from XRD experiments. The inset shows the corresponding half-life normalized traces. Bottom: Apparent activation energy as a function of extent of reduction calculated from the reduction traces in figure on top (Ressler et al., 2002). Reprinted from (Ressler et al., 2002), Copyright 2002, with permission from Elsevier.

function of the extent of reduction α. With increasing α at a given temperature, a transition from nucleation growth kinetics to a three-dimensional diffusion-controlled regime was observed. With decreasing temperature (less than about 650 K), a pronounced transition from a nucleation growth kinetics to a regime that is entirely controlled by oxygen diffusion in the MoO_3 lattice was found (Figure 58).

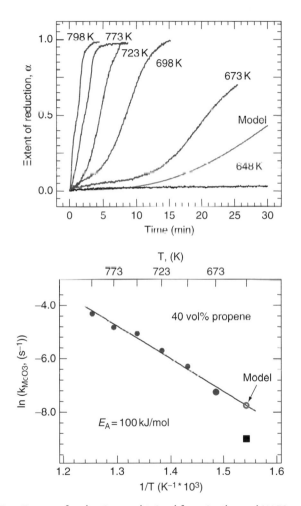

FIGURE 58   Top: Extent of reduction α obtained from isothermal XAFS experiments at various temperatures, including 648 K (square in bottom figure) in 40 vol % propene. The trace denoted "Model" (open circle in bottom figure) corresponds to a calculated extent of reduction assuming a rate of reduction denoted as "Model" in the bottom figure showing an Arrhenius diagram constructed from the rate constants determined for the extent of reduction traces in the top figure. At 673 K a change in the rate-limiting step is evident (Ressler et al., 2002). Reprinted from (Ressler et al., 2002), Copyright 2002, with permission from Elsevier.

On the basis of the results of the solid-state kinetics analysis, a schematic reaction mechanism for the reduction of MoO_3 in propene and the reoxidation in O_2 was proposed. The mechanism consists of (1) generation of oxygen vacancies at the (100) or (001) facets by reaction with propene, (2) vacancy diffusion in the MoO_3 bulk, (3) formation of "$Mo_{18}O_{52}$" type shear structures in the lattice, and (4) formation and growth of MoO_2 nuclei. The mechanism is in agreement with previous reports of the propene oxidation on MoO_3 as a structure-sensitive reaction.

In addition to investigating the reduction and oxidation properties of MoO_3, the evolution of characteristic defects in the bulk structure of MoO_3 under propene oxidation conditions by using time-resolved XAFS spectroscopy was investigated. Under the reaction conditions employed (273–773 K with a propene-to-O_2 ratio ranging from 1:1 to 1:5 (molar)), orthorhombic MoO_3 remained the only crystalline phase detected by XRD. The onset temperature of the reaction of propene and O_2 in the presence of MoO_3 coincided with the onset of reduction of MoO_3 in He, H_2, and propene (at $\sim$620 K) (Figure 59). At temperatures below $\sim$720 K, and independent of the atmosphere, partial reduction of MoO_3 was observed, resulting in the formation of "$Mo_{18}O_{52}$" type defects in the layer structure of α-MoO_3 (Figure 60). At temperatures above $\sim$720 K and in O_2 or in an oxidizing atmosphere, the "$Mo_{18}O_{52}$" type defects were reoxidized to MoO_3 (increasing the average molybdenum valence in shown in Figure 60). Evidently, the catalytically active molybdenum oxide phase develops under partial oxidation conditions at temperatures below 720 K and does not have the undisturbed ideal layer structure of orthorhombic α-MoO_3.

In the redox mechanism for the partial oxidation of propene on MoO_3, three stages are distinguished (Figure 61): (1) At temperatures below $\sim$600 K, the participation of oxygen from the MoO_3 bulk is negligible; (2) At temperatures of 600–700 K, oxygen vacancy diffusion in the bulk is sufficient to make a redox mechanism feasible. Because the complete reoxidation of the "$Mo_{18}O_{52}$" type shear structures is inhibited, a partially reduced MoO_3 with crystallographic shear (CS) planes in the lattice is obtained under reaction conditions; (3) At temperatures above approximately 700 K, sufficiently fast oxygen diffusion in the lattice combined with rapid formation and annihilation of CS permits the participation of a considerable amount of the lattice oxygen of MoO_3 in the partial oxidation of propene. The results clearly show the necessity and the large potential of bulk structural investigations of solid catalysts under reaction conditions. The bulk structure and particularly the type and amount of defects in the material (the "real" structure) markedly affect the catalytic properties. Hence, to rationally design an active solid catalyst, both structure and reactions of the surface and structure, defects, and reactions of the bulk need to be known in detail and considered carefully.

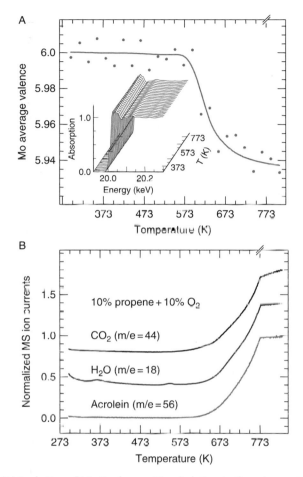

FIGURE 59   (A) Evolution of Mo K-edge position (relative to the edge position of MoO_3) during temperature-programmed reaction of propene and O_2 in the presence of MoO_3 (10% O_2 and 10% propene by volume in He) (temperature ramp from 300 to 773 K at 5 K/min, followed by a hold at 773 K) (inset shows evolution of XANES spectra during TPR). (B) Evolution of the corresponding gas-phase composition (CO_2 ($m/e = 44$), H_2O ($m/e = 18$), and acrolein ($m/e = 56$)) displaced for clarity during temperature-programmed reaction of propene and oxygen (Ressler et al., 2003). (Ressler et al., 2003) Copyright Wiley-VCH Verlag GmbH & Co. KGaA. Reproduced with permission.

5.8. Concluding Remarks regarding Time-Resolved Modes

The examples presented in this section clearly demonstrate the potential of time-resolved experiments to extend the suitability of XAFS spectroscopy for investigations of heterogeneous catalysis under dynamic conditions. XAFS experiments under stationary conditions indicate the structure of the working catalyst; however, correlations between the

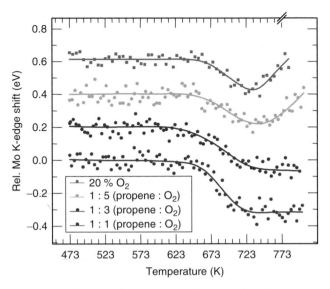

FIGURE 60 Evolution of Mo K-edge position (relative to the edge position of MoO_3 and displaced for clarity) during thermal treatment of MoO_3 in O_2 (20% in He) and during temperature-programmed reaction of propene and O_2 in the presence of MoO_3 (10% O_2 and 10% propene in He) (temperature ramp from 473 to 773 K at 10 K/min, followed by hold at 773 K). The three different ratios of O_2 to propene employed are indicated (Ressler et al., 2003). (Ressler et al., 2003) Copyright Wiley-VCH Verlag GmbH & Co. KGaA. Reproduced with permission.

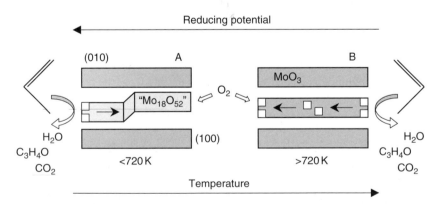

FIGURE 61 Schematic representation of the structure of MoO_3 under reaction conditions (propene and O_2) as a function of temperature and reducing potential of the gas phase. A; At temperatures < 720 K or in a reducing gas phase. B; At temperatures > 720 K and in an oxidizing gas phase. "$Mo_{18}O_{52}$" refers to crystallographic shear type defects in the MoO_3 structure (squares indicate oxygen vacancies) (Ressler et al., 2003). (Ressler et al., 2003) Copyright Wiley-VCH Verlag GmbH & Co. KGaA. Reproduced with permission.

structure of the catalyst and its catalytic properties can only be elucidated from monitoring the structural response of the catalyst to (rapidly or even periodically) changing reaction conditions ("chemical lock-in"). The examples presented illustrate that TR-XAFS spectroscopy is highly suitable to accomplishment of this task.

On one hand, TR-XAFS investigations that require the best time resolution available (such as isothermal reactions or rapid decompositions with half lives of the order of one minute) may be performed at an energy-dispersive XAFS station, with full advantage taken of the time resolution in the sub-second range. On the other hand, TR-XAFS investigations of processes with half lives of the order of several minutes may be performed in the QEXAFS mode, with advantage taken of the increased EXAFS data quality for a detailed structural analysis.

Although the experiments referred to here demonstrate the wealth of kinetics and structural data that can be obtained from TR-XAFS data, application of XAFS spectroscopy combined with complementary techniques provides unique and even more detailed information. This statement refers to the most elegant way of using XAFS spectroscopy simultaneously with other methods (e.g., XRD; Clausen, 1998; Clausen et al., 1993; Dent et al., 1995; Thomas et al., 1995), and it also refers to XAFS experiments complemented by experiments carried out under similar experimental conditions (e.g., laboratory techniques such as XRD, Raman spectroscopy, TG/DTA). More often than not, a detailed XAFS analysis is possible only when all additional data (characterizing phases, metal valences, and structure) representing the catalyst are available. Furthermore, the analysis of TR-XAFS data should aim at extracting as much information from the XANES part and the EXAFS part of a XAFS spectrum as possible.

Eventually, for time-resolved characterizations of solid–solid or solid–gas reactions in catalysis, a lower limit for the required time resolution exists, because of the required mass transport of reactants to the catalyst surface. In flow reactors, the characteristic time for gas-phase transport to the catalyst surface, after a rapid change in the gas atmosphere, usually amounts to several seconds. Hence, a time resolution in the range of $\sim$100 ms should be sufficient to resolve the changes in the bulk structure induced by the variation in gas composition or reaction temperature.

More important than pushing the time resolution into the microsecond range is measuring X-ray absorption data of superior quality. Therefore, to evaluate data quality and time resolution of a particular experimental station, "real" catalysts under reaction conditions should be compared (e.g., 3d and 4d metal oxides at elevated temperature measured to k equal to 16 A^{-1}, rather than metal foils). It must be kept in mind that kinetics data can also be obtained from complementary methods such as

TG/DTA, TPR, and complementary techniques for characterizing catalysts in the working state (e.g., XRD; Raman, IR, and UV–vis spectroscopies) can provide structural and metal valence information under reaction conditions. However, the capability of TR-XAFS spectroscopy to reveal *quantitative phase composition and average metal valence* together with the evolution of the *local structure* of a catalyst under varying (reaction) conditions, combined with a time resolution of 100 ms will continue to be a very powerful tool for kinetics investigations in solid-state chemistry and heterogeneous catalysis.

6. LOOK TO THE FUTURE

The field of *in situ* XAFS investigations of catalysts is relatively mature: many groups around the globe are conducting such studies, and significant new understanding continues to develop from them. As highlighted in this chapter, catalysts are being investigated in the presence of gas or liquid phases, and under all types of reaction conditions.

There are exciting new trends in the area that stand to revolutionize the spectroscopy. We start this discussion by borrowing a depiction of a potential roadmap for future spectroscopic studies of catalysts in reactive atmospheres from Weckhuysen (2003) (Figure 62). Many of the points raised in his monograph are relevant here. First, many spectroscopic observations that are claimed to be of working catalysts are not—often there is no link between the catalyst characterization data and any catalyst performance data. In the large majority of XAFS investigations that are claimed to represent working catalysts, there are no activity data presented, and when there are, they are often determined in a laboratory reactor and not online with the spectroscopic data. This limitation leads to the first point of Figure 62: *better measurement conditions*—if at all possible, the catalytic measurements should be made with the same catalyst at the same time as the spectroscopic measurements. (It is agreed that it is not essential in all cases for the activity data to be measured simultaneously with the XAFS data, but at a minimum some activity data should be presented for the same catalyst to show that the catalyst being characterized is actually active under the conditions of the spectroscopy).

There are multiple reasons why this criterion is not met on a routine basis. First, better spectroscopy–reactor cells need to be designed. Section 3 provides details of many of the innovative designs published, it is hoped that this will provide ideas for the much-needed improved designs. Second, XAFS investigations need to be performed at synchrotrons. This requirement necessitates that the catalysis researcher conduct the experiments at a remote location and typically with only a few days to

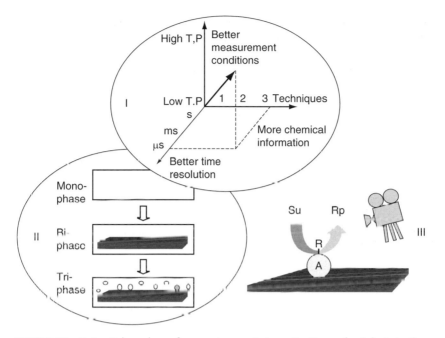

FIGURE 62 Potential roadmap for spectroscopic investigations of catalysts in the working state. Area I: measuring better, measuring faster, measuring more. Area II: extension of operational window from mono-phase to bi-phase to tri-phase reactions. Area III: single molecule and single-size spectroscopy-microscopy (Weckhuysen, 2003). Reproduced by permission of the PCCP Owner Societies.

complete the experiments. There are major impediments to adding the complexity of online analytical capabilities to beam lines.

It would be easier for many research groups if specialized catalysis facilities were supported at centralized synchrotron facilities. A recent example of this is the Synchrotron Catalysis Consortium (SCC) at the National Synchrotron Light Source at Brookhaven National Laboratory (http://www.yu.edu/scc/). The aim of this fledgling consortium is to promote the use of synchrotron techniques to perform cutting-edge catalysis science research with samples under reaction conditions by providing assistance and developing new techniques to the catalysis community through the following concerted efforts: (1) dedicated beam time on two XAFS spectroscopy beam lines, (2) dedicated facilities, including state-of-the-art reaction cells, gas-handling systems, and advanced detectors, (3) dedicated research staff to assist the experimental setup and data analysis, (4) training courses and help sessions, and (5) development and testing of new hardware and software for catalysis research.

The second point made in Figure 62 is *measuring faster*. This point refers to developing the field of time-resolved spectroscopy into the sub-second region. Quick-scanning XAFS spectroscopy has been available for many years (Section 5), and yet on most XAFS spectroscopy beam lines the data are still collected in a step-scan mode. The technology exists now to retrofit many of these beam lines with the quick-scanning capability. Of course, this may necessitate an improvement in the optics on the beam lines, with collimating and focusing mirrors to increase the flux on the sample.

The final point raised in Section 5 is also repeated. In QEXAFS spectroscopy it is not usual that the scans need to be collected faster, but that the data quality needs to be higher, so that detailed structural information can be derived. This goal will likely be achieved in the future by the improved design of beam line optics, more experiments being performed on undulator or wiggler beam lines, and improvements in detector technology.

The third point is *measuring more*. Weckhuysen implied by this statement that the researcher should combine two or more spectroscopic techniques in one cell, to produce multitechnique characterizations. Given the complexity of a typical solid catalyst, the use of multiple characterization techniques has always been appreciated in catalyst characterization science. The novelty suggested here is to combine these techniques so that the spectroscopic measurement is made on the same catalyst at the same time. Other chapters in this volume address this point.

Some examples of such multitechnique cells are presented in Section 3, with examples given in Section 5 (the usefulness for the combination of XAFS and XRD for many catalysts has been realized for many years—although for some reason not implemented on many beam lines worldwide). It is likely that we will see many more investigations of this type in the next few years.

In the field of XAFS spectroscopy of working catalysts, the three points raised by Weckhuysen are only a beginning. There are at least two other directions that could be added to this roadmap. Both have already started to show their potential value.

We call the first new direction *measuring complexity*, and the second *measuring spatially*.

By measuring complexity, we refer to the recent breakthroughs in measuring high energy-resolution fluorescence detection (HERFD) XAFS spectroscopy. A detailed description of HERFD is beyond the scope of this review, and the reader is referred to several key references for details (de Groot, 2001; de Groot et al., 2002, 2005; Hamalainen et al., 1991). In brief, the method relies on the detection of a single fluorescence X-ray emission channel. This is achieved experimentally by using a high energy resolution fluorescence spectrometer, typically using either a single or an array of germanium single crystals that are spherically bent at the appropriate radius to diffract only the wavelength of interest, and a solid state detector (Bergmann et al., 1998; Schulke et al., 1995; Stojanoff et al., 1992). A typical arrangement is depicted in Figure 63. These experiments demand the extremely high brightness of undulator beam lines at modern third-generation synchrotron sources. This field is only in its infancy and yet has spawned many different submethods, and the information content is truly remarkable.

The first example we discuss is the use of HERFD-XANES, which has been demonstrated to resolve different adsorption sites in supported metal catalysts (see, e.g, Safonova et al., 2006; Tromp et al., 2007; van Bokhoven et al., 2006). The critical property of HERFD-XANES utilized in these investigations is the enhanced spectral resolution, resulting from selecting an appropriate emission channel. The recorded spectrum is broadened only by the instrumental broadening and not dominated by the usually significantly greater core-hole broadening. This in turn results in sharper features in the XANES region of the spectrum.

An example of the enhanced resolution is presented in Figure 64, where the Au L$_3$-edge XANES of a gold foil is shown both in the normal transmission mode and in the HERFD mode (Safonova et al., 2006). This spectral sharpening was used to determine the actual adsorption site of CO on a small alumina-supported platinum cluster (Safonova et al., 2006). The measured spectra are shown in Figure 65. Clearly, the features in the XANES spectra recorded by the use of HERFD are richer than what is determined by using total fluorescence detection.

Although these spectra are remarkable in themselves, they would not mean much unless they could be interpreted—this is where there has been significant improvement in the last few years, and the advances promise to be even more significant in the next few years (e.g., moving

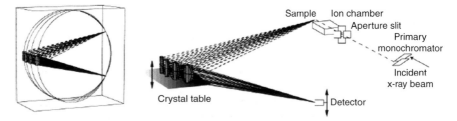

FIGURE 63 Schematic of a HERFD crystal array spectrometer, in this case using six spherically bent Ge(620) crystals to collect the Fe Kβ region with approximately 1.0 eV resolution. The inset (left) shows the Rowland circles for each analyzer crystal (Heijboer et al., 2004). Reprinted with permission from (Heijboer et al., 2004). Copyright 2004 American Chemical Society.

A

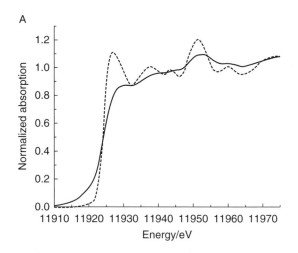

FIGURE 64 Au L$_3$-edge XANES measured in normal transmission mode (solid line) and by using HERFD mode (dotted) (Safonova et al., 2006). Reprinted with permission from (Safonova et al., 2006). Copyright 2006 American Chemical Society.

beyond the muffin-tin potential), by the use of *ab initio* self-consistent multiple scattering codes, such as the FEFF8.2 code (Ankudinov et al., 2002a; Ankudonov et al., 1998). When this code is used to calculate the XANES region of the XAFS spectrum, quantitative interpretation of the spectrum can be attained (Ankudinov et al., 2000, 2001a,b, 2002b,c).

In the study of Safonova et al. (2006) (Figure 65), FEFF8.2 calculations were performed on small platinum clusters (six atoms each) with the CO adsorbed in different adsorption sites. The double feature at the white line was reproduced only when the CO was adsorbed in an on-top configuration, and not in bridging or face-bridging sites. Thus, this proof-of-principle investigations shows that the CO adsorption site can be determined by using HERFD-XANES coupled with advanced calculations. Because the technique has all of the advantages of "regular" detection XAFS spectroscopy, it also carries all of its advantages: the measurement can be made under high pressures and temperatures, and it can be applicable to a wide range of samples. We eagerly await such investigations.

Another application of HERFD-XANES has been in the spectra of the preedge peaks of the K-edges of the first-row transition metals. The most investigated catalyst according to this approach has been the zeolite FeZSM-5, for which high-resolution Kβ-detected XANES spectroscopy has been used to determine both the average oxidation state of iron and the local bonding geometry around the Fe atoms in a series of catalysts (Battiston et al., 2003a–c; Heijboer et al., 2004, 2005). Izumi et al. (2005) also used this approach to selectively detect the vanadium K$_{\alpha 1}$ fluorescence of a series of V/TiO$_2$ catalysts.

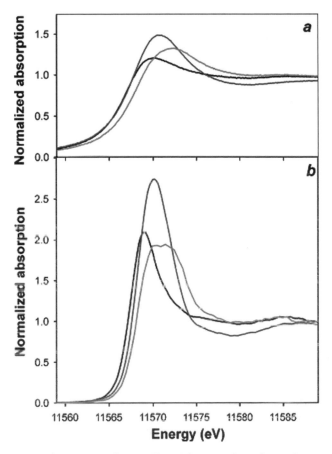

FIGURE 65 Pt L₃-edge XANES of a 5 wt% Pt/Al₂O₃ catalyst after reduction in the presence of various atmospheres: He (1), He(O₂) (2), and 1% CO/He (3) measured by using total fluorescence detection (top) and HERFD (bottom) (Safonova et al., 2006). Reprinted with permission from (Safonova et al., 2006). Copyright 2006 American Chemical Society.

The preedge peaks of K-edge XANES have been used for decades as a measure of the electronic structure of the element of interest. However, in K-edge XANES these features are weak and not clearly resolved form the rising background of the absorption edge itself. The preedge peak arises from transitions from a 1s electron to the lowest unoccupied electronic states that mainly have 3d character. In a centrosymmetric system, these transitions are dipole forbidden and therefore have very low intensity. On deviation from a centrosymmetric environment, mixing of 3d with 4p orbitals is allowed. As a consequence of this hybridization, some electric-dipole-allowed 1s to 4p character is added to the 3d-band transitions, resulting in enhanced intensity of the preedge structure. Therefore, changes in the preedge intensity reflect changes in the local geometry.

An example of the Fe K preedge feature measured using conventional fluorescence XANES and Kβ-detected XANES is shown in Figure 66. Using this technique, Heijboer et al. (2005) were able to greatly improve the accuracy of the interpretation of the preedge feature and quantitatively determine the degree of extraction of iron from the zeolite framework as a result of steaming of the sample. The measurement of the fluorescent Kβ transition with a high-resolution spectrometer is a new characterization technique. It is anticipated that this spectroscopy will become a standard tool to catalysis research similar to traditional XANES and EXAFS spectroscopy. Of course, what is needed is availability of experimental stations where these techniques can be readily applied to catalysis research, and in particular to developments of methodology for characterization of catalysts in the working state. New developments in this area are expected to continue.

Our final example of measuring complexity is the recent use of 1s2p resonant inelastic X-ray scattering spectroscopy (RIXS). This technique allows the collection of L-edge spectra of, for example, the first-row transition metals by use of hard X-rays. The L-edges of the 3d transition metals occur in the soft X-ray region between 400 and 1000 eV and are routinely measured under vacuum conditions. This requirement essentially precludes any investigations of functioning catalysts.

It was shown recently that by combining detailed crystal-field multiplet simulations together with the RIXS spectroscopy that the spectral shapes can be analyzed and explained for a series of iron oxides (de Groot

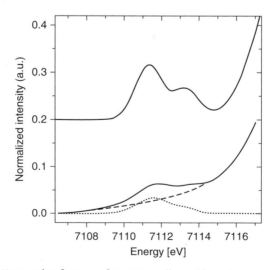

FIGURE 66 Fe K pre-edge feature of Fe$_2$SiO$_4$ collected by using conventional XANES (bottom) and Kβ-detected XANES (top) (Heijboer et al., 2004). Reprinted with permission from (Heijboer et al., 2004). Copyright 2004 American Chemical Society.

et al., 2005). The challenge of this spectroscopy is to combine high resolution (0.3 eV) with good sensitivity in order to make it practical to measure a complete RIXS experiment in a reasonable time, even for dilute samples and difficult conditions. Once again, work is in progress in this field and advancements are expected to continue.

Clearly, this field of high-resolution fluorescence detection is growing, and many new applications are being developed. In a recent paper by de Groot (2007), he projected an extension of this to concept of "selective" XAS. The concept is that if the detector is set at an X-ray emission line for a specific state, then only that state is observed. In this manner, selective extra absorption allows one to measure the L_3-edge of an element without observing the L_2-edge or the L_2-edge under the influence of the L_3 and L_1 edges (Glatzel et al., 2005).

Another important application of selective extra absorption is the use of chemical shifts in the various X-ray emission channels. This opportunity yields the possibility of measuring the valence-selective X-ray absorption spectra, both XANES and EXAFS. This technique has been applied successfully to Fe(III) as a test case (Glatzel et al., 2002). At least in principle, many applications of this methodology are possible. A metal will have a different X-ray emission spectrum than an oxide. This difference can be used to measure the X-ray absorption spectra of the metal and the oxide in a mixed metal oxide. Such mixtures are encountered frequently in heterogeneous catalysis.

The second extension of the Weckhuysen roadmap is *measuring spatially*. In catalysis research, little use has been made of micro-focus beam lines for collecting spectroscopic information spatially. Most often the beam is fixed in position on a catalyst, and great care it taken to ensure that the beam position is indeed stable. However, recently Grunwaldt et al. (2006) combined XANES spectroscopy with a charge coupled device (CCD) camera and used a suitable cell to demonstrate that the two-dimensional mapping of the rhodium oxidation state in the catalyst bed can be achieved during the oxidation of a methane on a Rh/Al_2O_3 catalyst. A schematic of the experimental arrangement is shown in Figure 67. The reader is referred to the publication for details of the experimental procedure.

Typical results obtained by this method are shown in Figure 68. The axial distribution of rhodium species in the capillary cell can be clearly mapped at a spatial resolution of 10 μm. This experiment represents an exciting advance in the application of spatially resolved XAFS spectroscopy. The authors stated that although the full field imaging with the 2-D detector had a spatial resolution of 4 μm associated with the optical aspects and counting statistics, the experimental resolution was limited to 10 μm. If these types of experiments are carried out at high-brightness beam lines at third-generation synchrotron sources and are coupled with

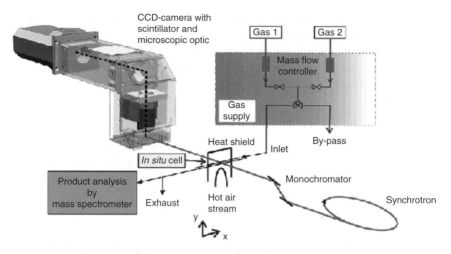

FIGURE 67 Schematic of the 2-D mapping of catalyst samples at the few μm scale (Grunwaldt et al., 2006). Reprinted with permission from (Grunwaldt et al., 2006). Copyright 2006 American Chemical Society.

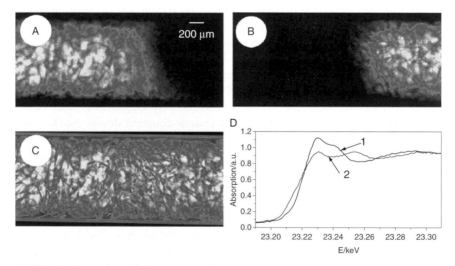

FIGURE 68 Measures of the amount of oxidized rhodium species in a Rh/Al₂O₃ catalyst (image (A) and spectrum 1 in (D)), reduced rhodium species (image (B) and spectrum (2) in (D)), and the distribution of other elements that show a featureless absorption spectrum (C). The original image recorded by the CCD camera was 3.0 × 1.5 mm, and the reaction gas mixture (6% CH₄/3%O₂/He) entered the capillary cell from the left (Grunwaldt et al., 2006). Reprinted with permission from (Grunwaldt et al., 2006). Copyright 2006 American Chemical Society.

high-resolution area detectors then significant improvements in spatial resolution are expected. Even today a resolution of approximately 1μm can be achieved. However, this approach of using 2-D CCD detectors does have a finite limitation of spatial resolution, and will be limited to the micron length scale.

Another approach is to use focused X-ray beams in the scanning mode. This technique will require specialized focusing optics and a fast monochromator. Such a nanoprobe beam line is currently being developed at the APS and is expected to have the spatial resolution of several tens of nanometers. At the proposed NSLS-II synchrotron, a beam line with a spatial resolution of ten nanometers is planned. With the development of these beam lines, it is expected that spatial imaging will be available for characterization of catalysts, and of course the hope is to do this with catalysts in reactive atmospheres.

A further development could be 3-D imaging of a catalyst bed. Already X-ray tomography experiments are being conducted at a spatial resolution of several μm, and the dream is to develop this field of 3-D imaging with a spatial resolution in the nanometer range.

The second area of Weckhuysen's roadmap—extending the operational window from mono-phasic to biphasic and triphasic systems—does not need much discussion here. With the design of specialized cells and appropriate detectors, this methodology is already being applied to the characterization of catalysts by XAFS spectroscopy (see some of the examples in Section 3).

The third area is the single molecule or single site aspect of characterization. This is clearly the most speculative and most challenging of the three areas. A dream in catalyst characterization is to monitor the catalytic transformation of a single reactant molecule on a particular active site. Today we are a long way from being able to do this; but there is some hope.

Roeffaers et al. (2006) adapted real-time monitoring of the chemical transformation of individual organic molecules by fluorescence microscopy to monitor reactions catalyzed by crystals of layered double hydroxide immersed in reagent solution. By using a wide-field microscope, they were able to map the spatial distribution of catalytic activity over the entire crystal by counting single turnover events. They hypothesized that reporter probes with different sizes and functionalities could be used to gain insight into steric, electronic, and polarity characteristics of active sites. They expect that such developments, when used in conjunction with the methods for characterizing the catalyst material, will provide better insight into challenging aspects of catalysis such as the structure-sensitive catalysis by supported metals.

Breakthroughs like this raise the hope that single-site or single-molecule imaging will be realized, and it is anticipated that the many advantages of XAS will play a role in this development.

The future trends in XAFS spectroscopy relevant to characterization of catalysts in reactive atmospheres will thus be a combination of μm–ns time-resolved XAFS spectroscopy, time-resolved and spatially resolved XAFS spectroscopy, and state-resolved XAFS observations of the local structures of working catalysts. These more precise and definitive measurements, when coupled with advances in theory, will lead to more reliable structural analysis of catalysts and the ability to definitively resolve the structures in mixed-phase catalysts. It is indeed an exciting and continuously evolving field.

7. CONCLUDING REMARKS

XAS is a powerful and widely applicable method for characterizing catalysts in the working state. The fact that it is a photon-in/photon-out technique means that it is relatively straightforward to probe the structure of the catalyst while it is in the presence of the reactants and products. The spectroscopy can be conducted at almost any temperature and pressure, for characterization of most elements in the periodic table, and with a wide range of elemental concentrations (tens of ppm to bulk)—and information can be gained about the structure of the catalyst during reaction in the presence of both gas-phase and liquid-phase reactants. In this review, we have attempted to provide some insight into this field of catalyst characterization, using examples of static and dynamic systems, and to provide a detailed review of the literature of the design of cells for characterization of catalysts in the functioning state needed to conduct this work. We are left with the conclusion that this remains a dynamic research field; with the many new developments in the spectroscopy, coupled with advancements in theory, and the construction of new synchrotron radiation sources, XAS will remain a key characterization technique for catalysts in their most important state—when they are functioning.

ACKNOWLEDGMENT

One of us (SRB) gratefully acknowledges Dr. Shelly Kelly for many enlightening discussions.

REFERENCES

Ahuja, S.P., Derrien, M.L., and Le Page, J.F., *Ind. Eng. Chem. Prod. Res. Develop.* **9**, 272 (1970).
Alexeev, O.S., Li, F., Amiridis, M.D., and Gates, B.C., *J. Phys. Chem. B* **109**(6), 2338 (2005).
Allen, P.G., Conradson, S.D., Wilson, M.S., Gottesfeld, S., Raistrick, I.D., Valerio, J., and Lovato, M., *J. Electroanal. Chem.* **384**(1–2), 99 (1995).
Als-Nielsen, J., Grubel, G., and Clausen, B.S., *Nucl. Instrum. Meth. Phys. Res. B* **97**(1–4), 522 (1995).

Amiridis, M.D., Zhang, T.J., and Farrauto, R.J., *Appl. Catal. B* **10**(1–3), 203 (1996).
Ankudinov, A.L., Bouldin, C.E., Rehr, J.J., Sims, J., and Hung, H., *Phys. Rev. B* **65**(10), 104107 (2002a).
Ankudinov, A.L., Rehr, J.J., and Bare, S.R., *Chem. Phys. Lett.* **316**(5–6), 495 (2000).
Ankudinov, A.L., Rehr, J.J., Low, J., and Bare, S.R., *Phys. Rev. Lett.* **86**(8), 1642 (2001a).
Ankudinov, A.L., Rehr, J.J., Low, J.J., and Bare, S.R., *J. Chem. Phys.* **116**(5), 1911 (2002b).
Ankudinov, A.L., Rehr, J.J., Low, J.J., and Bare, S.R., *J. Synchrotron Rad.* **8**, 578 (2001b).
Ankudinov, A.L., Rehr, J.J., Low, J.J., and Bare, S.R., *Top. Catal.* **18**(1–2), 3 (2002c).
Ankudonov, A.L., Ravel, B., Rehr, J.J., and Conradson, S.D., *Phys. Rev. B* **58**, 998 (1998).
Argo, A.M., and Gates, B.C., *J. Phys. Chem. B* **107**(23), 5519 (2003).
Argo, A.M., and Gates, B.C., *Langmuir* **18**(6), 2152 (2002).
Argo, A.M., Odzak, J.F., and Gates, B.C., *J. Am. Chem. Soc.* **125**(23), 7107 (2003).
Argo, A.M., Odzak, J.F., Goellner, J.F., Lai, F.S., Xiao, F.S., and Gates, B.C., *J. Phys. Chem. B* **110**(4), 1775 (2006).
Argo, A.M., Odzak, J.F., Lai, F.S., and Gates, B.C., *Nature* **415**(6872), 623 (2002).
Asakura, K., and Iwasawa, Y., *J. Phys. Chem.* **93**(10), 4213 (1989).
Bamwenda, G.R., Tsubota, S., Nakamura, T., and Haruta, M., *Catal. Lett.* **44**(1–2), 83 (1997).
Bare, S.R., Kelly, S.D., Sinkler, W., Low, J.J., Modica, F.S., Valencia, S., Corma, A., and Nemeth, L.T., *J. Am. Chem. Soc.* **127**(37), 12924 (2005).
Bare, S.R., Mickelson, G.E., Modica, F.S., Ringwelski, A.Z., and Yang, N.,*Rev. Sci. Instrum.* **77**(2), 023105/1 (2006).
Bare, S.R., Yang, N., Kelly, S.D., Mickelson, G.E., and Modica, F.S., *Catal. Today* **126**(1–2), 18 (2007).
Bart, J.C.J., *Adv. Catal.* **34**, 203 (1986).
Barton, D.G., Soled, S.L., Meitzner, G.D., Fuentes, G.A., and Iglesia, E., *J. Catal.* **181**(1), 57 (1999).
Battiston, A.A., Bitter, J.H., de Groot, F.M.F., Overweg, A.R., Stephan, O., van Bokhoven, J.A., Kooyman, P.J., van der Spek, C., Vanko, G., and Koningsberger, D.C., *J. Catal.* **213**(2), 251 (2003a).
Battiston, A.A., Bitter, J.H., Heijboer, W.M., de Groot, F.M.F., and Koningsberger, D.C., *J. Catal.* **215**(2), 279 (2003b).
Battiston, A.A., Bitter, J.H., and Koningsberger, D.C., *J. Catal.* **218**(1), 163 (2003c).
Bazin, D., *in* "X-ray Absorption Fine Structure for Catalysts and Surfaces" (Y. Iwasawa, Ed.) p. 113. World Scientific, Singapore, 1996.
Beale, A.M., van der Eerden, A.M.J., Jacques, S.D.M., Leynaud, O., O'Brien, M.G., Meneau, F., Nikitenko, S., Bras, W., and Weckhuysen, B.M., *J. Am. Chem. Soc.* **128**(38), 12386 (2006).
Beale, A.M., van der Eerden, A.M.J., Kervinen, K., Newton, M.A., and Weckhuysen, B.M., *Chem. Comm.*, 3015 (2005).
Becker, E., Carlsson, P.A., Gronbeck, H., and Skoglundh, M., *J. Catal.* **252**(1), 11 (2007).
Benfield, R.E., *J. Chem. Soc. Faraday Trans.* **88**(8), 1107 (1992).
Bergmann, U., Grush, M.M., Horne, C.R., DeMarois, P., Penner-Hahn, J.E., Yocum, C.F., Wright, D.W., Dube, C.E., Armstrong, W.H., Christou, G., Eppley, H.J., and Cramer, S.P., *J. Phys. Chem. B* **102**(42), 8350 (1998).
Betta, R.A.D., Boudart, M., Foger, K., Loffler, D.G., and Sanchezarrieta, J., *Rev. Sci. Instrum.* **55**(12), 1910 (1984).
Bhirud, V., Goellner, J.F., Argo, A.M., and Gates, B.C., *J. Phys. Chem. B* **108**(28), 9752 (2004).
Bond, G.C., and Thompson, D.T., *Gold Bulletin* **33**(2), 41 (2000).
Bottger, I., Schedel-Niedrig, T., Timpe, O., Gottschall, R., Havecker, M., Ressler, T., and Schlögl, R., *Chem. Eur. J.* **6**(10), 1870 (2000).
Breen, J.P., Burch, R., Hardacre, C., and Hill, C.J., *J. Phys. Chem. B* **109**(11), 4805 (2005).
Bressler, C., Saes, M., Chergui, M., Grolimund, D., Abela, R., and Pattison, P., *J. Chem. Phys.* **116**(7), 2955 (2002).

Brown, M., Peierls, R.E., and Stern, E.A., *Phys. Rev. B* **15**(2), 738 (1977).

Caballero, A., Morales, J.J., Cordon, A.M., Holgado, J.P., Espinos, J.P., and Gonzalez-Elipe, A. R., *J. Catal.* **235**(2), 295 (2005).

Castner, D.G., Watson, P.R., and Chan, I.Y., *J. Phys. Chem.* **94**(2), 819 (1990).

Cattaneo, R., Rota, F., and Prins, R., *J. Catal.* **199**(2), 318 (2001).

Cattaneo, R., Shido, T., and Prins, R., *J. Catal.* **185**(1), 199 (1999).

Cattaneo, R., Weber, T., Shido, T., and Prins, R., *J. Catal.* **191**(1), 225 (2000).

Chauvistre, R., Hormes, J., and Sommer, K., *Kautsch. Gummi Kunstst.* **47**(7), 481 (1994).

Chen, L.X., *J. Elec. Spectrosc. Relat. Phenom.* **119**(2–3), 161 (2001).

Chen, Y., Ciuparu, D., Lim, S., Yang, Y.H., Haller, G.L., and Pfefferle, L., *J. Catal.* **226**(2), 351 (2004b).

Chen, Y., Ciuparu, D., Lim, S.Y., Yang, Y.H., Haller, G.L., and Pfefferle, L., *J. Catal.* **225**(2), 453 (2004a).

Choy, J.H., Kim, Y.I., Yoon, J.B., and Choy, S.H., *J. Mat. Chem.* **11**(5), 1506 (2001).

Cimini, F., and Prins, R., *J. Phys. Chem. B* **101**(27), 5277 (1997).

Ciuparu, D., Chen, Y., Lim, S., Haller, G.L., and Pfefferle, L., *J. Phys. Chem. B* **108**(2), 503 (2004a).

Ciuparu, D., Chen, Y., Lim, S., Yang, Y.H., Haller, G.L., and Pfefferle, L., *J. Phys. Chem. B* **108** (40), 15565 (2004b).

Ciuparu, D., Haider, P., Fernandez-Garcia, M., Chen, Y., Lim, S., Haller, G.L., and Pfefferle, L., *J. Phys. Chem. B* **109**(34), 16332 (2005).

Clausen, B.S., *Catal. Today* **39**(4), 293 (1998).

Clausen, B.S., Grabaek, L., Steffensen, G., Hansen, P.L., and Topsøe, H., *Catal. Lett.* **20**(1–2), 23 (1993).

Clausen, B.S., Grabaek, L., Topsøe, H., Hansen, L.B., Stoltze, P., Nørskov, J.K., and Nielsen, O. H., *J. Catal.* **141**, 368 (1993).

Clausen, B.S., Steffensen, G., Fabius, B., Villadsen, J., Feidenhansl, R., and Topsøe, H., *J. Catal.* **132**(2), 524 (1991).

Clausen, B.S., and Topsøe, H., *Catal. Today* **9**(1–2), 189 (1991).

Clausen, B.S., Topsøe, H., Candia, R., Villadsen, J., Lengeler, B., Alsnielsen, J., and Christensen, F., *J. Phys. Chem.* **85**(25), 3868 (1981).

Clausen, B.S., Topsøe, H., Hansen, L.B., Stoltze, P., and Nørskov, J.K., *Catal. Today* **21**(1), 49 (1994).

Corma, A., Nemeth, L.T., Renz, M., and Valencia, S., *Nature* **412**(6845), 423 (2001).

Costello, C.K., Yang, J.H., Law, H.Y., Wang, Y., Lin, J.N., Marks, L.D., Kung, M.C., and Kung, H.H., *Appl. Catal. A* **243**(1), 15 (2003).

Coulston, G.W., Bare, S.R., Kung, H., Birkeland, K., Bethke, G.K., Harlow, R., Herron, N., and Lee, P.L., *Science* **275**(5297), 191 (1997).

Couves, J.W., Thomas, J.M., Waller, D., Jones, R.H., Dent, A.J., Derbyshire, G.E., and Greaves, G.N., *Nature* **354**(6353), 465 (1991).

Dacapito, F., Boscherini, F., Buffa, F., Vlaic, G., Paschina, G., and Mobilio, S., *J. Non-Cryst. Solids* **156**, 571 (1993).

Davis, R.J., *Science* **301**(5635), 926 (2003).

de Groot, F., *Chem. Rev.* **101**(6), 1779 (2001).

de Groot, F., *Coord. Chem. Rev.* **249**(1–2), 31 (2005).

de Groot, F.M.F., *AIP Conf. Proc.* **882**, 37 (2007).

de Groot, F.M.F., Glatzel, P., Bergmann, U., van Aken, P.A., Barrea, R.A., Klemme, S., Havecker, M., Knop-Gericke, A., Heijboer, W.M., and Weckhuysen, B.M., *J. Phys. Chem. B* **109**(44), 20751 (2005).

de Groot, F.M.F., Krisch, M.H., and Vogel, J., *Phys. Rev. B* **66**(19), 195112 (2002).

Deng, W., Carpenter, C., Yi, N., Flytzani-Stephanopoulos, M., *Top. Catal.* Vol. **44**(1–2), 199 (2007).

Dent, A.J., *Top. Catal.* **18**(1–2), 27 (2002).

Dent, A.J., Greaves, G.N., Roberts, M.A., Sankar, G., Wright, P.A., Jones, R.H., Sheehy, M., Madill, D., Catlow, C.R.A., Thomas, J.M., and Rayment, T., *Nucl. Instrum. Meth. Phys. Res. B* **97**(1–4), 20 (1995).

Dent, A.J., Oversluizen, M., Greaves, G.N., Roberts, M.A., Sankar, G., Catlow, C.R.A., and Thomas, J.M., *Physica B* **209**(1–4), 253 (1995).

Douillard, L., GautierSoyer, M., Duraud, J.P., Fontaine, A., and Baudelet, F., *J. Phys. Chem. Solids* **57**(4), 495 (1996).

Elam, W.T., Ravel, B.D., and Sieber, J.R., *Radiat. Phys. Chem.* **63**(2), 121 (2002).

Epple, M., Troger, L., and Hilbrandt, N., *J. Chem. Soc. Faraday Trans.* **93**(17), 3035 (1997).

Evans, J., Gauntlett, J.T., and Mosselmans, J.F.W., *Faraday Discuss.* (89), 107 (1990).

Evans, J., and Newton, M.A., *J. Mol. Catal. A Chem.* **182**(1), 351 (2002).

Evans, J., O'Neill, L., Kambhampati, V.L., Rayner, G., Turin, S., Genge, A., Dent, A.J., and Neisius, T., *J. Chem. Soc.-Dalton Trans.* (10), 2207 (2002).

Fiddy, S.G., Newton, M.A., Campbell, T., Dent, A.J., Harvey, I., Salvini, G., Turin, S., and Evans, J., *Phys. Chem. Chem. Phys.* **4**(5), 827 (2002).

Fierro-Gonzalez, J.C., and Gates, B.C., *J. Phys. Chem. B* **108**(11), 16999 (2004).

Fierro-Gonzalez, J.C., and Gates, B.C., *Langmuir* **21**(13), 5693 (2005).

Frahm, R., *Nucl. Instrum. Meth. A* **270**(2–3), 578 (1988).

Frahm, R., *Rev. Sci. Instrum.* **60**(7), 2515 (1989).

Frahm, R., Wong, J., Holt, J.B., Larson, E.M., Rupp, B., and Waide, P.A., *Phys. Rev. B* **46**(14), 9205 (1992).

Frenkel, A.I., *J. Synchrotron Rad.* **6**, 293 (1999).

Frenkel, A.I., Hills, C.W., and Nuzzo, R.G., *J. Phys. Chem. B* **105**(51), 12689 (2001).

Frenkel, A.I., Kleifeld, O., Wasserman, S.R., and Sagi, I., *J. Chem. Phys.* **116**, 9449 (2002).

Fuggle, J.C., and Ingelsfield, J.E., "Unoccupied Electronic States." Springer-Verlag, Berlin, 1992.

Fu, Q., Saltsburg, H., and Flytzani-Stephanopoulos, M., *Science* **301**(5635), 935 (2003).

Gamarra, D., Belver, C., Fernandez-Garcia, M., and Martinez-Arias, A., *J. Am. Chem. Soc.* **129** (40), 12064 (2007).

Geantet, C., Soldo, Y., Glasson, C., Matsubayashi, N., Lacroix, M., Proux, O., Ulrich, O., and Hazemann, J.L., *Catal. Lett.* **73**(2–4), 95 (2001).

Girardon, J.S., Khodakov, A.Y., Capron, M., Cristol, S., Dujardin, C., Dhainaut, F., Nikitenko, S., Meneau, F., Bras, W., and Payen, E., *J. Synchrotron Rad.* **12**, 680 (2005).

Glatzel, P., de Groot, F.M.F., Manoilova, O., Grandjean, D., Weckhuysen, B.M., Bergmann, U., and Barrea, R., *Phys. Rev. B* **72**(1), 014117 (2005).

Glatzel, P., Jacquamet, L., Bergmann, U., de Groot, F.M.F., and Cramer, S.P., *Inorg. Chem.* **41** (12), 3121 (2002).

Grandjean, D., Beale, A.M., Petukhov, A.V., and Weckhuysen, B.M., *J. Am. Chem. Soc.* **127**(41), 14454 (2005).

Greegor, R.B., and Lytle, F.W., *J. Catal.* **63**(2), 476 (1980).

Groothaert, M.H., van Bokhoven, J.A., Battiston, A.A., Weckhuysen, B.M., and Schoonheydt, R.A., *J. Am. Chem. Soc.* **125**(25), 7629 (2003).

Groppo, E., Prestipino, C., Cesano, F., Bonino, F., Bordiga, S., Lamberti, C., Thune, P.C., Niemantsverdriet, J.W., and Zecchina, A., *J. Catal.* **230**(1), 98 (2005).

Grunwaldt, J.D., and Baiker, A., *Phys. Chem. Chem. Phys.* **7**(20), 3526 (2005).

Grunwaldt, J.D., Caravati, M., and Baiker, A., *J. Phys. Chem. B* **110**(51), 25586 (2006).

Grunwaldt, J.D., Caravati, M., Hannemann, S., and Baiker, A., *Phys. Chem. Chem. Phys.* **6**(11), 3037 (2004).

Grunwaldt, J.D., Caravati, M., Ramin, M., and Baiker, A., *Catal. Lett.* **90**(3–4), 221 (2003b).

Grunwaldt, J.D., and Clausen, B.S., *Top. Catal.* **18**(1–2), 37 (2002).

Grunwaldt, J.D., Hannemann, S., Schroer, C.G., and Baiker, A., *J. Phys. Chem. B* **110**(17), 8674 (2006).

Grunwaldt, J.D., Kappen, P., Basini, L., and Clausen, B.S., *Catal. Lett.* **78**(1–4), 13 (2002).

Grunwaldt, J.D., Keresszegi, C., Mallat, T., and Baiker, A., *J. Catal.* **213**(2), 291 (2003a).

Grunwaldt, J.D., Kimmerle, B., Hannemann, S., Baiker, A., Boye, P., and Schroer, C.G., *J. Mat. Chem.* **17**(25), 2603 (2007).

Grunwaldt, J.D., Lutzenkirchen-Hecht, D., Richwin, M., Grundmann, S., Clausen, B.S., and Frahm, R., *J. Phys. Chem. B* **105**(22), 5161 (2001).

Grunwaldt, J.D., Molenbroek, A.M., Topsøe, N.Y., Topsøe, H., and Clausen, B.S., *J. Catal.* **194** (2), 452 (2000).

Grunwaldt, J.D., Ramin, M., Rohr, M., Michailovski, A., Patzke, G.R., and Baiker, A., *Rev. Sci. Instrum.* **76**(5), 054104/1 (2005).

Grzybowska-Swierkosz, B., *Top. Catal.* **11**(1–4), 23 (2000).

Guay, D., Tourillon, G., Dartyge, E., Fontaine, A., Mcbreen, J., Pandya, K.I., and O'Grady, W. E., *J. Electroanal. Chem.* **305**(1), 83 (1991).

Gunter, M.M., Ressler, T., Jentoft, R.E., and Bems, B., *J. Catal.* **203**(1), 133 (2001).

Guzman, J., and Gates, B.C., *J. Am. Chem. Soc.* **126**(9), 2672 (2004).

Guzman, J., and Gates, B.C., *J. Phys. Chem. B* **106**(31), 7659 2002).

Hagelstein, M., Hatje, U., Forster, H., Ressler, T., and Metz, W., *Zeolites and Related Microporous Materials: State of the Art 1994* **84**, 1217 (1994).

Hagelstein, M., SanMiguel, A., Fontaine, A., and Goulon, J., *J. de Physique IV* **7**(C2), 303 (1997).

Haider, P., Grunwaldt, J.D., Seidel, R., and Baiker, A., *J. Catal.* **250**(2), 313 (2007).

Haller, G.L., in "X-ray Absorption Fine Structure for Catalysts and Surfaces" (Y. Iwasawa, Ed.), World Scientific, Singapore, 1996.

Hamalainen, K., Siddons, D.P., Hastings, J.B., and Berman, L.E., *Phys. Rev. Lett.* **67**(20), 2850 (1991).

Hammer, B., and Nørskov, J.K., *Nature* **376**, 238 (1995).

Hannemann, S., Casapu, M., Grunwaldt, J.D., Haider, P., Trussel, P., Baiker, A., and Welter, E., *J. Synchrotron Rad.* **14**, 345 (2007).

Haruta, M., *Catal. Today* **36**(1), 153 (1997).

Haruta, M., and Date, M., *Appl. Catal. A* **222**(1–2), 427 (2001).

Haruta, M., Tsubota, S., Kobayashi, T., Kageyama, H., Genet, M.J., and Delmon, B., *J. Catal.* **144**(1), 175 (1993).

Hass, F., Bron, M., Fuess, H., and Claus, P., *Appl. Catal. A* **318**, 9 (2007).

Hatje, U., Ressler, T., Petersen, S., and Forster, H., *J. Phys. IV* **4**(C9), 141 (1994).

Hecht, D., Borthen, P., Frahm, R., and Strehblow, H.H., *J. de Physique IV* **7**(C2), 717 (1997).

Hecht, D., Frahm, R., and Strehblow, H.H., *J. Phys. Chem.* **100**(26), 10831 (1996).

Heijboer, W.M., Glatzel, P., Sawant, K.R., Lobo, R.F., Bergmann, U., Barrea, R.A., Koningsberger, D.C., Weckhuysen, B.M., and de Groot, F.M.F., *J. Phys. Chem. B* **108**(28), 10002 (2004).

Heijboer, W.M., Koningsberger, D.C., Weckhuysen, B.M., and de Groot, F.M.F., *Catal. Today* **110**(3–4), 228 (2005).

Hilbrandt, N., and Martin, M., *Solid State Ionics* **95**(1–2), 61 (1997).

Ichikawa, M., Fukushima, T., Yokoyama, T., Kosugi, N., and Kuroda, H., *J. Phys. Chem.* **90**(7), 1222 (1986).

Inada, Y., Hayashi, H., Funahashi, S., and Nomura, M., *Rev. Sci. Instrum.* **68**(8), 2973 (1997).

Iwamoto, M., Furukawa, H., Mine, Y., Uemura, F., Mikuriya, S.I., and Kagawa, S., *J. Chem. Soc. Chem. Comm.* (16), 1272 (1986).

Iwasawa, Y. (Ed.) "X-ray Absorption Fine Structure for Catalysts and Surfaces", World Scientific, Singapore, 1996.

Izumi, Y., Kiyotaki, F., Yagi, N., Vlaicu, A.M., Nisawa, A., Fukushima, S., Yoshitake, H., and Iwasawa, Y., *J. Phys. Chem. B* **109**(31), 14884 (2005).

Jentoft, R.E., Deutsch, S.E., and Gates, B.C., *Rev. Sci. Instrum.* **67**(6), 2111 (1996).

Jentoft, R.E., Hahn, A.H.P., Jentoft, F.C., and Ressler, T., *Phys. Chem. Chem. Phys.* **7**(14), 2830 (2005).

Jentoft, R.E., Hahn, A.H.P., Jentoft, F.C., and Ressler, T., *Physica Scripta* **T115**, 794 (2005).

Jentys, A., *Phys. Chem. Chem. Phys.* **1**(17), 4059 (1999).

Kabe, T., Qian, W.H., Funato, A., Okoshi, Y., and Ishihara, A., *Phys. Chem. Chem. Phys.* **1**(5), 921 (1999).

Kaminaga, U., Matsushita, T., and Kohra, K., *Jap. J. Appl. Phys.* **20**(5), L355 (1981).

Kampers, F.W.H., Maas, T.M.J., Vangrondelle, J., Brinkgreve, P., and Koningsberger, D.C., *Rev. Sci. Instrum.* **60**(8), 2635 (1989).

Kappen, P., Troger, L., Materlik, G., Reckleben, C., Hansen, K., Grunwaldt, J.D., and Clausen, B.S., *J. Synchrotron Rad.* **9**, 246 (2002).

Kawai, T., Bando, K.K., Lee, Y.K., Oyama, S.T., Chun, W.J., and Asakura, K., *J. Catal.* **241**(1), 20 (2006).

Keegan, M.B.T., Dent, A.J., Blake, A.B., Conyers, L., Moyes, R.B., Wells, P.B., and Whan, D.A., *Catal. Today* **9**(1–2), 183 (1991).

Kelly, S.D., Hesterberg, D., and Ravel, B., *in* "Analysis of Soils and Minerals Using X-ray Absorption Spectroscopy" (L. Ulery and L.R. Drees, Eds.), p. 367. Soil Science Society of America, Madison, 2008.

Kip, B.J., Duivenvoorden, F.B.M., Koningsberger, D.C., and Prins, R., *J. Catal.* **105**(1), 26 (1987).

Knop-Gericke, A., de Groot, F.M.F., van Bokhoven, J.A., and Ressler, T., *in* "In situ Spectroscopy of Catalysts" (B.M. Weckhuysen, Ed.), American Scientific, Stevenson Ranch, CA, 2004.

Koch, A., Hagelstein, M., San Miguel, A., Fontaine, A., and Ressler, T., *SPIE* **85**, 2461 (1995).

Koningsberger, D.C., and Prins, R., "X-ray Absorption: Principles, Applications, Techniques of EXAFS, SEXAFS and XANES" John Wiley & Sons, NY, 1988.

Lamberti, C., Bordiga, S., Bonino, F., Prestipino, C., Berlier, G., Capello, L., D'Acapito, F., Xamena, F.X.L.I., and Zecchina, A., *Phys. Chem. Chem. Phys.* **5**(20), 4502 (2003).

Lamberti, C., Prestipino, C., Bonino, F., Capello, L., Bordiga, S., Spoto, G., Zecchina, A., Moreno, S.D., Cremaschi, B., Garilli, M., Marsella, A., Carmello, D., et al., *Angew. Chem. Int. Ed.* **41**(13), 2341 (2002).

Lim, S., Ciuparu, D., Chen, Y., Yang, Y.H., Pfefferle, L., and Haller, G.L., *J. Phys. Chem. B* **109** (6), 2285 (2005).

Liu, D.J., and Robota, H.J., *J. Phys. Chem. B* **103**(14), 2755 (1999).

Louwers, S.P.A., Craje, M.W.J., Vanderkraan, A.M., Geantet, C., and Prins, R., *J. Catal.* **144**(2), 579 (1993).

Lutzenkirchen-Hecht, D., and Frahm, R., *J. Phys. Chem. B* **105**(41), 9988 (2001).

Lytle, F.W., *J. Catal.* **43**(1–3), 376 (1976).

Lytle, F.W., Sayers, D.E., and Moore, E.B., *Appl. Phys. Lett.* **24**(2), 45 (1974).

Lytle, F.W., Wei, P.S.P., Greegor, R.B., Via, G.H., and Sinfelt, J.H., *J. Chem. Phys.* **70**(11), 4849 (1979).

Malinowski, E.R., and Howery, D.G., "Factor Analysis in Chemistry." John Wiley and Sons, New York, 1980.

Mansour, A.N., Cook, J.W., and Sayers, D.E., *J. Phys. Chem.* **88**(11), 2330 (1984).

Maris, E.P., Ketchie, W.C., Oleshko, V., and Davis, R.J., *J. Phys. Chem. B* **110**(15), 7869 (2006).

Martinez-Arias, A., Fernandez-Garcia, M., Hungria, A.B., Iglesias-JueZ, A., and Anderson, J.A., *Catal. Today* **126**(1–2), 90 (2007).

Mavrikakis, M., Stoltze, P., and Nørskov, J.K., *Catal. Lett.* **64**(2–4), 101 (2000).

Mccaulley, J.A., *J. Phys. Chem.* **97**(40), 10372 (1993).

Meitzner, G., Bare, S.R., Parker, D., Woo, H., and Fischer, D.A., *Rev. Sci. Instrum.* **69**(7), 2618 (1998).

Meitzner, G., Via, G.H., Lytle, F.W., Fung, S.C., and Sinfelt, J.H., *J. Phys. Chem.* **92**(10), 2925 (1988).

Meitzner, G., Via, G.H., Lytle, F.W., and Sinfelt, J.H., *J. Phys. Chem.* **96**(12), 4960 (1992).

Millet, P., Durand, R., Dartyge, E., Tourillon, G., and Fontaine, A., *J. Electrochem. Soc.* **140**(5), 1373 (1993).

Moggridge, G.D., Schroeder, S.L.M., Lambert, R.M., and Rayment, T., *Nucl. Instrum. Meth. Phys. Res. B* **97**(1–4), 28 (1995).

Mol, J.C., and Moulijn, J.A., *Adv. Catal.* **24**, 131 (1975).

Moulijn, J.A., and Mol, J.C., *J. Mol. Catal.* **46**(1–3), 1 (1988).

Neils, T.L., and Burlitch, J.M., *J. Catal.* **118**(1), 79 (1989).

Newton, M.A., Burnaby, D.G., Dent, A.J., Diaz-Moreno, S., Evans, J., Fiddy, S.G., Neisius, T., Pascarelli, S., and Turin, S., *J. Phys.l Chem. A* **105**(25), 5965 (2001).

Newton, M.A., Burnaby, D.G., Dent, A.J., Diaz-Moreno, S., Evans, J., Fiddy, S.G., Neisius, T., and Turin, S., *J. Phys. Chem. B* **106**(16), 4214 (2002).

Newton, M.A., Dent, A.J., and Evans, J., *Chem. Soc. Rev.* **31**(2), 83 (2002).

Newville, M., *J. Synchrotron Rad.* **8**, 322 (2001).

Odzak, J.F., Argo, A.M., Lai, F.S., Gates, B.C., Pandya, K., and Feraria, L., *Rev. Sci. Instrum.* **72** (10), 3943 (2001).

Oudenhuijzen, M.K., Kooyman, P.J., Tappel, B., van Bokhoven, J.A., and Koningsberger, D. C., *J. Catal.* **205**(1), 135 (2002).

Overbury, S.H., Schwartz, V., Mullim, D.R., Yan, W.F., and Dai, S., *J. Catal.* **241**(1), 56 (2006).

Ovesen, C.V., Clausen, B.S., Schiotz, J., Stoltze, P., Topsøe, H., and Nørskov, J.K., *J. Catal.* **168** (2), 133 (1997).

Parvulescu, V.I., Grange, P., and Delmon, B., *Catal. Today* **46**(4), 233 (1998).

Pettiti, I., Gazzoli, D., Inversi, M., Valigi, M., De Rossi, S., Ferraris, G., Porta, P., and Colonna, S., *J. Synchrotron Rad.* **6**, 1120 (1999).

Pickup, D.M., Mountjoy, G., Holland, M.A., Wallidge, G.W., Newport, R.J., and Smith, M.E., *J. Phys. Condes. Matter* **12**(47), 9751 (2000).

Ravel, B., and Newville, M., *J. Synchrotron Rad.* **12**, 537 (2005).

Reed, C., Lee, Y.K., and Oyama, S.T., *J. Phys. Chem. B* **110**(9), 4207 (2006).

Rehr, J.J., and Albers, R.C., *Rev. Mod. Phys.* **72**(3), 621 (2000).

Reifsnyder, S.N., and Lamb, H.H., *J. Phys. Chem. B* **103**(2), 321 (1999).

Reitz, T.L., Lee, P.L., Czaplewski, K.F., Lang, J.C., Popp, K.E., and Kung, H.H., *J. Catal.* **199**(2), 193 (2001).

Renz, M., Blasco, T., Corma, A., Fornes, V., Jensen, R., and Nemeth, L., *Chem. Eur. J.* **8**(20), 4708 (2002).

Ressler, T., *Anal. Bioanal. Chem.* **376**(5), 584 (2003).

Ressler, T., *J. Synchrotron Rad.* **5**, 118 (1998).

Ressler, T., Hagelstein, M., Hatje, U., and Metz, W., *J. Phys. Chem. B* **101**(34), 6680 (1997).

Ressler, T., Jentoft, R.E., Wienold, J., Gunter, M.M., and Timpe, O., *J. Phys. Chem. B* **104**(27), 6360 (2000).

Ressler, T., and Timpe, O., *J. Catal.* **247**(2), 231 (2007).

Ressler, T., Timpe, O., Neisius, T., Find, J., Mestl, G., Dieterle, M., and Schlögl, R., *J. Catal.* **191** (1), 75 (2000).

Ressler, T., Wienold, J., and Jentoft, R.E., *Solid State Ionics* **141**, 243 (2001).

Ressler, T., Wienold, J., Jentoft, R.E., and Girgsdies, F., *Eur. J. Inorg. Chem.* **2**, 301 (2003).

Ressler, T., Wienold, J., Jentoft, R.E., and Girgsdies, F., *Nucl. Instrum. Meth. Phys. Res. B* **200**, 165 (2003).

Ressler, T., Wienold, J., Jentoft, R.E., and Neisius, T., *J. Catal.* **210**(1), 67 (2002).

Ressler, T., Wienold, J., Jentoft, R.E., Neisius, T., and Gunter, M.M., *Top. Catal.* **18**(1–2), 45 (2002).

Ressler, T., Wienold, J., Jentoft, R.E., Timpe, O., and Neisius, T., *Solid State Commun.* **119**(3), 169 (2001).

Ressler, T., Wong, J., Roos, J., and Smith, I.L., *Environ. Sci. & Techn.* **34**(6), 950 (2000).

Revel, R., Bazin, D., Seigneurin, A., Barthe, P., Dubuisson, J.M., Decamps, T., Sonneville, H., Poher, J.J., Maire, F., and Lefrancois, P., *Nucl. Instrum. Meth. Phys. Res. B* **155**(1–2), 183 (1999).

Richwin, M., Zaeper, R., Lutzenkirchen-Hecht, D., and Frahm, R., *Rev. Sci. Instrum.* **73**(3), 1668 (2002).

Rodriguez, J.A., Chaturvedi, S., Hanson, J.C., Albornoz, A., and Brito, J.L., *J. Phys. Chem. B* **102**(8), 1347 (1998).

Rodriguez, J.A., Chaturvedi, S., Hanson, J.C., and Brito, J.L., *J. Phys. Chem. B* **103**(5), 770 (1999).

Rodriguez, J.A., Hanson, J.C., Chaturvedi, S., Maiti, A., and Brito, J.L., *J. Chem. Phys.* **112**(2), 935 (2000).

Roeffaers, M.B.J., Sels, B.F., Uji-i, H., De Schryver, F.C., Jacobs, P.A., De Vos, D.E., and Hofkens, J., *Nature* **439**(7076), 572 (2006).

Rumpf, H., Janssen, J., Modrow, H., Winkler, K., and Hormes, J., *J. Solid State Chem.* **163**(1), 158 (2002).

Safonova, O.V., Deniau, B., and Millet, J.M.M., *J. Phys. Chem. B* **110**(47), 23962 (2006).

Safonova, O.V., Tromp, M., van Bokhoven, J.A., de Groot, F.M.F., Evans, J., and Glatzel, P., *J. Phys. Chem. B* **110**(33), 16162 (2006).

Saib, A.M., Borgna, A., de Loosdrecht, J.V., van Berge, P.J., and Niemantsverdriet, J.W., *Appl. Catal. A* **312**, 12 (2006).

Sanchez, R.M.T., Ueda, A., Tanaka, K., and Haruta, M., *J. Catal.* **168**(1), 125 (1997).

Sankar, G., Cao, E.H., and Gavriilidis, A., *Catal. Today* **125**(1–2), 24 (2007).

Sankar, G., Thomas, J.M., Rey, F., and Greaves, G.N., *J. Chem. Soc. Chem. Comm.* (24), 2549 (1995).

Sankar, G., Thomas, J.M., Waller, D., Couves, J.W., Catlow, C.R.A., and Greaves, G.N., *J. Phys. Chem.* **96**(19), 7485 (1992).

Sankar, G., Wright, P.A., Natarajan, S., Thomas, J.M., Greaves, G.N., Dent, A.J., Dobson, B.R., Ramsdale, C.A., and Jones, R.H., *J. Phys. Chem.* **97**(38), 9550 (1993).

Sayers, D., Stern, E., and Lytle, F.W., *Phys. Rev. Lett.* **27**, 1204 (1971).

Scheuring, E.M., Clavin, W., Wirt, M.D., Miller, L.M., Fischetti, R.F., Lu, Y., Mahoney, N., Xie, A.H., Wu, J.J., and Chance, M.R., *J. Phys. Chem.* **100**(9), 3344 (1996).

Schneider, S., Bazin, D., Dubuisson, J.M., Ribbens, M., Sonneville, H., Meunier, G., Garin, F., Maire, G., and Dexpert, H., *J. X-ray Sci. Technol.* **8**, 221 (2000).

Schulke, W., Kaprolat, A., Fischer, T., Hoppner, K., and Wohlert, F., *Rev. Sci. Instrum.* **66**(3), 2446 (1995).

Shannon, I.J., Maschmeyer, T., Sankar, G., Thomas, J.M., Oldroyd, R.D., Sheehy, M., Madill, D., Waller, A.M., and Townsend, R.P., *Catal. Lett.* **44**(1–2), 23 (1997).

Shibata, J., Shimizu, K., Takada, Y., Shichia, A., Yoshida, H., Satokawa, S., Satsuma, A., and Hattori, T., *J. Catal.* **227**(2), 367 (2004).

Shido, T., and Prins, R., *Curr. Opin. Solid State Mat. Sci.* **3**(4), 330 (1998).

Shido, T., Yamaguchi, A., Inada, Y., Asakura, K., Nomura, M., and Iwasawa, Y., *Top. Catal.* **18** (1–2), 53 (2002).

Sibeijn, M., Vanveen, J.A.R., Bliek, A., and Moulijn, J.A., *J. Catal.* **145**(2), 416 (1994).

Silversmit, G., Poelman, H., Sack, I., Buyle, G., Marin, G.B., and De Gryse, R., *Catal. Lett.* **107** (1–2), 61 (2006).

Sinfelt, J.H., "Bimetallic Catalysts", Wiley NY 1983.

Stanislaus, A., and Cooper, B.H., *Catal. Rev. Sci. Eng.* **36**, 75 (1994).

Stern, E.A., *Contemp. Phys.* **19**(4), 239 (1978).

Stern, E.A., and Heald, S.M., *in* "Basic Principles and Applications of EXAFS" (E.E. Koch, Ed.), Vol. 10. p. 995. North-Holland, 1983.

Stern, E.A., and Kim, K., *Phys. Rev. B* **23**(8), 3781 (1981).

Stojanoff, V., Hamalainen, K., Siddons, D.P., Hastings, J.B., Berman, L.E., Cramer, S., and Smith, G., *Rev. Sci. Instrum.* **63**(1), 1125 (1992).

Sun, M.Y., Burgi, T., Cattaneo, R., and Prins, R., *J. Catal.* **197**(1), 172 (2001).

Sun, M.Y., Burgi, T., Cattaneo, R., van Langeveld, D., and Prins, R., *J. Catal.* **201**(2), 258 (2001).

Teo, B.K., and Joy, D.C. (Eds.) "EXAFS Spectroscopy Techniques and Applications", Plenum, NY, 1981.

Teo, B.K., "EXAFS Spectroscopy Techniques and Applications", Springer, Berlin, 1986.

Thiel, D.J., Livins, P., Stern, E.A., and Lewis, A., *Nature* **363**(6429), 565 (1993).

Thomas, J.M., Greaves, G.N., Richard, C., and Catlow, C.R.A., *Nucl. Instrum. Meth. Phys. Res. B* **97**(1–4), 1 (1995).

Thomas, J.M., Greaves, G.N., Sankar, G., Wright, P.A., Chen, J.S., Dent, A.J., and Marchese, L., *Angew. Chem. Int. Ed.* **33**(18), 1871 (1994).

Tinnemans, S.J., Mesu, J.G., Kervinen, K., Visser, T., Nijhuis, T.A., Beale, A.M., Keller, D.E., van der Eerden, A.M.J., and Weckhuysen, B.M., *Catal. Today* **113**(1–2), 3 (2006).

Tohji, K., Udagawa, Y., Mizushima, T., and Ueno, A., *J. Phys. Chem.* **89**(26), 5671 (1985).

Tomov, I.V., Oulianov, D.A., Chen, P.L., and Rentzepis, P.M., *J. Phys. Chem. B* **103**(34), 7081 (1999).

Troger, L., Hilbrandt, N., and Epple, M., *J. de Physique IV* **7**(C2), 323 (1997).

Tromp, M., van Bokhoven, J.A., Safonova, O.V., de Groot, F.M.F., Evans, J., and Glatzel, P., *AIP Conf. Proc.* **882**, 651 (2007).

Valden, M., Lai, X., and Goodman, D.W., *Science* **281**(5383), 1647 (1998).

van Bokhoven, J.A., Louis, C., T Miller, J., Tromp, M., Safonova, O.V., and Glatzel, P., *Angew. Chem. Int. Ed.* **45**(28), 4651 (2006).

Van Nordstrand, R.A., *Adv. Catal.* **12**, 149 (1960).

Villinski, J.E., O'Day, P.A., Corley, T.L., and Conklin, M.H., *Environ. Sci. & Techn.* **35**(6), 1157 (2001).

Viswanathan, R., Hou, G.Y., Liu, R.X., Bare, S.R., Modica, F., Mickelson, G., Segre, C.U., Leyarovska, N., and Smotkin, E.S., *J. Phys. Chem. B* **106**(13), 3458 (2002).

Walton, R.I., Dent, A.J., and Hibble, S.J., *Chem. Mater.* **10**(11), 3737 (1998).

Walton, R.I., and Hibble, S.J., *J. Mat. Chem.* **9**(6), 1347 (1999).

Weckhuysen, B.M. (Ed.) "*In situ* Spectroscopy of Catalysts", American Scientific Publishers, Stevenson Ranch, 2004.

Weckhuysen, B.M., *Phys. Chem. Chem. Phys.* **5**(20), 4351 (2003).

Weiher, N., Beesley, A.M., Tsapatsaris, N., Delannoy, L., Louis, C., van Bokhoven, J.A., and Schroeder, S.L.M., *J. Am. Chem. Soc.* **129**(8), 2240 (2007).

Weiher, N., Bus, E., Delannoy, L., Louis, C., Ramaker, D.E., Miller, J.T., and van Bokhoven, J. A., *J. Catal.* **240**(2), 100 (2006).

Weiher, N., Bus, E., Gorzolnik, B., Moller, M., Prins, R., and van Bokhoven, J.A., *J. Synchrotron Rad.* **12**, 675 (2005).

Wienold, J., Jentoft, R.E., and Ressler, T., *Eur. J. Inorg. Chem.* **6**, 1058 (2003).

Wong, J., Lytle, F.W., Messmer, R.P., and Maylotte, D.H., *Phys. Rev. B* **30**(10), 5596 (1984).

XAFS IX Proceedings, *J. Phys. IV* (1997).

XAFS VII Proceedings, *Jpn. J. Appl. Phys.* (1993).

XAFS VIII Proceedings, *Physica B* (1995).

XAFS X Proceedings, *J. Synchrot. Radiat.* (1999).

XAFS XI Proceedings, *J. Synchrot. Radiat.* (2001).

XAFS XII Proceedings, *Physica Scripta* (2005).

XAFS XIII Proceedings, *AIP Conf. Proc.* Vol 882 (2007).

Xu, Y., and Mavrikakis, M., *J. Phys. Chem. B* **107**(35), 9298 (2003).

Yamaguchi, A., Shido, T., Inada, Y., Kogure, T., Asakura, K., Nomura, M., and Iwasawa, Y., *Catal. Lett.* **68**(3–4), 139 (2000).

Yamaguchi, A., Suzuki, A., Shido, T., Inada, Y., Asakura, K., Nomura, M., and Iwasawa, Y., *J. Phys. Chem. B* **106**(9), 2415 (2002).

Yang, N., Mickelson, G.E., Greenlay, N., Kelly, S.D., and Bare, R., *in "In Situ* EXAFS of Ni–W Hydrocracking Catalysts", (B. Hedman and P. Pianetta, Eds.) Vol. 882. p. 663. AIP Conf. Proc., Melville, NY, 2007.

Yoon, B., Hakkinen, H., Landman, U., Worz, A.S., Antonietti, J.M., Abbet, S., Judai, K., and Heiz, U., *Science* **307**(5708), 403 (2005).

Zaeper, R., Richwin, M., Wollmann, R., Lutzenkirchen-Hecht, D., and Frahm, R., *Nucl. Instrum. Meth. A* **467**, 994 (2001).

Zhang, Z., Chen, H., and Sachtler, W.M.H., *J. Chem. Soc. Faraday Trans.* **87**, 1413 (1991).

INDEX

A

Acidic site characterization
  benzhydrol, 22–23
  benzophenone, 21–22
  benzpinacol, 22
  Ti/Al binary oxide, 22–23
  triplet–triplet (T–T) absorption spectrum, 22
Acrylonitrile, 98–100
Alkane ammoxidation reactions,
      mixed-metal oxide catalysts
  antimony–vanadium mixed-metal oxides, 98
  Sb₂O₄ phase, 100
  VSbO₄ phase, 98, 100
Alkane oxidation reactions, mixed metal oxides
  ¹⁸O isotopic labeling, 94
  acrylic acid synthesis catalyst, 98
  bulk bismuth molybdate catalyst, 94
  molybdenum–vanadium–tungsten mixed
      oxide catalyst, 95
  MoO₃, 94, 96
  selectivity, 96–97
  turnover frequency (TOF), 94

B

Basic site characterization, 23
Brillouin zone center (BZC), 50

C

Catalytically active sites
  chromium-containing zeolites, 19
  copper-containing zeolites
    Cu(I)/APO-5, 18
    Cu(I)/SAPO-5, 18
    N₂O decomposition, 18
  magnesium oxide (MgO), model system
    2-methylbut-3-yn-2-ol (MBOH), 19
    excitation and emission process, 21
    luminescence fingerprint, 20
    oxidative coupling, methane, 19
    water interaction and hydrogen bonding, 21
  molybdenum-containing zeolites
    catalytic cycle, 16
    Mo–MCM-41, 14–16
    nitrogen yield, 15
    NO conversion, 14–15
  silver-containing zeolites
    Ag⁺-exchanged ZSM-5, 17

  photoluminescence bands, 17
  titanium-containing zeolites
    carbon dioxide addition, 6–7
    charge transfer (CT) excited state, 7
    Fourier transforms, 5
    methanol, 7–8
    N₂ formation, selectivity, 7
    physical parameters and photocatalytic
        properties, 6
    Ti–MCM-41 and Ti–MCM-48, 5
    titanium oxide, 5, 7
  vanadium-containing zeolites
    bond length, 10
    C-Hyd-V₀.₀₅Siβ and C-Hyd-V₁.₅Siβ, 8–11
    chemical shift, 13
    diffuse reflectance UV–vis spectrum, 8, 10
    photocatalytic activity, 13
    physicochemical parameters, 12
    quenching rate constant, 13–14
    vibrational energy, 10
Cell-autoclave reactor, 93–94
CO oxidation, gold cluster catalysts
  activity, 417
  Au/TiO₂, 416
  Au–O coordination number and distance, 416
  cationic and zerovalent gold, 414
  CO conversion, 415, 417
  high energy-resolution fluorescence detection
      (HERFD), 418
  MgO, 412
  structure elucidation, 418

D

Dehydrated supported metal oxide catalyst,
      Raman spectroscopy
  anatase nanocrystal, 71
  SiO₂, 70
  vanadia, 71
Diffuse reflectance spectroscopy
  combined optical elements, 159
  fiber optics
    fiber bundle and advantages, 159
    limiting angle, 158
  integrating spheres
    average reflectance, 154
    detector flux, 154
    sphere and wall section flux, 153–154
    sphere multiplier, 155

Diffuse reflectance spectroscopy (*cont.*)
  interpretation via Kubelka–Munk theory
    concentration, 144–145
    data quantification, 145
    limitation, 137–138
    photon fluxes and reflectance factor, 139
    remission function, 140–141
    scattering coefficient, 144
    zero absorption limit, 142–144
  mirror optics
    advantages and spectral results, 158
    design, 155–156
    Harrick DR accessory, 156–157
    vs. integrating spheres, 156
  optical accessory design
    diffuse illumination, 149
    regular illumination, 149–150
  standard materials
    polytetrafluoroethylene, 150–152
    precision and accuracy, 152

E

Electrostatic lens system
  gas flow scheme, 227–228
  infrared laser, 228
  limitation, 229
  molecular flow approach, 226
  photoelectron collection efficiency, 227
  photoelectron scattering, 225–227
  signal intensity parameters, 224
  X-ray transmission, 224–225
Extended X-ray absorption fine structure (EXAFS)
    spectroscopy, 283–284

F

Fiber optics, UV–vis spectroscopy
  alumina-supported chromia
    monitoring, 192–193
  design parameters, 160
  fiber bundle and advantages, 159
  H_2O_2 decomposition, 194
  limitations, 203–204
  limiting angle, 158
  NO and N_2O decomposition, 197
  reaction cells
    design, 164
    diode array spectrometer, 165
    NMR probe, 166
    XAFS spectroscopy, 165

H

Heterogeneous reactions and catalytic processes
  CO adsorption, Pd(111)
    C1s core-level spectra, 229–230
    coverage-dependent vibrational spectra, 231
    deconvolution, C1s spectra, 230

    Fe2p and C1s core-level spectra, 233–234
    LEED pattern characteristics, 232
    SFG spectra, 231–232
  CO oxidation, Ru(0001)
    mechanism, 257
    RuO_2 phase formation, 257–258
    subsurface oxygen, 256–257
    surface morphology, 265–266
    temperature dependence, 263–265
  ethene epoxidation, silver
    catalyst surface, 240–241
    Langmuir–Hinshelwood mechanism, 247
    oxygen species, 243–246
    PTRMS spectrum, 241–242
    spectroscopic characteristics, 244
    surface carbonates, 242–243
  methanol decomposition, Pd(111)
    C–O bond scission, 235–236
    C1s core-level spectra, 236–237
    SFG postreaction, 238–239
    TPRS investigations, 239–240
  methanol dehydrogenation, Pd(111), 234–235
  methanol oxidation, copper
    catalytic activity, 252–253
    Cu–O trilayer, 255–256
    depth-profiling measurement, 251–252
    methanol-to-oxygen ratio, 249–251
    NEXAFS experiment, 248
    reaction paths, 247
    UHV–X-ray investigations, 247–248
High energy-resolution fluorescence detection
    (HERFD), 448
High-pressure X-ray photoelectron spectroscopy
  differential pumping stages, 219–220
  thin film synthesis, 220
Hydrated supported metal oxide, Raman
    spectroscopy
  adsorbed metallates, 69
  MoO_3/Al_2O_3, 66
  phase diagram, 68
  point of zero charge (PZC), 67–68
Hydrogenation, XAFS spectroscopy
  catalytic activity, 407, 409
  ethene conversion, 407
  interatomic distance, 409
  $Ir_4/\gamma–Al_2O_3$, 407–408
  metal clusters, 406
  supported iridium clusters, 407
  toluene, 409–410

I

Intervalence charge transfer (IVCT), 83, 95, 190

K

Kubelka–Munk theory (also see Diffuse
  reflectance spectroscopy)

L

Langmuir–Hinshelwood mechanism, 247
Lanthana catalyst, 102
Ligand-to-metal charge transfer (LMCT), 81, 183, 190, 191, 193

M

Maleic anhydride production, vanadium phospho-oxides (VPO) catalyst (VO)$_2$P$_2$O$_7$, 101
   bulk lattice oxygen sites, 101
   n-butane, 100–101
   V$_2$O$_5$, 100
Metal catalysts (and oxidized forms), Raman spectroscopy
   AgO, 110
   ethene and methanol oxidation, 111
   isotopic oxygen, 110
   palladium, 111
   Pt–O vibration, 112
Mirror optics, UV–Vis spectroscopy
   advantages and spectral results, 158
   experimental limitation, 203–204
   Harrick DR accessory, 156–157
   independent calibration, 175
   reaction cells
      design, 160
      Harrick reaction cell, 163–164
      pressure range, 163
      temperature range, 164
   reference beam attenuation effect, 170
   vs. integrating spheres, 156, 166–167

N

Nanostructures and diffraction, XRD
   bulk-sensitivity, 303
   empirical method, 299
   EXAFS analyses, 301
   intensity function, 298
   microstructural parameters, 296–297
   Miller index, 298
   morphologies, 299–300
   pore system, 296
   Rietveldt method, 301
   Scherrer analysis, 299
   small angle X-ray scattering (SAXS), 302
   surface crystallography, 302–303
   systematic deviations, 297
   Williamson–Hall plot, 300
Nuclear magnetic resonance spectroscopy (NMR), 308

O

Oxygenates/hydrocarbons, reduction
   Raman bands, 86

V$_2$O$_5$/CeO$_2$ catalysts, 85–86
V$_2$O$_5$/Nb$_2$O$_5$ catalysts, 85

P

Photoluminescence (PL) spectroscopy
   acidic site characterization
      benzhydrol, 22–23
      benzophenone, 21–22
      benzpinacol, 22
      Ti/Al binary oxide, 22–23
      triplet–triplet (T–T) absorption spectrum, 22
   basic site characterization, 23
   catalytically active sites
      chromium-containing zeolites, 19
      copper-containing zeolites, 18
      magnesium oxide (MgO), model system, 19–21
      molybdenum-containing zeolites, 14–16
      silver-containing zeolites, 17
      titanium-containing zeolites, 4–8
      vanadium-containing zeolites, 8–14
   pressure effect
      high-pressure domain, 29–32
      physical vs. chemical quenching, 28–29
   temperature effect
      orange luminescence band, 33–34
      SiC wafer, 33
   TiO$_2$-containing photocatalysts
      electrons and holes, charge separation, 25
      gas-phase photocatalytic reaction, 24
      hole transfer, 25–26
      O$_2$ photodesorption, 24
      platinum particle, 24–25
      rutile and anatase, 24
      transient absorption spectra, 25–27
Polytetrafluoroethylene (PTFE), 150–152
Pressure effect, photoluminescence spectra
   high-pressure domain
      layered double-hydroxide intercalation compound, 29–30
      physical vs. chemical quenching, 28–29
Proton transfer reaction mass spectrometer (PTRMS)
   ethene epoxidation, silver, 241–242
   gas flow, 228

Q

Quenching rate constant, 13

R

Raman microspectroscopy, 112–113
Raman spectroscopy
   advantages, 54–55
   Brillouin zone center (BZC), 50
   under catalytic reaction conditions
      carbonaceous deposits, 88–89

Raman spectroscopy (*cont.*)
  DeNO$_x$ (SCR) and DeSO$_x$ reactions, 90–91
  H$_2$, reduction, 83–85
  MoO$_3$, 77–78
  NO$_x$-trap catalysts, 89–90
  oxides, dispersion and aggregation, 77
  oxides, resonance Raman effects, 80–83
  oxygenates/hydrocarbons, reduction, 85–88
  solid–solid wetting, 78
  spreading process, 77
  supercritical conditions and autoclave
      reactors, 91–92
  surface oxygen species, 79–80
catalyst preparation
  bulk iron molybdate, 72
  occluded surfactant molecule, 75–76
  PdCl$_2$ precursor, 76
  pH, 73–74
  phase transformation, 74
  propane ammoxidation, 76
  structural transformation, 74
  titanium oxide nanotube, 73
  vanadium antimonate catalyst, 76
  zeolites, 73
chemisorption, 65
CO$_2$ sequestration, 57
dehydrated supported metal oxide
    catalysts, 69–71
enhancement factors, 51
hydrated and dehydrated bulk metal oxide
    catalysts, 71–72
hydrated supported metal oxides, 66–69
imaging
  confocal microscopy, 55
  SNOM experiment, 56
limitations
  black-body radiation, 54
  fluorescence, 53
  laser heating, 52–53
molybdate catalyst, 57
MoO$_3$/Al$_2$O$_3$ catalysts, 66
phase transformation, 57
polymolybdates, 65
Rayleigh scattering, 49
reactor cells
  autoclave cell, 58–59
  catalyst structure, 62
  catalytic activity, 63
  cell design, 61–63
  conversion and selectivity, 62–63
  experimental apparatus setup, 58
  iron molybdate catalyst, 63
  oxide catalyst structural transformation, 64
  requirements, 60
  selective alkane oxidation, 61
  selective partial oxidation reactions, 64
  thermal radiation, 60
resonance effect, 50

scattering effect, 49
SERS effect, 51–52
simultaneous activity measurement
  alkane ammoxidation reactions, 98–100
  alkane oxidation reactions, 94–98
  bismuth molybdate catalyst, 92–93
  catalytic NO$_x$ decomposition, 92
  CCl$_4$ destruction, lanthana catalysts, 102
  cell–autoclave reactors, 93–94
  conductance cells, 92
  conversion and selectivity, 93
  gas-phase characterization, 93
  gas-phase reactant oxidation, acid
      catalysts, 101–102
  maleic anhydride production, 100–101
  metal catalysts, 110–112
  supported transition metal oxides, 102–110
  spectroscopic exclusion rule, 50
  surface acidity, 51
Rayleigh scattering, 49
Resonance Raman effects
  intervalence charge transfer (IVCT)
      transition, 83
  ligand-to-metal charge transfer (LMCT), 81
  MoO$_2$, 82
  transitions, resonance enhancement, 81
  TS-1, H$_2$O$_2$, 81
Resonant inelastic X-ray scattering spectroscopy
    (RIXS), 452

S

Scanning near-field optical microscopy
    (SNOM), 56
Scattering phenomena
  Bragg equation, 291
  chemical microstrain, 292
  electromagnetic waves, 289
  Ewald sphere intersects, 291–292
  gas–solid reactions, 294
  geometric analysis, 289
  methodologies, 296
  microstrain, 293–294
  schematic representation, 294
  Scherrer formula, 295–296
  translational symmetric lattices, 293
Selective catalytic reduction, 421–423
Stern–Volmer equation, 13
Strong metal–support interactions (SMSI), 278
Supported transition metal oxides
  ethane oxidation reaction, 103
  Mo=O bond, 107
  Raman bands, 105–106
  surface methoxy vibrations, 105
  V=O bond, 102, 105
  zirconia- and alumina-supported vanadia, 104
Surface enhanced Raman scattering (SERS), 51–52
Synchrotron catalysis consortium (SCC), 447

T

Temperature effect, photoluminescence spectra
orange luminescence band, 33–34
SiC wafer, 33
Temperature programmed reaction
spectroscopy, 239
Time-resolved X-ray absorption fine structure (TR-XAFS) spectroscopy
applications
catalyst sulfidation, 433
phase composition, 434
reaction intermediate, 433–434
reaction kinetics, 433
average metal valence and catalytic selectivity, correlation
average molybdenum valence, 435–436
$H_5[PV_2Mo_{10}O_{40}]$ oxidation catalyst, 434
oxygen ion diffusion, 438
reoxidation extent, 437
selectivity, 437
cells, 431
data analysis, 431–432
instrumentation
energy-dispersive design, 430
monochromator, 429
phase transition kinetics, 429
reduction and reoxidation, solid-state kinetics
Arrhenius diagram, 441
ion currents, 443
Mo K-edge spectra, 438
$Mo_{18}O_{52}$ type defects, 442
MoO_2 oxidation, O_2, 438, 442
MoO_3 reduction, propene, 438, 440
nucleation growth kinetics, 441
power rate law, 439
propene-to-O_2 ratio, 442
reaction kinetics, 439
time-resolved modes
catalyst structural response, 445
MoO_3 structure, 444
TiO_2-containing photocatalysts
electrons and holes charge separation, 25
gas-phase photocatalytic reaction, 24
hole transfer, 25–26
O_2 photodesorption, 24
rutile and anatase, 24
transient absorption spectra, 25–27
Turnover frequency (TOF), 94, 102, 104

U

Ultraviolet–Visible–Near Infrared (UV–vis–NIR)
spectroscopy
catalyst treatment
spectroscopic measurements, 184–189
correlation with other spectra
ATR (IR) spectra, 203
homogeneous oxidation, 201

propane dehydrogenation, 200–201
data acquisition and analysis
background correction, 167–171
sample preparation, 171–172
spectral representation, 173–175
diffuse reflectance spectroscopy
combined optical elements, 159
fiber optics, 158–159
integrating spheres, 153–155
mirror optics, 155–158
optical accessory design, 149–150
standard materials, 150–152
independent calibration, 175–176
Kubelka–Munk function
concentration, 144–145
data quantification, 145
limitation, 137–138
photon fluxes and reflectance factor, 139
remission function, 140–141
scattering coefficient, 144
vs. absorbance, 141–142
zero absorption limit, 142–144
reaction cells
design parameters, 160–161
fiber optics, 164–165
integrating spheres, 161–163
mirror optics, 163–164
transmission mode spectroscopy, 161
spectrometer requirements
background correction, 166–167
sample measurement, 166–167
step-like artifacts and time resolution, 167
spectroscopic and catalytic data acquisition
alkane isomerization, 197–198
butane dehydrogenation, 196–197
electrocatalysis, 198–200
ethene transformation, 195–196
H_2S with O_2 oxidation, 195
methanol and ethanol oxidation, 196
transmission
Lambert–Beer law, 135–137
vs. reflection spectroscopy, 134

X

XAFS of catalysts in reactive atmospheres, catalyst
characterization by XAFS
electronic structure
d-electron states, 365–366
nano clusters, 368
Pt L_2-edge data, 368–369
rhenium oxides, 367–368
XANES spectroscopy, 367
metal cluster size and shape
cubooctahedral platinum, 364–365
Pt L_3-edge EXAFS, 364, 366
refractory oxides, 362–363
single-shell analysis, 364

XAFS of catalysts in reactive atmospheres, catalyst
characterization by XAFS (*cont.*)
oxidation state of metal
bulk rhenium, 360–361
Mo K-edge feature, 362
Re L_3-edge XANES, 361, 363
species transformation
bimetallic clusters, 355
hydrocracking catalyst, 355–356
W L_3-edge and Ni K-edge EXAFS, 356–359
XANES spectra, 360
structure determination
EXAFS analysis, 350
framework species *vs.* nonframework, 352
Re L_3-edge EXAFS data, 353–354
rhenium oxide, 353–354
scattering path, 354–355
SiO_4 tetrahedra, 353
Sn-β zeolite, 350–352
zeolite framework, 349–350
X-ray absorption fine structure (XAFS)
spectroscopy
absorption process, 346–347
catalyst structure, 404
catalytic reactions
Ag/Al_2O_3 catalyst, 422
Ag_4^{2+}, 423
benzyl alcohol oxidation, 425–426
catalytic activity, 426–427
catalytic partial oxidation (CPO), 427
cobalt clusters, 420
cobalt MCM-41 catalyst, 421
Cu/ZrO_2 catalyst, 421–422
Cu–ZSM-5 catalyst, 423–424
CuCl reoxidation, 420
$CuCl_2/Al_2O_3$ catalysts, 419
ethene catalytic oxychlorination, 419
ethene polymerization, 421
H_2 prereduction, 420
O_2 conversion, 419–420
Pd(0) and Pd(II), 426
Pd/Al_2O_3, 425
selective catalytic reduction, 421–423
single walled carbon nanotubes (SWNT), 420
catalytic reactors
advantages, 392
beryllium windows, 379
boron nitride, 389–390
cartridge heaters, 396–397
cell design, 369–370
Dalla Betta cell, 375
DeNOx catalysis, 387–388
disadvantage, 385
fluorescence measurements, 380
functions, 372–373
Lytle cells, 373–374
metal and bimetallics, 374
Neils and Burlitch design, 376
Ni–Cr wire, 386

Odzak design, 391
photon energy, 347
plug-flow reactor, 388–389
polyether–ether–ketone (PEEK), 393–394
pore diffusion, 392–393
quartz capillary, 380–381
schematic descriptions, 385–386
screw-cap design, 398
signal/noise ratio (S/N), 382
solid catalyst, 373
spectra and catalytic activity, 381
spectroscopic data, 394
supercritical fluids, 395
tank reactor, 377–378
transmission geometry, 399
UV spectroscopy, 400
versatile cell, 397–398
wafers and powders, 373–374
Weiher XAFS cell, 396
window materials, 371
CO oxidation, gold cluster catalysts
activity, 417
Au/TiO_2, 416
Au–O coordination number and
distance, 416
cationic and zerovalent gold, 414
CO conversion, 415, 417
high energy-resolution fluorescence
detection (HERFD), 418
MgO, 412
reaction rate, 417
structure elucidation, 418
Fourier transform (FT), 343
hydrogenation
catalytic activity, 407, 409
ethene conversion, 407
interatomic distance, 409
$Ir_4/\gamma-Al_2O_3$, 407–408
metal clusters, 406
supported iridium clusters, 407
toluene, 409–410
mean square disorder, 404–405
methane, partial oxidation, 404
monograph, 345
multitechnique cells, 448
potential roadmap, 446–447
reactive atmospheres
electronic structure, 365–369
metal cluster size and shape, 362–365
oxidation state, 359–362
species transformation, 355–359
structure determination, 349–355
SciFinder, 343
scope and outline, 345–346
sigma squared term exponential
decay, 405
thermal and structural disorder, 348
XAFS data analysis, 344
X-ray absorption near edge structure (XANES)

charge coupled device (CCD) camera, 453
CO adsorption site, 450
Fe₂SiO₄, 452
FeZSM-5, 450
HERFD mode, 449
Pt/Al₂O₃ catalyst, 451
Rh/Al₂O₃ catalyst, 454
X-ray diffraction (XRD)
  applications
    fluid–solid reactions, 306
    functional analysis, 308
    gas–solid reactions, 306–307
    nitride phases, 307
  auxiliary information
    functional analysis, 308–309
    metal-oxide transitions, 309
  case studies
    catalytic conditions, 315–327
    methodologies, 314, 328
    nonperiodic structure, 328
    proxy atmospheres, 329
    redox reactions, 329
    single-crystal approach, 328
    structure–function correlations, 328–329
    synchrotron-based experiment, 314
    working catalysts, 330
  catalyst characterization, (also see specific
      characterization methods)
    ammonia iron, 282
    bulk-sensitive techniques, 275
    calcination processes, 283
    coherent scattering methods, 277–278
    crystallites and crystallinity, 277
    defects classification, 279–280
    EXAFS spectroscopy, 283–284
    iron ammonia synthesis, 280–281
    nanocrystalline materials, 279
    nanostructure, 275
    structural transformations, 282
    sub-stoichiometric phases, 280
  instrumentation
    auxiliary equipment, 309
    Bragg–Brentano reflection, 310
    cell designs technique, 312–313
    RbNO₃, 313
    steady-state flow conditions, 310
    STOE diffractometer, 311
    STOE theta–theta goniometer, 312
  nanostructures
    bulk-sensitivity, 303
    EXAFS analyses, 301
    microstructural parameters, 296–297
    Miller index, 298
    morphologies, 299–300
    pore system, 296
    Rietveldt method, 301
    Scherrer analysis, 299
    small angle X-ray scattering (SAXS), 302
  surface crystallography, 302–303
  Williamson–Hall plot, 300
  powder diffraction
    advantages, 288–289
    aim and scope, 289
    physics and data analysis, 290
  reactive atmospheres, 287–288
  scattering phenomena
    Bragg equation, 291
    chemical microstrain, 292
    electromagnetic waves, 289
    Ewald sphere intersects, 291–292
    gas–solid reactions, 294
    geometric analysis, 289
    methodologies, 296
    microstrain, 293–294
    schematic representation, 294
    Scherrer formula, 295–296
    translational symmetric lattices, 293
  static vs. dynamic analysis
    complementary measurements, 287
    functional compartments, 285
    nanostructure, 284
    single-crystal model, 284
    solid-state conversions, 287
    structural adaptations, 286
    structure–function correlations, 284–285
  working catalysts
    core–shell structure, 306
    electron microscopy, 304–305
    microscopic methods, 305–306
    phase cooperation, 303–304
    TEM image, 306
X-ray photoelectron spectroscopy (XPS)
  CO adsorption, Pd(111)
    C1s core-level spectra, 229–230
    coverage-dependent vibrational spectra, 231
    deconvolution, C1s spectra, 230
    Fe2p and C1s core-level spectra, 233–234
    LEED pattern characteristics, 232
    palladium carbonyls, 232–233
    SFG spectra, 231–232
  CO oxidation, Ru(0001)
    mechanism, 257
    pressure gap, 256
    RuO₂ phase formation, 257–258
    subsurface oxygen, 256–257
    surface morphology, 265–266
    temperature dependence, 263–265
  differential pumping
    application and concept, 218–219
    CO adsorption, 220
    complex gas adsorption, 220–221
    high-pressure XPS, 219–220
  electrostatic lens system
    gas flow scheme, 227–228
    infrared laser, 228
    limitation, 229

X-ray photoelectron spectroscopy (XPS) (*cont.*)
    molecular flow approach, 226
    photoelectron collection efficiency, 227
    photoelectron scattering, 225–227
    signal intensity parameters, 224
    X-ray transmission, 224–225
  equipment, 218
  ethene epoxidation, silver
    catalyst surface, 240–241
    Langmuir–Hinshelwood mechanism, 247
    oxygen species, 243–246
    PTRMS spectrum, 241–242
    spectroscopic characteristics, 244
    surface carbonates, 242–243
  gas cell
    disadvantages, 222–223
    preparation chamber, 221–222

    pressure difference, 222
methanol decomposition, Pd(111)
  C–O bond scission, 235–236
  C1s core-level spectra, 236–237
  process and model, 237
  SFG postreaction, 238–239
  TPRS investigations, 239–240
methanol dehydrogenation,
    Pd(111), 234–235
methanol oxidation, copper
  catalytic activity, 252–253
  Cu–O trilayer, 255–256
  depth-profiling measurement,
    251–252
  methanol-to-oxygen ratio, 249–251
  NEXAFS experiment, 248
  reaction paths, 247